Forest Plant, Soil, Microorganisms and Their Interactions

Forest Plant, Soil, Microorganisms and Their Interactions

Guest Editors

Fuxi Shi
Xianwei Wang
Yang Zhang
Jiusheng Ren

MDPI

Basel • Beijing • Wuhan • Barcelona • Belgrade • Novi Sad • Cluj • Manchester

Guest Editors
Fuxi Shi
College of Forestry
Jiangxi Agricultural
University
Nanchang
China

Xianwei Wang
Northeast Institute of
Geography and Agroecology
Chinese Academy of Sciences
Changchun
China

Yang Zhang
College of Forestry
Jiangxi Agricultural
University
Nanchang
China

Jiusheng Ren
School of Water Resources
and Environmental
Engineering
East China University of
Technology
Nanchang
China

Editorial Office
MDPI AG
Grosspeteranlage 5
4052 Basel, Switzerland

This is a reprint of the Special Issue, published open access by the journal *Forests* (ISSN 1999-4907), freely accessible at: https://www.mdpi.com/journal/forests/special_issues/20TF288224.

For citation purposes, cite each article independently as indicated on the article page online and as indicated below:

Lastname, A.A.; Lastname, B.B. Article Title. *Journal Name* **Year**, *Volume Number*, Page Range.

ISBN 978-3-7258-4451-7 (Hbk)
ISBN 978-3-7258-4452-4 (PDF)
https://doi.org/10.3390/books978-3-7258-4452-4

Contents

forests

MDPI

Editorial

Updates on Plants, Soil, Microorganisms, and Their Interactions in Forest Ecosystems

Fuxi Shi [1,2,3], Jiusheng Ren [4] and Yang Zhang [2,*]

1 Jiangxi Key Laboratory of Watershed Soil and Water Conservation, Jiangxi Provincial Academy of Water Resources Sciences, Nanchang 330029, China; shifuxi2008@163.com
2 Key Laboratory of National Forestry and Grassland Administration on Forest Ecosystem Protection and Restoration of Poyang Lake Watershed, College of Forestry, Jiangxi Agricultural University, Nanchang 330045, China
3 Matoushan National Observation and Research Station of Chinese Forest Ecosystem, Zixi 335300, China
4 Jiangxi Provincial Key Laboratory of Genesis and Remediation of Groundwater Pollution, School of Water Resources and Environmental Engineering, East China University of Technology, Nanchang 330013, China; renjiusheng256@ecut.edu.cn
* Correspondence: zhangyang0558@163.com

Forests, covering one-third of the global landmass, are the world's most vital terrestrial ecosystem [1]. They offer multiple ecosystem services [2], such as food production, biodiversity conservation, sandstorm prevention, water and soil conservation, nutrient cycling, carbon (C) sequestration, climate mitigation, culture and recreation, etc. [3–8]. Generally, plants, soil, and microbes do not exist alone in an ecosystem; rather, they are tightly linked to each other [9–12]. Their complex coupling interrelationships are crucial for driving ecological processes and ecosystem functions in forest ecosystems [13].

Aboveground plants provide essential resources, such as organic carbon and nutrition, mainly through litter and root exudates, maintaining soil microbial activity and functional performance [12]. In turn, soil microorganisms (e.g., bacteria and fungi) decompose dead plant material, thereby influencing soil nutrient availability, plant growth, and the composition of aboveground communities [10]. Therefore, these interdependencies among plants, soil, and microorganisms play a very important role in determining the key ecosystem processes. Any disruption in these linkages caused by human-driven environmental changes, including climate warming, increased atmospheric nitrogen (N) deposition, and land use changes, can result in imbalances in nutrient cycling [14,15], which could ultimately affect the overall ecological functioning and stability of the forest ecosystems [16]. In these circumstances, delving deeper into the relationships among plants, soil, microorganisms, and their responses to environmental changes can enhance our comprehension of the internal mechanisms in forest ecosystems. This Special Issue selected 15 papers to discuss the new knowledge and different perspectives on plant–soil microbe interactions, aiming to promote forest management.

Forest succession is recognized as a key biotic factor that plays a significant role in shaping above-ground ecological processes and material cycling. For example, Yuan et al. [17] discovered that forest aging can lead to divergent changes in the stoichiometric characterizations of C, N, and P among plants, litter, soil, and microorganisms. This finding contributes to a better comprehension of the nutrient utilization strategies and regulatory mechanisms in *Pinus taiwanensis* plantation forest ecosystems. Qiu et al. [18] demonstrated that variations in fungal functional structure intensify during the succession process of *Pinus tabulaeformis* plantations in the North Warm Temperate Zone. This variation is primarily influenced by soil pH, dry matter content, C/N ratio, and N content. Ren et al. [19]

investigated the shifts in soil microbial communities and assembly processes during vegetation succession in a subtropical forest. Their study revealed that the community assembly of soil bacteria transitions from a deterministic process to a stochastic process and back to a deterministic process as forest succession progresses from abandoned land to evergreen broad-leaved forests. In contrast, stochastic processes largely govern the community assembly of soil fungal communities during all succession stages. Moreover, soil organic carbon (SOC) is responsible for the change in bacterial communities, whereas SOC, total N (TN), C:N ratio, and pH collectively regulate the changes in fungal community structure. Li et al. [20] found that the introduction of broad-leaved tree species to a Chinese fir plantation not only increased the soil nutrient content, but also enhanced the diversity of soil fungal communities, resulting in the microbial communities of mixed forests being more diverse. Therefore, for the sustainable development of Chinese fir plantations, it is crucial to direct more focus on the middle and late stages of their growth rather than just the early stages. These results imply that mixed forest models should be prioritized as the primary forestry management approach.

Climate warming, atmospheric N deposition, and land use changes are three important human-driven abiotic factors that may impact soil nutrient cycling and soil microorganisms in forest ecosystems. For instance, Shu et al. [21] observed that soil N mineralization was more influenced by warming in the subtropical *Castanopsis hystrix* plantation than in the temperate *Quercus aliena* natural forest, which may be attributed to differences in the soil nutrient availability, fine root biomass, and microbial biomass of forests in the two climatic zones. Hou et al. [22] showed that after four years of the continuous addition of N, soil microbial communities in subtropical planted coniferous forests displayed diverse responses that were dependent on the afforestation tree species. Long-term high-level N additions consistently lead to detrimental effects on soil microbial communities (especially fungi), highlighting that soil pH and N availability are crucial factors influencing the diversity of soil bacterial and fungal communities in these forest ecosystems. Li et al. [23] studied soil microbial communities in *Pseudotsuga sinensis* forests with different degrees of rocky desertification in the Karst Region of Southwest China. They found that the richness and diversity of microbial communities decreased with an increase in the degree of rocky desertification, primarily influenced by soil bulk density, pH, SOC, available N, and available P. Xiao et al. [24] found that relatively heavy thinning can accelerate soil P-bioavailable turnover by stimulating multiple nutrient cycles, highlighting the essential roles of microbial biomass turnover and soil nutrient supply in the plant-available P mechanism in *Eucalyptus* coppice plantations.

Elevation is a significant natural factor that plays a crucial role in influencing plant functional traits, soil characteristics, soil microbes, and their relationships within forest ecosystems. For example, Zhang et al. [25] discovered a notable increase in aboveground vegetation productivity across diverse altitudinal vegetation belts in the Chinese Tianshan Mountains since the early 21st century. Huang et al. [26] found that the differences in soil water availability and soil development due to different altitude habitats could significantly impact leaf-scale water use efficiency and nutrient status on the western slope of Wuyi Mountain in southern China. In the Fanjing Mountain Forest ecosystem, Wang et al. [27] found that the soil properties and enzyme activities were the main factors that drove the elevational distribution of soil microbial communities. Jin et al. [28] explored the differences in soil properties and root zone bacteria communities at different altitudes within the distribution range *Davidia involucrata* in the Sichuan Province of China. They also found a significant link between soil parameters and soil bacterial communities in root zones of mature trees across varying altitude gradients.

Moreover, some research was conducted on the application of soil microbial inoculants in agroforest ecosystems. For example, Lai et al. [29] found that ribosome-inactivating proteins, such as curcin protein, exhibit broad-spectrum antifungal properties. Ribosome proteins could significantly impact soil fungi and thereby affect the soil ecosystem in Jatropha plantations. Deng et al. [30] isolated an endophytic bacterium (*Pseudomonas* sp. En3) from the leaf of *Populus tomentosa*. These plant growth-promoting endophytic bacteria have a notable growth-promoting impact on poplar seedlings. de Oliveira et al. [31] found that, when inoculated with plant growth-promoting bacteria (*Exiguobacterium sibiricum*), the plant of *Corymbia citriodora* exhibited a taller height, higher chlorophyll *b* content, larger shoot and total dry mass, and greater ability to colonize the roots, leading to the production of higher-quality seedlings.

In summary, the 15 publications in this Special Issue represent a small sample of current scientific research, offering new insights into plants, soil, microorganisms, and their interactions in various forest ecosystems. These findings may have implications for forest conservation, restoration, and management. Moving forward, there is a need to delve deeper into the relationships among plants, soil, and microorganisms, particularly within the framework of constructing a global ecological civilization.

Conflicts of Interest: The authors declare that they have no conflicts of interest.

References

1. FAO. *Global Forest Resources Assessment 2020—Main Report*; FAO: Rome, Italy, 2020.
2. Millennium Ecosystem Assessment. *Ecosystems and Human Wellbeing: Current State and Trends: Findings of the Condition and Trends Working Group*; Island Press: Washington, WA, USA, 2005.
3. Lal, R. Forest soils and carbon sequestration. *Forest. Ecol. Manag.* **2005**, *220*, 242–258. [CrossRef]
4. Peura, M.; Burgas, D.; Eyvindson, K.; Repo, A.; Mönkkönen, M. Continuous cover forestry is a cost-efficient tool to increase multifunctionality of boreal production forests in Fennoscandia. *Biol. Conserv.* **2018**, *217*, 104–112. [CrossRef]
5. Felipe-Lucia, M.R.; Soliveres, S.; Penone, C.; Manning, P.; van der Plas, F.; Boch, S.; Prati, D.; Ammer, C.; Schall, P.; Gossner, M.M.; et al. Multiple forest attributes underpin the supply of multiple ecosystem services. *Nat. Commun.* **2018**, *9*, 4839. [CrossRef]
6. Snäll, T.; Triviño, M.; Mair, L.; Bengtsson, J.; Moen, J. High rates of short-term dynamics of forest ecosystem services. *Nat. Sustain.* **2021**, *4*, 951–957. [CrossRef]
7. Nocentini, S.; Travaglini, D.; Muys, B. Managing mediterranean forests for multiple ecosystem services: Research progress and knowledge gaps. *Curr. For. Rep.* **2022**, *8*, 229–256. [CrossRef]
8. Pan, Y.D.; Birdsey, R.A.; Phillips, O.L.; Houghton, R.A.; Fang, J.Y.; Kauppi, P.E.; Keith, H.; Kurz, W.A.; Ito, A.; Lewis, S.L.; et al. The enduring world forest carbon sink. *Nature* **2024**, *631*, 563–569. [CrossRef] [PubMed]
9. Wardle, D.; Bardgett, R.; Klironomos, J.; Setälä, H.; van der Putten, W.H.; Wall, D. Ecological linkages between aboveground and belowground biota. *Science* **2004**, *304*, 1629–1633. [CrossRef]
10. Hättenschwiler, S.; Gasser, P. Soil animals alter plant litter diversity effects on decomposition. *Proc. Natl. Acad. Sci. USA* **2005**, *102*, 1519–1524. [CrossRef] [PubMed]
11. van der Heijen, M.G.A.; Bardgett, R.D.; van Straalen, N.M. The unseen majority: Soil microbes as drivers of plant diversity and productivity in terrestrial ecosystems. *Ecol. Lett.* **2008**, *11*, 296–310. [CrossRef] [PubMed]
12. Lange, M.; Eisenhauer, N.; Sierra, C.A.; Bessler, H.; Engels, C.; Griffiths, R.I.; Mellado-Vázquez, P.G.; Malik, A.A.; Roy, J.; Scheu, S.; et al. Plant diversity increases soil microbial activity and soil carbon storage. *Nat. Commun.* **2015**, *6*, 6707. [CrossRef] [PubMed]
13. Topanotti, L.R.; Fuchs, J.M.; Albert, M.; Schick, J.; Penanhoat, A.; Lu, J.Z.; Pérez, C.A.R.; Foltran, E.C.; Appleby, S.; Wildermuth, B.; et al. Enhancing economic multifunctionality without compromising multidiversity and ecosystem multifunctionality via forest enrichment. *Sci. Adv.* **2024**, *10*, eadp6566. [CrossRef]
14. Schmidt, M.W.I.; Torn, M.S.; Abiven, S.; Dittmar, T.; Guggenberger, G.; Janssens, I.A.; Kleber, M.; Kögel-Knabner, I.; Lehmann, J.; Manning, D.A.C.; et al. Persistence of soil organic matter as an ecosystem property. *Nature* **2011**, *478*, 49–56. [CrossRef] [PubMed]
15. Hooper, D.U.E.; Adair, C.; Cardinale, B.J.; Byrnes, J.E.K.; Hungate, B.A.; Matulich, K.L.; Gonzalez, A.; Duffy, J.E.; Gamfeldt, L.; O'Connor, M.I. A global synthesis reveals biodiversity loss as a major driver of ecosystem change. *Nature* **2012**, *486*, 105–108. [CrossRef]
16. Yu, Q.S.; He, C.Q.; Anthony, M.A.; Schmid, B.; Gessler, A.; Yang, C.; Zhang, D.H.; Ni, X.F.; Feng, Y.H.; Zhu, J.L.; et al. Decoupled responses of plants and soil biota to global change across the world's land ecosystems. *Nat. Commun.* **2024**, *15*, 10369. [CrossRef]

17. Yuan, M.; Wang, Y.R.; Wang, Y.; Wang, Y.; Wang, S.W.; Pan, Y.; Zhou, W.M.; Xiang, X.Y.; Tong, Y.W. Ecological Stoichiometric Characteristics of C, N, and P in *Pinus taiwanensis* Hayata Needles, Leaf Litter, Soil, and Microorganisms at Different Forest Ages. *Forests* **2024**, *15*, 1954. [CrossRef]
18. Qiu, Z.L.; Liu, H.; Chen, C.L.; Liu, C.C.; Shu, J. Environmental Driving Mechanism and Response of Soil's Fungal Functional Structure to Near-Naturalization in a Warm Temperate Plantation. *Forests* **2024**, *15*, 1540. [CrossRef]
19. Ren, J.S.; Huang, K.X.; Xu, F.F.; Zhang, Y.; Yuan, B.S.; Chen, H.M.; Shi, F.X. The Changes in Soil Microbial Communities and Assembly Processes along Vegetation Succession in a Subtropical Forest. *Forests* **2024**, *15*, 242. [CrossRef]
20. Li, W.Y.; Sun, H.M.; Cao, M.M.; Wang, L.Y.; Fang, X.H.; Jiang, J. Diversity and Structure of Soil Microbial Communities in Chinese Fir Plantations and *Cunninghamia lanceolata–Phoebe bournei* Mixed Forests at Different Successional Stages. *Forests* **2023**, *14*, 1977. [CrossRef]
21. Shu, W.W.; Wang, H.; Liu, S.R.; Liu, Y.C.; Min, H.L.; Li, Z.Y.; Dell, B.; Chen, L. Differential Responses of Soil Nitrogen Forms to Climate Warming in *Castanopsis hystrix* and *Quercus aliena* Forests of China. *Forests* **2024**, *15*, 1570. [CrossRef]
22. Hou, Z.; Zhang, X.H.; Chen, W.; Liang, Z.Q.; Wang, K.Q.; Zhang, Y.; Song, Y.L. Differential Responses of Bacterial and Fungal Community Structure in Soil to Nitrogen Deposition in Two Planted Forests in Southwest China in Relation to pH. *Forests* **2024**, *15*, 1112. [CrossRef]
23. Li, W.J.; He, B.; Feng, T.; Bai, X.L.; Zou, S.; Chen, Y.; Yang, Y.R.; Wu, X.F. Soil Microbial Communities in Pseudotsuga sinensis Forests with Different Degrees of Rocky Desertification in the Karst Region, Southwest China. *Forests* **2024**, *15*, 47. [CrossRef]
24. Xiao, X.S.; Ali, I.; Du, X.; Xu, Y.Y.; Ye, S.M.; Yang, M. Thinning Promotes Soil Phosphorus Bioavailability in Short-Rotation and High-Density *Eucalyptus grandis* × *E. urophylla* Coppice Plantation in Guangxi, Southern China. *Forests* **2023**, *14*, 2067. [CrossRef]
25. Zhang, Y.; An, C.B.; Jiang, L.; Zheng, L.Y.; Tan, B.; Lu, C.; Zhang, W.S.; Zhang, Y.Z. Increased Vegetation Productivity of Altitudinal Vegetation Belts in the Chinese Tianshan Mountains despite Warming and Drying since the Early 21st Century. *Forests* **2023**, *14*, 2189. [CrossRef]
26. Huang, K.X.; Xue, Z.J.; Wu, J.C.; Wang, H.; Zhou, H.Q.; Xiao, Z.B.; Zhou, W.; Cai, J.F.; Hu, L.W.; Ren, J.S.; et al. Water Use Efficiency of Five Tree Species and Its Relationships with Leaf Nutrients in a Subtropical Broad-Leaf Evergreen Forest of Southern China. *Forests* **2023**, *14*, 2298. [CrossRef]
27. Wang, J.C.; Xiao, S.Y.; Hayat, K.; Liao, X.F.; Chen, J.Z.; Zhang, L.Y.; Xie, Y.G. Investigating the Effects of Elevation on Microbial Communities and Soil Properties at Fanjing Mountain, China. *Forests* **2024**, *15*, 1980. [CrossRef]
28. Jin, Y.; Li, X.; Hu, Y.; Huang, J.Z.; Chen, Y.; Kou, Y.P.; Li, X.L.; Dong, M.; Deng, D.Z.; Li, Y. Diversity Patterns of Bacteria in the Root Zone of *Davidia involucrata* Along an Altitudinal Gradient. *Forests* **2024**, *15*, 1920. [CrossRef]
29. Lai, Z.P.; Zhang, B.B.; Niu, X.F.; Ma, R.; Wang, T.; Cheng, C.; Ren, Y.Y.; Wang, X.Y.; Hu, N.; Jiang, N.; et al. The Effect of Curcin Protein and Jatropha Plantation on Soil Fungi. *Forests* **2023**, *14*, 2088. [CrossRef]
30. Deng, B.Y.; Wu, L.; Xiao, H.J.; Cheng, Q. Characterization of *Pseudomonas* sp. En3, an Endophytic Bacterium from Poplar Leaf Endosphere with Plant Growth-Promoting Properties. *Forests* **2023**, *14*, 2203. [CrossRef]
31. de Oliveira, A.M.; de Abreu, C.M.; Grazziotti, P.H.; de Andrade, G.F.P.; Gomes, J.V.; Avelino, N.R.; Menezes, J.F.S.; Barroso, G.M.; dos Santos, J.B.; da Costa, M.R. Production of Seedlings of Corymbia citriodora Inoculated with Endophytic Bacteria. *Forests* **2024**, *15*, 905. [CrossRef]

forests

Article

Investigating the Effects of Elevation on Microbial Communities and Soil Properties at Fanjing Mountain, China

Juncai Wang [1,2], Shengyang Xiao [1,2], Kashif Hayat [3], Xiaofeng Liao [1,2,4], Jingzhong Chen [5], Lanyue Zhang [1,2] and Yuangui Xie [1,2,*]

1 Guizhou Academy of Sciences, Guiyang 550001, China; wangjuncai12@126.com (J.W.); xiaoshengyang918@163.com (S.X.); lxfnsd@163.com (X.L.); zlyue_0122@163.com (L.Z.)
2 The Land Greening Remediation Engineering Research Center of Guizhou Province, Guiyang 550001, China
3 Key Laboratory of Pollution Exposure and Health Intervention, Interdisciplinary Research Academy, Zhejiang Shuren University, Hangzhou 310015, China
4 College of Resources and Environmental Engineering, Guizhou University, Guiyang 550025, China
5 College of Pharmacy, Guizhou University of Traditional Chinese Medicine, Guiyang 550025, China; chenjingzhong022@gzy.eud.cn
* Correspondence: xieyuangui12@126.com

Abstract: Elevation is one of the most influential factors affecting soil characteristics and microbial communities in forest ecosystems. Nevertheless, there is no consensus on how soil characteristics, soil microbials, and their relationships response to the elevation of the mountain ecosystem. We investigated the soil physicochemical characteristics, the activity of soil enzymes, and the microbial community at elevational sites from 600 to 2400 m above sea level (asl) in the western slopes of the Fanjing Mountain ecosystem, China. The soil microbial communities were determined by high throughput 16S rRNA and ITS amplicon sequencing. The results demonstrated that soil total nitrogen (TN) showed a slight decrease, whereas total phosphorus (TP) and total potassium (TK) gradually tended to increase with increasing elevation. The large macroaggregates (>2 mm) accounted for the largest proportion of the aggregate fraction (66.23%–76.13%) in the 0–10 cm soil layer with elevation. The average values of the soil electrical conductivity (EC), soil organic carbon (SOC), and cation exchange capacity (CEC) concentration in the 0–60 cm layer undulated with increasing elevation, and the highest values were observed at 1500–1800 m asl and 1800–2100 m asl, respectively. The activities of soil urease, sucrase, acid phosphatase, and catalase clearly differed ($p < 0.05$) with increasing elevation, and the minimum values were found at 2100–2400 m asl. Interesting, with increasing soil depth, the values of these factors tended to decrease, indicating surface aggregation. In addition, the soil microbial (bacterial and fungal) community diversity exhibited a single-peak pattern with elevation. Our results also revealed that the soil bacterial and fungal communities varied significantly at different elevation sites. The bacterial communities were dominated by the phyla Acidobacteria, Pseudomonadota, and Chloroflexi, and the phyla Basidiomycota and Ascomycota dominated the fungal communities. The Pearson and redundancy analyses revealed that the SOC, TP, four soil enzymes, and soil aggregates were significant factors influencing the soil microbial community. In conclusion, soil properties and enzyme activities jointly explained the elevational pattern of the soil microbial community in the Fanjing Mountain. The results of this study provide insights into the influence of elevation on soil characteristics, microbial communities, and their relationships in the Fanjing Mountain ecosystem.

Keywords: soil physiochemical properties; extracellular enzymes; high-throughput sequencing; soil microbial community; elevation

1. Introduction

Soil microorganisms are fundamental biological units of soil ecosystems, and their community structure, diversity, and functioning can be utilized to evaluate their health [1,

2]. As the link between soil and plants, soil microorganisms play important roles in the storage of various types of soil carbon, soil organic matter (SOM) decomposition, mineralization, plant community assembly, nutrient immobilization, and other biochemical processes [3,4]. In recent years, changes in soil characteristics and microbial community structures according to variety environmental factors have attracted considerable interest for novel perspectives on the influence of environments on biological communities [3,5,6], especially in various forest soils [7]. Elevation is a critical aspect of mountain topography and can lead to spatial heterogeneity of biological and abiotic elements by influencing ecological factors such as temperature, precipitation, and light [7,8]. For example, several studies have demonstrated that elevation is the main factor affecting the diversity and abundance of soil microbial communities in various forest soils because it can cause variations in precipitation and temperature, which in turn will influence different forest soil microbial community structures and species diversity [9–11]. Nevertheless, the response of the soil microbial community and diversity differ across the elevation of different forest ecosystems, in which the climate, forest type, and azonal soil properties covary even over small spatial distances [12–14]. Moreover, there is no decisive pattern response of soil microbial diversity along elevation; it is essential to study the distribution and abundance of the soil microbial community and its relationships with soil property factors at different elevation sites, which may provide insight into the effects of elevation on the soil properties and microbial community composition in montane ecosystems.

Elevation has been taken as a strong "natural laboratory" to explore the elevational patterns of soil microorganism communities and their properties because of the dramatic climate variation at small spatial scales [15–17]. Compared with the elevational patterns of plant and animal communities, microbial diversity displays a different trend at the same increasing elevation owing to the greater sensitivity of microorganisms to environmental variations [18,19]. Currently, although diverse modes of elevation-associated soil microbial community and diversity have been extensively studied, including monotonously linear increasing [8], monotonously linear decreasing [9], humpback [20], concave [21,22], or no regular change distribution patterns related to [23], a consensus remains complicated. The lack of accordance potentially suggests that the elevational pattern of microbial diversity and community composition is site-dependent, confirming the conclusions of a global synthesis study [9,24]. Additionally, numerous studies have demonstrated that greater diversity and richness of understory vegetation are beneficial to litter decomposition and matter cycling and that the litter decomposition rate is controlled by the types and activities of soil enzymes, and higher soil enzyme activity in turn accelerates plant growth and development [24,25]. Thus, soil enzyme activity can be a crucial indicator of soil health status and microbial activity because of its functional properties and susceptibility to environmental stress [26,27]. The relationships among elevation, soil physicochemical characteristics, enzyme activities, and microbial communities still need to be explored further, which is beneficial for enhancing our comprehension of the spatial changes in active soil substances in mountain ecosystems. In conclusion, the elevational patterns of soil microbial community structure and enzyme activity were closely correlated with geographical distributions, which verified the importance of exploring the elevational patterns of microbial diversity and enzyme activity in soils from more habitats.

The Fanjing Mountains have the most biodiversity-rich terrestrial zones at the same latitude in the world and are typical representatives of global hotspots for biodiversity conservation [28,29]. The vertical zone spectrum of vegetation in Fanjing Mountains has the general structure of the vertical band spectrum of humid subtropical mountain vegetation in eastern China; for example, it is an evergreen broad-leaved forest belt below 1300 and 1400 m, a mixed evergreen and deciduous broad-leaved mixed forest belt between 1400 and 2000 m, and a subalpine coniferous and broad-leaved mixed forest and shrub meadow belt between 2200 and 2570 m [29,30]. Although many researchers have reported that the soil characteristics can be used to evaluate the elevation-dependent variations in the microbial community structure of mountain ecosystems [30,31], our understanding of the impact of elevation on soil characteristics and the microbial community in Fanjing Mountain is limited. The aims of this work were to (1) investigate the response of soil physicochemical characteristics to elevation on the western slope of Fanjing Mountain; (2) explore the changes in soil microbial diversity and communities at different elevation sites; and (3) identify the possible environmental driver factors that could change soil microbial communities among the determined physicochemical factors of soil. We hypothesized that soil physicochemical characteristics, the activities of soil enzymes, soil microbial communities, and diversity respond differently to the elevation. Consequently, we examined the alterations in soil properties, enzyme activities, and the microbial community at six elevation sites on Fanjing Mountain (600–2400 m asl) and to identify the possible environmental drivers of elevational changes in the soil microbial community. This research provides important basic information and novel understandings of the formerly substituted soil properties and microbiomes of Fanjing Mountain and will help us better understand the effects of elevation variation on belowground microbial communities.

2. Materials and Methods

2.1. Study Area Description

The Fanjing Mountain is located in the northeastern part of Guizhou Province in China (27°49′50″–28°1′30″ N, 108°45′55″–108°48′30″ E, Figure 1) [28,29]. Fanjing Mountain, with 96.5% forest coverage, has a subtropical humid monsoon climate. The forests in the study area are largely mountainous deciduous broad-leaved secondary forests and evergreen broadleaf forests. The highest peak elevation is 2570.0 m asl, and the relative height difference of the mountain body is more than 2000 m. The annual precipitation is 1100–2600 mm and is mainly concentrated in April–October. The average temperature in the hottest month (July) is 25.3 °C, the coldest month (February) average temperature is 2 °C, the annual frost-free period is 270–278 d, and the annual average relative humidity is 80% [28,29]. There is an evergreen broad-leaved forest belt below 1300~1400 m, an evergreen and deciduous broad-leaved mixed forest belt between 1400 and 2000 m, and a subalpine coniferous and broad-leaved mixed forest and shrub meadow belt between 2200 and 2570 m in Fanjing Mountains [29,30].

Additionally, the parent material of soil is mainly the residual and slope deposits of Proterozoic Banxi group metamorphic rock weathering, and the predominant soil types include yellow soil and yellow-brown soil, alongside pronounced soil thickness, clay content, total iron, and free iron-oxide content, all of which exhibited a reduction with increasing elevation in Fanjing Mountain [29,30]. Meanwhile, the annual average temperature decline rate is 0.5–0.56 °C/100 m, and the rainfall increases with the increase of elevation in Fanjing Mountain, which has obvious vertical zonal mountain climate attributes [30,31].

2.2. Study Plots and Soil Sample

In Sept 2022, the experimental sites were established at various elevations, from west to east, following the Niuwei River on the western slopes of Fanjing Mountain (Figure 1). The elevation sites were set at intervals of 300 m and split into an elevation range from the bottom to the top of 600–2400 m asl (Figure 1). At each elevation site, three replicate soil sampling plots (20 m × 20 m) more than 100 m apart were established at the same elevation along an "S" path drawn at each site. Soil samples were obtained from the 0–10 cm, 10–20 cm, 20–40 cm, and 40–60 cm layers at each elevation. A total of 72 (3 replicate samples × 6 elevations × 4 soil layers) soil samples were obtained from all elevation locations. All the soil samples were brought back to the laboratory and separated into two parts: one part was air-dried, sieved through 2.0 and 0.25 mm meshes, and stored at room temperature for the determination of soil physicochemical characteristics, and the other part was sieved through 2.0 mm mesh immediately stored at −80 °C for the determination of soil microorganisms. The soil profiles of different elevations are shown in Figure S1.

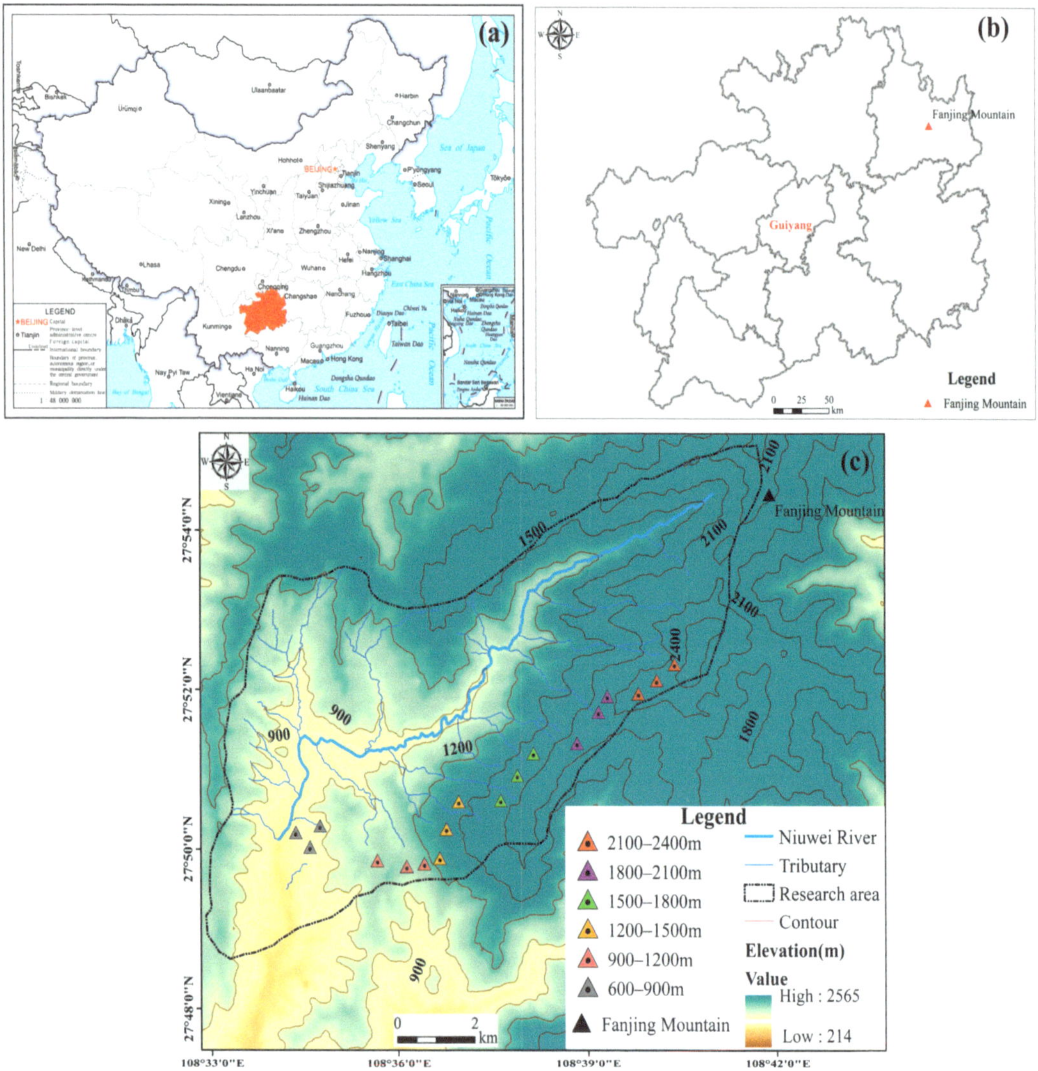

Figure 1. (**a**) Map of Guizhou in China. (**b**) Map of Fanjin Mountain in Guizhou, China. (**c**) Map of the study area in Fanjing Mountain, China.

2.3. Soil Physiochemical Property Determination

The soil pH was measured in a 1:2.5 (mass:volume) suspension of soil to water for 1–2 min via a pH meter (PHSJ-6L, shanghai yanfeng Electronic technology Co., Ltd., Shanghai, China) [7]. The soil cation exchange capacity (CEC) was evaluated using the method of ammonium chloride [21]. The soil electric conductivity (EC) was determined via a conductivity meter (DDSJ-318 T, INESA, Shanghai, China) [32]. The soil aggregates were separated and analyzed using the dry-sieving method with different particle sizes of >5, 5–2, 2–1, 1–0.5, 0.5–0.25, and <0.25 mm [33]. Soil organic carbon (SOC) was quantified by the oil bath heating potassium dichromate-volumetric method, soil total nitrogen (TN) was determined by heating digestion-alkaline hydrolysis and the diffusion-semimicro Kelvin's method, soil total phosphorus (TP) was measured by the molybdenum antimony colorimetric method, and soil total potassium (TK) was measured by the heating digestion-flame photometry method [21,32,33]. In addition, soil urease (SUE) activity was determined by the sodium phenol-sodium hypochlorite colorimetric method; soil sucrase (SS) activity was measured using the 3,5-dinitrosalicylic acid colorimetric method; the soil catalase (SCAT) activity was determined by titration with the KMnO4 method; and the soil acid phosphatase (SACP) activity was assessed via p-nitrophenyl phosphate colorimetry [20,21,25]. The deionized water was used to analyze the above indicators.

2.4. Soil DNA Extraction and High-Throughput Sequencing

The microbial DNA was extracted from 0.5 g of fresh weight soil using soil DNA extraction kits (Omega, Norcross, GA, USA), following the manufacturer's instructions. The purity and quality of the extracted total DNA were examined via 1% agarose gel electrophoresis using a NanoDrop2000 spectrophotometer (Thermo Fischer Scientific, Waltham, MA, USA). For bacteria, the V3–V4 regions of the 16S rRNA gene were amplified using the universal primers 338F (5′-ACTCCTACGGGAGGCAGCA-3′) and 806R (5′-GGACTACHVGGGTWTCTAAT-3′). For fungi, the universal primers ITS1 (5′-CTTGGTCA TTTAGAGGAAGTAA-3′) and ITS2 (5′-TGCGTTCTTCATCG ATGC-3′) were used to amplify the internal transcribed spacer (ITS) regions of fungal rRNA [34]. Polymerase chain reactions (PCRs) for bacteria and fungi were performed in detail [7]. The Illumina MiSeq platform (Illumina, San Diego, CA, USA) at Microeco Tech Co., Ltd. (Shenzhen, China) was used for high-throughput sequencing. The raw sequencing data were deposited into the NCBI Sequence Read Archive (SRA) database (BioProject ID: PRJNA1093166).

2.5. Sequencing Analysis

Reads were quality-filtered and demultiplexed to acquire available sequences via the QIIME (version 1.9.1) pipeline and based on the following criteria: (1) a minimum length of 150 bp, (2) Phred quality scores over 20, (3) no ambiguous bases, and (4) mononucleotide repeats <8 bp. The valid sequences were assigned to operational taxonomic units (OTUs) at 97% similarity using UPARSE software (version 7.0.1). The SILVA (release 128) and UNITE (version 7.2) databases were utilized for annotation of bacterial and fungal species information, respectively. The microbial taxonomic composition was drawn via R package version 3.2. The soil microbial alpha diversity was evaluated by the α- diversity (Chao 1 index, Shannon and Simpson index) indices via QIIME [11,34,35].

2.6. Statistical Analysis

One-way analysis of variance (ANOVA) and Tukey's multiple comparisons test were employed to determine the obvious differences in soil physicochemical characteristics (such as soil pH, TN, TK, and soil enzyme activities, etc.), and the significance level at $p < 0.05$ was determined via SPSS 23.0 (IBM, Armonk, NY, USA). The data are presented as the means $\pm$ standard deviations of triplicate samples. Pearson correlation analysis was implemented to estimate the relationships between the soil physicochemical characteristics and the soil microbial composition. All the graphs were generated via Origin 2022 (Origin Lab, Northampton, MA, USA). Redundancy analysis (RDA) was performed to analyze the relationships between physicochemical characteristics and the dominant genera in the soil microbial communities via Canoco 5.0 software [9]. The network visualization was carried out using Wekemo Bioincloud (https://www.bioincloud.tech/), an online platform for data analysis.

3. Results

3.1. Elevation-Induced Influences on Soil Properties at Different Soil Depth

The measured soil TN, TP, and TK concentrations varied with elevation from 600 m to 2400 m asl (Figures 2a,b and S2a). The soil TN concentration slightly decreased, whereas the TP and TK concentrations gradually tended to increase with elevation (Figures 2a,b and S2a). At the same elevation, the concentrations of TN and TP decreased with increasing soil depth, while the trend of the TK concentration was the opposite. The soil N concentration significantly differed among the different soil layers at the same elevation ($p < 0.05$). There were no significant differences between different elevation sites in the same soil layer at low and middle elevations (<1800 m asl), except at high elevations (>1800 m asl). The TP concentration did not vary significantly among the different soil layers at the same elevation when the elevation was less than 1500 m asl, whereas the opposite results were observed with increasing elevation (>1500 m asl). Similarly, there was a marked difference between high and low elevations within the same soil layer. In addition, the TK concentration of the same soil layer at high elevations (>1800 m asl) was remarkably greater than that at low and middle elevations (<1800 m asl). Interestingly, the TK concentration in the topsoil layer (0–10 cm) was lower than that in the deep soil layer (20–60 cm) at the same elevation. The above results revealed that the changes in the TN, TP, and TK concentrations in the surface soil and at high elevation in the Fanjing Mountain were more drastic than those at low elevations and in the deep soil layer.

We subsequently measured and analyzed the distribution of the changes in the soil aggregate size fraction with increasing elevation (Figure S2b). The main soil aggregates of Fanjing Mountain varied from microaggregates (<0.25 mm) to macroaggregates (>0.25 mm) in the same soil layer with increasing elevation, and the large macroaggregates (>2 mm) accounted for the largest proportion of the aggregate fraction (66.23%–76.13%) in the 0–10 cm soil layer. Moreover, the main soil aggregates changed significantly from macroaggregates (>0.25 mm) to microaggregates (<0.25 mm) with increasing soil depth at the same elevation. For example, the number of microaggregates (<0.25 mm) significantly increased from 10.79% to 32.64%, 4.87% to 7.89%, and 5.37% to 13.57% at 600–900 m asl, 1500–1800 m asl, and 2100–2400 m asl, respectively. Overall, with increasing elevation, the ratio of microaggregates in the same soil layer decreased and the proportion of macroaggregates increased, whereas the opposite trend occurred with increasing soil depth at the same elevation.

The soil pH was determined at different elevations 600–2400 m asl below 5.5, with the lowest soil pH being 4.09 in the topsoil layer (0–10 cm) at the 1800–2100 m asl (Figure S2c). Specifically, increased elevation did not significantly change the pH, but the soil pH was slightly acidic and generally increased with increasing soil depth. The average soil EC, SOC, and CEC concentrations in the 0–60 cm layer fluctuated with increasing elevation; the highest values were observed at 1500–1800 m asl or 1800–2100 m asl, and the maximum value was 1.64–3.36 times greater than the minimum value (Figures S2d and 2c,d). Moreover, the soil depth had a notable effect on the soil EC, SOC, and CEC, and their values in the topsoil layer (0–10 cm) were significantly greater than those in the subsurface layer ($p < 0.05$), indicating surface aggregation.

Our results revealed that the activities of soil urease (S-UE), sucrase (S-S), acid phosphatase (S-ACP), and catalase (S-CAT) clearly differed with variation at different elevation sites, and the enzyme activities were significantly greater in the topsoil layer (0–10 cm) than in the other soil layers ($p < 0.05$) (Figure 3a–d). Figure 3a shows the activity of soil urease detected at each elevation. The activity of soil urease was notably greater at sites 1200–1500 m asl and 1500–1800 m asl than at other elevations, and the lowest soil urease activity was found at 2100–2400 m asl, followed by 1800–2100 m asl. The urease activity of the surface soil layer was the highest at 1200–1500 m asl, which was 99.65% higher than that at 2100–2400 m asl. In contrast, the highest soil sucrase activity was observed at 600–900 m asl, and the lowest value was observed at 1800–2100 m asl, which linearly decreased with elevation (Figure 3b). Similar to the soil acid phosphatase and catalase activities, the lowest values were found at 2100–2400 m asl, and the variations in the two enzymatic activities were not significant at elevations less than 2100 m (Figure 3c,d). These results indicated that soil enzyme activity was significantly affected by elevation and soil layer.

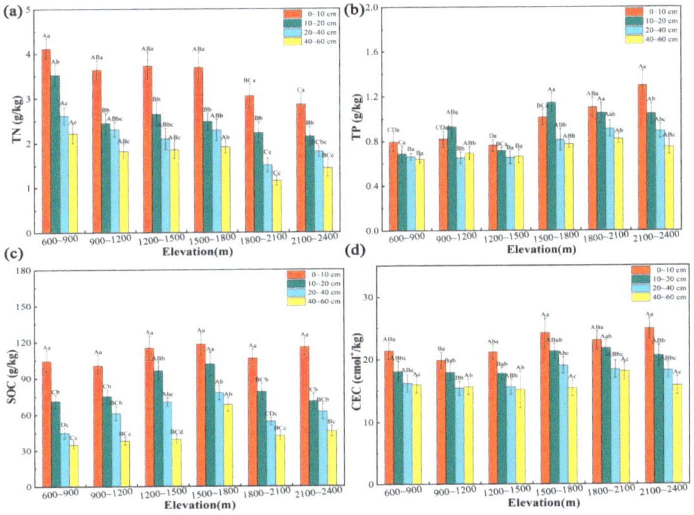

Figure 2. Soil properties, e.g., (**a**) TN, (**b**) TP, (**c**) SOC and (**d**) CEC under the different elevation sites. TN: total nitrogen; SOC: soil organic carbon; CEC: cation exchange capacity. Different capital letters indicate significant differences between different elevations in the same soil depth according to Tukey's HSD test. Different lowercase letters indicate significant differences between different soil depths at the same elevations. Bars indicate standard errors (mean ± SE, *n* = 3). The number in the upper right corner of the figure represents the soil depth.

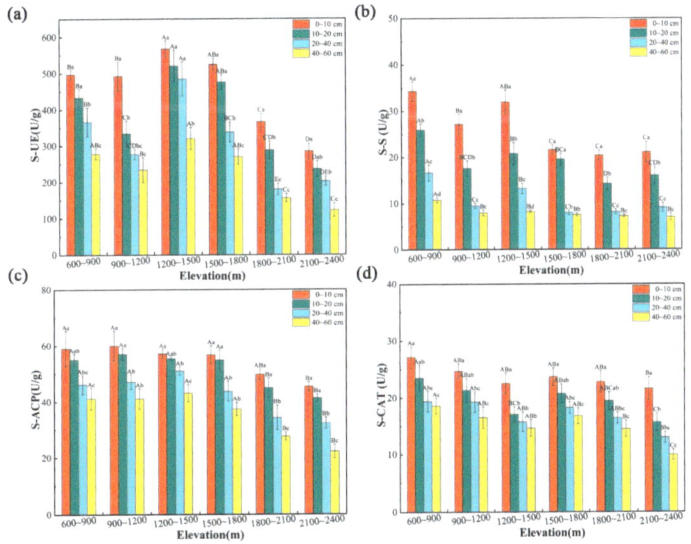

Figure 3. Soil enzyme activities, e.g., (**a**) soil urease, (**b**) soil sucrase, (**c**) soil acid phosphatase and (**d**) soil catalase along the elevation sites. S-UE: soil urease; S-S: soil sucrase; S-ACP: soil acid phosphatase; S-CAT: soil catalase. Different capital letters indicate significant differences between different elevations at the same soil depth according to Tukey's HSD test. Different lowercase letters indicate significant differences between different soil depths at the same elevation. Bars represent standard errors (mean ± SE, *n* = 3). The number in the upper right corner of the figure represents the soil depth.

3.2. Influence of Elevation on Microbial Composition

High-throughput pyrosequencing was used to characterize the influences of elevation on the alpha diversity of soil microbial communities at different depths in the soil. With respect to the topsoil layer bacterial richness (Chao1) and α-diversity (Shannon and Simpson indices), three indices generally first increased and then decreased with increasing elevation (Figure 4a–c). The highest bacterial Chao1, Shannon and Simpson index values were found at 1500–1800 m, at 1928.9, 9.60, and 0.9963, and the lowest values of the three indices were found at 2100–2400 m asl, at 1425.51, 9.03, and 0.9945, respectively. However, the Shannon and Simpson indices did not obviously differ at the different elevations. For fungi, the Chao1, Shannon and Simpson indices significantly decreased at 2100–2400 m asl compared with those 600–900 m asl and 900–1200 m asl and decreased with increasing elevation (Figure 4f,h). Additionally, the α-diversity of the fungal community was more strongly influenced by elevation than was that of the bacterial community.

The dominant classified bacterial and fungal phyla and genera were chosen to explore variations in the soil microbial communities at different elevations. Figure 4d shows that the main bacterial phyla were Acidobacteria, Pseudomonadota, Chloroflexi, Actinobacteria, and Verrucomicrobia, accounting for 42.51%–50.30%, 29.78%–40.74%, 1.5%–10.89%, 2.52%–5.24%, and 2.18%–4.55%, respectively, of the relative abundance of bacterial phyla at each spot, indicating that the relative abundance of the main bacterial phyla fluctuated pattern with increasing elevation. At the genus level, the primary bacterial species were *Candidatus Solibacter, Rhodoplanes,* and *Candidatus Koribacter*, which accounted for more than 22% of the bacterial sequences (Figure 4e). The relative abundances of *Rhodoplanes* and *Burkholderia* first increased but then decreased with increasing elevation. In addition, the soil fungal communities at all elevations were dominated by *Basidiomycota* (ranging from 31.63% to 69.96%), Ascomycota (ranging from 9.78% to 40.35%), unclassified (ranging from 5.66% to 32.09%), and Mortierellomycota (ranging from 3.51% to 11.20%) at the phylum level (Figure 4i). Specifically, the relative abundances of Basidiomycota and Mortierellomycota first increased but then decreased with increasing elevation. At the genus level, the dominant fungi were *Mortierella* among Mortierellomycota and *Russula, Sebacina,* and *Inocybe* among Basidiomycota (Figure 4j).

3.3. Relationships Among the Microbial Community, Enzymatic Activity, and Characteristics of the Soil

The partial Mantel test was conducted to evaluate the main variables driving the changes in the soil microorganism structure, soil enzyme activity, and soil physicochemical properties (Figure 5a). We found that the TK, SC, SACP, and SCAT were the determining factors of the soil bacterial communities at the different elevation sites (Figure 5a). Although these soil physicochemical characteristics had no remarkable influence on the soil fungal communities, pH, EC, SA5, and SOC had a marginal influence on the fungal community structure. Furthermore, Pearson correlation analysis was performed on the soil properties, revealing that pH, SUE, SC, SACP, SCAT, and $SA_{0.25}$ were positively correlated with TN; TK, EC, CEC, and SA_{21} were positively correlated with TP; and SOC was positively correlated with CEC (Figure 5a). Interestingly, there were positive correlations between SUE, SC, SACP, and SCAT. Moreover, TN was strongly negatively correlated with TP, TK, EC, and SA_{21}; EC was notably negatively correlated with pH, SUE, SC, and $SA_{0.5}$; and CEC was strongly negatively correlated with the activities of the four soil enzymes. In addition, pH, SUE, SC, SACP, and $SA_{0.25}$ were significantly negatively correlated with TP.

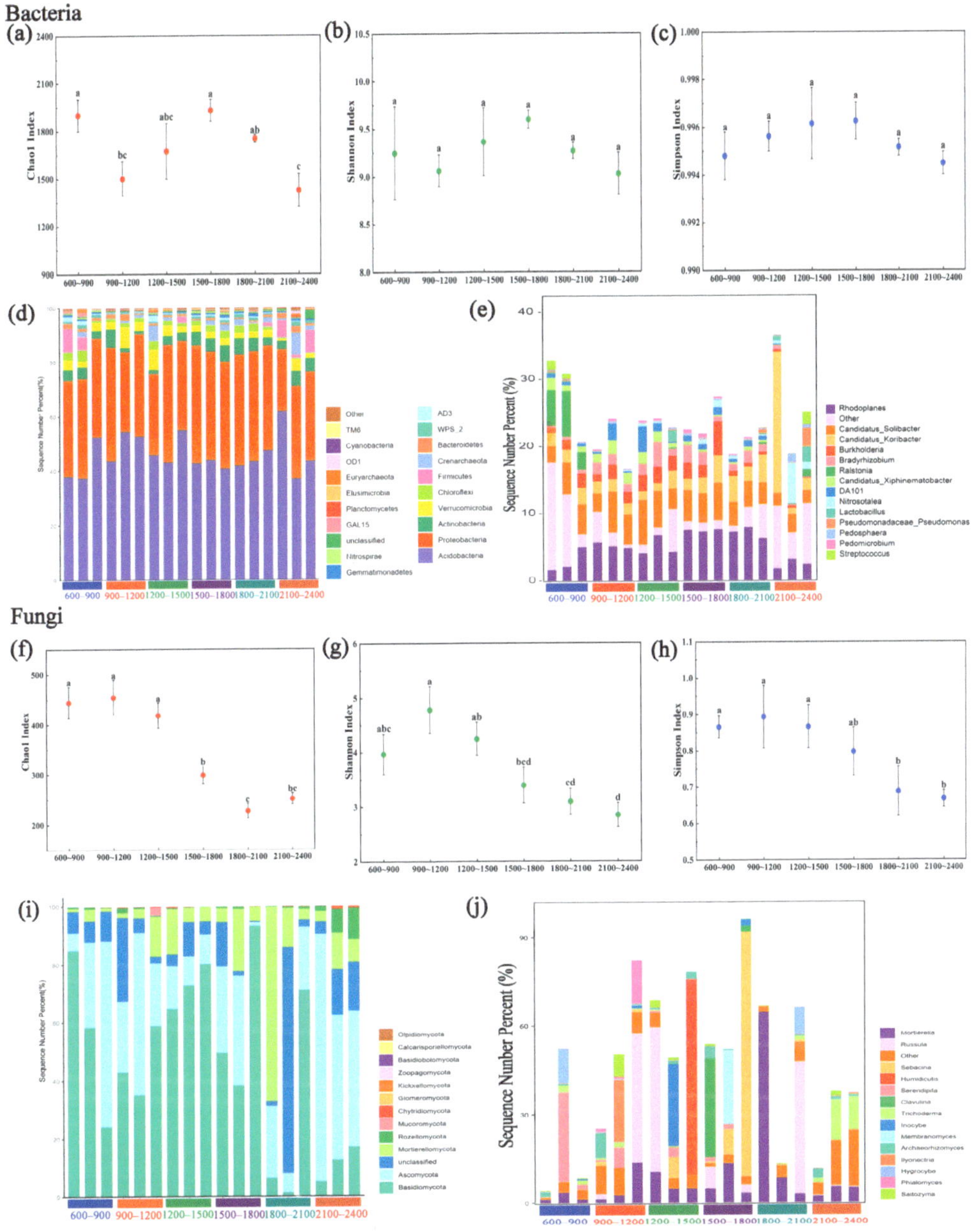

Figure 4. Soil microbial community structure at different elevations. Comparison of bacterial alpha diversity at different elevation sites based on the Chao 1 (**a**), Shannon (**b**), and Simpson (**c**) diversity indices. Relative abundances of bacteria at the phylum level (**e**) and genus level (**d**). Comparison of fungal alpha diversity among different elevation sites based on the Chao 1 (**f**), Shannon (**g**), and Simpson (**h**) diversity indices. Relative abundances of fungi at the phylum level (i) and genus level (**j**). Significant differences were measured by Tukey's HSD test, and different lowercase letters above the point graph denote significant differences at $p < 0.05$ under different elevation sites.

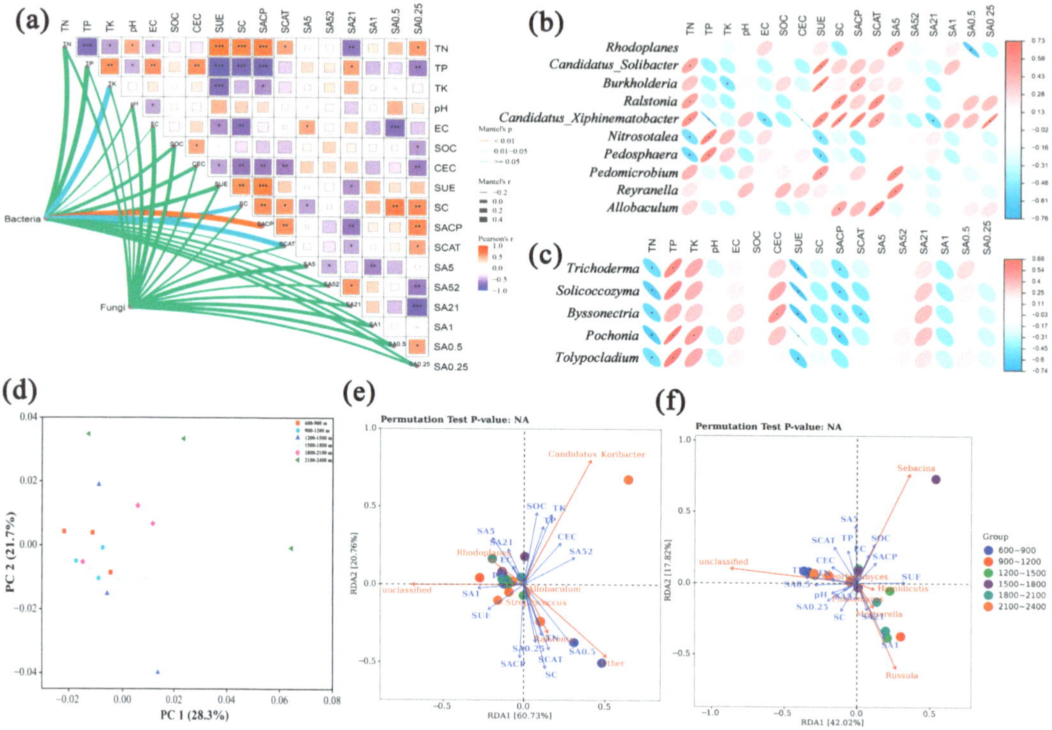

Figure 5. Mantel test of soil bacteria and fungi with respect to soil physicochemical characteristics at different elevations. (**a**). Pearson correlations between predominant bacterial (**b**) and fungal (**c**) genus and soil physicochemical characteristics at different elevations. PCA analysis of soil bacteria community in different elevations (**d**). RDA between elevation and soil properties (blue arrows) and the predominant (relative abundance) bacterial (**e**) and fungal (**f**) genus (red arrows) in Fanjing Mountain. Note: SA_5, soil aggregate (>5 mm); SA_{52}, soil aggregate (5–2 mm); SA_{21}, soil aggregate (2–1 mm); SA_1, soil aggregate (1–0.5 mm); $SA_{0.5}$, soil aggregate (0.5–0.25 mm); $SA_{0.25}$, soil aggregate (<0.25 mm); SUE, soil urease; SC, soil sucrase; SACP, soil acid phosphatase; SCAT, soil catalase. * $p < 0.05$, ** $p < 0.01$, *** $p < 0.001$.

The Pearson correlations between the soil microbial communities at the genus level and the soil property indices are shown in Figure 5b,c. For the bacterial communities (Figure 5b), we observed that TN was significantly positively correlated with the relative abundances of 3 bacterial genera, *Candidatus Solibacter*, *Ralstonia*, and *Candidatus Xiphinematobacter*; TK was positively correlated with *Nitrosotalea* and *Pedosphaer*; and TN was negatively correlated with these genera. *Ralstonia*, *Candidatus Xiphinematobacter,* and *Allobaculum* were positively related to SC and SCAT. Similarly, the SUE was positively correlated with *Candidatus Solibacter*, *Burkholderia*, *Candidatus Xiphinematobacter,* and *Pedomicrobium* but negatively correlated with *Nitrosotalea* and *Pedosphaer.* However, *Candidatus Xiphinematobacter* was negatively correlated with TP, EC, CEC, and SA_{21}. TK was negatively correlated with the relative abundance of *Burkholderia.* In addition, for the fungal communities (Figure 5c), TN, SUE, and SACP were negatively correlated with *Trichoderma*, *Solicoccozyma*, *Byssonectria*, and *Pochonia*, whereas TP was significantly positively related. *Pochonia* was positively correlated with TK, and CEC was positively correlated with *Byssonectria.* In general, some soil physiochemical factors, such as TN, pH, SA1, etc., were positively correlated with soil enzyme activity, whereas other factors, such as TP, EC, SOC, etc., were negatively correlated, and they were both positively and negatively correlated with soil microorganisms.

Further insight into the difference of soil microbial community at different elevations was gained through principal component analysis (PCA); two principal components of soil bacteria and fungus were extracted, and the total interpretation was 50.0% and 50.9%, respectively (Figures 5d and S3). PCA analysis results showed that soil microbial community structure was more dispersed at high elevation than at low elevation. In addition, to further evaluate the remarkable environmental factors affecting the dominant genera in the microbial communities, a redundancy analysis (RDA) was conducted (Figure 5d,e). According to the RDA, the first axis was closely related to pH, SA_1, SUE, EC, and SA_{52}, which explained 60.73% of the total variation in the soil bacterial community structure. However, the second axis was related primarily to SACP, SOC, SA0.25, and SC and explained 20.76% of the variation. SC, SOC, TK, $SA_{0.5}$, and SACP were significant factors influencing the bacterial community (Figure 5e). Meanwhile, axis 1 and axis 2 represented 42.02% and 17.82% of the total variation in the soil fungal community composition, respectively, and SA_1, SA_5, SUE, TK, SOC, and SCAT had substantial influences on the fungal communities (Figure 5f).

4. Discussions

The Fanjing Mountain is in a representative karst rocky desertification region in Southwest China [28,31] and is a considerable part of terrestrial ecosystems. In the present study, we investigated how elevation influences soil physiochemical characteristics, soil enzyme activities, and corresponding changes in soil bacterial and fungal community composition on the western slopes of Fanjing Mountain at elevations ranging from 600–2400 m. These changes obviously differed with the alteration in spatial scale.

4.1. Soil Characteristics and Enzymatic Activities with Elevation

Soil characteristics are shaped primarily by climate, organisms, parent material, and elevation, probably by regulating soil biogeochemical cycles, which influence plant growth and development, and plant-associated microbiota [1,23,25,36–38]. This study revealed a significant decrease in the soil TN concentration with increasing elevation on Fanjing Mountain, which may be related to elevation changes in soil temperature and precipitation, affecting nitrogen mineralization and inorganic nitrogen production [5,19]. The soil TP and EC concentrations had generally tended to increase with increasing elevation. Moreover, the soil TK and CEC concentrations did not markedly change with elevation. This difference might be associated with changes in the natural environment of Fanjing Mountain, which could influence litter decomposition, the degree of soil weathering, and other inputs into the soil to some degree [21,39]. In addition, we found that the soil pH and SOC usually exhibited a single-peaked model with increasing elevation. The changes in the characteristics of pH and SOC could be ascribed to one reason: the study area is situated in a mountainous area, where the plants are flourishing and the vegetation coverage rate is high, and a thick litter, as the primary source of soil nutrients, accumulates largely in the topsoil. A decrease in temperature because elevation decreases soil evaporation and the soil respiration rate, leading to changes in pH and SOC concentrations [3,5,40]. Importantly, we found that the aforementioned results decreased with increasing soil depth, except for the soil pH. These differences might be associated with differences in organic matter input, nutrient availability, climate type, and litter decomposition [41]. Previous studies have reported inconsistent soil physiochemical property patterns at various elevations [20,21,39]. Overall, these results might reveal a different influence of elevation on soil characteristic factors.

The activity of soil enzymes is an important indicator of the health status and microbial activity of a soil, and these enzymes are prone to fast turnover and high susceptibility to quick responses to variations in elevation and the soil environment [42–44]. In this study, we observed a significant effect of elevation on S-UE, S-S, S-ACP, and S-CAT, with the lowest values occurring from 2100–2400 m asl, whereas there was no significant difference in S-ACP and S-CAT and a slight increase in S-UE at elevations ranging from 600–2100 m asl. The results are consistent with this hypothesis, and these results have both similarities and differences with those of previous studies [42,45,46]. This discrepancy could be caused by the diverse vegetation cover, plant root exudates, and climatic regimes at specific elevations because soil enzyme activities are very susceptible to environmental signals [17,22,24,27]. In addition, we observed that soil enzyme activity decreased with increasing depth in the soil layer at the same elevation, which was in accordance with the results of previous studies [27,46]. This might be because of the abundant litter return, good hydrothermal, and aeration conditions at the soil layer surface, which provide an appropriate living environment and many nutrient substances for microorganism growth; meanwhile, microbial metabolism is vigorous, leading to a relatively high active matter concentration [46,47]. Additionally, elevation directly affects the changes in soil enzyme activities by affecting soil characteristics, such as pH, SOC, and soil aggregates. In summary, elevation has a profound effect on soil physicochemical characteristics and enzyme activity in Fanjing Mountain, and this effect also plays a crucial role in the formation of soil microbial diversity and the shaping of ecological functions in Fanjing Mountain.

4.2. Differential Responses of Microbial Structures Across Elevation Sites

Here, we researched soil microbial diversity and community structure at different elevation sites. Although the trends in bacterial and fungal abundance and diversity exhibit single-peak patterns with elevation, the trends in bacterial diversity were not significant at all elevations, which might indicate that elevation does not directly affect bacterial diversity but has an indirect influence via other environmental factors [5,12]. These results support our hypothesis that there is a strong elevational influence on microbial diversity, especially fungal diversity. Our results were similar to those of several previous studies, for example, at different elevation sites on Mount Kilimanjaro in East Africa and Sygera Mountain (Tibet, China) [30,35]. However, Yang et al. [41] reported that the alpha-diversity and fungal richness of soil bacteria first decreased and then increased with elevation in southern Himalaya. This is likely because the relative contributions of elevational influences on microbial diversity rely primarily on their relative effects on the soil environment. Additionally, the soil bacterial diversity was greater than the fungal diversity. These findings are in line with those of Yang et al. [41], who suggested that the different microbial niches in the Fanjing Mountain differ with elevation. In brief, the inconsistency of soil microbial diversity patterns at different elevation sites may be due to varying environmental factors induced by elevation.

The dominant phyla of the soil bacterial communities along the elevational gradient were the same, consisting of Acidobacteria, Pseudomonadota, and Chloroflexi. This finding is basically consistent with those of previous studies [3,7,41]. Several previous studies have shown that Acidobacteria are oligotrophic bacteria that can degrade complex lignin and cellulose and provide nutrients for the soil [34]. Pseudomonadota contain a variety of metabolic nitrogen-fixing bacterial species and are involved in a variety of biochemical cycles [35]. These results suggested that the bacterial populations of these phyla have strong adaptability. In addition, for fungal communities, the dominant fungal phyla were Ascomycota and Basidiomycota, with Ascomycota accounting for greater relative abundance at higher elevations and Basidiomycota accounting for greater relative abundance at lower elevations. Previous studies have demonstrated that Basidiomycota have the capacity to degrade aliphatic/protein and Ascomycota are common saprophytic bacteria in soil that can decompose refractory organic matter and play a considerable role in the soil nutrient cycling process and the function and stability of soil microorganisms [3,35]. We speculate that the advantage of Basidiomycota at lower elevations could be correlated with their capacity to degrade complicated lignocellulose components because the vegetation species structure at relatively low elevations is more complicated and the soil litter layer is thicker in Fanjing Mountain. In contrast, Ascomycota can utilize more resources and are more resistant to environmental pressures, increasing its survival benefit at high elevations [5,35]. Together, these results indicated that, to a large extent, elevation determined the soil microbial diversity and community composition, which could be one of the reasons for the complex biodiversity of Fanjing Mountain.

4.3. Effects of Soil Properties on the Soil Microbial Community

Many studies have investigated the relationships among soil physiochemical characteristics, soil enzyme activity, and soil microbial communities, but there is no consensus on this pattern or a consistent understanding of the factors driving it [21,31,45]. On the basis of the results of the Mantel test and RDA, TK, SC, and SACP strongly influenced the bacterial community composition, and TK, SA_1, SUE, and SOC substantially influenced the fungal community at the different elevation sites. Several previous studies have shown the alterations in TK and SOC could be affected by litter input and vegetation productivity, which likely shape the distinct microbial communities by supplying various resources for microbial growth [5,34,35]. In addition, our results also demonstrated that soil enzyme activity and soil aggregates can influence microbial community composition, to some extent, which is in line with the results of previous studies [11,42,48].

Generally, elevation strongly influences climatic properties (e.g., temperature and precipitation), therefore indirectly affecting soil microbial community composition [6,12,48]. Moreover, the variation in soil nutrients attracted by an increase in elevation may influence the soil microbial community structure indirectly [49,50]. Moreover, temperature is closely connected with soil enzyme activities and litter decomposition; thus, the influences of elevation on soil microbial diversity and communities can be reflected indirectly by alterations in soil characteristics (e.g., SOC, soil nutrients, and enzyme activity). Our findings provide a theoretical basis for protecting Fanjing Mountain ecology, vegetation restoration, and soil management. However, the primary limitation of the present work is that the narrow observable results obtained from this small-scale elevational study might not be applicable for large-scale elevational studies; thus, further investigations are needed.

5. Conclusions

This study revealed the different responses of soil properties and the microbial community along a continuous elevational gradient (600–2400 m asl) on Fanjing Mountain. Elevation significantly affects soil characteristics and enzyme activities, especially at the highest elevation. With increasing soil depth, they tended to decrease, resulting in surface aggregation. In addition, the soil bacterial and fungal community diversity presented a single-peak pattern with elevation. The predominant bacterial species were the phyla Acidobacteria, Pseudomonadota, and Chloroflexi and the genus *Candidatus Solibacter*, and *Rhodoplanes*. The phyla Basidiomycot and Ascomycota and the genus *Mortierella*, *Russula*, and *Sebacina* dominated the fungal communities. The soil SOC, TK, and enzyme activities (SC, SUE) were the main factors that drove the elevational distribution of the microbial communities. The soil microorganism community composition varied greatly among the elevations. The results of this study contribute to the understanding of the interactions among soil properties and microorganisms with increasing elevation, which provides insights into the influence of elevation on soil characteristics, microbial diversity, and communities in the Fanjing Mountain Forest ecosystem. In the future, we need to study how climate change will affect Fanjing Mountain ecosystems and continue to track the relationship between biological indicators and environmental change.

Supplementary Materials: The following supporting information can be downloaded at: https://www.mdpi.com/article/10.3390/f15111980/s1, Figure S1: Soil profile at different elevations; Figure S2: Soil properties under the different elevation gradient. TK: Total potassium, EC: Electrical conductivity. Different capital letters indicate significant differences between different elevations in the same soil depth according to Tukey's HSD test. Different lowercase letters indicate significant differences between different soil depth in the same elevations. Bars indicate standard errors (±SE, $n = 3$); Figure S3: PCA analysis of soil fungi community in different elevations.

Author Contributions: J.W.: Writing-original draft, Writing-review and editing, Data curation. S.X.: Data curation, Investigation. K.H.: Writing–review and editing, Methodology. X.L.: Writing–review and editing, Methodology. J.C.: Data curation, Investigation. L.Z.: Formal analysis, Methodology. Y.X.: Conceptualization, Funding acquisition, Methodology, Supervision. All authors have read and agreed to the published version of the manuscript.

Funding: This work was supported by the Guizhou Provincial Science and Technology Projects (QKH-ZK [2022] 276, QKH-ZK [2023] 231, QKH-ZK [2021] 100), Guizhou Academy of Sciences Projects (No. QKY [R] [2023] 04 and QKY [R] [2023] 05), Guizhou Provincial Development and Reform Investment Project (No. Qian-Development-Reform-Investment [2019] 886), and Innovative Talent Team Project of Guizhou Academy of Sciences (No. QKY-Talent [2019] 07).

Data Availability Statement: Data will be made available on request.

Acknowledgments: We would like to thank the Fanjingshan Forest Ecological Station, Guizhou Academy of Sciences, for their help during sampling.

Conflicts of Interest: The authors declare that they have no known competing financial interests or personal relationships that could have appeared to influence the work reported in this paper.

References

1. Cui, H.J.; Wang, G.X.; Yang, Y.; Yang, Y.; Chang, R.Y.; Ran, F. Soil microbial community composition and its driving factors in alpine grasslands along a mountain elevational gradient. *J. Mt. Sci.* **2016**, *13*, 1013–1023. [CrossRef]
2. Ivashchenko, K.; Sushko, S.; Selezneva, A.; Ananyeva, N.; Zhuravleva, A.; Kudeyarov, V.; Makarov, M.; Blagodatsky, S. Soil microbial activity along an altitudinal gradient: Vegetation as a main driver beyond topographic and edaphic factors. *Appl. Soil Ecol.* **2021**, *168*, 104197. [CrossRef]
3. Yang, Y.; Qiu, K.; Xie, Y.; Li, X.; Zhang, S.; Liu, W.; Huang, Y.; Cui, L.; Wang, S.; Bao, P. Geographical, climatic, and soil factors control the altitudinal pattern of rhizosphere microbial diversity and its driving effect on root zone soil multifunctionality in mountain ecosystems. *Sci. Total Environ.* **2023**, *904*, 166932. [CrossRef] [PubMed]
4. Wang, Y.; Xue, D.; Hu, N.; Lou, Y.; Zhang, Q.; Zhang, L.; Zhu, P.; Gao, H.; Zhang, S.; Zhang, H.; et al. Post-agricultural restoration of soil organic carbon pools across a climate gradient. *Catena* **2021**, *200*, 105138. [CrossRef]
5. Wang, S.; Heal, K.V.; Zhang, Q.; Yu, Y.; Tigabu, M.; Huang, S.; Zhou, C. Soil microbial community, dissolved organic matter and nutrient cycling interactions change along an elevation gradient in subtropical China. *J. Environ. Manag.* **2023**, *345*, 118793. [CrossRef]
6. Shen, C.; He, J.-Z.; Ge, Y. Seasonal dynamics of soil microbial diversity and functions along elevations across the tree line. *Sci. Total Environ.* **2021**, *794*, 148644. [CrossRef]
7. Xiang, J.; Gu, J.; Wang, G.; Bol, R.; Yao, L.; Fang, Y.; Zhang, H. Soil pH controls the structure and diversity of bacterial communities along elevational gradients on Huangshan, China. *Eur. J. Soil Biol.* **2024**, *120*, 103586. [CrossRef]
8. Yu, C.; Han, F.; Fu, G. Effects of 7 years experimental warming on soil bacterial and fungal community structure in the Northern Tibet alpine meadow at three elevations. *Sci. Total Environ.* **2019**, *655*, 814–822. [CrossRef]
9. Feng, E.; Zhang, L.; Kong, Y.; Xu, X.; Wang, T.; Wang, C. Distribution Characteristics of Active Soil Substances along Elevation Gradients in the Southern of Taihang Mountain, China. *Forests* **2023**, *14*, 370. [CrossRef]
10. Jia, W.; Huang, P.; Zhu, K.; Gao, X.; Chen, Q.; Chen, J.; Ran, Y.; Chen, S.; Ma, M.; Wu, S. Zonation of bulk and rhizosphere soil bacterial communities and their covariation patterns along the elevation gradient in riparian zones of three Gorges reservoir, China. *Environ. Res.* **2024**, *249*, 118383. [CrossRef]
11. Zhao, Y.; Zhou, Y.; Jia, X.; Han, L.; Liu, L.; Ren, K.; Ye, X.; Qu, Z.; Pei, Y. Soil characteristics and microbial community structure on along elevation gradient in a *Pinus armandii* forest of the Qinling Mountains, China. *For. Ecol. Manag.* **2022**, *503*, 119793. [CrossRef]
12. Liu, J.J.; Jin, L.; Shan, Y.X.; Burgess, K.S.; Ge, X.J. Elevation explains variation in soil microbial diversity and community composition under experimental warming and fertilization treatments in mountain meadows. *Appl. Soil Ecol.* **2022**, *171*, 104311. [CrossRef]
13. Ma, T.; Zhang, X.; Wang, R.; Liu, R.; Shao, X.; Li, J.; Wei, Y. Linkages and key factors between soil bacterial and fungal communities along an altitudinal gradient of different slopes on mount Segrila, Tibet, China. *Front. Microbiol.* **2022**, *13*, 1024198. [CrossRef] [PubMed]
14. Zhang, J.; Xu, M.; Zou, X.; Chen, J. Structural and functional characteristics of soil microbial community in a *Pinus massoniana* forest at different elevations. *PeerJ* **2022**, *10*, e13504. [CrossRef]
15. Murugan, R.; Djukic, I.; Keiblinger, K.; Zehetner, F.; Bierbaumer, M.; Zechmeister-Bolternstern, S.; Joergernsen, R.G. Spatial distribution of microbial biomass and residues across soil aggregate fractions at different elevations in the Central Austrian Alps. *Geoderma* **2019**, *339*, 1–8. [CrossRef]
16. Hu, L.; Xiang, Z.; Wang, G.; Rafique, R.; Liu, W.; Wang, C. Changes in soil physicochemical and microbial properties along elevation gradients in two forest soils. *Scand. J. For. Res.* **2016**, *31*, 242–253. [CrossRef]

17. Zhu, B.; Li, C.; Wang, J.; Li, J.; Li, X. Elevation rather than season determines the assembly and co-occurrence patterns of soil bacterial communities in forest ecosystems of Mount Gongga. *Appl. Microbiol. Biotechnol.* **2020**, *104*, 7589–7602. [CrossRef]
18. Guo, G.; Kong, W.; Liu, J.; Zhao, J.; Du, H.; Zhang, X.; Xia, P. Diversity and distribution of autotrophic microbial community along environmental gradients in grassland soils on the Tibetan Plateau. *Appl. Microbiol. Biotechnol.* **2015**, *99*, 8765–8776. [CrossRef]
19. Li, N.; Du, H.; Li, M.H.; Na, R.; Dong, R.; He, H.S.; Zong, S.; Huang, L.; Wu, Z. Deyeuxia angustifolia upward migration and nitrogen deposition change soil microbial community structure in an alpine tundra. *Soil Biol. Biochem.* **2023**, *180*, 109009. [CrossRef]
20. Liu, Z.; Fang, J.; Song, B.; Yang, Y.; Yu, Z.; Hu, J.; Dong, K.; Takahashi, K.; Adams, J.M. Stochastic processes dominate soil arbuscular mycorrhizal fungal community assembly along an elevation gradient in central Japan. *Sci. Total Environ.* **2023**, *855*, 158941. [CrossRef]
21. Luo, H.; Wang, C.; Zhang, K.; Ming, L.; Chu, H.; Wang, H. Elevational changes in soil properties shaping fungal community assemblages in terrestrial forest. *Sci. Total Environ.* **2023**, *900*, 165840. [CrossRef] [PubMed]
22. Abbott, K.M.; Quirk, T.; Fultz, L.M. Soil microbial community development across a 32-year coastal wetland restoration time series and the relative importance of environmental factors. *Sci. Total Environ.* **2022**, *821*, 153359. [CrossRef] [PubMed]
23. Whitaker, J.; Ostle, N.; Nottingham, A.T.; Ccahuana, A.; Salinas, N.; Bardgett, R.D.; Meir, P.; McNamara, N.P. Microbial community composition explains soil respiration responses to changing carbon inputs along an Andes-to-Amazon elevation gradient. *J. Ecol.* **2014**, *102*, 1058–1071. [CrossRef]
24. Chen, W.; Wang, J.; Chen, X.; Meng, Z.; Xu, R.; Duoji, D.; Zhang, J.; He, J.; Wang, Z.; Chen, J.; et al. Soil microbial network complexity predicts ecosystem function along elevation gradients on the Tibetan Plateau. *Soil Biol. Biochem.* **2022**, *172*, 108766. [CrossRef]
25. Bhople, P.; Djukic, I.; Keiblinger, K.; Zehetner, F.; Liu, D.; Bierbaumer, M.; Zechmeister-Boltenstern, S.; Joergensen, R.G.; Murugan, R. Variations in soil and microbial biomass C, N and fungal biomass ergosterol along elevation and depth gradients in Alpine ecosystems. *Geoderma* **2019**, *345*, 93–103. [CrossRef]
26. Ni, B.; Zhao, W.; Zuo, X.; You, J.; Li, Y.; Li, J.; Du, Y.; Chen, X. Deyeuxia angustifolia Kom. encroachment changes soil physicochemical properties and microbial community in the alpine tundra under climate change. *Sci. Total Environ.* **2022**, *813*, 152615. [CrossRef]
27. Ren, C.; Zhou, Z.; Guo, Y.; Yang, G.; Zhao, F.; Wei, G.; Han, X.; Feng, L.; Feng, Y.; Ren, G. Contrasting patterns of microbial community and enzyme activity between rhizosphere and bulk soil along an elevation gradient. *Catena* **2021**, *196*, 104921. [CrossRef]
28. Chen, H.; Zhou, Y.; Fei, Y.; Wei, Q. Differences of Moss Mites Communities at Different Vegetation Succession Stages in Subalpine Wetland (Jiulongchi, Fanjing Mountain), Southwest China. *Forests* **2023**, *14*, 332. [CrossRef]
29. Luo, W.; Liu, Y.; Liu, Y.; Mu, G.; Wu, X. A Study on the Evolutionary Characteristics of Soil Properties and Their Drivers in Central Subtropical Forests: The Case of Fanjing Mountains in Southwest China. *Pol. J. Environ. Stud.* **2024**, *33*, 1813–1822. [CrossRef]
30. Xiong, H.; Yu, F.; Gu, X.; Wu, X. Biomass, net production, carbon storage and spatial distrubution features of different forest vegetation in Fanjing Mountains. *Ecol. Environ. Sci.* **2021**, *30*, 264–273.
31. Zhang, Z.; Wu, X.; Zhang, J.; Liu, Y.; Luo, W.; Mou, G. Community assembly of bacterial generalists and specialists and their network characteristics in different altitudinal soils on Fanjing Mountain in Southwest China. *Catena* **2024**, *238*, 107863. [CrossRef]
32. Bao, S.D. *Soil and Agricultural Chemistry Analysis*; Agriculture Press of China: Beijing, China, 2000.
33. Lv, Z.; Rønn, R.; Liao, H.; Rensing, C.; Chen, W.; Huang, Q.; Hao, X. Soil aggregates affect the legacy effect of copper pollution on the microbial communities. *Soil Biol. Biochem.* **2023**, *182*, 109048. [CrossRef]
34. Huang, K.; Xiang, J.; Ma, Y.; Cheng, J.; Gu, J.; Hu, M.; Yang, Y.; Fang, Y.; Wang, G.; Zhang, H. Response of Soil Microbial Communities to Elevation Gradient in Central Subtropical *Pinus taiwanensis* and *Pinus massoniana* Forests. *Forests* **2023**, *14*, 772. [CrossRef]
35. Zhang, Y.; Heal, K.V.; Shi, M.; Chen, W.; Zhou, C. Decreasing molecular diversity of soil dissolved organic matter related to microbial community along an alpine elevation gradient. *Sci. Total Environ.* **2022**, *818*, 151823. [CrossRef] [PubMed]
36. Frindte, K.; Pape, R.; Werner, K.; Löffler, J.; Knief, C. Temperature and soil moisture control microbial community composition in an arctic-alpine ecosystem along elevational and micro-topographic gradients. *ISME J.* **2019**, *13*, 2031–2043. [CrossRef] [PubMed]
37. Gao, J.; Muhanmmad, S.; Yue, L.; He, Y.; Tsechoe, D.; Zhang, X.; Kong, W. Changes in CO_2-fixing microbial community characteristics with elevation and season in alpine meadow soils on the northern Tibetan Plateau. *Shengtai Xuebao* **2018**, *38*, 3816–3824.
38. Li, X.; Xie, J.; Zhang, Q.; Lyu, M.; Xiong, X.; Liu, X.; Lin, T.; Yang, Y. Substrate availability and soil microbes drive temperature sensitivity of soil organic carbon mineralization to warming along an elevation gradient in subtropical Asia. *Geoderma* **2020**, *364*, 114198. [CrossRef]
39. Aqeel, M.; Khalid, N.; Noman, A.; Ran, J.; Manan, A.; Hou, Q.; Dong, L.; Sun, Y.; Deng, Y.; Lee, S.S.; et al. Interplay between edaphic and climatic factors unravels plant and microbial diversity along an altitudinal gradient. *Env. Res.* **2024**, *242*, 117711. [CrossRef]
40. Yang, L.; Lyu, M.; Li, X.; Xiong, X.; Lin, W.; Yang, Y.; Xie, J. Decline in the contribution of microbial residues to soil organic carbon along a subtropical elevation gradient. *Sci. Total Environ.* **2020**, *749*, 141583. [CrossRef]

41. Yang, N.; Li, X.; Liu, D.; Zhang, Y.; Chen, Y.; Wang, B.; Hua, J.; Zhang, J.; Peng, S.; Ge, Z.; et al. Diversity patterns and drivers of soil bacterial and fungal communities along elevational gradients in the Southern Himalayas, China. *Appl. Soil Ecol.* **2022**, *178*, 104563. [CrossRef]
42. Liu, S.; Xu, G.; Chen, H.; Zhang, M.; Cao, X.; Chen, M.; Chen, J.; Feng, Q.; Shi, Z. Contrasting responses of soil microbial biomass and extracellular enzyme activity along an elevation gradient on the eastern Qinghai-Tibetan Plateau. *Front. Microbiol.* **2023**, *14*, 974316. [CrossRef] [PubMed]
43. Martin-Sanz, J.P.; Valverde-Asenjo, I.; de Santiago-Martin, A.; Quintana-Nieto, J.R.; Gonzalez-Huecas, C.; Lopez-Lafuente, A.L.; Dieguez-Anton, A. Enzyme activity indicates soil functionality affectation with low levels of trace elements. *Environ. Pollut.* **2018**, *243*, 1861–1866. [CrossRef] [PubMed]
44. Zhou, F.; Cui, J.; Zhou, J.; Yang, J.; Li, Y.; Leng, Q.; Wang, Y.; He, D.; Song, L.; Gao, M.; et al. Increasing atmospheric deposition nitrogen and ammonium reduced microbial activity and changed the bacterial community composition of red paddy soil. *Sci. Total Environ.* **2018**, *633*, 776–784. [CrossRef] [PubMed]
45. Zhou, J.; Sun, Y.; Blagodatskaya, E.; Berauer, B.J.; Schuchardt, M.; Holz, M.; Shi, L.; Dannenmann, M.; Kiese, R.; Jentsch, A.; et al. Response of microbial growth and enzyme activity to climate change in European mountain grasslands: A translocation study. *Catena* **2024**, *239*, 107956. [CrossRef]
46. Zuo, Y.; Zhang, H.; Li, J.; Yao, X.; Chen, X.; Zeng, H.; Wang, W. The effect of soil depth on temperature sensitivity of extracellular enzyme activity decreased with elevation: Evidence from mountain grassland belts. *Sci. Total Environ.* **2021**, *777*, 146136. [CrossRef]
47. Sun, W.; Xiao, E.; Pu, Z.; Krumins, V.; Dong, Y.; Li, B.; Hu, M. Paddy soil microbial communities driven by environment- and microbe-microbe interactions: A case study of elevation-resolved microbial communities in a rice terrace. *Sci. Total Environ.* **2018**, *612*, 884–893. [CrossRef]
48. Ma, L.; Liu, L.; Lu, Y.; Chen, L.; Zhang, Z.; Zhang, H.; Wang, X.; Shu, L.; Yang, Q.; Song, Q.; et al. When microclimates meet soil microbes: Temperature controls soil microbial diversity along an elevational gradient in subtropical forests. *Soil Biol. Biochem.* **2022**, *166*, 108566. [CrossRef]
49. Ma, T.; Zhan, Y.; Chen, W.; Hou, Z.; Chai, S.; Zhang, J.; Zhang, X.; Wang, R.; Liu, R.; Wei, Y. Microbial traits drive soil priming effect in response to nitrogen addition along an alpine forest elevation gradient. *Sci. Total Environ.* **2024**, *907*, 167970. [CrossRef]
50. Merino-Martín, L.; Hernández-Cáceres, D.; Reverchon, F.; Angeles-Alvarez, G.; Zhang, G.; Dunoyer de Segonzac, D.; Dezette, D.; Stokes, A. Habitat partitioning of soil microbial communities along an elevation gradient: From plant root to landscape scale. *Oikos* **2023**, *2023*, e09034. [CrossRef]

forests

MDPI

Article

Ecological Stoichiometric Characteristics of C, N, and P in *Pinus taiwanensis* Hayata Needles, Leaf Litter, Soil, and Micro-Organisms at Different Forest Ages

Meng Yuan, Yurong Wang, Yang Wang, Yi Wang, Shiwen Wang, Yang Pan, Wangming Zhou, Xiaoyan Xiang and Yuewei Tong *

Engineering Technology Research Center for Aquatic Organism Conservation and Water Ecosystem Restoration of Anhui Province, College of Life Science, Anqing Normal University, 1318, Jixian Road, Yixiu District, Anqing 246133, China; yuanmeng@stu.aqnu.edu.cn (M.Y.); wangyr@stu.aqnu.edu.cn (Y.W.); wangyang@stu.aqnu.edu.cn (Y.W.); wangyi24@mails.ucas.ac.cn (Y.W.); wangsw@stu.aqnu.edu.cn (S.W.); panyang@aqnu.edu.cn (Y.P.); zhouwangming@aqnu.edu.cn (W.Z.); xiaoyanxiang@aqnu.edu.cn (X.X.)

* Correspondence: tongyuewei@aqnu.edu.cn; Tel.: +86-18654064606

Abstract: The ecological stoichiometric characterization of plant and soil elements is essential for understanding the biogeochemical cycles of ecosystems. Based on three forest ages of *Pinus taiwanensis* Hayata (*P. taiwanensis*) plantations in the Gujingyuan National Nature Reserve (i.e., young (16 years), middle-aged (32 years), and mature forests (50 years)), we conducted a field experiment to analyzed C, N, and P stoichiometry and the relationships between needles, litter, soil, and micro-organisms in *P. taiwanensis* plantations. We intended to elucidate the nutritional characteristics and stability mechanisms of the artificial *P. taiwanensis* forest ecosystem. The results showed that the C contents of live needles, leaf litter, soil, and micro-organisms in *P. taiwanensis* plantation forests of the three forest ages were 504.17–547.05, 527.25–548.84, 23.40–35.85, and 0.33–0.54 g/kg, respectively; the respective N contents were 11.02–13.35, 10.71–11.76, 1.42–2.56, and 0.08–0.12 g/kg; and the respective P contents were 0.82–0.91, 0.60–0.74, 0.19–0.36, and 0.03–0.06 g/kg. Forest age significantly influenced both the C, N, and P contents in live needles, leaf litter, soil, and micro-organisms as well as stoichiometric characteristics ($p < 0.05$). Furthermore, although the litter N:P content was comparable to that of needles, the ratios of C:N and C:P in the litter were notably higher compared to those in needles. Soil C:P and N:P ratios were the highest in mature forests while microbial C:P and N:P ratios continuously decreased. Stoichiometric analyses of our findings suggest that forest stand age can influence divergent changes in element cycling among plants, soil, and micro-organisms. The presented results can aid in further understanding nutrient utilization strategies and regulatory mechanisms for *P. taiwanensis* plantation forest systems.

Keywords: *Pinus taiwanensis* Hayata; stand age; ecological stoichiometry; plant–soil continuum; Dabie Mountains

1. Introduction

Ecological stoichiometry is an emerging ecological tool for studying the balance and cycling of multiple elements, providing an integrated approach for studying the coupling of C, N, P, and other elements in ecosystem processes [1,2]. Combining biological studies from different fields, taxa, and scales has been widely utilized to reveal the functions and roles of nutrient ratios and the regulatory mechanisms of ecosystem components for the assessment of nutrient availability in various ecosystems [3,4]. For the forest ecosystem, studies on ecological stoichiometry coupling have focused on various study areas, forest types, successional stages, or the above- or below-ground aspects (e.g., plants, litter, and soils) of nutrient cycling [5–9]. For example, the leaf nutrient content status better reflects the capacity of the soil to provide essential elements to plants and gradually return them to the soil as litter. In contrast, the soil influences nutrient uptake and leaf stoichiometric characteristics through nutrient availability, transformation, and release [5,10–12]. The ecological stoichiometric characteristics of soil microbial biomass intuitively reflect microbial retention and the utilization of soil nutrients [13]. Meanwhile, the ratios of C, N, and P in plants and soils are pivotal in determining the composition and structure of microbial communities [14]. The interactions between plant nutrient demand, soil nutrient supply, plant self-regulation, and nutrient return during litter decomposition increase the complexity of nutrient content studies along the plant–litter–soil continuum [15]. Therefore, it can be observed that the systematic study of plant–litter–soil–microbial stoichiometry is of paramount importance in elucidating the processes and mechanisms underlying nutrient cycling within terrestrial ecosystems.

Previous studies have demonstrated that the forest microenvironment is continually evolving with age [16]. As the age of a forest changes, its internal environment, compositional structure, and soil properties are altered, thus affecting patterns of nutrient partitioning [17,18]. Numerous scholars have examined the associations among forest C, N, and P contents; their respective stoichiometric characteristics; and stand age, yielding innumerable findings. The effects of stand age and stand structure on soil micro-organisms, enzyme activities, and nutrient contents have been reported for Mediterranean black pine (*Pinus nigra Ar.* ssp. *salzmannii*) [19]. In a study, the leaf, root, and soil stoichiometric properties were characterized across different shrub ages of *Ammopiptanthus mongolicus*, and it was found that P may act as a restrictive element for the growth of plants and recovery processes in ecosystems inhabited by *Ammopiptanthus mongolicus* populations [20]. The dynamics of C, N, and P concerning stand age and spatial distribution contribute to the spatial and temporal patterns of C–N–P stoichiometric characteristics in forest ecosystems [12]. However, the precise impact of forest stand age on plant–soil–microbial C:N:P stoichiometry changes remains elusive. Therefore, employing the principles and methodologies of ecological stoichiometry to examine the stoichiometric ratios of C, N, and P in plant and soil components within forest ecosystems, along with an exploration of the sources of plant nutrients within the soil nutrient cycle and the equilibrium constraints between them, is crucial for enhancing the precision of local forestry management practices [21,22].

Pinus taiwanensis Hayata is an essential tree species distributed in the mountainous regions of the southeast coast and the subtropical eastern regions of China, including Henan, Jiangxi, Zhejiang, and Anhui. This species possesses significant economic, ecological, and social value [23]. As a dominant tree species at high altitudes, *P. taiwanensis* fulfills an ecological successional function that is unparalleled among other coniferous forests in harsh climatic and impoverished soil environments [24]. Numerous studies have been conducted regarding the classification, geographic distribution, genetics, population ecology, physiology, biochemistry, and wood anatomy of *P. taiwanensis* [25–27]. In recent years, eco-chemometric studies of the plant–soil continuum under artificial management have emerged as a research hotspot at the regional scale [28]. Some scholars have compared the characteristics of soil nutrient distributions and the primary drivers of soil nutrients in *P. taiwanensis* forests located at varying altitudes [29,30]. However, relatively few studies have concentrated on the plant–soil–microbial C:N:P nutrient content and stoichiometry of *P. taiwanensis* at varying ages. Therefore, we chose the young, middle-aged, and mature stands of *P. taiwanensis* in the Gujingyuan National Nature Reserve, in the Dabie Mountains, and analyzed the C, N, and P contents and their stoichiometric ratios in needles, litter, soil, and micro-organisms to deepen our insight on the strategies of nutrient utilization of C, N, and P in *P. taiwanensis* plantation ecosystems of varying stand ages, thereby providing a scientific basis for their sustainable management. The research questions we aimed to address were as follows: (1) How do the C, N, and P nutrient contents vary across forest ages? (2) Do the biochemical stoichiometric characteristics and nutrient contents of the four fractions of C, N, and P align with changes in the stand age?

2. Materials and Methods

2.1. Overview of the Study Area

The research site was positioned in the Gujingyuan National Nature Reserve, Yuexi County, Anqing City, Anhui Province (Figure 1), in the southeast of the Dabie Mountains, with an altitude of 210 to 1465 m. The geographic coordinates are 116°25′02″–116°33′23″ E and 30°56′43″–31°07′37″ N, spanning a north–south width of 20.02 km and an east–west length of 13.25 km, covering a total area of 7904.3 hectares. The study region is typified by a northern subtropical mesic monsoon regime, exhibiting a mean annual thermometric mean of 14.4 °C, a median annual precipitation of 1445.82 mm, an average annual evaporation of 1444 mm, a normative annual frequency of precipitation events at 147 days, and a vernalization-free period ranging from 180 to 220 days. The soil matrix is primarily composed of hemp gravel, hemp sand, and hemp clay. Pinus taiwanensis is the most prevalent species in the region, which is often mixed with broad-leaved forests to form mixed conifer and broad-leaved forests [31].

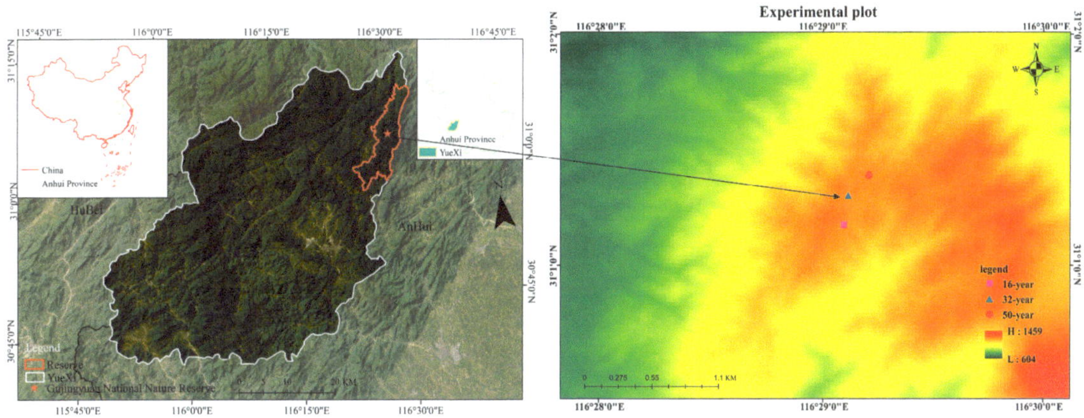

Figure 1. Geographic location of the study area. The study region is located in Gujingyuan National Nature Reserve, Dabie Mountainous Region, China. Different shapes represent different *P. taiwanensis* stand age groups.

2.2. Sample Site Settings

In early July 2023, based on a field survey, three forest age classes of *P. taiwanensis* forests were selected in the Gujingyuan National Nature Reserve, namely, young (16 years), middle-aged (32 years), and mature (50 years) forests, and three 20 m × 20 m sample plots were established corresponding to each forest age class. The distance between the sample plots was not less than 1 km. The per-tree checking was conducted on the *P. taiwanensis* trees in the sample plots with the average tree height, DBH, and stand density measured, and also, the vegetation status under three forests was investigated (Figure 2). As *P. taiwanensis* is typically encountered in arid and barren mountainous regions, thriving in diverse environments across high-elevation gradients, the understory vegetation is seldom herbaceous. Coniferous forests possess smaller canopies, allowing sunlight to penetrate the understory more directly, fostering a conducive environment for shrub growth. The survey revealed that the young forest sample maintained an open structure, abundant with shrub plants and ample light. Conversely, mature forests exhibited a lower richness of understory vegetation. The main tree species in the understory include *Lindera glauca*, *Lindera reflexa*, and *Quercus serrata*, among others. Basic information on the three forest age classes is shown in Table 1.

(a) (b) (c)

Figure 2. Conditions of three experimental sites. (**a–c**) show the three stand profiles of young, middle-aged, and mature-age forests, respectively.

Table 1. Basic information and variables of standard plots in different age groups.

Forest Type							Soil Layer (0–10 cm)	
Forest Type	**Age**	**Altitude (m)**	**Mean DBH (cm)**	**Mean Height (m)**	**Mean Stand Density (trees·ha^{-1})**	**Mean Canopy Density**	**pH**	**Moisture (%)**
Young Forest	16	1208.8	11.12 ± 0.33	9.44 ± 0.40	533	0.83 ± 0.03	5.43 ± 0.17	25.72 ± 1.53
Middle-aged Forest	32	1234.5	18.06 ± 0.17	13.39 ± 0.48	500	0.64 ± 0.04	5.67 ± 0.16	25.39 ± 0.99
Mature Forest	50	1241.5	28.22 ± 0.55	15.31 ± 0.62	470	0.59 ± 0.03	5.60 ± 0.18	26.29 ± 1.12

Mean values ± standard errors are shown.

2.3. Sample Collection and Determination

Five average trees were chosen as the standard trees in each mature-age plot. Healthy and mature live needles were collected from the middle and upper crowns of each tree, evenly distributed across the east, west, south, and north directions, using high-branching scissors. In three plots of each forest age class, five 2 × 2 m litter traps were set up at the four corners and centers to collect litter. The collected samples were thoroughly mixed and placed in a paper bag. Firstly, live needle samples were killed at 105 °C for 30 min to inactivate the enzymes in the needles for a short period of time, then the temperature was reduced to 70 °C to remove the water from the needles until a stable weight was attained. Subsequently, the needles and litter samples were ground into a powder with a particle size of 0.15 mm to ascertain the C, N, and P contents. Concurrently, soil samples were collected beneath the same trees where coniferous litter was gathered; in particular, 5 soil specimens were gathered from the uppermost 0 to 10 cm of the topsoil stratum using a soil drill in each square. These soil samples were thoroughly mixed. One part was air-dried and screened with a 0.25 mm sieve to determine soil C, N, and P contents while the other part was sieved and stored at 4 °C to assess soil microbial C, N, and P.

The total plant (needles and litter) and soil organic C contents were assessed with a revised Walkley–Black acid dichromate ($FeSO_4$) exothermic titration method [32]. Plant total N and soil total N contents were determined according to the Kjeldahl method after digestion with H_2SO_4 and H_2O_2. The P content was also ascertained through the molybdenum anti-colorimetric method [33]. Moreover, the chloroform fumigation extraction method was employed to evaluate the soil microbial biomass C, N, and P contents. The microbial C and N contents were derived through comparing the unfumigated and fumigated soil samples. A colorimetric determination of microbial P was carried out using UV spectrophotometry [34].

2.4. Data Analysis

Excel 2019 was utilized to organize the experimental data. One-way ANOVAs were conducted for the live needle, litter, soil, and soil microbial C, N, and P contents, along with the stoichiometric ratios of *P. taiwanensis* across different ages using SPSS 27.0 (IBM Corp., Armonk, NY, USA). The least significant difference (LSD) test was employed at a 95% confidence interval ($p < 0.05$) to assess whether the sample means differed significantly from a normal distribution, and the mean ± standard deviation values have been utilized in the graphical representations. Pearson correlation coefficient assessment was conducted to ascertain the interrelationships among C, N, and P in plant and soil samples and their respective ecological stoichiometric proportions. The results were depicted graphically utilizing Origin 2021 software (OriginLab Corp., Northampton, MA, USA).

3. Results

3.1. C, N, and P Contents in Live Needle, Leaf Litter, Soil, and Microbial Communities Across Different Forest Ages

The C, N, and P contents of *P. taiwanensis* forests exhibited a pattern of live needle > leaf litter > soil > micro-organisms, except for the leaf litter C content being greater than that of needle C (Table 2). Firstly, the live needle, leaf litter, soil, and microbial C contents in different *P. taiwanensis* plantations ranged from 504.17 to 547.05, 527.25 to 548.84, 23.40 to 35.85, and 0.33 to 0.54 g/kg, respectively. An analysis of variance (ANOVA) revealed that the C contents of all the four fractions varied significantly across different stand ages ($p < 0.05$). With the advancement of forest age, the needle and litter C contents reduced first and then increased and exhibited an inverted U-shaped trend. However, the soil C content continued to decline, and the microbial C content was also highest in the young forests.

Table 2. Chemical properties of C, N, and P contents in live needles, leaf litter, soil, and micro-organisms of *P. taiwanensis* plantations across different age groups.

Chemical Properties (g/kg)	Different Fractions	Young Forest	Middle-Aged Forest	Mature Forest
C	Live needle	504.17 ± 5.72^{b}	547.05 ± 6.01^{a}	543.84 ± 5.76^{a}
	Leaf litter	535.66 ± 4.94^{b}	548.84 ± 8.00^{a}	527.25 ± 5.98^{b}
	Soil	35.85 ± 1.45^{a}	25.16 ± 2.02^{b}	23.40 ± 0.58^{b}
	Micro-organisms	0.54 ± 0.018^{a}	0.33 ± 0.012^{c}	0.36 ± 0.007^{b}
N	Live needle	11.02 ± 0.46^{b}	12.26 ± 0.29^{a}	13.34 ± 0.59^{a}
	Leaf litter	10.71 ± 0.45^{b}	11.03 ± 0.35^{b}	11.76 ± 0.35^{a}
	Soil	2.56 ± 0.12^{a}	1.42 ± 0.22^{b}	1.53 ± 0.11^{b}
	Micro-organisms	0.12 ± 0.004^{a}	0.09 ± 0.014^{b}	0.08 ± 0.019^{b}
P	Live needle	0.85 ± 0.02^{b}	0.92 ± 0.15^{a}	0.82 ± 0.07^{b}
	Leaf litter	0.74 ± 0.04^{a}	0.60 ± 0.01^{b}	0.70 ± 0.01^{a}
	Soil	0.36 ± 0.022^{a}	0.23 ± 0.025^{b}	0.19 ± 0.014^{b}
	Micro-organisms	0.03 ± 0.002^{b}	0.05 ± 0.009^{a}	0.06 ± 0.002^{a}

Different superscript lowercase characters denote statistically significant ($p < 0.05$) variations in the C, N, and P contents within needles, litter, soil, and micro-organisms across various stand age categories. Live needles, leaf litter, and soil were collected from each sample plot for each of the three stand ages, with five replicates per sample plot.

Then, the N contents of *P. taiwanensis* plantations were in the ranges of 11.02 to 13.34, 10.71 to 11.76, 1.42 to 2.56, and 0.08 to 0.12 g/kg for needles, litter, soil, and micro-organisms, respectively. The N contents of needles, leaf litter, and micro-organisms differed significantly among forest ages ($p < 0.05$). The age of the forest exerted a comparable influence on the N content in the needles and litter of *P. taiwanensis*; mature forests showed the highest level of N in their needles and litter. However, soil N content exhibited a U-shaped trend with an increasing stand age, and the highest levels of soil and microbial N contents were found in the young forests.

Moreover, the needle, litter, soil, and microbial P contents of three *P. taiwanensis* plantations in the study area were 0.82 to 0.92, 0.60 to 0.74, 0.19 to 0.36, and 0.03 to 0.06 g/kg, respectively. The P contents of needles, litter, soil, and micro-organisms differed significantly among forest ages ($p < 0.05$). In particular, the needle P content was also showed an inverted U-shaped trend, with the needle P content being higher in middle-aged forests compared to young and mature ones. Leaf litter and soil P contents were highest in young forests while microbial P content was highest in mature forests.

3.2. Ecological Stoichiometric Ratios Among Needles, Leaf Litter, Soil, and Micro-Organisms in Different Stand Ages

The C:N, C:P, and N:P ratios across the four distinct fractions revealed the order to be leaf litter > live needles > soil > micro-organisms, and the differences between the four pools were significant ($p < 0.05$). The effects of forest age on the four fractions of C:N, C:P, and N:P were not uniform. The values of C:N for live needles, litter, soil, and microbes in *P. taiwanensis* plantations were 40.81 to 45.79, 44.86 to 50.08, 14.01 to 17.91, and 3.53 to 4.58, respectively (Figure 3). The C:N ratios of needles and litter in mature forests differed significantly from those in young and middle-aged forests ($p < 0.05$); both litter and needle C:N decreased with an increasing stand age. Meanwhile, variations in soil C:N ratios were observed across the three forest age classes, and soil C:N reached its highest value and microbial C:N reached its lowest value in the middle-aged forest.

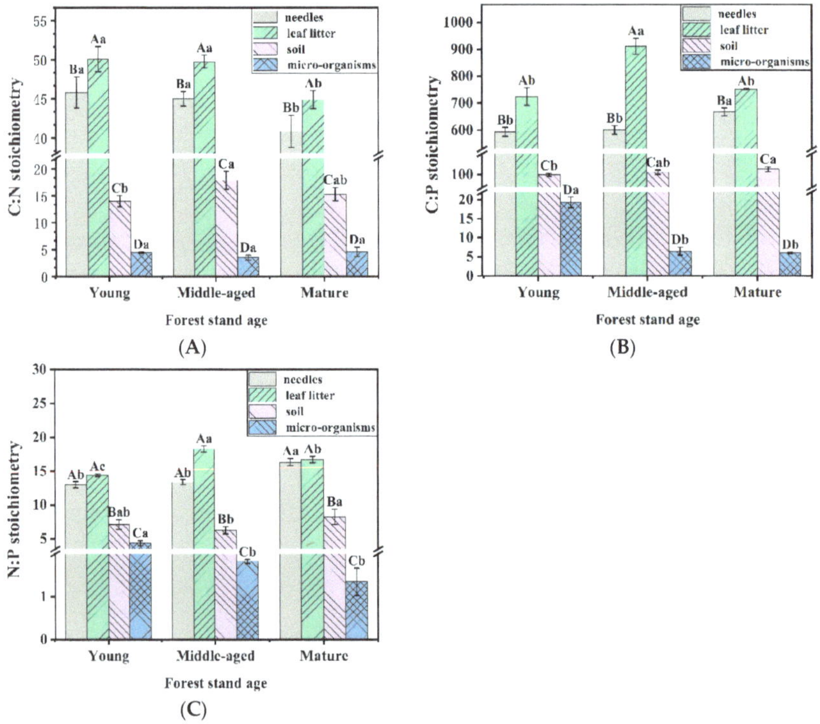

Figure 3. Stoichiometry of live needle, leaf litter, soil, and microbial C:N (**A**), C:P (**B**), and N:P (**C**) in *P. taiwanensis* plantations of different ages. Capital letters (A, B, C, D) denote substantial variations ($p < 0.05$) among needles, leaf litter, soil, and micro-organisms at the same stand age; lowercase letters (a, b, c) signify significant differences ($p < 0.05$) across various stand ages within the same research component.

The needle, litter, soil, and microbial C:P ratios of *P. taiwanensis* plantations within the study region were in the ranges of 593.97 to 667.24, 724.15 to 911.71, 94.70 to 125.56, and 5.99 to 19.26 g/kg, respectively. The effects of forest age on needle, litter, soil, and microbial C:P ratios were all significant ($p < 0.05$); leaf litter C:P was significantly higher than needle C:P; microbial C:P decreased with an increasing forest age and had a highest level in young forests compared to middle and mature forests.

The N:P ratios of the needle, litter, soil, and micro-organisms of *P. taiwanensis* plantations were 12.98 to 16.36, 14.45 to 18.32, 6.27 to 7.14, and 1.35 to 4.36 g/kg, respectively. The effects of different forest ages on needle, litter, soil, and microbial N:P ratios were also significant ($p < 0.05$). The needle N:P ratio continued to increase while stand age advanced, and the microbial N:P diminished. The N:P ratio in needles was highest in the mature forest stage while it was highest in the middle-aged forest for the leaf litter.

3.3. Correlation of Factors in Different Fractions of *P. taiwanensis* Plantation Forests

The ecological stoichiometry of C, N, and P in the needles, litter, soil, and microbiota of *P. taiwanensis* plantations exhibited strong correlations (Figure 4). As the results showed, the needle C and N contents exhibited a strong, significant correlation with soil and microbial C, N, and P values. The leaf litter N showed a significant correlation with microbial and soil P, and leaf litter P was significantly correlated with microbial C content and soil N. Meanwhile, there was a highly significant correlation between soil C, N, and P contents with microbial C, N, and P contents ($p < 0.01$). For the ecological stoichiometric ratios, the needle C:N was significantly and positively correlated with leaf litter C:N ($p < 0.05$), and the needle C:P and N:P ratios displayed highly significant correlations ($p < 0.01$) with the litter C:N ratio. Additionally, there was a notably inverse relationship between litter N:P and microbial C:P and N:P ($p < 0.01$), and the litter C:N ratio was significantly associated with soil N:P, while the litter C:P ratio showed a significant correlation with both microbial and soil C:N ratios ($p < 0.05$).

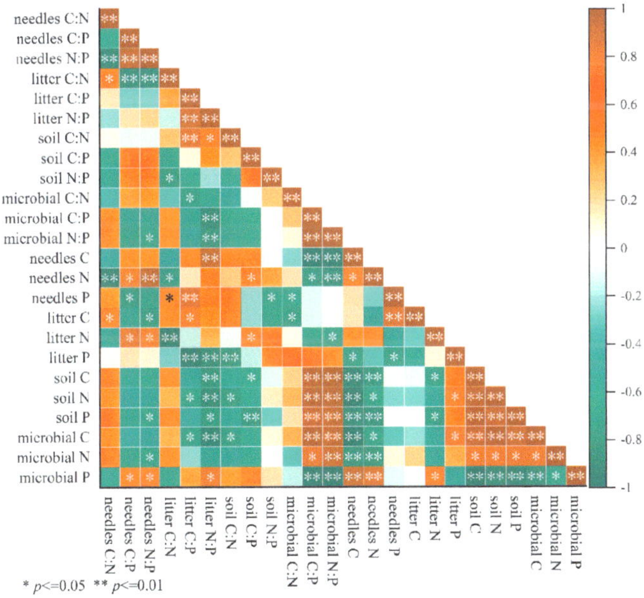

Figure 4. Stoichiometric relationships in the plant-soil continuum of four components: needles, leaf litter, soil, and micro-organisms. Correlations between live needle, leaf litter, soil, and microbial C, N, and P contents and ecological stoichiometric ratios in *P. taiwanensis* plantations are shown. Symbols * and ** indicate correlations of significance at the 0.05 and 0.01 probability levels, respectively.

4. Discussion

4.1. Changes in Live Needle, Leaf Litter, Soil, and Microbial CNP Contents with Forest Age

The circulation of elements C, N, and P is a crucial element influencing the functionality of forest ecosystems [3]. Most previous studies have shown that the C, N, and P contents of plants and their ecological stoichiometric ratios vary with forest age, with different patterns of change being influenced by forest age [18]. In this research, needle C, N, and P contents were significantly influenced by the stand age, suggesting variations in the ecological adaptations of the plants across different developmental stages [35]. The mean value of C content in *P. taiwanensis* needles in this study area exceeded the global mean C content for terrestrial plant leaves [36], indicating that the content of organic compounds synthesized by *P. taiwanensis* needles was high. Meanwhile, the mean value of N content in *P. taiwanensis* needles was significantly lower than the mean value in leaves of terrestrial plants in China [37] and the mean value in leaves of plants globally [36]. Furthermore, the mean value of P content in live needles was significantly lower than the mean value in land plant leaves [37]. All these results indicated that the distribution pattern in the live needle of *P. taiwanensis* was "high C, low N and P", consistent with the findings of Ashfaq [38]. This pattern may be related to the characteristics of the tree species as conifers typically have a high C content and low N and P contents [39]. The growth rate of *P. taiwanensis* gradually increases in the middle-aged stage, thus increasing dry matter synthesis, and more rRNA is required for protein synthesis [40]. In stands ranging from young to mature, leaf C contents in middle-aged forests were higher, indicating that middle-aged stands have higher leaf organic matter content and greater C storage, compared to other stand ages. The reason for this phenomenon may have been the influence of plant age on photosynthesis [41]. On the other hand, litter acts as the hub connecting vegetation and soil and is the primary way in which plants return nutrients to the ecosystem [42]. The litter P content was lower than the global averages for senescent leaves while the C and N contents were higher [43]. The trends in leaf litter content and live needle C and N contents with an increasing stand age were consistent, indicating a close relationship between the *P. taiwanensis* litter and needles. However, the influences of forest age on the needle and litter C, N, and P contents were not uniform, suggesting that forest age had a non-uniform influence on their variability. Nevertheless, both are significantly influenced by the age of the forest.

The C, N, and P contents of the soil are affected by changes in soil physicochemical properties and the return of apoplastic nutrients [44]. According to the stoichiometric values in this study, C and N contents in 0–10 cm soils were high, compared to national levels, while the P content was low [45]. This imbalance further led to soil C:N:P stoichiometry alterations as the C and N accumulation rate is usually faster than that of P [46]. Additionally, there was a general decline in soil C, N, and P contents as the forest age progressed. This suggests a specific degree of soil degradation by *P. taiwanensis* in this area, consistent with the study by Zeng et al. on *Larix gmelinii* (Rupr.) [47]. The reason may be that as a plant grows, the roots draw nutrients from the soil and store them in the trunk and leaves [48]. The vegetation canopy, plant biodiversity, topographical features, soil textural attributes, and substrate lithology are pivotal in modulating the pools of soil organic carbon and essential nutrients [49,50]. Soil N is primarily derived from litter decomposition and atmospheric deposition [51]. Young forest stands are distinguished by their low canopy density, rich shrubs and herbaceous plants, abundant light, and high air temperatures. Consequently, soil microbial activity is heightened, facilitating the decomposition of organic matter and the conversion of N into soil nutrients, thus leading to higher soil C and N contents in young forests [52]. Soil P is derived from rock weathering and the decay of organic matter such as decomposing plant material [53]. The pattern of change in P levels of soil is analogous to that of N, and similar findings have been observed across diverse mixed forests located in the Qinling Mountains region of China [54].

Soil micro-organisms are the main agents responsible for the decomposition of soil organic matter and the turnover of nutrients [4]. The microbial C content can indicate the abundance of living soil micro-organisms [55]. The microbial N content comprehensively reflects the nitrogen mineralization and fixation process by microbial bodies while microbial P is a crucial source of available P for plants, with a fast turnover rate and susceptibility to environmental influences [56]. The responses of microbial C and N to stand age in *P. taiwanensis* plantations in this study were consistent with the soil C and N. This response reflects the characteristics of the soil microbial biomass, affected by soil nutrients [57]. The depletion of soil C and N in middle-aged and mature forests affected the activities of soil micro-organisms, leading to a reduction in soil microbial C and N levels. Microbial P gradually increased, favoring the conversion and release of soil-available P, similar to the results obtained by Fang et al. when considering *Quercus serrata* forests of different ages [58].

4.2. Changes in Live Needle, Litter, Soil, and Microbial CNP Stoichiometric Ratios with Stand Age

The C:N, C:P, and N:P ratios of live needles, leaf litter, soil, and micro-organisms represent the competition between different components to maintain ecological balance and their efforts to adapt to the environment in order to meet their own needs, thus characterizing the changes in total productivity at the elemental level [59]. Needle N:P serves as an indicator for assessing plant limitations in terms of N and P. Notably, the N:P ratio in needles varied significantly across different stand ages. *P. taiwanensis* was found to be limited by different nutrients at various stand ages. The observed increase in needle N:P ratios with age suggests that *P. taiwanensis* forests progressively become more P-limited [60]. As important physiological indices of plants, C:N and C:P can indicate both a plant's capacity for carbon assimilation as well as the effectiveness of nutrient acquisition and utilization [61]. The mean values of C:N and C:P in *P. taiwanensis* needles at different forest ages were higher than the global mean values for plant leaves [36], suggesting that *P. taiwanensis* exhibits an efficient use of N and P along with a significant capacity for carbon sequestration. The mature forest stand exhibited elevated leaf C:P and N:P ratios compared to the other stands, suggesting a greater capacity for C assimilation at this stage.

The conditions of soil nutrients and the absorption and utilization of nutrients by plants significantly influence the regulation of the ecological stoichiometric properties of C, N, and P in litter. When the soil nutrient supply is inadequate, plants may resort to the re-uptake of nutrients from the litter, thereby altering its ecological stoichiometric characteristics [62]. The N and P contents of leaf litter were lower than those in live needles while the C:N and C:P ratios significantly exceeded those found in live needles, which, as in the study by Wang et al., indicated that *P. taiwanensis* is characterized by nutrient reabsorption [63]. Studies have shown that the litter N:P ratio is a crucial factor influencing litter decomposition and nutrient cycling and negatively correlates with the litter decomposition rate [64]. The litter N:P ratio of *P. taiwanensis* initially rose before declining as the forest age advanced, indicating that the rate of litter decomposition initially slowed down before accelerating as the forest age increased, similar to the results of the study by Jiang et al. on *Pinus tabulaeformis* forests of different ages [65].

Previous studies have indicated that the soil C:N ratio is indicative of the rate at which organic matter undergoes mineralization, the C:P ratio demonstrates the magnitude of P potential released through the microbial degradation of soil organic matter into inorganic forms, and the N:P ratio reflects soil P availability [66]. The mean values of the C:N, C:P, and N:P ratios in *P. taiwanensis* soils under different forest ages were higher than the respective national mean values for terrestrial soils [45]. This indicates that the conversion of soil organic matter into minerals in *P. taiwanensis* forests in this region is slow, which is conducive to the accumulation of soil organic C and enhancement of the function of soil C pools in *P. taiwanensis* forests. The slow decomposition of N and P is not conducive to the release of N and P, easily restricting the growth of *P. taiwanensis* [67]. The C:P ratio in the soil is vital for the development of plants and development, indicating the potential of plants to release P from the environment and to absorb and utilize it [68]. A higher soil C:P ratio reflects low soil P availability [69]. In this study, soil C:P ratios changed minimally with an increasing forest age, suggesting that the trend of soil P availability remained relatively stable during the subsequent growth of *P. taiwanensis* forests. Furthermore, numerous studies have proven that a high microbial C:N ratio is associated with a high proportion of soil fungi in the microbial community [70]. The differences in microbial C:N ratios among the three stands in this study were not significant, indicating that the abundance of soil microbial fungi in *P. taiwanensis* plantations did not vary significantly among the three stands. The microbial C:P ratio continuously decreased with an increasing forest age, suggesting an enhancement in the ability of micro-organisms to fix P, which was the strongest in mature forests [71]. The demand for and utilization of resources by micro-organisms are motivated by the biochemical ratios in their biomass and their capacity for growth; however, they may also be influenced by the relative supply of available resources [72].

4.3. Coupling of Needle, Litter, Soil, and Microbial CNP and Stoichiometric Ratios

Correlation analyses revealed significant correlations among the C, N, and P stoichiometric characteristics of *P. taiwanensis* live needles, leaf litter, soil, and micro-organisms. These results indicated that C, N, and P elements were transported and transformed among the four components (i.e., plants, litter, soil, and micro-organisms) in forest ecosystems [73]. The soil microbial biomass is pivotal in regulating soil nutrient cycling and energy flow [74]. Furthermore, the soil environment is a crucial factor influencing the growth of micro-organisms as soil is intimately linked to micro-organisms [75]. This study also revealed a substantial association among soil-based metrics and microbial stoichiometric ratios. The highly significant correlation between soil C, N, and P contents and needle C and N contents indicates that the stoichiometric balance of soil C, N, and P significantly affects leaf nutrient uptake. Therefore, it is crucial to explore the profound connections between C, N, and P within plants, soils, and micro-organisms [76]. In this study, there was a notable correlation between the needle C:P and N:P ratios and the litter C:N ratio. Additionally, leaf litter N:P ratio exhibited a significant negative correlation with the microbial C:P and N:P ratios. The litter C:P and N:P ratios showed a strong, positive correlation with the soil C:N ratio, indicating that live needles are (directly or indirectly) correlated with each indicator of litter, soil, and micro-organisms [77,78]. This result demonstrates the significant coupling relationships between the ecological stoichiometric characteristics of the four components, namely, needles, leaf litter, soil, and micro-organisms. This suggests the mutual transportation and transfer of C, N, and P among these four components during the growth of *P. taiwanensis*.

5. Conclusions

Consequently, the current study examined *P. taiwanensis* plantations of young, middle-aged, and mature forests in the Gujingyuan National Nature Reserve, a representative region in northern subtropical China. By analyzing C, N, and P contents in lived needles, litter, soil, and micro-organisms, we investigated the patterns of changes in the stoichiometric ratios of C, N, and P across different forest ages. The results showed that the nutrient requirements of the plantations varied with the stand age. Specifically, significant differences in the stoichiometric characteristics of C, N, and P in needles, leaf litter, soil, and micro-organisms were observed between different stand ages of *P. taiwanensis* plantation forests. Litter can serve as a nutrient hub between leaf and soil systems. According to the needle N:P reabsorption theory, the N:P ratio increased with forest age, suggesting a transition from N to P limitation as the forest matured. At the same time, soil C and P contents decreased under all three stand ages, suggesting a certain degree of soil degradation in the area. Microbial C:P values decreased as stand age advanced, implying that the microbial capacity to fix P increased. The understory soil of *P. taiwanensis* in the study area is sandy and low in nutrients, as well as in microbial C, N, and P contents. *P. taiwanensis* fulfills an ecological successional function unparalleled among coniferous forests in harsh climatic and impoverished soil environments. Overall, the C, N, and P contents of *P. taiwanensis* needles, litter, soil, and micro-organisms were closely related to their ecological stoichiometric ratios. Forest age is an important factor influencing elemental cycling among plants, litter, soil, and micro-organisms.

Author Contributions: Formal analyses, surveys, and writing—original manuscript, M.Y.; data compilation, analysis, and investigation, Y.W. (Yi Wang) and Y.W. (Yurong Wang); conceptualization and methodology, X.X. and W.Z.; investigation, Y.W. (Yang Wang), S.W. and Y.P.; writing—reviewing and editing, Y.T. All authors have read and agreed to the published version of the manuscript.

Funding: This research was financially supported by the Project of the Natural Science Foundation of Anhui Province (2208085QC72 and 1908085MC58) and the Scientific Research Projects of Universities in Anhui Province (2022AH051052).

Data Availability Statement: Data are available on request from the authors.

Acknowledgments: The authors sincerely thank the Gujingyuan National Nature Reserve in Anhui Province for supporting and permitting this study. The authors are thankful to the staff of the reserve for their help.

Conflicts of Interest: The authors declare no conflicts of interest.

References

1. Elser, J.J.; Dobberfuhl, D.R.; MacKay, N.A.; Schampel, J.H. Organism size, life history, and N:P stoichiometry: Toward a unified view of cellular and ecosystem processes. *BioScience* **1996**, *46*, 674–684. [CrossRef]
2. Sistla, S.A.; Schimel, J.P. Stoichiometric flexibility as a regulator of carbon and nutrient cycling in terrestrial ecosystems under change. *New Phytol.* **2012**, *196*, 68–78. [CrossRef] [PubMed]
3. Leroux, S.J.; Wal, E.V.; Wiersma, Y.F.; Charron, L.; Ebel, J.D.; Ellis, N.M.; Hart, C.; Kissler, E.; Saunders, P.W.; Moudrá, L.; et al. Stoichiometric distribution models: Ecological stoichiometry at the landscape extent. *Ecol. Lett.* **2017**, *20*, 1495–1506. [CrossRef] [PubMed]
4. Van de Waal, D.B.; Elser, J.J.; Martiny, A.C.; Sterner, R.W.; Cotner, J.B. Editorial: Progress in ecological stoichiometry. *Front. Microbiol.* **2018**, *9*, 1957. [CrossRef] [PubMed]
5. Wang, Z.; Zheng, F. Ecological stoichiometry of plant leaves, litter and soils in a secondary forest on China's Loess Plateau. *PeerJ* **2020**, *8*, e10084. [CrossRef]
6. Wang, W.; Peng, Y.; Chen, Y.; Lei, S.; Wang, X.; Farooq, T.H.; Liang, X.; Zhang, C.; Yan, W.; Chen, X. Ecological stoichiometry and stock distribution of C, N, and P in three forest types in a Karst region of China. *Plants* **2023**, *12*, 2503. [CrossRef]
7. Yu, M.F.; Tao, Y.; Liu, W.; Xing, W.; Liu, G.; Wang, L.; Ma, L. C, N, and P stoichiometry and their interaction with different plant communities and soils in subtropical riparian wetlands. *Environ. Sci. Pollut. Res. Int.* **2020**, *27*, 1024–1034. [CrossRef]
8. Asada, K.; Kanda, T.; Yamashita, N.; Asano, M.; Eguchi, S. Interpreting stoichiometric homeostasis and flexibility of soil microbial biomass carbon, nitrogen, and phosphorus. *Ecol. Model.* **2022**, *470*, 110018. [CrossRef]
9. Jiang, J.; Lu, Y.; Chen, B.; Ming, A.; Pang, L. Nutrient resorption and C:N:P stoichiometry responses of a *Pinus massoniana* plantation to various thinning intensities in Southern China. *Forests* **2022**, *13*, 1699. [CrossRef]
10. Zhang, G.; Zhang, P.; Peng, S.; Chen, Y.; Cao, Y. The coupling of leaf, litter, and soil nutrients in warm temperate forests in northwestern China. *Sci. Rep.* **2017**, *7*, 11754. [CrossRef]
11. Manivannan, S.; Chadha, K.L. Standardization of time of sampling for leaf nutrient diagnosis on Kinnow mandarin in North-West India. *J. Plant Nutr.* **2011**, *34*, 1820–1827. [CrossRef]
12. Bui, E.N.; Henderson, B.L.C. N:P stoichiometry in Australian soils concerning vegetation and environmental factors. *Plant Soil* **2013**, *373*, 553–568. [CrossRef]
13. He, X.; Hou, E.; Veen, G.F.; Ellwood, M.D.F.; Dijkstra, P.; Sui, X.; Zhang, S.; Wen, D.; Chu, C. Soil microbial biomass increases along elevational gradients in the tropics and subtropics but not elsewhere. *Glob. Ecol. Biogeogr.* **2020**, *29*, 345–354. [CrossRef]
14. Yang, Y.; Liu, H.; Yang, X.; Yao, H.; Deng, X.; Wang, Y.; An, S.; Kuzyakov, Y.; Chang, S.X. Plant and soil elemental C:N:P ratios are linked to soil microbial diversity during grassland restoration on the Loess Plateau, China. *Sci. Total Environ.* **2022**, *806*, 150557. [CrossRef] [PubMed]
15. Agren, G.I.; Bosatta, E. Theoretical ecosystem ecology: Understanding element cycles. *J. Ecol.* **1998**, *18*, 1025–1026.
16. Ren, Y.; Gao, G.L.; Ding, G.D.; Zhang, Y.; Zhao, P.S. Patterns and environmental drivers of C, N, and P stoichiometry in the leaf–litter–soil system associated with Mongolian pine forests. *Ecol. Evol.* **2024**, *14*, e11172. [CrossRef]
17. Inagaki, Y.; Miura, S.; Kohzu, A. Effects of forest type and stand age on litterfall quality and soil N dynamics in Shikoku District, southern Japan. *For. Ecol. Manag.* **2004**, *202*, 107–117. [CrossRef]
18. Shen, J.; Fan, S.; Zhang, J.; Liu, G. Stoichiometric homeostasis of N and P in the leaves of different–aged *Phyllostachys edulis* after bamboo forest expansion in Subtropical China. *Forests* **2024**, *15*, 1181. [CrossRef]
19. Lucas-Borja, M.E.; Hedo, J.; Cerdá, A.; Candel-Pérez, D.; Viñegla, B. Unravelling the importance of forest age stand and forest structure driving microbiological soil properties, enzymatic activities and soil nutrients content in Mediterranean Spanish black pine (*Pinus Nigra Ar. Ssp. Salzmannii*) Forest. *Sci. Total Environ.* **2016**, *562*, 145–154. [CrossRef]

20. Dong, X.; Xu, D.; Wang, D.; Han, C.; Huang, Y.; Zhang, J. Leaf-root-soil stoichiometric characteristics in different shrub ages of *Ammopiptanthus mongolicus*. *Plants* **2023**, *12*, 3103. [CrossRef]
21. Wang, Y.; Zhang, Y.; Wang, L.; Jing, X.; Yu, L.; Liu, P. Response of leaf biomass, leaf and soil C:N:P stoichiometry characteristics to different site conditions and forest ages: A case of *Pinus Tabuliformis* plantations in the temperate mountainous area of China. *Front. Plant Sci.* **2022**, *13*, 1060406. [CrossRef] [PubMed]
22. Gong, Z.; Sheng, M.; Zheng, X.; Zhang, Y.; Wang, L. Ecological stoichiometry of C, N, P, and Si of Karst Masson Pine Forests: Insights for the forest management in southern China. *Sci. Total Environ.* **2024**, *912*, 169490. [CrossRef] [PubMed]
23. Li, Q.Z. Population Structure and Dynamics of *Pinus taiwanensis* Hayata at Songyang County, Zhejiang Province, China. *Plant Ecol.* **1990**, *86*, 119–129.
24. Su, S.J.; Liu, J.F.; Lan, S.R.; Hong, W.; Li, W.Z. A review of *Pinus taiwanensis* studies (1960–2014) and the knowledge domain analysis. *Fujian J. Agric. For. Univ.* **2015**, *44*, 478–486. (In Chinese)
25. Chu, M.H.; Hsiao, S.W.; Kao, Y.C.; Yin, H.W.; Kuo, Y.H.; Lee, C.K. Cytotoxicity effect of constituents of *Pinus taiwanensis* Hayata Twigs on B16–F10 Melanoma Cells. *Molecules* **2022**, *27*, 2731. [CrossRef]
26. Lyu, M.; Sun, M.; Peñuelas, J.; Sardans, J.; Sun, J.; Chen, X.; Zhong, Q.; Cheng, D. Thermal acclimation of foliar Carbon Metabolism in *Pinus taiwanensis* along an elevational gradient. Front. *Front. Plant Sci.* **2021**, *12*, 778045.
27. Fang, M.F.; Wang, Y.J.; Zu, Y.M.; Dong, W.L.; Wang, R.N.; Deng, T.T.; Li, Z.H. The complete Chloroplast Genome of the Taiwan Red Pine *Pinus taiwanensis* (Pinaceae). *Mitochondrial DNA Part A* **2016**, *27*, 2732–2733. [CrossRef]
28. Xiang, Y.; Pan, P.; Ouyang, X.; Zang, H.; Rao, J. The chemical stoichiometry characteristics of plant-soil carbon and nitrogen in subtropical *Pinus massoniana* natural forests. *Sci. Rep.* **2024**, *14*, 5031. [CrossRef]
29. Jiang, M.; He, Y.; Liu, J.; Xing, H.; Gu, Y.J.; Wei, X.; Zhu, Z. Elevation gradient altered soil C, N, and P stoichiometry of *Pinus taiwanensis* forest on Daiyun Mountain. *Forests* **2019**, *10*, 1089. [CrossRef]
30. Chen, B.; Chen, L.; Jiang, L.; Zhu, J.; Chen, J.; Huang, Q.; Liu, J.; Xu, D.; He, Z.C. C:N:P stoichiometry of plant, litter and soil along an elevational gradient in subtropical forests of China. *Forests* **2022**, *13*, 372. [CrossRef]
31. Wang, X.J. Study on the distribution and biodiversity of the plant communities in Gujingyuan national nature reserve. *Anhui For. Sci. Technol.* **2019**, *45*, 10–14. (In Chinese)
32. Bao, S.D. *Soil and Agricultural Chemistry Analysis*; China Agriculture Press: Beijing, China, 2000. (In Chinese)
33. Kuo, S. Phosphorus. In *Methods of Soil Analysis*; John Wiley & Sons, Ltd.: Hoboken, NJ, USA, 1996; pp. 869–919.
34. Wu, J.; Joergensen, R.G.; Pommerening, B.; Chaussod, R.; Brookes, P.C. Measurement of soil microbial biomass C by fumigation-extraction-an automated procedure. *Soil Biol. Biochem.* **1990**, *22*, 1167–1169. [CrossRef]
35. Li, H.; Crabbe, M.J.C.; Xu, F.; Wang, W.; Ma, L.; Niu, R.; Gao, X.; Li, X.; Zhang, P.; Ma, X.; et al. Seasonal variations in carbon, nitrogen and phosphorus concentrations and C:N:P stoichiometry in different organs of a *Larix Principis-Rupprechtii* Mayr. plantation in the Qinling Mountains, China. *PLoS ONE* **2017**, *12*, e0185163. [CrossRef] [PubMed]
36. Elser, J.J.; Fagan, W.F.; Denno, R.F.; Dobberfuhl, D.R.; Folarin, A.; Huberty, A.; Interlandi, S.; Kilham, S.S.; Mccauley, E.; Schulz, K.L. Nutritional constraints in terrestrial and freshwater food webs. *Nature* **2000**, *408*, 578–580. [CrossRef]
37. Han, W.; Fang, J.Y.; Reich, P.B.; Woodward, F.I.; Wang, Z.H. Biogeography and variability of eleven mineral elements in plant leaves across gradients of climate, soil and plant functional type in China. *Ecol. Lett.* **2011**, *14*, 788–796. [CrossRef]
38. Ashfaq, A.; Hussain, M.; Ali, S.; Akhtar, K.; Mamoona, W.M.; Zamir, A. Ecological stoichiometry in *Pinus massoniana* L. plantation: Increasing nutrient limitation in a 48–Year chronosequence. *Forests* **2022**, *13*, 469. [CrossRef]
39. Stierle, A.A.; Stierle, D.B. Bioactive secondary metabolites produced by the fungal endophytes of conifers. *Nat. Prod. Commun.* **2015**, *10*, 1671–1682. [CrossRef]
40. Matamouros, S.; Gensch, T.; Cerff, M.; Sachs, C.C.; Abdollahzadeh, I.; Hendriks, J.; Horst, L.; Tenhaef, N.; Tenhaef, J.; Noack, S.; et al. Growth-rate dependency of ribosome abundance and translation elongation rate in *Corynebacterium glutamicum* differs from that in *Escherichia coli*. *Nat. Commun.* **2023**, *14*, 5611. [CrossRef]
41. Bielczynski, L.W.; Łącki, M.K.; Hoefnagels, I.; Gambin, A.; Croce, R. Leaf and plant age affects photosynthetic performance and photoprotective capacity. *Plant Physiol.* **2017**, *175*, 1634–1648. [CrossRef]
42. Hu, T.; Xiong, K.; Yu, Y.; Wang, J.; Wu, Y. Ecological stoichiometry and homeostasis characteristics of plant-litter-soil system with vegetation restoration of the karst desertification control. *Front. Plant Sci.* **2023**, *14*, 122469131. [CrossRef]
43. Yuan, Z.; Chen, H.Y.H. Global trends in senesced-leaf nitrogen and phosphorus. *Glob. Ecol. Biogeogr.* **2009**, *18*, 532–542. [CrossRef]
44. Shaver, G.R.; Chapin, E.S., III. Long–term responses to factorial, NPK fertilizer treatment by Alaskan wet and moist tundra sedge species. *Ecography* **1995**, *18*, 259–275. [CrossRef]
45. Tian, H.; Chen, G.; Zhang, C.; Melillo, J.M.; Hall, C.A.S. Pattern and variation of C:N:P ratios in China's soils: A synthesis of observational data. *Biogeochemistry* **2010**, *98*, 139–151. [CrossRef]
46. Zhang, W.; Qiao, W.; Gao, D.; Dai, Y.; Deng, J.; Yang, G.; Han, X.; Ren, G. Relationship between soil nutrient properties and biological activities along a restoration chronosequence of *Pinus tabulaeformis* plantation forests in the Ziwuling Mountains, China. *Catena* **2017**, *161*, 85–95. [CrossRef]
47. Zeng, F.P.; Chi, G.Y.; Chen, X.; Shi, Y. The stoichiometric characteristics of C, N and P in soil and root of larch (*Larix* spp.) plantation at different stand ages in mountainous region of eastern Liaoning Province, China. *Chin. J. Ecol.* **2016**, *35*, 1819–1825. (In Chinese)

48. Chang, Y.; Zhong, Q.; Yang, H.; Xu, C.; Hua, W.; Li, B. Patterns and driving factors of leaf C, N, and P stoichiometry in two forest types with different stand ages in a mid–subtropical Zone. *For. Ecosyst.* **2022**, *9*, 100005. [CrossRef]
49. Aponte, C.; Marañón, T.; García, L.V. Microbial C, N and P in soils of Mediterranean oak forests: Influence of season, canopy cover and soil depth. *Biogeochemistry* **2010**, *101*, 77–92. [CrossRef]
50. Deng, W.; Wang, X.; Hu, H.; Zhu, M.; Chen, J.; Zhang, S.; Cheng, C.; Zhu, Z.; Wu, C.; Zhu, L. Variation characteristics of soil organic carbon storage and fractions with stand age in north subtropical *Quercus acutissima* Carruth. Forest in China. *Forests* **2022**, *13*, 1649. [CrossRef]
51. Chen, Y.; Song, Q.; Pan, L.; Jia, M.; Li, C.; Hu, B.; Wu, G. Trace metals, organic carbon and nutrients in the Beidagang Wetland Nature Reserve, northern China. *PLoS ONE* **2018**, *13*, e0204812. [CrossRef]
52. Chen, X.; Chen, H.Y.H. Plant mixture balances terrestrial ecosystem C:N:P stoichiometry. *Nat. Commun.* **2021**, *12*, 4562. [CrossRef]
53. Gong, Y.; Lv, G.; Guo, Z.; Chen, Y.; Cao, J. Influence of aridity and salinity on plant nutrients scales up from species to community level in a desert ecosystem. *Sci. Rep.* **2017**, *7*, 6811. [CrossRef] [PubMed]
54. Pang, Y.; Tian, J.; Zhao, X.; Chao, Z.; Wang, Y.; Zhang, X.; Wang, D. The linkages of plant, litter and soil C:N:P stoichiometry and nutrient stock in different secondary mixed forest types in the Qinling Mountains, China. *PeerJ* **2020**, *8*, e9274. [CrossRef] [PubMed]
55. Liu, S.; Plaza, C.; Ochoa-Hueso, R.; Trivedi, C.; Wang, J.; Trivedi, P.; Zhou, G.; Piñeiro, J.; Martins, C.S.C.; Singh, B.K.; et al. Litter and soil biodiversity jointly drive ecosystem functions. *Glob. Chang. Biol.* **2023**, *29*, 6276–6285. [CrossRef] [PubMed]
56. Gavazov, K.; Ingrisch, J.; Hasibeder, R.; Mills, R.T.E.; Buttler, A.; Gleixner, G.; Pumpanen, J.; Bahn, M. Winter ecology of a subalpine grassland: Effects of snow removal on soil respiration, microbial structure and function. *Sci. Total Environ.* **2017**, *590*, 316–324. [CrossRef] [PubMed]
57. Xu, X.; Hui, D.; King, A.W.; Song, X.; Thornton, P.E.; Zhang, L. Convergence of microbial assimilations of soil carbon, nitrogen, phosphorus, and sulfur in terrestrial ecosystems. *Sci. Rep.* **2015**, *5*, 17445. [CrossRef]
58. Fang, P.; Hu, H.; Wang, X. Soil ecological stoichiometry characteristics of different aged *Quercus acutissima* plantations. *J. Southwest For. Univ.* **2022**, *42*, 39–47. (In Chinese)
59. Ghaley, B.B.; Sandhu, H.S.; Porter, J.R. Relationship between C:N/C:O stoichiometry and ecosystem services in managed production systems. *PLoS ONE* **2015**, *10*, e0123869. [CrossRef]
60. Koerselman, W.; Meuleman, A.F.M. The vegetation N:P ratio: A new tool to detect the nature of nutrient limitation. *J. Appl. Ecol. UK* **1996**, *33*, 1441–1450. [CrossRef]
61. Zhang, G.; Han, X.; Elser, J.J. Rapid top-down regulation of plant C:N:P stoichiometry by grasshoppers in an Inner Mongolia grassland ecosystem. *Oecologia* **2011**, *166*, 253–264. [CrossRef]
62. Franklin, O.; Ågren, G.I. Leaf senescence and resorption as mechanisms of maximizing photosynthetic production during canopy development at N limitation. *Funct. Ecol.* **2002**, *16*, 727–733. [CrossRef]
63. Weiqi, W.; Linglin, X.U.; Congsheng, Z.; Chuan, T.; Linhai, Z. Carbon, nitrogen and phosphorus ecological stoichiometric ratios among live plant-litter-soil systems in estuarine wetland. *Acta Ecol. Sin.* **2011**, *31*, 7119–7124.
64. Barantal, S.; Schimann, H.; Fromin, N.; Hättenschwiler, S. C, N and P fertilization in an Amazonian rainforest supports stoichiometric dissimilarity as a driver of litter diversity effects on decomposition. *Proc. Biol. Sci.* **2014**, *281*, 20141682. [CrossRef] [PubMed]
65. Jiang, P.; Cao, Y.; Chen, Y.; Wang, F. Variation of C, N, and P stoichiometry in plant tissue, litter, and soil during stand development in *Pinus tabulaeformis* plantation. *ACTA Ecol. Sin.* **2016**, *36*, 6188–6197. (In Chinese)
66. Quan, Q.; Wang, C.; He, N.; Zhang, Z.; Wen, X.; Su, H.; Wang, Q.; Xue, J. Forest type affects the coupled relationships of soil C and N mineralization in the temperate forests of northern China. *Sci. Rep.* **2014**, *4*, 6584. [CrossRef] [PubMed]
67. Liu, A.N.; Zhang, Y.; Hou, Z.F.; Hui Lü, G. Allometric scaling of biomass with nitrogen and phosphorus above- and below-ground in herbaceous plants varies along Water-Salinity gradients. *AoB Plants* **2021**, *13*, plab030. [CrossRef]
68. Jiang, W.; Gong, L.; Yang, L.; He, S.; Liu, X. Dynamics in C, N, and P stoichiometry and microbial biomass following soil depth and vegetation types in low mountain and hill region of China. *Sci. Rep.* **2021**, *11*, 19631. [CrossRef]
69. Awoonor, J.K.; Dogbey, B.F.; Salis, I. Human-induced land use changes and phosphorus limitation affect soil microbial biomass and ecosystem stoichiometry. *PLoS ONE* **2023**, *18*, e0290687. [CrossRef]
70. Fanin, N.; Fromin, N.; Buatois, B.; Hättenschwiler, S.; Cleland, E. An experimental test of the hypothesis of non-homeostatic consumer stoichiometry in a plant litter-microbe system. *Ecol. Lett.* **2013**, *16*, 764–772. [CrossRef]
71. Pokharel, P.; Chang, S.X. Biochar decreases and nitrification inhibitor increases phosphorus limitation for microbial growth in a Wheat-Canola rotation. *Sci. Total Environ.* **2023**, *858*, 159773. [CrossRef]
72. Elser, J.J.; Acharya, K.; Kyle, M.; Cotner, J.B.; Sterner, R.W. Growth rate-stoichiometry couplings in diverse biota. *Ecol. Lett.* **2003**, *6*, 936–943. [CrossRef]
73. Mcgroddy, M.E.; Daufresne, T.; Hedin, L.O. Scaling of C:N:P stoichiometry in forests worldwide: Implications of terrestrial Redfield-type ratios. *Ecology* **2004**, *85*, 2390–2401. [CrossRef]
74. Yang, G.; Ma, Y.; Ma, X.; Wang, X.; Lu, C.; Xu, W.; Luo, J.; Guo, D. Changes in soil organic carbon components and microbial community following spent mushroom substrate application. *Front. Microbiol.* **2024**, *15*, 1351921. [CrossRef] [PubMed]
75. Wolińska, A.; Podlewski, J.; Słomczewski, A.; Grządziel, J.; Gałązka, A.; Kuźniar, A. Fungal indicators of sensitivity and resistance to long-term Maize Monoculture: A culture-independent approach. *Front. Microbiol.* **2022**, *12*, 799378. [CrossRef] [PubMed]

76. Chang, Y.; Liu, W.; Mao, Y.; Yang, T.; Chen, Y. Biochar addition alters C: N: P stoichiometry in Moss Crust-Soil continuum in Gurbantünggüt Desert. *Plants* **2022**, *11*, 814. [CrossRef] [PubMed]
77. Kück, P.; Hita Garcia, F.; Misof, B.; Meusemann, K. Improved phylogenetic analyses corroborate a plausible position of *Martialis heureka* in the ant tree of life. *PLoS ONE* **2011**, *6*, e21031. [CrossRef]
78. Ahemad, M. Phosphate-solubilizing bacteria-assisted phytoremediation of metalliferous soils: A review. *3 Biotech* **2015**, *5*, 111–121. [CrossRef]

forests

Article

Diversity Patterns of Bacteria in the Root Zone of *Davidia involucrata* Along an Altitudinal Gradient

Yang Jin [1], Xin Li [1], Yu Hu [1], Junzhong Huang [2], Yan Chen [1], Yongping Kou [3], Xinlei Li [1], Ming Dong [1], Dongzhou Deng [4,*] and Yan Li [1,*]

[1] Ecological Security and Protection Key Laboratory of Sichuan Province, Mianyang Normal University, Mianyang 621000, China
[2] Xiaohegou Nature Reserve Administrative Office, Mianyang 622550, China
[3] CAS Key Laboratory of Mountain Ecological Restoration and Bioresource Utilization & Ecological Restoration and Biodiversity Conservation Key Laboratory of Sichuan Province, Chengdu Institute of Biology, Chinese Academy of Sciences, Chengdu 610041, China
[4] Sichuan Academic of Forestry, Chengdu 610081, China
* Correspondence: dongzhoud@163.com (D.D.); leeleehi@163.com (Y.L.)

Abstract: *Davidia involucrata* has an ancient origin, representing a remnant from the paleotropical flora that thrived during the Tertiary period. Altitudinal gradient acts as a natural testing ground for studying climate change, and research on the distribution patterns of microorganisms along altitudinal gradients is crucial in understanding the adaptability of *D. involucrata* to climate change. In our study, we examined sample sites ranging from 1600 to 2200 m in elevation, which are part of the primary habitat zone for *Davidia involucrata* within the Xuebaoding National Nature Reserve. In 2021, field surveys were conducted across four altitudinal gradients (1600 m, 1800 m, 2000 m and 2200 m) of the *D. involucrata* distribution in the nature reserve. The sampling plots were set in each altitudinal gradient, and three representative and healthy mature trees were selected as sample trees for each plot. Rhizosphere soils were used to test the soil stoichiometry characteristics and root zone microbial communities. Our findings indicated pronounced differences in soil total carbon (TC) and total phosphorus (TP) content and C:P and N:P ratios between the four altitude sites ($p < 0.05$). Analysis of the bacterial communities revealed higher richness (PD and Chao1 indexes) at ASL2000 and ASL2200 (high altitude) compared to ASL1600 and ASL2000 (low altitude) ($p < 0.05$). Non-metric multidimensional scaling analysis demonstrated a distinct clustering of bacterial communities between the high and low altitudes ($p < 0.01$). At the phylum level, Proteobacteria and Acidobacteria were predominant at high altitudes, while Actinobacteria and Chloroflexi dominated at low altitudes. The core microbiome, shared among all altitudes, comprised 377 genes. The analysis of differential abundance revealed notable disparities in the prevalence of certain bacterial genera with altitude, with *Arthrobacter* and *Acidothermus* experiencing the most pronounced shifts ($p < 0.05$). This confirmed that environmental factors significantly influenced bacterial community structure and abundance. Spearman's rank correlations revealed that both Chao1 and PD indices were positively correlated with elevation, TC, and TN, with Chao1 showing stronger relationships. Both indices were negatively correlated with MAT, while only Chao1 exhibited a significant negative correlation with pH. Linear regression analysis further confirmed the significant associations between Chao1 index and elevation, TN, MAT, and pH. Furthermore, redundancy analysis demonstrated that altitude (ASL) and TN were the primary factors shaping soil bacterial community composition, explaining 21.32% and 30.70% of the variance, respectively. Altitude significantly influenced microbial community structure ($p = 0.003$). Distinct microbial taxa showed specific associations with environmental gradients, suggesting niche specialization in response to soil conditions. These findings suggest that altitude influences soil nutrient characteristics and microbial community composition in the *D. involucrata* habitat, offering insights into the ecological factors affecting this endangered species.

Keywords: *Davidia involucrata*; altitude; bacterial community structure; soil stoichiometry characteristics

1. Introduction

The altitude pattern of biodiversity is one of the classic problems in ecological research, and can offer important insights into latitudinal-induced biodiversity resulting from climate changes, as well as anticipating the prospective effects of climate change on different gradients [1]. Approximately 270 years ago, the study of biodiversity patterns across altitudinal gradients begin to receive attention [2]. However, previous studies have predominantly concentrated on the patterns of plant and animal biodiversity changes along altitudinal gradients driven by temperature differences. Most studies suggest that biodiversity exhibits either a unimodal pattern or a decrease with increasing altitude [2,3].

The study of microbial distribution across altitudes and the driving factor analysis constitutes a foundation for biogeographical research. As a crucial component of ecosystems, microorganisms are closely linked to ecosystem functional integrity [4]. Bacteria and fungi are critical for soil biodiversity, with their community structure and diversity shaped by various ecological factors [5]. However, research on the patterns of microorganisms along altitude gradients, especially small-scale altitudinal gradients within the species distribution range are not yet well understood, and the mechanisms that drive these patterns remain largely unknown [2]. Elucidating the altitude patterns of microorganisms can provide a more comprehensive and systematic understanding of biodiversity responses and ecosystem functioning under climate changes.

Despite their central role in ecosystem functioning, due to technical limitations, altitudinal patterns in microbial biodiversity have only begun to be studied in the past decade or so [2]. Over these years, encouraging progress has been made in exploring patterns along large-scale altitudinal gradients [6]. The distinct biogeographical distributions observed among microbial communities across altitudinal gradients are significantly influenced by a multitude of factors [7]. Empirical investigations into altitudinal gradients of bacterial diversity have identified two forms: monotonic declines and hump-shaped models [8–10]. However, inconsistent trends has also been reported [11,12]. Altitude gradients across short geographical distances can result in dramatic changes in climate and biological replacement [1]. The altitudinal responses of bacterial and fungal diversity, as well as their community compositions, may exhibit heterogeneous characteristics, as they demonstrate divergent responses to climatic and soil characteristics [13]. The reasons for the observed disparities in research outcomes may be intricately associated with a variety of factors, including the sampling scale, plant community structure, bedrock typology [8], climatic conditions, the specific microbial taxa under investigation, and differences in microbial species [14]. For instance, as elevation increases, there is a pronounced variation in soil nutrient parameters, such as total nitrogen, total organic carbon, moisture content, and urease activity. These alterations are correlated with the composition and metabolic activity of the soil microbial communities [15]. Additionally, vegetation can exert a direct influence on microbial community composition through rhizospheric activities. Furthermore, the type and structure of plant communities may transform with changes in elevation, which in turn can also impact the soil microbial diversity [15].

The soil microbial community composition is also modulated by the specific vegetation type. Past investigations into the elevational patterns of soil microbes have predominantly involved the collection of community soils across varying altitude gradients, exposing them to the compounding effects of plant community composition and soil physicochemical properties [16]. The function of soil microorganisms in individual species has been poorly studied [16]. To elucidate the intricate dynamics between soil ecological processes and plant growth, it is necessary to study the altitude pattern of soil microorganisms of a single species. To elucidate the complex interactions between soil ecological processes and plant growth, conducting research on the altitudinal pattern of soil microorganisms related to a specific plant species is of paramount importance.

Davidia involucrata Baill. is a deciduous tree of the genus *Davidia* in the Nyssaceae family, endemic to China [17]. It has an ancient origin, representing a remnant from the paleotropical flora that thrived during the Tertiary period [18]. It was once widely distributed in temperate and subtropical regions, and now only exists in parts of central and southwestern of China [19]. This species is restricted to relatively narrow altitudinal ranges. In the eastern distribution area, *D. involucrata* occurs from 700 to 2000 m in altitude, while in the western region, it is more commonly found from 1000 to 2800 m. Meanwhile, the optimal distribution altitude in the eastern and western region is within a span of 1200 to 2200 m [20]. Although *D. involucrata* has high research value, its ecological adaptation to altitude is still poorly understood.

Grasping the distribution of microbial diversity across altitudinal gradients and identifying the determinants of these patterns is essential for elucidating the mechanism underlying plant–soil interactions. This knowledge is crucial for enhancing our comprehension of how relic species adapt to the challenges posed by climate change [21]. This study addresses several questions: (1) What is the pattern of soil bacteria diversity in root zone soil of *D. involucrata* along altitudinal gradients? and (2) Which climate and soil factors drive the altitude patterns of soil microbes? To answer these questions, we set up four sample plots at 1600, 1800, 2000, and 2200 m, and ascertained variations in bacterial diversity and community composition across a small-scale altitudinal range in order identify the principal factors influencing these trends.

2. Materials and Methods

2.1. Study Site and Sampling

The Xuebaoding Nature Reserve, in Pingwu County, Mianyang City, Sichuan Province, China (31°59′31″~33°02′41″ N, 103°50′31″~104°58′13″ E) served as the study site (Figure 1). Characterized by a temperate, semi-humid continental monsoon climate, this region has a mean yearly temperature of 10.1 °C and receives over 1300 mm of precipitation annually. Yellow-brown soil is the predominant soil type in the area. The soil layer is thick and contains a lot of gravel fragments.

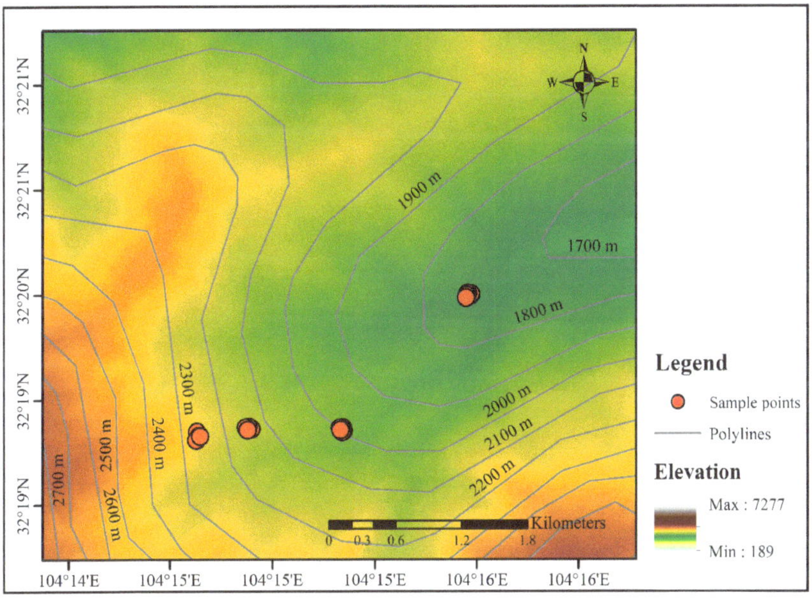

Figure 1. Map of the research area.

The research sites extended from 1600 to 2200 m above sea level (m a.s.l.) and represented the concentrated distribution area along the altitudinal gradient in the Xuebaoding national nature reserve. In 2021, field surveys were conducted in two extreme altitude gradients of the *D. involucrata* distribution in the nature reserve. The study was conducted during the growing season of *D. involucrata* at four altitudinal sampling plots (1600 m a.s.l., 1800 m a.s.l., 2000 m a.s.l., and 2200 m a.s.l.). Three representative and healthy mature trees with comparable growth vigor, a similar diameter at breast height (35~45 cm), and freedom from diseases and pests were selected as sample trees in each plot.

Around each sample tree, three soil samples were collected using a soil corer, and the collected soil cores were immediately placed in a cooler to maintain freshness. Upon returning to the laboratory, soil physical properties were determined. The surface litter and cover materials were removed from the base of the sample tree. Prior to sampling, all sampling tools were disinfected and sterilized using 75% alcohol. Surface soil samples from the top 10 cm layer were collected using the sterilized tools. These samples underwent sieving through a 20-mesh screen, after which they were split into two parts. One portion, sealed in bags, was transported to the lab for physicochemical analysis. The other, stored in sterile centrifuge tubes within a liquid nitrogen container, was designated for soil microbial community structure analysis. The soil samples for soil characteristic measurements were shade-dried naturally for pH determination and the analysis of soil element content.

The soil samples for soil microbial diversity analysis, upon returning to the laboratory, were immediately placed in liquid nitrogen and sent to a contracted company for sequencing.

2.2. Soil Physicochemical Properties Analyses

Soil mean annual temperature (MAT) and mean annual humidity (MAH) were measured using a TMS-4 temperature and moisture meter (TOMST, Prague, Czech Republic) at each altitude.

For the determination of soil pH, a fresh soil sample equivalent to 5 g of dry soil was taken, combined with 25 mL of CO_2-free water, placed on a magnetic stirrer for 30 min, and then measured with a pH meter (FE20-Five Easy TM pH, Mettler Toledo, Greifensee, Germany). Determination of total carbon (TC), total nitrogen (TN), and total phosphorus (TP) was performed as follows: soil was dried naturally, ground finely, and sieved through a 100 mesh sieve, and contents were then determined using a carbon and nitrogen analyzer (Vario MAX, Elementar, Langenselbold, Germany); total phosphorus was determined using the molybdenum blue colorimetric method.

2.3. Soil DNA Exrtraction, PCR Amplification and Illumina Miseq Sequencing

Root zone soil (0.5 g dry weight) underwent DNA isolation using the E.Z.N.A.® soil DNA extraction kit (Omega Bio-tek, Norcross, GA, USA). PCR amplification targeted the V3-V4 hypervariable regions of bacterial 16S rRNA genes, employing primers 338F (5′-ACTCCTACGGGAGGCAGCA-3′) and 806R (5′-GGACTACHVGGGTWTCTAAT-3′). After separation on 2% agarose gel, the amplicons were extracted and purified with the AxyPrep DNA Gel Extraction Kit (Axygen Biosciences, Union City, CA, USA). QuantiFluor™-ST (Promega, Madison, WI, USA) facilitated DNA quantification. The resulting products then underwent paired-end sequencing (2 × 300 bp) on an Illumina MiSeq platform (Illumina, San Diego, CA, USA).

2.4. Data Quality Control and Processing

The original sequences were multiplexed and subjected to quality filtering using Trimmomatic. The paired-end reads were then merged using FLASH [22]. Subsequently, the operational taxonomic units (OTUs) were clustered at a 97% similarity threshold using the UPARSE software (version 7.1, http://drive5.com/uparse), and chimeric sequences were identified and removed using UCHIME [23] (accessed on 15 October 2024).

Following that, the classification of each 16S rRNA gene sequence was performed using the SILVA r138 database [24]. To account for variations in sequence numbers that could affect the composition of bacterial communities, the sequence numbers for each soil sample were standardized to the minimum sequence number after removing singletons. In this study, a total of 1,255,896 bacterial sequences were obtained.

2.5. Statistical Analyses

This text utilized OTU table analysis to examine the α and β diversity of soil microorganisms at different altitudes. Statistical analyses of the data were conducted using Excel 2019 and R 4.2.1 software. ANOVA was employed to assess the differences in soil nutrient characteristics of *D. involucrata* and α diversity indexes of root zone soil bacteria across four altitudes, and a *t*-test was employed to assess the differences in soil nutrient characteristics of *D. involucrata* and α diversity indexes of root zone soil bacteria across high and low altitudes. Non-metric multidimensional scaling (NMDS) was used to analyze the differences in community composition between altitudes. Spearman correlation analysis and redundancy analyses (RDA) were employed to analyze the correlation between soil microbial communities and environmental factors.

3. Results

3.1. Community Composition of Soil Bacteria in Root Zone of D. involucrata

The effects of altitude on microbial community diversity were examined across four elevations ranging from 1600 m a.s.l. to 2200 m a.s.l. (Figure 2A). The Faith's Phylogenetic Diversity (PD) index (Figure 2A) did not exhibit statistically significant changes with altitude ($p > 0.05$), but revealed significant differences between low (L_ASL) and high (H_ASL) altitude sites ($p = 0.017$), with higher values observed at H_ASL. In contrast, the Chao1 index, representing species richness, demonstrated significant differences between lower (1600–1800 m) and higher (2000–2200 m) altitudes ($p < 0.05$; Figure 2A). H_ASL exhibited significantly greater Chao1 index values compared to L_ASL ($p = 0.001$).

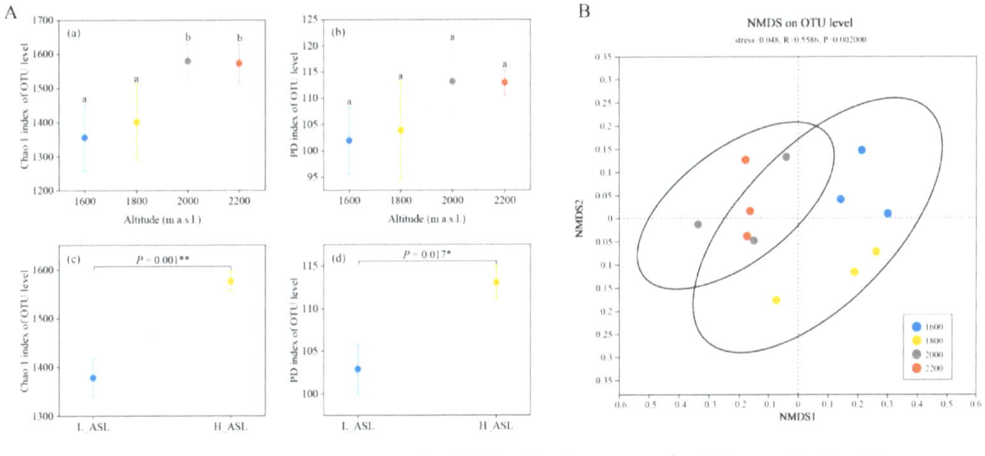

Figure 2. Soil bacterial communities at the OTU level in the root zone at different altitudes. (**A**) α diversity metrics for bacterial communities based on 16sRNA regions in the soil of *Davidia involucrata* at different altitudes (**a**) Chao 1 index. (**b**) PD index. (**c**) Chao 1 index and (**d**) PD index comparison between low (L_ASL) and high altitude (H_ASL) levels. Each point represents the mean value, and error bars indicate standard error. Different lowercase letters denote significant differences among altitudes ($p < 0.05$). * $p < 0.05$, ** $p < 0.01$. (**B**) NMDS of soil bacterial communities at the OTU level in the root zone at different altitudes. Each point represents a sample, with colors indicating different altitudes: blue (1600 m), yellow (1800 m), grey (2000 m), and red (2200 m). The ellipses represent 95% confidence intervals for the high and low elevation groupings. The stress value of 0.048 indicates a good fit of the NMDS ordination. ANOSIM analysis (R = 0.5586, $p = 0.002$) revealed significant differences in bacterial community composition among altitudes.

Non-metric multidimensional scaling (NMDS) analysis of bacterial communities at the OTU level demonstrated clear separation between lower (1600 m and 1800 m) and higher (2000 m and 2200 m) altitude samples (Figure 2B). ANOSIM confirmed significant differences in bacterial community composition among altitudes ($R = 0.5586$, $p = 0.002$).

At the phylum level, the dominant bacterial groups at both high and low altitudes were Proteobacteria (39.01%, 32.50%), Actinobacteria (22.01%, 26.07%), Acidobacteria (15.88%, 13.39%), and Chloroflexi (6.82%, 8.27%) (Figure 3A). High altitude samples showed higher relative abundances of Proteobacteria and Acidobacteria, while low altitude samples had higher proportions of Actinobacteria and Chloroflexi.

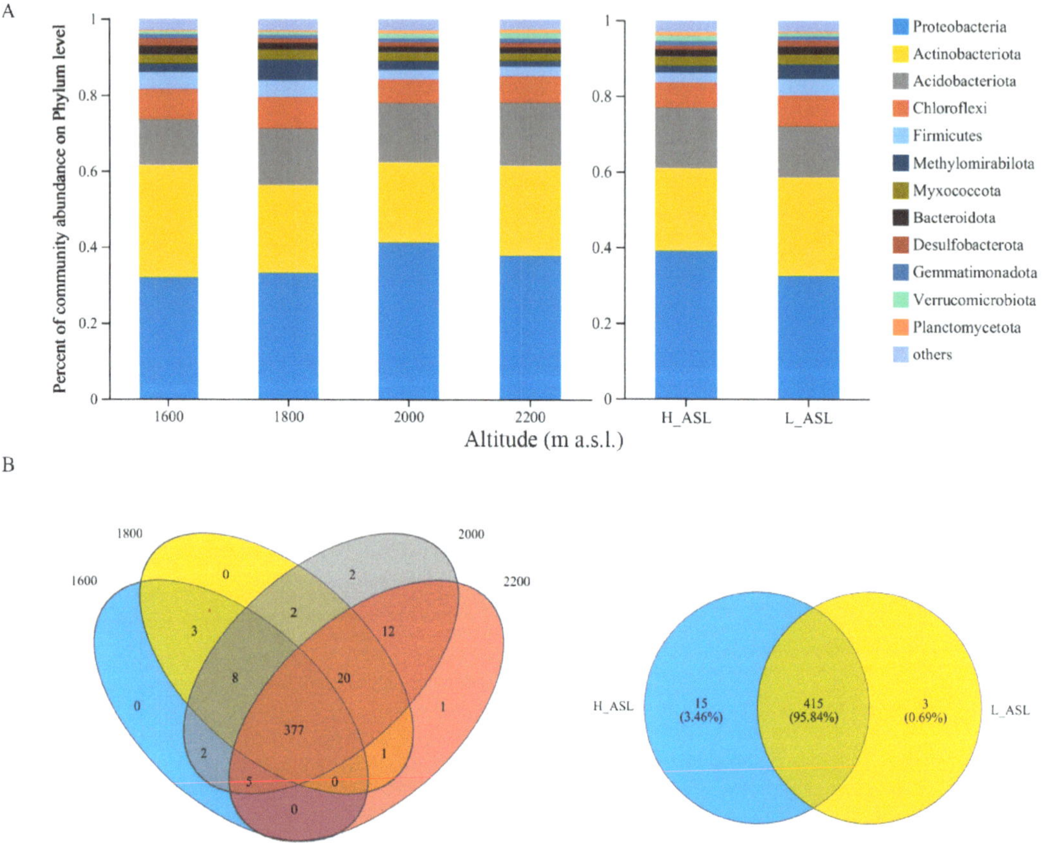

Figure 3. (**A**) Proportions of dominant bacteria in the root zone of different altitudes at the phylum level. Phyla representing less than 1% of the total reads are grouped as "others". Stacked bar plot shows the relative abundance of bacterial phyla at four specific altitudes (1600 m, 1800 m, 2000 m, and 2200 m), and across low altitude (L_ASL) and high altitude (H_ASL) groups. (**B**) Venn diagram showing the distribution of genus among four altitudes in soil samples. The numbers in each section represent the number of genera. The overlapping areas indicate shared genera at different altitudes.

The core microbiome, shared among all altitudes, comprised 377 genera (Figure 3B). Differential abundance analysis revealed significant differences in bacterial genera between H_ASL and L_ASL groups, with *Arthrobacter* and *Acidothermus* showing the most pronounced changes (Figure 4).

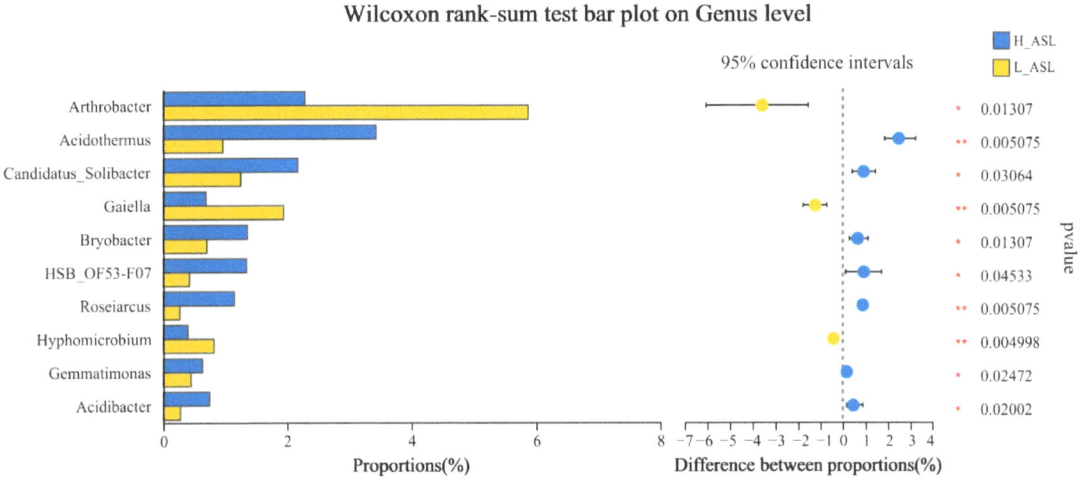

Figure 4. Differential abundance analysis of bacterial genera between high and low altitude based on Wilcoxon rank-sum test. The left panel shows the relative abundance (proportions) of each genus in high altitude (H_ASL, blue) and low altitude (L_ASL, yellow) samples. Positive values in the right panel indicate higher abundance in high altitude samples, while negative values indicate higher abundance in low altitude samples, at 95% confidence intervals. Genera are ordered by their statistical significance. Blue dots indicate higher abundance at a high altitude, while yellow dots represent higher abundance at a low altitude. The vertical dashed line at 0 represents no difference between altitudes. p-values are shown on the far right, with significance levels indicated by (* $p < 0.05$, ** $p < 0.01$). Only genera with significant differences ($p < 0.05$) are shown.

3.2. Soil Physicochemical Properties Along the Altitudinal Gradient

Soil physicochemical properties varied significantly along the altitudinal gradient (1600–2200 m a.s.l.). TC, TN, and TP concentrations increased with elevation, reaching maxima at 2000–2200 m (Figure 5A–C). The C:N ratio peaked at 2000 m and was lowest at 1600 m, while C:P and N:P ratios rose sharply above 1800 m (Figure 5D–F). MAT decreased steadily with elevation, whereas MAH peaked at 1800 m before declining (Figure 5G,H). Soil pH showed a slight downward trend, reaching its lowest at 2000 m (Figure 5I). Significant differences were observed between high (H_ASL) and low (L_ASL) altitude sites (Table 1): TC and TN were higher at H_ASL ($p < 0.01$), TP was higher at L_ASL ($p = 0.04$), C:P and N:P ratios were higher at H_ASL ($p < 0.01$), and pH was lower at H_ASL ($p < 0.01$). MAT was significantly lower at H_ASL ($p < 0.01$).

Table 1. Environment elements in the soil of *Davidia involucrata* at high altitude and low altitude.

	TC	TN	TP	C:N	C:P	N:P	MAT	MAH	pH
L_ASL	53.50 ± 4.32 a	4.91 ± 0.10 a	1.29 ± 0.22 b	10.89 ± 0.67 a	42.06 ± 3.78 a	3.89 ± 0.59 a	10.49 ± 1.22 b	0.33 ± 0.11 a	5.74 ± 0.17 b
H_ASL	71.24 ± 3.75 b	6.36 ± 0.11 b	1.03 ± 0.14 a	11.20 ± 0.46 a	70.06 ± 6.00 b	6.28 ± 0.77 b	8.38 ± 0.26 a	0.37 ± 0.06 a	5.27 ± 0.25 a
p-value	<0.01	<0.01	0.04	0.37	<0.01	<0.01	<0.01	0.42	<0.01

Notes: Values followed by different small letters represent significant differences between high altitude and low altitude ($p < 0.05$). H_ASL: high altitude; L_ASL: low altitude.

3.3. The Relationship Between Environmental Factors and Bacterial Communities

Spearman's rank correlations indicated that both the Chao1 and PD indices were positively correlated with elevation, TC, and TN. The Chao1 index showed stronger correlations ($p < 0.01$) compared to the PD index ($p < 0.05$). Both indices exhibited negative correlations with MAT, with Chao1 showing a stronger relationship ($r = -0.73$, $p < 0.01$) than PD ($r = -0.62$, $p < 0.05$). Interestingly, only the Chao1 index showed a significant negative correlation with pH (r = −0.59, $p < 0.05$), while the correlation between PD and pH was not statistically significant (r = −0.48, $p > 0.05$) (Table 2). Specifically, the Chao1 index showed positive correlations with elevation (R^2_{adj} = 0.555, p = 0.003), TC (R^2_{adj} = 0.613, p = 0.002), TN (R^2_{adj} = 0.669, $p < 0.001$), and C:P (R^2_{adj} = 0.518, p = 0.005), and negative correlations with MAT (R^2_{adj} = 0.452, $p < 0.01$) and pH (R^2_{adj} = 0.320, p = 0.003) (Figure 6).

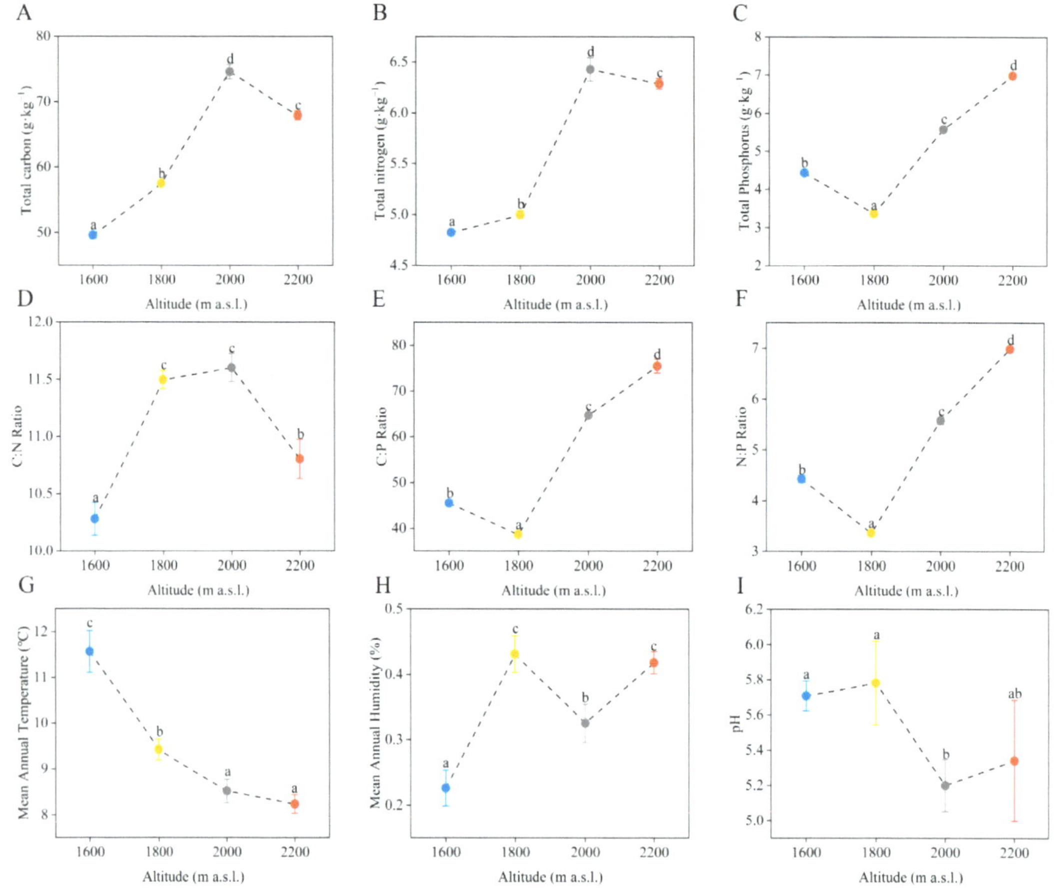

Figure 5. Environment elements in the soil of *Davidia involucrata* at different altitudes: (**A**) total carbon, (**B**) total nitrogen, (**C**) total phosphorus, (**D**) C:N ratio, (**E**) C:P ratio, (**F**) N:P ratio, (**G**) Mean annual temperature, (**H**) Mean annual humidity, and (**I**) Soil pH. Each point represents the mean value, and error bars indicate standard error. Different lowercase letters denote significant differences among altitudes ($p < 0.05$).

Table 2. Spearman correlation coefficients between environmental factors and alpha diversity indices (Chao1, Faith's PD).

	ASL	TC	TN	TP	C:N	C:P	N:P	MAT	MAH	pH
Chao1	0.75 **	0.80 **	0.80 **	−0.11	0.43	0.58 *	0.55	−0.73 **	0.42	−0.59 *
PD	0.58 *	0.63 *	0.59 *	−0.10	0.31	0.48	0.43	−0.62 *	0.40	−0.48

Note: Asterisks denote significance levels: ** $p < 0.01$, * $p < 0.05$.

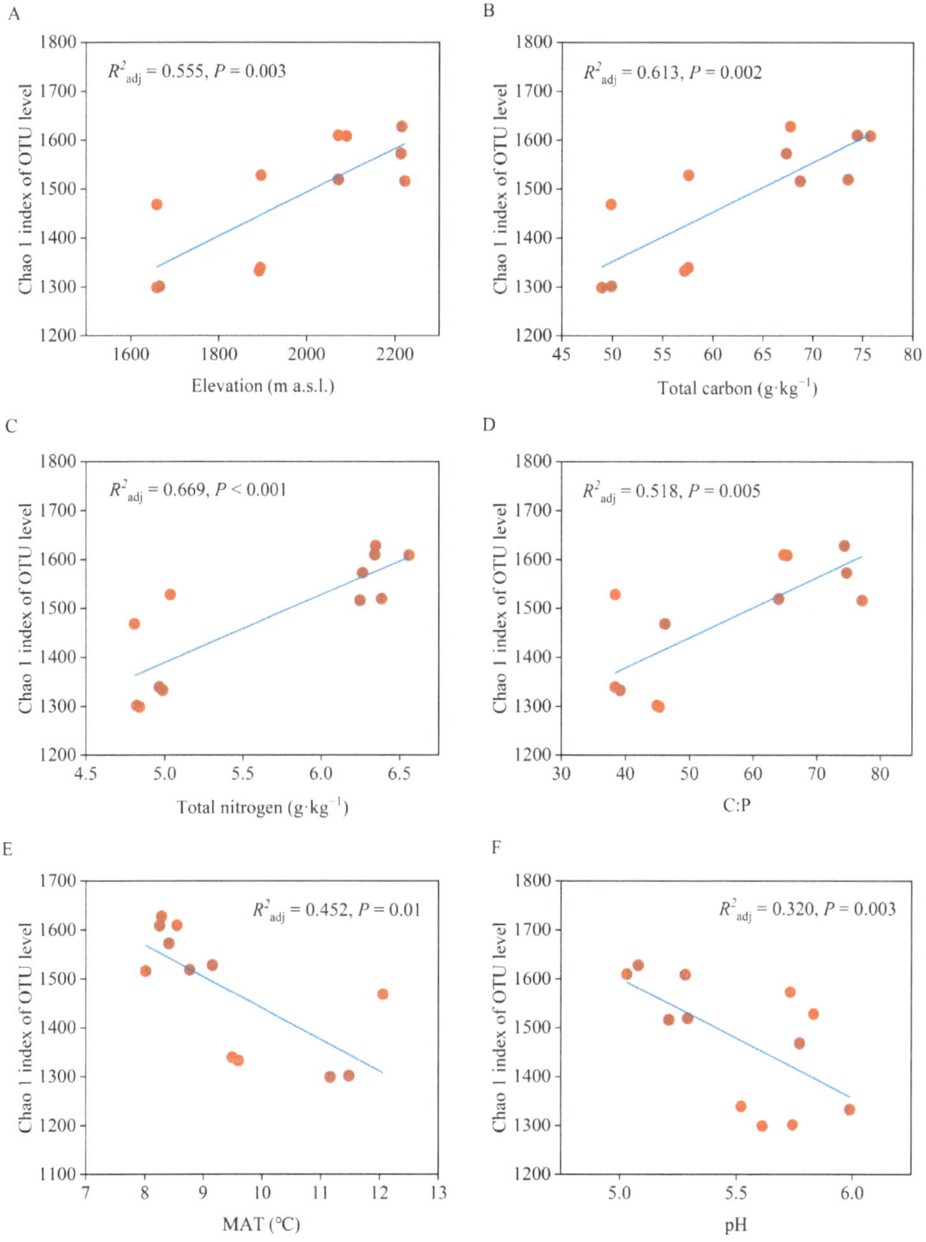

Figure 6. Relationships between Chao1 index of operational taxonomic units (OTUs) and environmental factors: (**A**) elevation, (**B**) total carbon, (**C**) total nitrogen, (**D**) C:P ratio, (**E**) MAT, and (**F**) pH. The blue lines represent the fitted linear regression models, and the grey shaded areas indicate the 95% confidence intervals. R^2_{adj} and *p*-values are shown for each relationship.

Redundancy analysis (RDA) revealed that the first two axes explained 51.83% of the total variance in microbial community composition (Figure 7). Environmental factors showed varying influences, with TN and ASL exhibiting the strongest correlations with community structure. Variance partitioning indicated that TN accounted for the highest proportion of explained variance (30.70%), followed by ASL (21.32%) and TC (15.93%). However, only ASL showed a statistically significant effect ($F = 13.3587$, $p = 0.003$) (Table 3). Distinct microbial genus-level associations were observed along environmental gradients, with *Arthrobacter* negatively correlated with TN and TC, while *Roseiarcus* and *Acidothermus* showed positive correlations.

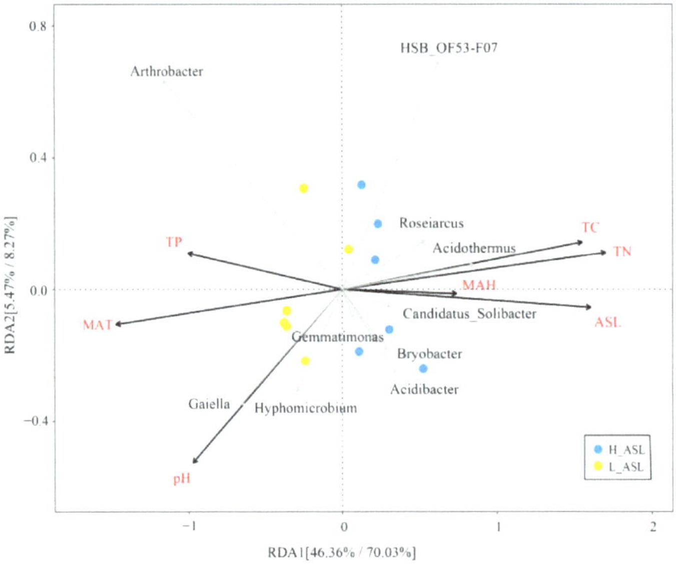

Figure 7. Redundancy analysis (RDA) of soil microbial community composition in relation to environmental factors. The first two RDA axes explain 51.83% of the total constrained variation (RDA1: 46.36%, RDA2: 5.47%).

Table 3. Influence of environmental factors on soil microbial community composition.

Environmental Factor	Explained Variance (%)	F-Value	*p*-Value	Significance
TN	30.7	2.271	0.18	-
ASL	21.32	13.3587	0.003	**
TC	15.93	0.8642	0.419	-
pH	12.74	1.7887	0.249	-
MAT	12.14	0.6543	0.521	-
TP	5.62	1.0991	0.326	-
MAH	1.5	0.6706	0.533	-

Notes: The explained variance represents the proportion of variation in microbial community composition accounted for by each factor. *p*-values were obtained from permutation tests (999 permutations). ** indicates statistical significance at $p < 0.01$.

4. Discussion

Microorganisms are integral to ecosystem functioning, exerting a significant influence on the regulation of key processes, such as the cycling of nutrients and materials. Microbial diversity and abundance are intricately linked to the strength and characteristics of ecological functions in ecosystems. Altitude significantly influences the diversity and abundance of soil microorganisms [25]. The α-diversity of soil bacteria in this research did not vary much between altitudes, but notable differences were observed in the community composition between high and low altitudes. In our research, the soil bacterial richness of the root zone was remarkably higher at a high altitude (ASL_{2000} and ASL_{2200}) than at a low altitude (ASL_{1600} and ASL_{1800}). This pattern deviates from the trend observed in numerous other organism groups; for example, in one study, it was shown that as the altitude gradient increased from 2400 m to 3400 m, the richness of bacterial taxa and their phylogenetic diversity exhibited a consistent decline [1,26]. Meanwhile, other studies on Gongga (3000 m to 4100 m) and Changbai Mountain (2000 m to 2500 m) in China and the Rocky Mountains (2460 m to 3380 m) in the United States have demonstrated the same pattern [1,27,28]. Although the trend of the results of our research reflects that bacterial diversity increases with an increase in altitude, the altitude range of the present study is from 1600 m to 2200 m, which belongs to the mid-mountainous zone as compared to the above mentioned studies, and is similar to the results of a previous study performed on Mt. Fuji; that is, bacterial diversity exhibits a hump-shaped pattern in relation to altitude (1000 m to 3700 m). Across a montane altitudinal gradient in eastern Peru, previous research determined that there was no substantial variation in bacterial diversity associated with changes in altitude (1800 m to 3200 m) [29]. Therefore, the variation in bacterial diversity with altitude differs at different altitude spans. However, over a wider altitude span (440~3250 m), bacterial diversity has been found to not be significantly related to altitude [5]. Our study reveals the altitudinal distribution of soil bacteria diversity in *D. involucrata* (1600~2200 m) in mountain ecosystems. This helps to understand the distribution patterns of bacterial diversity of *D. involucrata* across its altitudinal range. On the other hand, there are fewer research studies on the altitudinal patterns of soil microorganisms in the root zone of single woody plants. This is the first study to reveal the soil bacterial diversity, community structure, and spatial distribution of microorganisms in the root zone of *D. involucrata* trees in the Xuebaoding National Nature Reservation.

While the geographic spread of bacterial communities is predominantly influenced by climatic factors [30,31], soil properties exert a direct impact on the regulation of these communities [32,33]. Specifically, ecological niche environmental filtering processes related to soil carbon and nitrogen content are likely to be the primary determinants in shaping the structure of bacterial communities across altitudinal gradients [28]. Shen et al. (2015) indicated that soil C and N content is an essential ecological determinant affecting the vertical distribution of bacteria [28]. The bacterial assemblages exhibited substantial variation with changes in altitude, and there was a pronounced correlation between the community structure and the levels of total C and N in the soil within the Changbai Mountain tundra ecosystem [28]. Furthermore, the balance of resource stoichiometry is pivotal for the function of microorganisms [32]. Reed et al. (2011) highlighted that the ratio of resource elements, such as the nitrogen to phosphorus ratio, can often provide a more accurate forecast of microbial activity than considering a single nutrient in isolation [34]. Eisele et al. (1989) suggested a possible link between nitrogen fixation rates and the proportion of available nitrogen to accessible phosphorus [34,35]. Wan et al. (2015) discovered that the soil microbial community structure changed in response to C:N ratio variations [36]. Therefore, using ecological stoichiometry may help to reveal how soil factors impact microbial communities. In our research, we observed significant changes in the MAT, pH, STC, STN, STP content, C:P, and N:P between high and low altitudes. In addition, altitude, soil physicochemical properties, and chemical characteristics also showed a high correlation with bacterial diversity in our results. These findings emphasize the significant impact of altitude-related changes in temperature, pH, and soil organic matter content on microbial community diversity within root zone soil. The complex interplay between elevation and soil physicochemical properties seems to be a primary force shaping the patterns of microbial diversity within this mountain ecosystem. The abundance and composition of bacterial communities in soil is directly associated with soil characteristics, and this also offers insight into the greater soil bacterial richness at higher altitudes as opposed to lower altitudes [37,38]. The altitudinal variation trend of nutrients in soil may be the main influencing factor driving the emergence of bacterial patterns in *D. involucrata* along the altitudinal gradient.

In addition to plant communities, the balance of C and N within the litter-soil-microbe continuum also influences the shifts in bacterial composition and diversity [32,33]. Furthermore, due to the distinct nutrient requirements and dosages of diverse functional groups, the stoichiometry of carbon and nitrogen in the soil continuum may exert both individual and combined influences on bacterial communities [33,39]. For example, in nitrogen-rich soil environments, the functional groups associated with N cycling having a high assemblage, and similarly, in carbon-rich soils, there is a high abundance of bacteria that participate in the carbon cycle [32,33]. The main bacterial phyla in the root zone of *D. involucrata* include Proteobacteria, Actinobacteria, Acidobacteria, and Chloroflexi, with the functions of these groups frequently connected to the cycling processes of carbon and nitrogen. Acidobacteria taxa are frequently linked to the carbon cycle, whereas Proteobacteria, Chloroflexi, and Actinobacteria are primarily engaged in nitrogen cycling activities [32,33,40]. In addition, the distribution of different genera of *Arthrobacter* and *Acidothermus* was significantly or very significantly observed in the rhizospheric soils at high altitude (H_ASL) and low altitude (L_ASL). Our results also revealed significant correlations between environmental factors and bacterial diversity. Specifically, we detected a positive association between bacteria diversity and altitudes, TC, TN, and C:P, and negative correlation between bacteria diversity and pH, as well as MAT. This suggests that the differences in environmental conditions along the altitude gradient have led to distinct patterns in bacteria distribution and community composition, and altitude acts as a key factor driving the divergence of bacteria communities and their relationships with environmental variables. Therefore, consistent with other research, the interaction between soil carbon, nitrogen, and phosphorus is indicated to exert a possible influence on the distribution characteristics of bacterial communities in the root zone soil of *D. involucrata* [41].

5. Conclusions

This study explored the differences in microclimatic characteristics, soil properties, and root zone bacteria communities between altitudes within the *D. involucrata* distribution range in the Xuebaoding Nature Reserve, Sichuan Province, China. In this study, the results exhibited a notable disparity in soil bacteria community characteristics in root zones of mature trees, especially in terms of richness and relative abundance. The composition and diversity of bacterial communities were different at high and low altitudes. There were also differences in soil pH and stoichiometric characteristics among altitudes. There was a significant linkage between environmental parameters and soil bacterial communities. TC, TN, C:P, pH, and MAT values were significantly correlated with bacterial diversity, while ASL and TN were key factors in the formation of bacterial community diversity across the studied altitudes. Our results aid in understanding the soil bacteria pattern of *D. involucrata* within its altitudinal distribution range, and the associated driving factors.

Author Contributions: Conceptualization, Y.K.; methodology, Y.C.; formal analysis, Y.J., X.L. (Xin Li) and X.L. (Xinlei Li); investigation, Y.J., X.L. (Xin Li) and J.H.; resources, M.D.; data curation, Y.H.; writing-original draft, Y.J.; writing-review and editing, D.D.; supervision, M.D.; project administration, D.D. and Y.L.; funding acquisition, Y.L. All authors have read and agreed to the published version of the manuscript.

Funding: This work was supported by the open project of CAS Key Laboratory of Mountain Ecological Restoration and Bioresource Utilization and Ecological Restoration and the Biodiversity Conservation Key Laboratory of Sichuan Province, Chengdu Institute of Biology, Chinese Academy of Sciences (kxysws2002), the Natural Science Foundation of Sichuan Province (23NSFSC0142), the Scientific Research Initiation Project of Mianyang Normal University (QD2023A01), the Research Program of the Ecological Security and Protection Key Lab of Sichuan Province (ESP201906).

Data Availability Statement: The original contributions presented in the study are included in the article. Further inquiries can be directed to the corresponding author.

Conflicts of Interest: The authors declare no conflicts of interest.

References

1. Bryant, J.A.; Lamanna, C.; Morlon, H.; Kerkhoff, A.J.; Enquist, B.J.; Green, J.L. Colloquium paper: Microbes on mountainsides: Contrasting elevational patterns of bacterial and plant diversity. *Proc. Natl. Acad. Sci. USA* **2008**, *105* (Suppl. S1), 11505–11511. [CrossRef] [PubMed]
2. Wang, J.; Meier, S.; Soininen, J.; Casamayor, E.O.; Pan, F.; Tang, X.; Yang, X.; Zhang, Y.; Wu, Q.; Zhou, J.; et al. Regional and global elevational patterns of microbial species richness and evenness. *Ecography* **2016**, *40*, 393–402. [CrossRef]
3. Rahbek, C. The role of spatial scale and the perception of large-scale species-richness patterns. *Ecol. Lett.* **2004**, *8*, 224–239. [CrossRef]
4. Yang, T.; Adams, J.M.; Shi, Y.; Sun, H.; Cheng, L.; Zhang, Y.; Chu, H. Fungal community assemblages in a high elevation desert environment: Absence of dispersal limitation and edaphic effects in surface soil. *Soil Biol. Biochem.* **2017**, *115*, 393–402. [CrossRef]
5. Fierer, N.; Jackson, R.B. The diversity and biogeography of soil bacterial communities. *Proc. Natl. Acad. Sci. USA* **2006**, *103*, 626–631. [CrossRef]
6. Liu, X.; Yang, T.; Shi, Y.; Zhu, Y.; He, M.; Zhao, Y.; Adams, J.M.; Chu, H. Strong partitioning of soil bacterial community composition and co-occurrence networks along a small-scale elevational gradient on Zijin Mountain. *Soil Ecol. Lett.* **2021**, *3*, 290–302. [CrossRef]
7. Lauber, C.L.; Hamady, M.; Knight, R.; Fierer, N. Pyrosequencing-based assessment of soil pH as a predictor of soil bacterial community structure at the continental scale. *Appl. Environ. Microbiol.* **2009**, *75*, 5111–5120. [CrossRef]
8. Singh, D.; Takahashi, K.; Kim, M.; Chun, J.; Adams, J.M. A Hump-Backed Trend in Bacterial Diversity with Elevation on Mount Fuji, Japan. *Microb. Ecol.* **2011**, *63*, 429–437. [CrossRef]
9. He, L.B.; Sun, X.Y.; Li, S.Y.; Zhou, W.Z.; Yu, J.T.; Zhao, G.Y.; Chen, Z.; Bai, X.T.; Zhang, J.S. Depth effects on bacterial community altitudinal patterns and assembly processes in the warm-temperate montane forests of China. *Sci. Total Environ.* **2024**, *914*, 169905. [CrossRef]
10. Ni, Y.Y.; Yang, T.; Zhang, K.P.; Shen, C.C.; Chu, H.Y. Fungal communities along a small-scale elevational gradient in an alpine tundra are determined by soil carbon nitrogen rations. *Front. Microbiol.* **2018**, *9*, 1815. [CrossRef]
11. Singh, D.; Lee-Cruz, L.; Kim, W.-S.; Kerfahi, D.; Chun, J.-H.; Adams, J.M. Strong elevational trends in soil bacterial community composition on Mt. Halla, South Korea. *Soil Biol. Biochem.* **2014**, *68*, 140–149. [CrossRef]
12. Shen, C.; Xiong, J.; Zhang, H.; Feng, Y.; Lin, X.; Li, X.; Liang, W.; Chu, H. Soil pH drives the spatial distribution of bacterial communities along elevation on Changbai Mountain. *Soil Biol. Biochem.* **2013**, *57*, 204–211. [CrossRef]
13. Zhou, Y.N.; Meng, F.F.; Ochieng, B.; Xu, J.N.; Zhang, L.; Kimirei, I.A.; Feng, M.H.; Zhu, L.F.; Wang, J.J. Climate and environmental variables drive stream biofilm bacterial and fungal diversity on tropical mountainsides. *Microb. Ecol.* **2024**, *87*, 28. [CrossRef] [PubMed]
14. Shen, C.; Gunina, A.; Luo, Y.; Wang, J.; He, J.Z.; Kuzyakov, Y.; Hemp, A.; Classen, A.T.; Ge, Y. Contrasting patterns and drivers of soil bacterial and fungal diversity across a mountain gradient. *Environ. Microbiol.* **2020**, *22*, 3287–3301. [CrossRef]
15. Tang, M.Z.; Li, L.; Wang, X.L.; You, J.; Li, J.N.; Chen, X. Elevational is the main factor controlling the soil microbial community structure in alpine tundra of the Changbai Mountain. *Sci. Rep.* **2020**, *10*, 12442. [CrossRef]
16. Wu, D.; Yang, T.; Lin, C.; Fu, X.; Zhao, Y.; Chu, H. Elevational Distribution of Soil Diazotrophic Community in Root Zone of Ginkgo biloba in Tianmu Mountain. *Soils* **2022**, *54*, 958–967.
17. Li, H. *Davidia* as the type of a new family *Davidaceae*. *Lloydia* **1954**, *17*, 329–331.
18. Fu, L.K.; Jin, J.M. *Red Data Book of Chinese Plants*; Science Press: Beijing, China, 1991. (In Chinese)
19. Tang, C.Q.; Ohsawa, M. Tertiary relic deciduous forests on a humid subtropical mountain, Mt. Emei, Sichuan, China. *Folia Geobot.* **2002**, *37*, 93–106. [CrossRef]
20. Su, Z.; Zhang, S. The reproductive phenology ang the influencing factors of *Davidia involucrata* population. *J. Sichuan Teach. Coll. (Nat. Sci.)* **1999**, *4*, 4–9. [CrossRef]
21. Zhang, Y.G.; Cong, J.; Lu, H.; Li, G.L.; Xue, Y.D.; Deng, Y.; Li, H.; Zhou, J.Z.; Li, D.Q. Soil bacterial diversity patterns and drivers along an elevational gradient on Shenlongjia Mountain, China: Soil bacterial elevational pattern. *Microb. Biotechnol.* **2015**, *8*, 739–746. [CrossRef]
22. Magoč, T.; Salzberg, S.L. FLASH: Fast length adjustment of short reads to improve genome assemblies. *Bioinformatics* **2011**, *27*, 2957–2963. [CrossRef] [PubMed]
23. Blaxter, M.; Mann, J.; Chapman, T.; Thomas, F.; Whitton, C.; Floyd, R.; Abebe, E. Defining operational taxonomic units using DNA barcode data. *Philos. Trans. R. Soc. Lond. B Biol. Sci.* **2005**, *360*, 1935–1943. [CrossRef]
24. Quast, C.; Pruesse, E.; Yilmaz, P.; Gerken, J.; Schweer, T.; Yarza, P.; Peplies, J.; Glöckner, F.O. The SILVA ribosomal RNA gene database project: Improved data processing and web-based tools. *Nucleic Acids Res.* **2013**, *41*, D590–D596. [CrossRef]
25. Zhao, Y.H.; Li, T.; Shao, P.S.; Sun, J.K.; Xu, W.J.; Zhang, Z.H. Variation in bacterial community structure in rhizosphere and bulk soils of different halophytes in the yellow river delta. *Front. Ecol. Evol.* **2022**, *9*, 816918. [CrossRef]
26. Hu, L.; Xiang, Z.Y.; Wang, G.X.; Rafique, R.; Liu, W.; Wang, C.T. Changes in soil physicochemical and microbial properties along elevation gradients in two forest soils. *Scand. J. For. Res.* **2016**, *31*, 242–253. [CrossRef]
27. Wang, Y.; Li, C.; Shen, Z.; Rui, J.; Jin, D.; Li, J.; Li, X. Community assemblage of free-living diazotrophs along the elevational gradient of Mount Gongga. *Soil Ecol. Lett.* **2019**, *1*, 136–146. [CrossRef]

28. Shen, C.; Ni, Y.; Liang, W.; Wang, J.; Chu, H. Distinct soil bacterial communities along a small-scale elevational gradient in alpine tundra. *Front. Microbiol.* **2015**, *6*, 582. [CrossRef]
29. Fierer, N.; McCain, C.M.; Meir, P.; Zimmermann, M.; Rapp, J.M.; Silman, M.R.; Knight, R. Microbes do not follow the elevational diversity patterns of plants and animals. *Ecology* **2011**, *92*, 797–804. [CrossRef] [PubMed]
30. Han, L.-L.; Wang, Q.; Shen, J.-P.; Di, H.J.; Wang, J.-T.; Wei, W.-X.; Fang, Y.-T.; Zhang, L.-M.; He, J.-Z. Multiple factors drive the abundance and diversity of the diazotrophic community in typical farmland soils of China. *FEMS Microbiol. Ecol.* **2019**, *95*, fiz113. [CrossRef]
31. Shay, P.-E.; Winder, R.S.; Trofymow, J.A. Nutrient-cycling microbes in coastal Douglas-fir forests: Regional-scale correlation between communities, in situ climate, and other factors. *Front. Microbiol.* **2015**, *6*, 1097. [CrossRef]
32. Zhang, X.; Liang, C.; Song, J.; Ye, Z.; Wu, W.; Hu, B. Transcriptome analyses suggest a molecular mechanism for the SIPC response of *Amphibalanus amphitrite*. *Biochem. Biophys. Res. Commun.* **2020**, *525*, 823–829. [CrossRef] [PubMed]
33. Wang, S.; Zhao, S.; Yang, B.; Zhang, K.; Fan, Y.; Zhang, L.; Yang, X. The carbon and nitrogen stoichiometry in litter-soil-microbe continuum rather than plant diversity primarily shapes the changes in bacterial communities along a tropical forest restoration chronosequence. *CATENA* **2022**, *213*, 106202. [CrossRef]
34. Reed, S.C.; Cleveland, C.C.; Townsend, A.R. Functional Ecology of Free-Living Nitrogen Fixation: A Contemporary Perspective. *Annu. Rev. Ecol. Evol. Syst.* **2011**, *42*, 489–512. [CrossRef]
35. Eisele, K.A.; Schimel, D.S.; Kapustka, L.A.; Parton, W.J. Effects of available P and N:P ratios on non-symbiotic dinitrogen fixation in tallgrass prairie soils. *Oecologia* **1989**, *79*, 471–474. [CrossRef]
36. Wan, X.; Huang, Z.; He, Z.; Yu, Z.; Wang, M.; Davis, M.R.; Yang, Y. Soil C:N ratio is the major determinant of soil microbial community structure in subtropical coniferous and broadleaf forest plantations. *Plant Soil* **2014**, *387*, 103–116. [CrossRef]
37. Mirza, B.S.; Potisap, C.; Nüsslein, K.; Bohannan, B.J.M.; Rodrigues, J.L.M. Response of free-living nitrogen-fixing microorganisms to land use change in the Amazon rainforest. *Appl. Environ. Microbiol.* **2014**, *80*, 281–288. [CrossRef]
38. Limmer, C.; Drake, H.L. Effects of carbon, nitrogen, and electron acceptor availability on anaerobic N2-fixation in a beech forest soil. *Soil Biol. Biochem.* **1998**, *30*, 153–158. [CrossRef]
39. Lu, M.; Wang, S.; Zhang, Z.; Chen, M.; Li, S.; Cao, R.; Cao, Q.; Zuo, Q.; Wang, P. Modifying effect of ant colonization on soil heterogeneity along a chronosequence of tropical forest restoration on slash-burn lands. *Soil Tillage Res.* **2019**, *194*, 104329. [CrossRef]
40. Pfister, C.A.; Meyer, F.; Antonopoulos, D.A. Metagenomic profiling of a microbial assemblage associated with the California mussel: A node in networks of carbon and nitrogen cycling. *PLoS ONE* **2010**, *5*, e10518. [CrossRef]
41. Zheng, M.; Chen, H.; Li, D.; Luo, Y.; Mo, J. Substrate stoichiometry determines nitrogen fixation throughout succession in southern Chinese forests. *Ecol. Lett.* **2019**, *23*, 336–347. [CrossRef]

forests
MDPI

Article

Differential Responses of Soil Nitrogen Forms to Climate Warming in *Castanopsis hystrix* and *Quercus aliena* Forests of China

Weiwei Shu [1,2], Hui Wang [3], Shirong Liu [3], Yanchun Liu [4], Huilin Min [1], Zhaoying Li [1], Bernard Dell [5,6,*] and Lin Chen [1,*]

[1] Experimental Center of Tropical Forestry, Chinese Academy of Forestry, Pingxiang 532600, China; shuweiwei_@outlook.com (W.S.); minhuilin828@163.com (H.M.); lzying1877@sina.com (Z.L.)
[2] Guangxi Youyiguan Forest Ecosystem Research Station, Youyiguan Forest Ecosystem Observation and Research Station of Guangxi, Pingxiang 532600, China
[3] Key Laboratory of Forest Ecology and Environment of National Forestry and Grassland Administration, Ecology and Nature Conservation Institute, Chinese Academy of Forestry, No. 2 Dongxiaofu, Haidian District, Beijing 100091, China; wanghui@caf.ac.cn (H.W.); liusr@caf.ac.cn (S.L.)
[4] International Joint Research Laboratory for Global Change, School of Life Sciences, Henan University, Kaifeng 475004, China; yanchunliu@henu.edu.cn
[5] Agriculture and Forest Sciences, Murdoch University, Murdoch, WA 6150, Australia
[6] Ecology and Nature Conservation Institute, Chinese Academy of Forestry, Beijing 100091, China
* Correspondence: b.dell@murdoch.edu.au (B.D.); chenlin-ectf@hotmail.com (L.C.)

Abstract: Climate warming impacts soil nitrogen cycling in forest ecosystems, thus influencing their productivity, but this has not yet been sufficiently studied. Experiments commenced in January 2012 in a subtropical *Castanopsis hystrix* Hook. f. and Thomson ex A. DC. plantation and in May 2011 in a temperate *Quercus aliena* Blume forest, China. Four treatments were established comprising trenching, artificial warming (up to 2 °C), artificial warming + trenching, and untreated control plots. The plots were 2 × 3 m in size. In 2021 and 2022, soil nitrogen mineralization, soil nutrient availability, fine root biomass and microbial biomass were measured at 0–20 cm soil depth in 6 replicate plots per treatment. Warming significantly increased soil temperature in both forests. In the *C. hystrix* plantation, warming significantly increased available phosphorus (AP) and fine root biomass (FRB), but it did not affect soil microbial biomass carbon (MBC), microbial biomass nitrogen (MBN), microbial biomass phosphorus (MBP) and their ratios. Warming depressed the net mineralization rate (NMR) and net ammonification rate (NAR) of the *C. hystrix* plantation, probably because the competition for nitrogen uptake by fine roots and microorganisms increased, thus decreasing substrates for nitrogen mineralization and ammonification processes. Trenching and warming + trenching increased the net nitrification rate (NNR), which might be related to decreased NH_4^+-N absorption of trees in the trenched plots and the increased microbial activity involved in soil nitrification. In the *Q. aliena* forest, warming significantly increased NH_4^+-N, MBC/MBN, Root C/N, Root N/P, and decreased pH, MBN, MBN/MBP and Root P; and there was no effect of trenching. Notably, the NAR, NNR and NMR were largely unaffected by long-term warming. We attributed this to the negative effect of increasing NH_4^+-N and decreasing MBN/MBP offsetting the positive effect of soil warming. This study highlights the vulnerability of subtropical forest stands to long-term warming due to decreased soil N mineralization and increased NO_3^--N leaching. In contrast, the soil N cycle in the temperate forest was more resilient to a decade of continuous warming.

Keywords: experimental warming; fine roots; soil microorganisms; soil nitrogen mineralization

1. Introduction

According to the World Meteorological Organization, the earth's temperature in 2023 was about 1.4 °C higher than in the early industrialization period of 1850–1900 [1]. However, there are significant regional differences, and ecosystems at middle and high latitudes and elevations are likely to encounter more serious warming [2]. It is projected that the global temperature will further increase by 1.0–3.7 °C by the end of the 21st century, and pose a significant threat to global sustainable development and human livelihood [2,3]. Forests play a critical role in sequestering carbon, maintaining biodiversity and providing other ecosystem services. Nitrogen is a key driver of forest growth and therefore nitrogen provides a useful index to evaluate forest ecosystem response to global warming [4,5]. Soil nitrogen pool size and flux, nitrogen utilization by plants and N_2O emissions in terrestrial ecosystems are influenced by soil nitrogen mineralization, which is highly sensitive to warming [6].

Many warming control experiments have been carried out in forest ecosystems [7], and these have increased our understanding of the response of forests to climate warming. In particular, soil nitrogen mineralization is mainly influenced by soil organic matter, nitrogen availability [8], as well as soil temperature and humidity [9]. Increasing soil temperature changes nitrogen conversion by soil microbes and roots, and accelerates nitrogen mineralization by promoting the activity of extracellular enzymes involved in nitrogen circulation [10,11]. However, forest ecosystems differ greatly in species composition, community structure and soil types. In addition, tree species may differ in root morphology, physiology and nutrient requirements [12]. As a result, the response modes and feedback mechanisms to warming may differ across forest ecosystems. For example, in a mixed deciduous temperate forest, the annual average soil net nitrogen mineralization (0–10 cm) increased by 45% in a 5 °C ambient warming experiment [13]. These authors proposed that the soil net nitrogen rate accelerated due to the comprehensive actions of the decreased unstable organic carbon, increased NH_4^+-N and enhanced microbial metabolism in the soil. Furthermore, experimental warming significantly increased rates of net nitrogen mineralization, as well as net nitrification and denitrification in soil under *Picea asperata* Mast. and *Abies faxoniana* var. faxoniana (Rehder and Wilson) Tang S. Liu in plateau-climate zones in most sampling periods [12]. This is because increasing temperature and nutrient availability altered the morphology (root length and type) and activity (e.g., root exudates) of plant roots, thus modulating soil nitrogen mineralization. By contrast, the soil net nitrogen mineralization rates in superficial and deep soil layers of a temperate *Pinus tabuliformis* plantation decreased by 52% (0–10 cm) and 51% (10–20 cm), respectively, under warming conditions [14]. This was attributed to warming decreasing soil moisture content, reducing soil microbial activity, and restricting the diffusion of soluble carbon and nitrogen. Surprisingly, in a subtropical *Cunninghamia lanceolata* (Lamb.) Hook. plantation, increasing soil temperature by 4 °C had no significant effect on the in situ soil net nitrogen mineralization rate, the net nitrification rate or the in situ soil N_2O emission rate [15]. This may be partly due to increasing temperature promoting the activity of nitrogen-transforming microorganisms but at the same time decreasing microbial immobilization of nitrogen. The above range of responses of soil nitrogen mineralization in field warming experiments highlights uncertainty for climate change predictions in forest ecosystems and the necessity for more comprehensive studies to be undertaken.

Several meta-analysis studies discussed different responses of soil nitrogen mineralization of forest ecosystems to warming [6,16,17], and concluded that there remains significant uncertainty due to inconsistent warming techniques and observation periods. Our previous research showed that during the first two years of warming, soil warming increased heterotrophic respiration and inhibited root respiration in a *C. hystrix* forest, leading to an overall decrease in topsoil SOC content of about 16.9% [18]. After 5 years of warming, the labile C in the trenched plots decreased significantly, while the effect of warming on SOC content leveled off during the latter three years of the study [19]. However, in a *Q. aliena* forest, soil warming substantially elevated soil respiration and autotrophic respiration [20]. Furthermore, 5 years of continuous soil warming caused a decline in nitrogen availability in the *Q. aliena* forest, while microbial biomass-specific nitrogen-acquisition enzyme activity increased significantly [21]. However, how warming affects the soil nitrogen transformation process and its regulatory mechanisms in different climatic zones is still not clear. Further field warming studies in forest ecosystems in different climatic zones are conducive to improving our understanding of the response mechanism of nutrient cycling to climate change. In particular, they can provide reference data for the parameterization of large-scale climate-nitrogen cycling models.

Soil warming experiments established in January 2012 in a south subtropical *C. hystrix* plantation [18] and in May 2011 in a temperate *Q. aliena* natural forest [20] provide an opportunity to identify ten-year responses of forest soil nitrogen mineralization in two climatic regions to soil warming. In this study, it is hypothesized that: (1) soil warming will increase soil temperature and promote fine root growth and microbial activity, thus increasing the soil nitrogen mineralization rate; and (2) the soil net nitrogen mineralization rate in the temperate forest will be more sensitive to warming than in the subtropical forest.

2. Materials and Methods

2.1. Study Area Sites

There were two field experiments, one study was conducted in a south subtropical *C. hystrix* monoculture plantation and the other was in a temperate *Q. aliena* pure natural forest. The *C. hystrix* plantation is located in the Guangxi Youyiguan Forest Ecosystem Research Station of Experimental Center for Tropical Forestry, Chinese Academy of Forestry (22°10′ N, 106°50′ E), at an elevation of 550 m. The area is part of the south subtropical monsoon climate zone, with 1200–1500 mm annual average precipitation, mainly from April to September, 1200–1400 mm annual evaporation, 80%–84% relative humidity, and 20.5–21.7 °C average annual temperature. In the 2012 survey, the average diameter at breast height of *C. hystrix* was 25.7 cm, and the tree density was 333 trees ha^{-1}. The forest soil type is Oxisol, and the soil depth is >80 cm.

The *Q. aliena* forest is located in the Positioning Research Station of the Henan Neixiang Baotianman Forest Ecosystem, National Forestry and Grassland Administration (111°47′–112°04′ E, 33°20′–33°36′ N), with an elevation of 1400 m. It experiences a temperate monsoon climate and has four distinctive seasons. The average annual precipitation is 894 mm, mainly from June to August, and the average annual temperature is 15.1 °C. In 2012, the average diameter at breast height of *Q. aliena* was 20.1 cm, and the tree density was 1270 trees ha^{-1}. The forest soil type is Hapludalfs, with a soil depth of 20–60 cm [20].

2.2. Experimental Design

A split plot experiment was established at each site with warming as the main plot, and trenching as the split plot. The four treatments were: control, warming, trenching, and warming + trenching. Each treatment had six replications, forming 24 split plots. Six pairs of 4 × 3 m plots were randomly established on each site over a 30 × 70 m area, each pair containing a plot that was subject to warming using an infrared heater and a control plot that was not subject to warming. Each plot was further divided into two 2 × 3 m subplots, one of which was randomly assigned to be trenched while the other remained non-trenched. The trenches were dug 1 m deep to minimize the influence of roots entering the subplots. The warming experiments commenced in January 2012 (*C. hystrix*) and May 2011 (*Q. aliena*), respectively, and have been ongoing continuously since then, 24 hours a day. More detailed information on site preparation and treatment protocols is provided by Wang et al. [18] and Liu et al. [20].

2.3. Soil Collection and Analysis

Soil nitrogen mineralization rates were determined at the sites using the PVC tube method described by Tsui and Chen [22] in two incubation periods (20 July–20 August 2021, 22 July–22 August 2022). At the beginning of each incubation, three sample points were chosen randomly and the surface litter was removed in each split-plot. At each sample point, a pair of tubes was inserted 20 cm vertically into the soil using a rubber mallet, leaving 8 cm above the ground. The PVC tubes had an inner diameter of 5 cm and height of 28 cm, and there were four holes near the top for gas exchange. The tops were sealed with PVC caps to prevent the loss of nutrients by precipitation and to exclude litter fall. Three tubes were taken out immediately, the content was partitioned into 0–10 cm and 10–20 cm layers, and the layers from the three tubes were combined and mixed thoroughly providing two samples for determining the initial NH_4^+-N and NO_3^--N concentrations. The remaining tubes with closed caps and cotton gauze bottoms to exclude roots were left in situ and collected at the end of each incubation period for the measurement of the final NH_4^+-N and NO_3^--N concentrations. Meanwhile, temperatures of soil layers 0–10 cm and 10–20 cm were measured with a digital thermometer (Harvesting Science and Technology Co., Ltd., Beijing, China) at three positions (upper, middle and lower) near each incubation point.

The formulae used to calculate the soil net nitrogen mineralization rates are as follows:

$$\text{Net ammonification rate (NAR, mg N kg}^{-1}\text{ d}^{-1}\text{) = (ammonium nitrogen after incubation} - \text{ammonium nitrogen before incubation)/days of incubation} \quad (1)$$

$$\text{Net nitrification rate (NNR, mg N kg}^{-1}\text{ d}^{-1}\text{) = (nitrate nitrogen after incubation} - \text{nitrate nitrogen before incubation)/days of incubation} \quad (2)$$

$$\text{Net nitrogen mineralization rate (NMR, mg N kg}^{-1}\text{ d}^{-1}\text{) = NAR + NNR} \quad (3)$$

Soil samples collected at the same soil depth in each plot were combined on-site, placed in plastic bags and transported to the laboratory on ice. A total of 48 mixed soil samples (24 split plots × 2 soil layers) were collected from each forest type at each sampling time. After removing stones and coarse roots, each soil sample was sieved through a 2 mm sieve and divided into two parts. One part was air dried and the physical and chemical properties were determined: soil pH (soil:water mass ratio 1:2.5), soil organic matter (SOC) concentration (acid potassium dichromate oxidation), total nitrogen (TN) concentration (Kjeldahl method), total phosphorus (TP) concentration (alkali fusion-molybdenum blue method), and available phosphorus (AP) concentration (Olsen P method). The other was stored at 4 °C for 48 h before measuring the soil moisture (SM, after drying at 105 °C for 48 h), carbon (C), nitrogen (N) and phosphorus (P) concentrations in soil microorganisms, as well as NH_4^+-N and NO_3^--N concentrations (SEAL Auto Analyzer 3, SEAL, Norderstedt, Germany). We calculated soil bulk density (BD) by the following equation: BD (g cm^{-3}) = soil fresh weight × (1-soil moisture)/soil volume.

Soil microbial biomass carbon (MBC) and microbial biomass nitrogen (MBN) concentrations were obtained by the chloroform fumigation-potassium sulphate extraction method [23]. The dissolved organic carbon concentrations in fumigated and non-fumigated filtrates were measured with a total organic carbon analyser (TOC, VCPH/CPN, Kyoto, Japan). The total soluble nitrogen concentration in the filtrate was measured using a continuous flow analyser (Skalar San + +, Skalar, Delft, The Netherlands). Microbial biomass phosphorus (MBP) concentration was obtained using the chloroform fumigation-$NaHCO_3$ extraction method [24] followed by spectrophotometry at 880 nm. These parameters were calculated using the following formulas:

$$\text{MBC} = \Delta E_C / K_C \quad (4)$$

$$\text{MBN} = \Delta E_N / K_N \quad (5)$$

$$\text{MBP} = \Delta E_P / K_P / K_{Pi} \quad (6)$$

where ΔEC is the difference in dissolved organic carbon between fumigated and non-fumigated soils, and K_C is the transformation coefficient (0.45). ΔE_N is the difference in total soluble nitrogen content between fumigated and non-fumigated soils, and K_N is the transformation coefficient (0.57). ΔE_P is the difference in AP between fumigated and non-fumigated soils, and K_P is the transformation coefficient (0.4). KPi = (test value of soil added with KH_2PO_4 solution − test value of non-fumigated soil)/25 × 100%.

2.4. Fine Root Collection and Analysis

The roots obtained during soil collection were taken back to the laboratory for cleaning. The fine root (diameter ≤ 2 mm) fraction was obtained and morphological traits were analyzed using the root analysis system WinRHIZO 2013 (Regent Instruments Inc., Quebec, Canada). Next, they were dried in an oven at 65 °C to constant weight (48 h) and weighed to determine fine root biomass. The dried roots were ground in a ball mill, and the powder was used for elemental analysis. The C and N concentrations were obtained using an element analyser (Elementar Vario EL III, Elementar, Germany). The P concentration was measured using a continuous flow analyser (Skalar San + +, Skalar, Delft, The Netherlands) after acid digestion ($HClO_4$, H_2SO_4).

2.5. Statistical Analysis

We conducted two identical experiments using the same methods and timing, one in a *C. hystrix* plantation and the other in a *Q. aliena* forest. Thus, the statistical analysis was the same for both forests. In each forest, differences in the indexes at the same soil layer in the two years under different treatments were explored through one-way analysis of variance (one-way ANOVA). These indexes included soil physical and chemical properties, MBC, MBN, MBP, fine root biomass and nutrient concentrations, and net nitrogen transformation rates. In addition, differences in these indexes among different soil layers under the same warming and trenching treatments were analyzed. Furthermore, effects of warming, trenching, soil depth and interaction on soil physical and chemical properties, MBC, MBN, MBP, fine root biomass and nutrient concentrations, and net nitrogen transformation rates were examined by multi-way analysis of variance (multi-way ANOVA). We used Duncan's multiple range test for comparing means across treatments. The statistical analysis was carried out using SPSS 25.0 software (SPSS Inc., Chicago, IL, USA). Images were plotted using the R 4.4.0 software.

To quantify the effects of soil variables, microbial variables, and fine root variables on NAR, NNR and NMR in soil, we employed the linear mixture model using the "lme" function in the R package "nlme". The fixed effect terms included soil, microbial and fine root variables. All variables were normalized before modeling so that each variable had a mean of zero and a standard deviation of one. To reduce the effects of multicollinearity, a variance inflation factor < 3 was used to identify multicollinearity variables in multiple regression models [25]. We calculated the variance inflation factor using the R package "car". The pseudo-R2 was calculated using the function "r.squared GLMM" from the R package "MuMIn" to represent the variance of the fixed effects interpretation in a linear mixed model. The effect size of the fixed factor was measured by the regression coefficient in the linear mixed model.

3. Results

3.1. Soil Physicochemical Properties

The influences of warming, trenching, soil depth and interaction on soil physical and chemical properties varied with tree species (Table 1). For the *C. hystrix* plantation, the ST (2.1%) and AP (21.3%) increased significantly at 10–20 cm soil depth after warming compared to the unheated control ($p < 0.05$, Table 2). However, the difference between different warming and trenching treatments in the same soil layer was not significant for the other soil properties ($p > 0.05$, Table 2). In addition, the BD (32.5%–44.0%) and pH (3.0%–4.2%) of the 10–20 cm soil layer increased significantly compared to the 0–10 cm soil layer irrespective of warming or trenching treatments. Constantly, SOC, TN, TP, AP, C/P and N/P ratios of the 10–20 cm soil layer decreased sharply to different extents compared to the 0–10 cm soil layer for all treatments ($p < 0.05$, Tablc 2).

Table 1. Multi-way ANOVA results for the effect of warming (W), trenching (T), soil depth (D) and their interaction on soil physicochemical properties in the *Castanopsis hystrix* (CH) plantation and *Quercus aliena* (QA) forest.

Forest	Source of Variation	ST	SM	BD	pH	SOC	TN	AN	NN	TP	AP	C/N	C/P	N/P
	W	0.000 **	0.758	0.987	0.367	0.670	0.730	0.851	0.344	0.007 **	0.055	0.331	0.395	0.131
	T	0.477	0.943	0.963	0.679	0.657	0.825	0.728	0.678	0.845	0.631	0.149	0.563	0.859
	D	0.000 **	0.008 **	0.000 **	0.000 **	0.000 **	0.000 **	0.001 **	0.217	0.000 **	0.000 **	0.185	0.000 **	0.000 **
CH	T × D	0.733	0.777	0.549	0.371	0.569	0.422	0.955	0.867	0.651	0.907	0.517	0.727	0.474
	W × D	0.144	0.762	0.665	0.546	0.738	0.302	0.978	0.684	0.470	0.684	0.592	0.255	0.085
	W × T	0.516	0.944	0.395	0.332	0.785	0.584	0.725	0.302	0.242	0.274	0.934	0.387	0.285
	W × T × D	0.814	0.907	0.412	0.815	0.534	0.868	0.944	0.510	0.335	0.721	0.539	0.853	0.795
	W	0.000 **	0.644	0.560	0.000 **	0.736	0.599	0.055	0.550	0.291	0.338	0.783	0.846	0.878
	T	0.192	0.684	0.054	0.567	0.068	0.147	0.080	0.813	0.908	0.382	0.445	0.220	0.329
	D	0.001 **	0.140	0.000**	0.000 **	0.000 **	0.000 **	0.023 *	0.598	0.011 *	0.000 **	0.000 **	0.000 **	0.000 **
QA	T × D	0.943	0.704	0.414	0.946	0.968	0.597	0.231	0.100	0.411	0.306	0.221	0.598	0.945
	W × D	0.830	0.194	0.407	0.267	0.557	0.352	0.026 *	0.046 *	0.373	0.038 *	0.458	0.412	0.225
	W × T	0.245	0.068	0.393	0.457	0.210	0.091	0.369	0.174	0.781	0.529	0.212	0.327	0.101
	W × T × D	0.793	0.357	0.386	0.305	0.717	0.895	0.394	0.190	0.689	0.991	0.460	0.727	0.910

ST: soil temperature; SM: soil moisture; BD: bulk density; SOC: soil organic carbon; TN: total nitrogen; AN: ammonium nitrogen; NN: nitrate nitrogen; TP: total phosphorus; AP: available phosphorus. C/N: soil organic carbon/total nitrogen; C/P: soil organic carbon/total phosphorus; N/P: total nitrogen/total phosphorus. Significant probabilities are denoted by * and ** for $p < 0.05$ and 0.01, respectively.

Table 2. Effect of warming, trenching, and soil depth on soil physicochemical properties in the *Castanopsis hystrix* (CH) plantation and *Quercus aliena* (QA) forest in 2021 and 2022.

Forest	Index	Depth	ST (°C)	SM (%)	BD (g cm^{-3})	pH	SOC (g kg^{-1})	TN (g kg^{-1})	AN (mg kg^{-1})	NN (mg kg^{-1})	TP (g kg^{-1})	AP (mg kg^{-1})	C/N	C/P	N/P
CH	C	0–10	25.2 ± 0.1 Aa	29.2 ± 1.6 A	1.7 ± 0.1 B	4.1 ± 0.0 B	44.6 ± 2.8 A	2.8 ± 0.2 A	11.6 ± 1.7	1.1 ± 0.5	0.4 ± 0.0 A	1.8 ± 0.1 Aa	16.1 ± 0.3	130.0 ± 8.2 A	8.1 ± 0.6 A
		10–20	24.8 ± 0.1 Bb	25.3 ± 0.9 B	2.2 ± 0.1 A	4.2 ± 0.0 A	23.2 ± 1.9 B	1.5 ± 0.0 B	7.2 ± 1.1	0.9 ± 0.3	0.3 ± 0.0 B	0.8 ± 0.0 Bb	15.0 ± 0.5	79.1 ± 5.6 B	5.2 ± 0.3 B
	W	0–10	25.9 ± 0.1 Aa	29.5 ± 2.1 A	1.7 ± 0.1 B	4.1 ± 0.0 B	46.8 ± 3.0 A	2.8 ± 0.2 A	11.0 ± 1.4	1.8 ± 0.6	0.4 ± 0.0 A	2.0 ± 0.2 Aa	16.6 ± 0.5	129.9 ± 5.3 A	7.9 ± 0.3 A
		10–20	25.4 ± 0.1 Ba	26.1 ± 1.1 A	2.4 ± 0.1 A	4.2 ± 0.0 A	25.6 ± 1.3 B	1.7 ± 0.1 B	7.0 ± 0.7	1.5 ± 0.4	0.3 ± 0.0 B	0.9 ± 0.0 Ba	15.5 ± 0.6	84.7 ± 2.9 B	5.5 ± 0.2 B
	T	0–10	25.2 ± 0.1 Aa	29.0 ± 2.2 A	1.7 ± 0.1 B	4.1 ± 0.0 B	46.1 ± 2.3 A	2.9 ± 0.2 A	11.5 ± 1.6	1.6 ± 0.6	0.3 ± 0.0 A	1.8 ± 0.1 Aa	16.5 ± 0.7	136.4 ± 5.2 A	8.5 ± 0.5 A
		10–20	24.8 ± 0.1 Bb	25.6 ± 1.3 A	2.3 ± 0.1 A	4.2 ± 0.0 A	24.3 ± 2.3 B	1.5 ± 0.1 B	6.6 ± 0.6	1.1 ± 0.3	0.3 ± 0.0 B	0.8 ± 0.1 Bab	16.2 ± 0.8	84.9 ± 6.9 B	5.2 ± 0.4 B
	WT	0–10	25.8 ± 0.1 Aa	28.8 ± 2.3 A	1.7 ± 0.1 B	4.1 ± 0.0 A	45.9 ± 1.9 A	2.7 ± 0.1 A	10.8 ± 1.4	1.8 ± 0.4	0.4 ± 0.0 A	1.9 ± 0.1 Aa	17.0 ± 0.6	127.9 ± 5.1 A	7.6 ± 0.3 A
		10–20	25.3 ± 0.1 Bab	26.3 ± 1.1 A	2.3 ± 0.1 A	4.2 ± 0.0 A	24.3 ± 1.1 B	1.5 ± 0.1 B	7.0 ± 0.8	1.4 ± 0.3	0.3 ± 0.0 B	0.8 ± 0.0 Bab	16.1 ± 0.4	80.8 ± 3.9 B	5.0 ± 0.2 B

Table 2. *Cont.*

Forest	Index	Depth	ST (°C)	SM (%)	BD (g cm^{-3})	pH	SOC (g kg^{-1})	TN (g kg^{-1})	AN (mg kg^{-1})	NN (mg kg^{-1})	TP (g kg^{-1})	AP (mg kg^{-1})	C/N	C/P	N/P
QA	C	0–10	19.7 ± 0.4 b	27.8 ± 0.5	1.9 ± 0.1 B	5.0 ± 0.1 Aa	37.7 ± 2.0 A	3.1 ± 0.2 A	30.3 ± 4.8 Ab	1.1 ± 0.2 A	0.5 ± 0.0 A	3.5 ± 0.4 A	12.5 ± 0.4 A	78.8 ± 5.0 A	6.4 ± 0.4 A
		10–20	18.8 ± 0.4 b	27.3 ± 0.5	2.1 ± 0.1 A	5.1 ± 0.0 Aa	22.8 ± 2.0 B	1.9 ± 0.1 B	26.1 ± 1.3 Aa	1.4 ± 0.4 A	0.4 ± 0.0 B	2.1 ± 0.3 B	11.7 ± 0.3 A	57.3 ± 3.7 B	4.9 ± 0.3 B
	W	0–10	21.0 ± 0.4 ab	29.9 ± 1.4	1.9 ± 0.1 B	4.8 ± 0.0 Ab	41.2 ± 2.3 A	3.4 ± 0.2 A	38.7 ± 3.1 Aa	0.8 ± 0.2 A	0.5 ± 0.0 A	3.2 ± 0.2 A	12.1 ± 0.3 A	84.4 ± 4.5 A	7.0 ± 0.4 A
		10–20	20.0 ± 0.4 ab	29.7 ± 0.9	2.1 ± 0.1 A	4.9 ± 0.1 Ab	23.7 ± 2.0 B	2.1 ± 0.1 B	29.7 ± 2.8 Ba	0.9 ± 0.3 A	0.4 ± 0.0 B	2.5 ± 0.2 B	11.4 ± 0.3 A	56.2 ± 2.2 B	5.0 ± 0.2 B
	T	0–10	19.9 ± 0.4 b	28.5 ± 1.0	1.9 ± 0.1 A	5.0 ± 0.0 Ba	37.3 ± 2.7 A	3.0 ± 0.2 A	28.6 ± 1.7 Ab	0.9 ± 0.2 A	0.5 ± 0.0 A	3.0 ± 0.2 A	12.5 ± 0.3 A	80.6 ± 4.9 A	6.5 ± 0.4 A
		10–20	18.8 ± 0.4 b	28.8 ± 1.3	2.2 ± 0.1 A	5.2 ± 0.0 Aa	21.5 ± 1.9 B	2.0 ± 0.2 B	24.1 ± 3.4 Aa	1.1 ± 0.3 A	0.4 ± 0.0 B	2.0 ± 0.1 B	10.8 ± 0.3 B	54.1 ± 3.3 B	5.1 ± 0.3 B
	WT	0–10	21.6 ± 0.3 a	29.2 ± 1.1	1.9 ± 0.1 B	4.9 ± 0.1 Bab	36.3 ± 2.0 A	3.0 ± 0.2 A	33.8 ± 2.6 Aab	1.7 ± 0.3 A	0.5 ± 0.0 A	2.9 ± 0.2 A	12.4 ± 0.4 A	78.6 ± 3.8 A	6.4 ± 0.3 A
		10–20	20.7 ± 0.4 a	26.1 ± 1.5	2.4 ± 0.1 A	5.1 ± 0.1 Aab	19.9 ± 1.6 B	1.8 ± 0.1 B	23.3 ± 2.8 Ba	0.7 ± 0.1 B	0.5 ± 0.1 A	2.7 ± 0.4 A	11.4 ± 0.4 A	49.4 ± 4.3 B	4.3 ± 0.3 B

Control: C; Warming: W; Trenching: T; Warming and Trenching: WT. ST: soil temperature; SM: soil moisture; BD: bulk density; SOC: soil organic carbon; TN: total nitrogen; AN: ammonium nitrogen; NN: nitrate nitrogen; TP: total phosphorus; AP: available phosphorus; C/N: soil organic carbon/total nitrogen; C/P: soil organic carbon/total phosphorus; N/P: total nitrogen/total phosphorus. Within a column, different lowercase letters indicate significant differences at $p < 0.05$ between treatments for the same soil depth, and different uppercase letters indicate significant differences between soil depths for the same treatment, based on one-way ANOVA and Duncan's multiple range test. Values are means with standard errors.

For the *Q. aliena* forest, the ST was significantly increased by 6.3% (0–10 cm) and 6.5% (10–20 cm) after warming, and also increased by 10.2% (0–10 cm) and 10.2% (10–20 cm) in the warming + trenching treatment. In addition, the NH_4^+-N concentration of the *Q. aliena* forest increased significantly by 27.6% (0–10 cm) after warming, but the pH decreased sharply by 4.4% (0–10 cm) and 4.5% (10–20 cm) ($p < 0.001$, Table 2). However, there were no significant differences in the other soil properties between any warming and trenching treatments in the same soil layer ($p > 0.05$, Table 2). Comparing the two soil layers, the BD (12.8%–23.3%) and pH (3.8%–4.9%) of the 10–20 cm soil layer increased significantly compared to those of 0–10 cm soil layer regardless of warming or trenching treatments. However, SOC, TN, TP, AP, C/P and N/P ratios of the 10–20 cm soil layer decreased sharply to different extents compared to those of the 0–10 cm soil layer in all treatments ($p < 0.05$, Table 2). In addition, there were significant interactive effects of warming and soil depth on NH_4^+-N, NO_3^--N and AP ($p < 0.05$, Table 1).

3.2. Soil Microbial Biomass

The effects of warming, trenching, soil depth and their interactions on MBC, MBN and MBP, and their ratios were not obvious in the *C. hystrix* plantation (Table 3). Only warming significantly affected the MBC concentration ($p < 0.05$, Table 3), while soil depth and trenching did not influence MBC, MBN and MBP, or their ratios ($p > 0.05$, Table 3). In addition, the interaction of warming and trenching significantly influenced the MBC/MBN ratio in the *C. hystrix* plantation ($p < 0.05$, Table 3). Compared with the control, warming, and warming + trenching slightly increased the MBC concentration by 14.3% and 63.3%, respectively, but this was not significant (0–10 cm, $p > 0.05$, Figure 1a). In comparison with the 0–10 cm soil layer, the MBC concentration at 10–20 cm decreased by 48.3% after warming + trenching ($p < 0.05$, Figure 1a).

Table 3. Multi-way ANOVA results for the effect of warming(W), trenching (T), soil depth (D), and their interaction on the concentration and ratios of microbial biomass carbon (MBC), microbial biomass nitrogen (MBN), and microbial biomass phosphorus (MBP) in the *Castanopsis hystrix* (CH) plantation and *Quercus aliena* (QA) forest.

Forest	Source of Variation	MBC	MBN	MBP	MBC/MBN	MBC/MBP	MBN/MBP
CH	W	0.019 *	0.152	0.860	0.741	0.244	0.353
	T	0.706	0.855	0.488	0.718	0.353	0.365
	D	0.067	0.853	0.886	0.264	0.266	0.385
	T × D	0.346	0.680	0.121	0.581	0.521	0.377
	W × D	0.233	0.302	0.377	0.303	0.284	0.329
	W × T	0.349	0.201	0.550	0.042 *	0.495	0.348
	W × T × D	0.418	0.530	0.654	0.976	0.498	0.380
QA	W	0.089	0.001 **	0.816	0.008 **	0.759	0.049 *
	T	0.167	0.921	0.078	0.435	0.157	0.960
	D	0.000 **	0.218	0.008 **	0.200	0.507	0.368
	T × D	0.004 **	0.168	0.293	0.615	0.011 *	0.157
	W × D	0.000 **	0.097	0.292	0.059	0.357	0.615
	W × T	0.858	0.849	0.122	0.479	0.383	0.297
	W× T × D	0.210	0.005 **	0.040 *	0.362	0.358	0.061

Significant values are denoted by * and ** for $p < 0.05$ and 0.01, respectively.

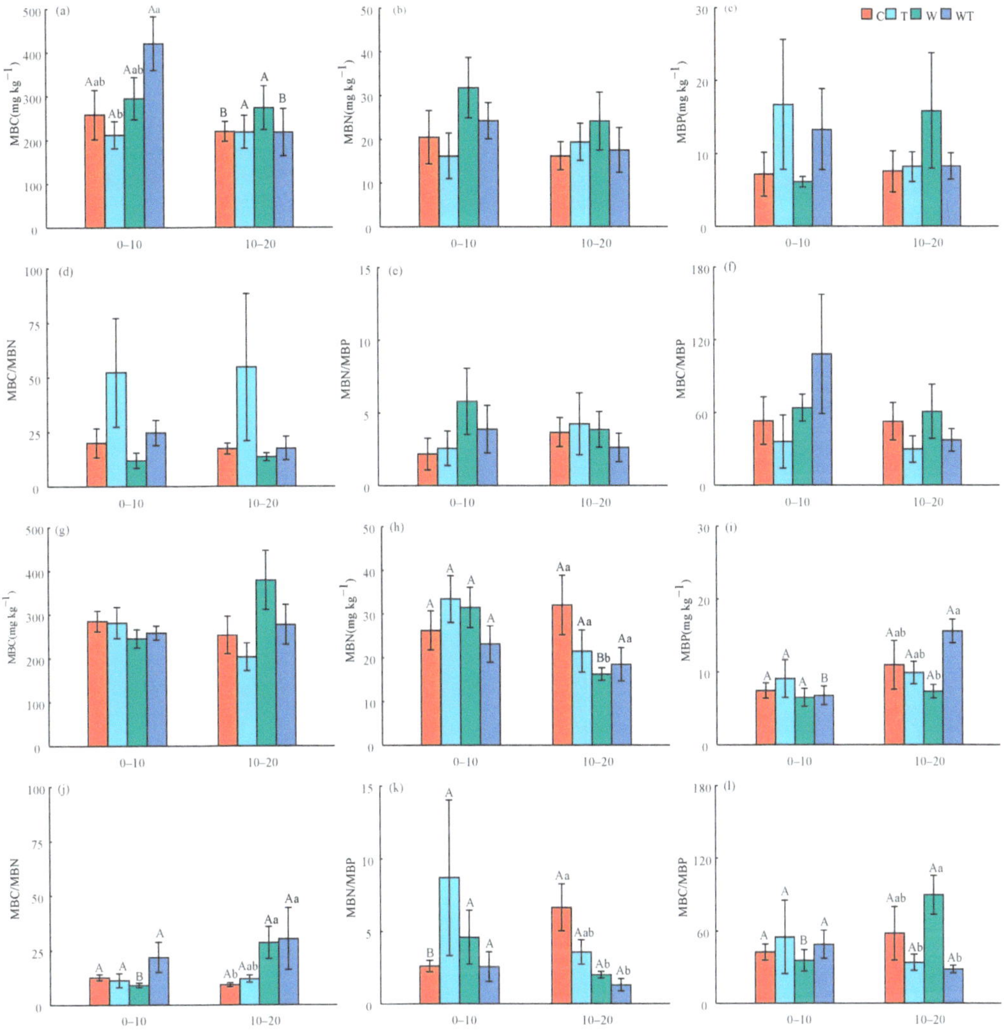

Figure 1. Concentrations and ratios of microbial biomass carbon (MBC), microbial biomass nitrogen (MBN), microbial biomass phosphorus (MBP) under four treatments (Control: C; Trenching: T; Warming: W; Warming and Trenching: WT) in the *Castanopsis hystrix* plantation (**a–f**) and *Quercus aliena* forest (**g–l**). Different lowercase letters indicate significant differences at $p < 0.05$ between treatments for the same soil depth, and different uppercase letters indicate significant differences between soil depths for the same treatment. Bars indicate standard errors of means.

However, in the *Q. aliena* forest, warming greatly influenced the MBN, MBN/MBP and MBC/MBN; soil depth significantly affected the MBC and MBP ($p < 0.05$, Table 3); but trenching did not alter the MBC, MBN, MBP and their ratios ($p > 0.05$, Table 3). We also found that there were significant interaction effects on the MBC, MBN, MBP concentrations and the MBC/MBP ratio ($p < 0.05$, Table 3). The MBN concentration in the warming and warming + trenching treatments decreased by 49.4% and 42.4%, respectively, compared to the control (Figure 1h). This resulted in the MBC/MBN being increased by 207.6% and 226.2%, and the MBN/MBP being decreased by 69.6% and 80.1%, respectively (10–20 cm, $p < 0.05$, Figure 1j,k). Compared to the 0–10 cm soil layer, the MBN concentration at

10–20 cm decreased by 48.5% after warming ($p < 0.05$, Figure 1h), while MBC/MBN and MBC/MBP at 10–20 cm increased by 214.7% and 149.9%, respectively ($p < 0.05$, Figure 1j,l). In addition, the MBN/MBP of the 10–20 cm soil layer in the control plot increased by 152.5% compared to those at 0–10 cm ($p < 0.05$, Figure 1k).

3.3. *Fine Root Biomass and Nutrients*

Warming, trenching, soil depth and their interaction effects on the fine root biomass and nutrient concentrations varied with tree species (Table 4). In the *C. hystrix* plantation, warming significantly increased the FRB of the 10–20 cm soil layer by 37.2% compared to the control ($p < 0.05$, Figure 2b), but had no apparent influence on the Root C, Root N, Root P concentration, and their ratios ($p > 0.05$, Figure 3a–f). In addition, trenching depressed fine root biomass of *C. hystrix*, although there was no statistical difference between trenching and the control ($p > 0.05$, Figure 2a,b). Overall, the FRB, Root N and Root P concentrations all decreased with soil depth, especially in the control and warming + trenching treatments (Figures 2 and 3).

Table 4. Multi-way ANOVA results for the effect of warming (W), trenching (T), depth (D) and their interaction on the concentration and ratios of root carbon (Root C), root nitrogen (Root N), root phosphorus (Root P) in the *Castanopsis hystrix* (CH) plantation and *Quercus aliena* (QA) forest, based on FRB: fine root biomass.

Forest	Source of Variation	FRB	Root C	Root N	Root P	Root C/N	Root C/P	Root N/P
CH	W	0.335	0.568	0.468	0.777	0.608	0.825	0.479
	T	0.000 **	0.179	0.357	0.787	0.231	0.648	0.736
	D	0.000 **	0.608	0.000 **	0.001 **	0.000 **	0.004 **	0.299
	T × D	0.840	0.752	0.863	0.552	0.995	0.570	0.442
	W × D	0.936	0.905	0.547	0.928	0.361	0.693	0.766
	W × T	0.476	0.727	0.192	0.294	0.320	0.445	0.721
	W ×T × D	0.231	0.355	0.099	0.229	0.604	0.706	0.908
QA	W	0.536	0.971	0.004 **	0.000 **	0.009 **	0.000 **	0.004 **
	T	0.001 **	0.418	0.748	0.985	0.687	0.640	0.828
	D	0.000 **	0.355	0.000 **	0.000 **	0.000 **	0.000 **	0.975
	T × D	0.312	0.488	0.662	0.403	0.879	0.679	0.487
	W × D	0.924	0.861	0.565	0.581	0.137	0.111	0.695
	W × T	0.560	0.592	0.770	0.465	0.741	0.276	0.316
	W × T × D	0.560	0.822	0.954	0.352	0.873	0.372	0.283

Significant probabilities are denoted by ** for $p < 0.01$.

In the *Q. aliena* forest, warming significantly influenced Root N, Root P, Root C/N, Root C/P and Root N/P, while trenching apparently influenced the fine root biomass ($p < 0.05$, Table 4). The Root P concentration at 10–20 cm soil layer in the warming and the warming + trenching treatments decreased significantly by 26.2% and 23.8%, respectively, compared to the control ($p < 0.05$, Figure 3i). As a result, the Root C/P (37.2% and 27.4%) and Root N/P ratios (18.9% and 13.3%) increased significantly ($p < 0.05$, Figure 3k,l). Additionally, Root N (21.4%–28.8%) and Root P concentrations (17.7%–31.1%) were significantly lower in the 10–20 cm soil layer than in 0–10 cm soil layer ($p < 0.05$, Figure 3h,i). Consequently, the Root C/N (31.3%–42.2%) and Root C/P ratios (21.1%–43.0%) were far higher in the 10–20 cm soil layer than in the 0–10 cm soil layer ($p < 0.05$, Figure 3j,l).

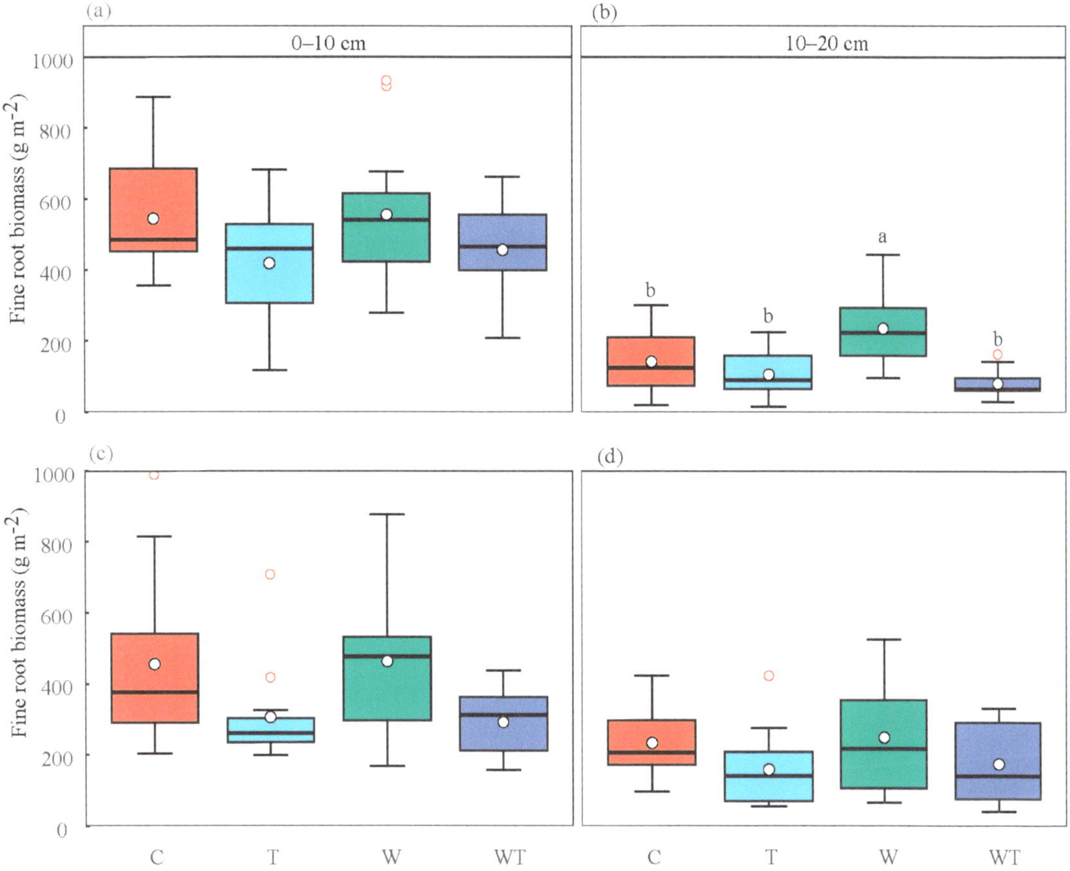

Figure 2. Fine root biomass at two soil depths under four treatments (Control: C; Trenching: T; Warming: W; Warming and Trenching: WT) in the *Castanopsis hystrix* plantation (**a**,**b**) and *Quercus aliena* forest (**c**,**d**). Boxplots display the 25% and 75% quartiles, the median (horizontal line within the box), and the maximum and minimum observed values within each dataset. Mean values are shown in white circles and outliers are shown in red circles. Different lowercase letters indicate a significant difference at $p < 0.05$ between treatments.

3.4. Soil Net Nitrogen Mineralization Rate

In the *C. hystrix* plantation, compared to the control, the NAR and NMR of the 0–10 cm soil layer decreased by 100.0% and 69.8% after warming ($p < 0.05$, Figure 4a,c). However, the NNR of the 0–10 cm soil layer in the trenching and warming + trenching treatments increased by 100.0% and 106.3%, respectively ($p < 0.05$, Figure 4b). In addition, the NAR, NNR and NMR of the 10–20 cm soil layer decreased significantly by 24.3%–75.7%, 75.0%–87.5%, and 24.3%–75.7%, respectively, in comparison with those of the 0–10 cm soil layer ($p < 0.05$, Figure 4a–c). We also observed that there was an interactive effect of trenching and depth on the NNR; and the interactive effect of warming and soil depth on the NAR and NMR ($p < 0.01$, Table 5).

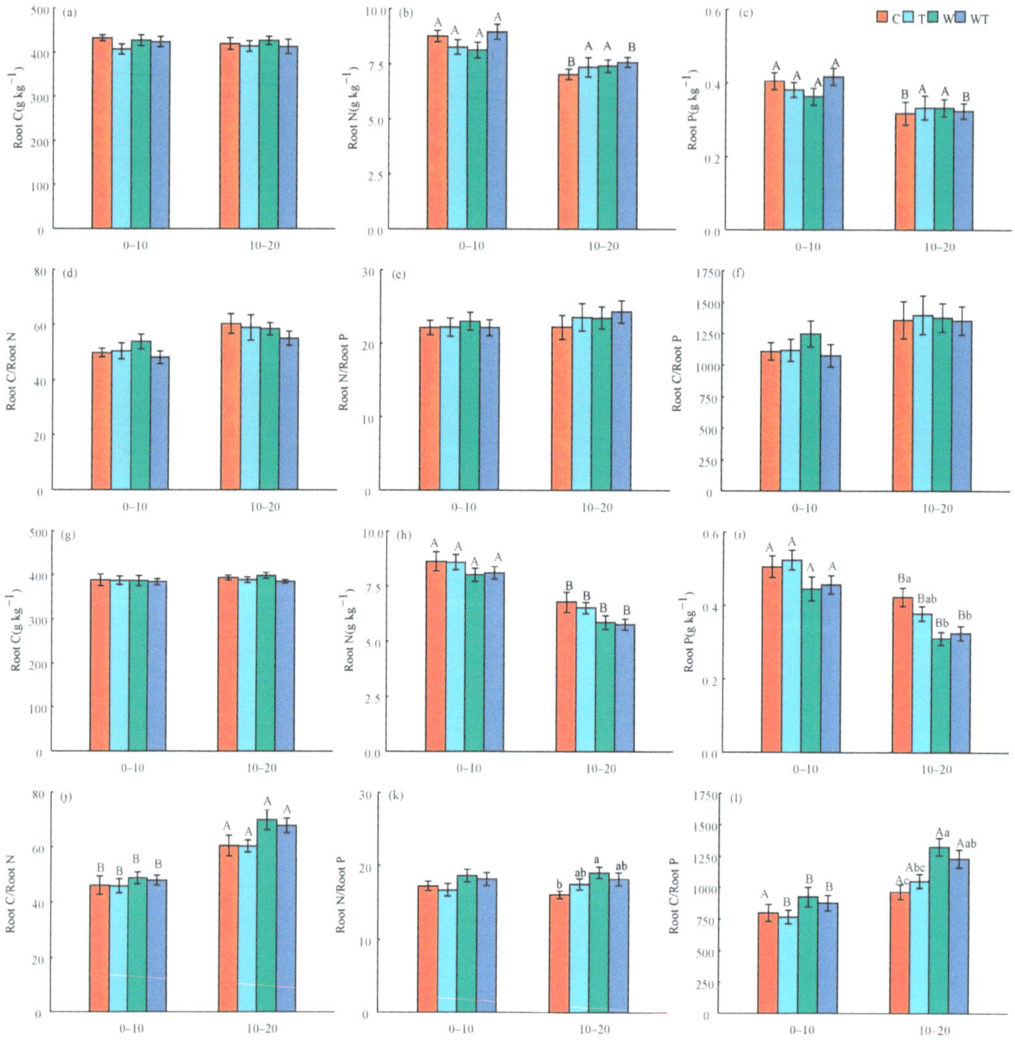

Figure 3. Concentrations and ratios of root carbon (Root C), root nitrogen (Root N), and root phosphorus (Root P) under four treatments (Control: C; Trenching: T; Warming: W; Warming and Trenching: WT) in the *Castanopsis hystrix* plantation (**a–f**) and *Quercus aliena* forest (**g–l**). Different lowercase letters indicate significant differences at $p < 0.05$ between treatments for the same soil depth, and different uppercase letters indicate significant differences between soil depths for the same treatment. Bars indicate standard errors of means.

However, only soil depth had a significant influence on the NAR, NNR and NMR in the *Q. aliena* forest ($p < 0.001$, Table 5). For example, the NAR (83.3%–125.5%), NNR (82.4%–91.2%) and NMR (85.5%–107.3%) of the 10–20 cm soil layer significantly decreased compared to those of the 0–10 cm soil layer ($p < 0.05$, Figure 4d–f).

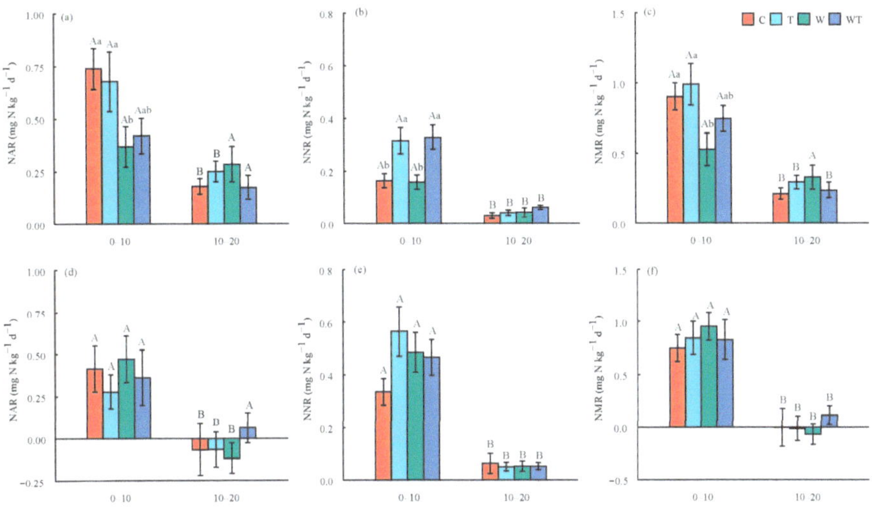

Figure 4. Rates of net mineralization (NMR), net nitrification (NNR) and net ammonification (NAR) under four treatments (Control: C; Trenching: T; Warming: W; Warming and Trenching: WT) in the *Castanopsis hystrix* plantation (**a–c**) and *Quercus aliena* forest (**d–f**). Different lowercase letters indicate significant differences at $p < 0.05$ between treatments for the same soil depth, and different uppercase letters indicate significant differences between soil depths for the same treatment. Bars indicate standard errors of means.

Table 5. Multi-way ANOVA results for the effect of warming (W), trenching (T), depth (D) and their interaction on soil net nitrogen transformation rates in the *Castanopsis hystrix* (CH) plantation and *Quercus aliena* (QA) forest.

Forest	Source of Variation	NAR	NNR	NMR
CH	W	0.015 *	0.604	0.030 *
	T	0.846	0.000 **	0.244
	D	0.000 **	0.000 **	0.000 **
	T × D	0.900	0.000 **	0.216
	W × D	0.008 **	0.736	0.009 **
	W × T	0.783	0.731	0.875
	W × T × D	0.227	0.912	0.232
QA	W	0.538	0.795	0.528
	T	0.860	0.203	0.723
	D	0.000 **	0.000 **	0.000 **
	T × D	0.225	0.151	0.607
	W × D	0.855	0.706	0.757
	W × T	0.555	0.132	0.945
	W × T × D	0.659	0.092	0.304

NAR: net ammonification rate; NNR: net nitrification rate; NMR net nitrogen mineralization rate. Significant probabilities are denoted by * and ** for $p < 0.05$ and 0.01, respectively.

3.5. Factors Influencing the Net Nitrogen Transformation Rate

The 54%, 50% and 52% of the variation in NAR, NNR and NMR could be explained by the predictor in the *C. hystrix* plantation. Specifically, ST had a significant negative effect on NAR (standardized coefficient = -0.23 ± 0.09, $p < 0.05$), while Root P and FRB had a significant positive effect on NAR (standardized coefficient = 0.5 ± 0.2, $p < 0.01$; 0.4 ± 0.1, $p < 0.05$, Figure 5a). The MBN had a significant negative effect on NNR and NMR (standardized coefficient = -0.4 ± 0.2, -0.3 ± 0.1, $p < 0.05$, Figure 5b,c), while the FRB had a significant positive effect on NMR (standardized coefficient = 0.4 ± 0.2, $p < 0.05$).

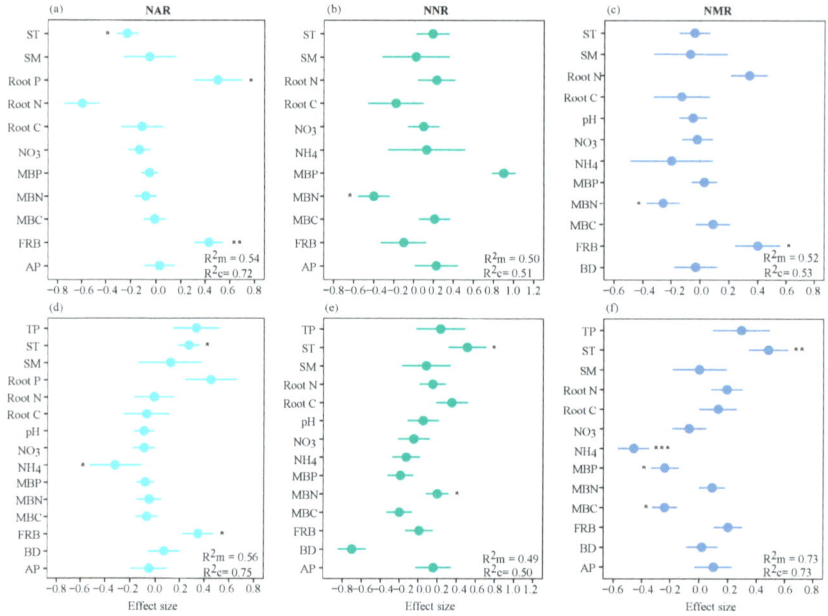

Figure 5. Summary of the linear mixed-effect modeling for multiple biotic and abiotic factors on net ammonification rate (NAR), net nitrification rate (NNR) and net nitrogen mineralization rate (NMR) in the *Castanopsis hystrix* plantation (**a–c**) and *Quercus aliena forest* (**d–f**). The dots represent the coefficients of the estimated effect size, and the lines represent standard errors. R^2m represents marginal R^2, and R^2c represents conditional R^2. * $p < 0.05$; ** $p < 0.01$; *** $p < 0.001$. ST: soil temperature; SM: soil moisture; BD: bulk density; NH_4: ammonium nitrogen; NO_3: nitrate nitrogen; TP: total phosphorus; AP: available phosphorus; MBC: microbial biomass carbon; MBN: microbial biomass nitrogen; MBP microbial biomass phosphorus; FRB: fine root biomass; Root C: fine root carbon; Root N: fine root nitrogen; Root P: fine root phosphorus.

In the *Q. aliena* forest, 56%, 49% and 73% of the variation in NAR, NNR and NMR, respectively, could be explained by the predictor. Among them, the ST had significant positive effects on NAR, NNR and NMR (standardized coefficient = 0.3 ± 0.1, 0.5 ± 0.2, $p < 0.05$, 0.5 ± 0.1, $p < 0.01$), and FRB had significant positive effects on NAR (standardized coefficient = 0.4 ± 0.1, $p < 0.05$), while NH_4^+-N had a negative effect on NAR (standardization coefficient = −0.3 ± 0.2, $p < 0.05$, Figure 5d). The MBN had a significant positive effect on NNR (standardization coefficient = 0.2 ± 0.1, $p < 0.05$, Figure 5e); The influences of NH_4^+-N, MBP and MBC on NMR were negative (standardized coefficient = −0.5 ± 0.1, $p < 0.001$, −0.2 ± 0.1, −0.1 ± 0.1, $p < 0.05$, Figure 5f).

4. Discussion

4.1. Warming Decreased Soil NAR and NMR in the Castanopsis Hystrix Plantation

In this study, the fine root biomass in the 10–20 cm soil layer in the *C. hystrix* plantation increased significantly after continuous soil warming for 11 years, indicating that trees probably increased the proportion of deep roots to cope with decreasing soil moisture content caused by the long-term warming process [26]. This was also observed in the studies of Wang et al. [27] and Zhao et al. [28] who found that fine root biomass had a stronger response to warming in deeper soil layers. Nevertheless, the fine root biomass at 0–10 cm in the *C. hystrix* plantation was not affected in the fourth and seventh years of warming, suggesting that short-term mild soil warming (+1.5 °C) had minimal influence on fine root growth in this soil horizon [18]. Previous studies have demonstrated that increasing the temperature by 4 °C had no influence on fine root biomass of 46-year-old *Picea abies* (L.) H. Karsten, but biomass declined sharply when the ambient temperature range was between +4 °C to +6 °C [29,30]. It is evident that the growth of fine roots is closely related to the duration and magnitude of temperature increases.

Soil microbial biomass or community structure plays a crucial role in soil nitrogen cycling [31]. In this study, the AP concentration in the 10–20 cm soil layer increased significantly after warming for 11 years, indicating that long-term climate warming will increase phosphorus availability in the south subtropical *C. hystrix* plantation. However, MBC, MBN, MBP and their ratios remained unchanged after warming, indicating that soil microbes did not benefit from the higher nutrient availability [32]. Unstable carbon consumption caused by long-term warming [19] might be a major reason for limited soil microbial biomass and nutrients as well as limited microbial reaction [33]. Previous studies demonstrated that the responses of MBC and MBN contents in soil to warming might be positive or negative. A meta-analysis study showed that MBC and MBN contents in soils had strong positive responses to mid-term (3–4 years) or short-term (1–2 years) warming, but their responses to long-term (5 years) warming tended to be weakened [34]. In this study, the soil microbes may have adapted to the warming episode up to 11 years, so they did not respond. In addition, the MBC content in soils shows relatively consistent positive responses to different increased amplitudes of temperature. In contrast, MBN's response to a low increase amplitude of temperature (<1 °C) was negative, and its response to a high increase amplitude of temperature (>2 °C) was positive. This might be related to warming-induced changes in soil moisture and substrate supply [34].

Generally, the NMR of soil increases with an increase in temperature [35] because high temperatures enhance the activity of soil microbes [36] and extracellular enzymes, and accelerate the transformation of macromolecular organic matter into forms easily assimilated by plants and microorganisms [10]. However, Gao et al. [37] found that long-term warming (14 years) decreased soil nitrogen cycling of Mediterranean-climate annual grassland due to a reduction in the abundance of relevant functional genes. In the present study, the NAR of the 0–10 cm soil layer in the *C. hystrix* plantation also decreased significantly after 11 years of warming, which was opposite to our first hypothesis. On the one hand, the Root P and fine root biomass (FRB) had significant positive effects on NAR (Figure 5a). Specifically, warming increased the underground distribution of plant biomass by enhancing the growth and turnover of fine roots (Figure 2a,b) which secreted more carbon into the soil [38], stimulating the growth of nitrogen cycling-related microorganisms [39]. In this case, the microorganisms produce more extracellular enzymes, thus accelerating the soil NAR. On the other hand, ST had a significant negative impact on NAR (Figure 5a). The increase in temperature also would have accelerated the decomposition of labile carbon, further accelerating microbial nitrogen immobilization by heterotrophic microbes. This is consistent with the increase in MBN concentration in the 0–20 cm soil layer after warming (Figure 1b), which would have decreased the substrate for soil ammonification and lowered the NAR of soils [40]. Overall, the negative effect of ST on NAR was greater than the positive effect of Root P and FRB, resulting in a significant decrease in NAR for the 0–10 cm *C. hystrix* soil horizon under warming. The NMR and NAR of the 0–10 cm soil layer fell sharply, while the NNR remained unchanged under warming conditions.

However, the NNR in the *C. hystrix* plantation increased by 100.0% and 106.3% after trenching and warming + trenching treatments (Figure 4b). This suggests that the risk of soil NO_3^--N leaching loss during the wet season may increase after trenching, and it is further intensified by long-term warming. In other studies on carbon input control based on ditching, the NNR of soils also increased significantly after trenching was applied in coniferous forests dominated by 27-year-old *Pinus taeda* L. [41], *Pseudotsuga menziesii* (Mirb.) Franco and *Tsuga heterophylla* (Raf.) Sarg. [42]. The increase in NNR was explained as follows: First, the NH_4^+-N absorption capacity of plants decreased after roots were cut, while the residual substrate for nitrification (NH_4^+-N in soils) increased to accelerate the NNR of soils [43]. Second, after plant roots were broken, dead roots and root exudates decreased as carbon sources for heterotrophic microorganisms, resulting in a decrease in microbial biomass and the immobilization of NH_4^+ by soil microorganisms [44]. Furthermore, root severing would cause an increase in NH_4^+ available to nitrifying bacteria and an increase in the net nitrification rate due to the negative correlation between MBN and NNR (Figure 5b).

4.2. Warming Did Not Change the NMR of Soils in the Quercus Aliena Forest

It could be expected that fine root biomass in temperate forests might show stronger responses to warming than in subtropical or tropical forests [28,45]. However, we found that warming had little effect on the fine root biomass in the temperate *Q. aliena* forest. According to our early findings, the fine root biomass at 0–10 cm increased greatly in the first four years of warming, but such positive effects began to be weakened in the fifth year [20,21]. Short-term warming promotes fine root biomass, which may be caused by a variety of factors. First, warming can extend the growing season of plants [46], which has a positive impact on plant roots; Moreover, the experimental warming will increase evapotranspiration and decrease soil water [20], leading to an increase in fine root biomass to enhance water uptake. That fine root biomass was hardly affected in the 11th year of warming in this study demonstrates that the warming effect fine on root biomass gradually disappears as warming continues [27]. Warming can increase the availability of plant-usable nitrogen in the soil (e.g., NH_4^+-N, Table 2). As a result, plants may not need to allocate additional photosynthetic resources to fine roots for nutrient acquisition, limiting the growth of fine roots under warming conditions [47].

We found that the MBN concentration in the 10–20 cm soil layer of the *Q, aliena* forest decreased significantly after 11 years of warming, thus increasing the MBC/MBN but decreasing the MBN/MBP ratios. However, the MBC and MBN of the 0–10 cm soil layer increased significantly in the first three years of warming [21]. This suggests that the positive response of soil microbial biomass also weakens gradually as warming continues. Short-term warming-induced increases in air and soil temperature and a fresh C supply from the plants' above- and below-ground parts might have enhanced microbial growth [48], resulting in an increase in MBC and MBN. However, water losses due to long-term rising temperatures may inhibit microbial growth [49], which eventually decreases the MBN but not the MBC. The maintenance of MBC is probably because root-related C input had offset losses of carbohydrates [12]. The sharp reduction in soil pH (Table 2) might be a major reason for the decreased MBN in the 10–20 cm soil layer [50]. It has been suggested that climate warming in temperate zones will inhibit the nitrogen fixation rate of soil microbes and decrease soil microbial reserves [51], thus increasing the inorganic nitrogen content of soils [52]. The considerable increase in NH_4^+-N concentration in the *Q. aliena* forest soils (Table 2) supports this view. In our study, the sharp reduction in Root P and sharp increase in Root N/P (Figure 3i,k) under warming conditions are consistent with previous research findings [27,28]. Ferric oxide, hydroxide and P adsorption by clay constituents in soils increases with temperature, while low soil pH (Table 2) enhances the fixation of soil P [53]. Thus, further long-term warming is likely to further decrease P absorption in the *Q. aliena* forest [54].

Warming usually improves the physiological activity of soil microbes, thus facilitating the decomposition and release of soil organic nitrogen and increasing the nitrogen content in soils [55]. Similarly, we found that NH_4^+-N concentration in soils increased significantly after 11-year warming. However, warming had no significant influence on the NAR, NNR and NMR of the *Q. aliena* forest, indicating that nitrogen transformation of soil microbes in the temperate *Q. aliena* forest was less temperature sensitive than in the *C. hystrix* plantation. This disagrees with our second hypothesis. On the one hand, the increase in soil temperature promotes NH_4^+-N concentration due to enhanced microbial metabolism and enzyme activity [56], thereby stimulating NMR and NAR. On the other hand, warming induces a reduction in SM which can hinder substrate accessibility [19], thus counteracting the temperature-induced stimulation of biochemical processes [51]. It seems that the negative effect of an increased NH_4^+-N concentration offsets the positive effect of ST on the NMR and NAR of the *Q. aliena* forest, thus preventing significant fluctuation in NMR and NAR after warming.

The NNR of *Q. aliena* forest soils also did not change greatly after warming. The optimal temperature of nitrification is about 25–30 °C [57], which is higher than the soil temperature during the warming treatment of the two-year growing season (18.7–21.6 °C) in our site. Therefore, warming may promote NNR in the *Q. aliena* forest because it brings the soil temperature close to the optimum. However, a significantly lower pH not only leads to significant reductions in MBN (Figure 1h), but may also negatively affect the growth and activity of nitrifying bacteria [53], thereby reducing NNR. Overall, the positive effect of increased ST offsets the negative effect of decreased MBN, and therefore, the NNR was not changed under warming conditions in the *Q. aliena* forest.

It is worth mentioning that there might be seasonal differences in the influences of warming on soil nitrogen cycling of forests in different climatic zones. However, this was not examined in our study as we limited our investigation to the wet seasons in two successive years. Hence, further longitudinal studies concerning the influences of seasonal changes on soil nitrogen cycling of forests in different climatic zones are needed.

5. Conclusions

This study found that soil nitrogen mineralization was more affected by long-term warming in the subtropical *C. hystrix* plantation than in the temperate *Q. aliena* natural forest. The sharp reduction in NAR and NMR of the *C. hystrix* plantation was mainly because the increasing temperature promoted the nitrogen immobilization of heterotrophic microorganisms and resulted in fewer substrates for NAR. The sharp rise in NNR caused by trenching and warming + trenching was mainly related to the reduced NH_4^+-N absorption of trenched plants and the reduced immobilization of NH_4^+ by soil microorganisms. It was unexpected that NAR, NNR and NMR did not change in the *Q. aliena* forest after warming. This is because warming not only stimulates soil N mineralization by increasing NH_4^+-N and reducing MBN, but also induces a reduction in SM which can hinder substrate accessibility, thus counteracting the temperature-induced stimulation of N mineralization. Therefore, the negative effect of increased NH_4^+-N and decreased MBN would have offset the positive effect of ST on the NMR and NAR of *Q. aliena* forest. Overall, the different responses of soil nitrogen mineralization to warming could be attributed to differences in soil nutrient availability, fine root biomass and microbial biomass of forests in the two climatic zones.

Author Contributions: Conceptualization, S.L.; methodology, L.C. and Y.L.; software, W.S.; formal analysis, W.S. and L.C.; investigation, L.C., Y.L., H.M. and Z.L.; writing—original draft preparation, W.S. and L.C.; writing—review and editing, B.D.; funding acquisition, H.W. All authors have read and agreed to the published version of the manuscript.

Funding: This research was funded by the Chinese Academy of Forestry, CAFYBB2020ZA001, and National Key Research and Development Program of China, 2023YFE0105100-1.

Data Availability Statement: The data that support the findings of this study are available from the corresponding authors upon reasonable request.

Acknowledgments: We would like to thank Jihui Zhang and Junxu Ma for their sampling work in the field and laboratory. We also thank Xianqiang Zhang and Dewei Huang for the maintenance of heating facilities in plots.

Conflicts of Interest: The authors declare no conflicts of interest.

References

1. Jones, N. When will global warming actually hit the landmark 1.5 C limit? *Nat. Geosci.* **2023**, *618*, 20. [CrossRef]
2. IPCC. Summary for policymakers. In *Climate Change 2021: The Physical Science Basis. Contribution of Working Group I to the Sixth Assessment Report of the Intergovernmental Panel on Climate Change*; Cambridge University Press: Cambridge, UK; New York, NY, USA, 2021.

3. Fernández-Martínez, M.; Sardans, J.; Chevallier, F.; Ciais, P.; Obersteiner, M.; Vicca, S.; Canadell, J.G.; Bastos, A.; Friedlingstein, P.; Sitch, S. Global trends in carbon sinks and their relationships with CO_2 and temperature. *Nat. Clim. Change* **2019**, *9*, 73–79. [CrossRef]
4. Liu, T.Z.; Nan, Z.B.; Hou, F.J. Grazing intensity effects on soil nitrogen mineralization in semi-arid grassland on the Loess Plateau of northern China. *Nutr. Cycl. Agroecosystems* **2011**, *91*, 67–75. [CrossRef]
5. Lebauer, D.S.; Treseder, K.K. Nitrogen limitation of net primary productivity in terrestrial ecosystems is globally distributed. *Ecology* **2008**, *89*, 371–379. [CrossRef] [PubMed]
6. Dai, Z.M.; Yu, M.J.; Chen, H.H.; Zhao, H.C.; Xu, J.M. Elevated temperature shifts soil N cycling from microbial immobilization to enhanced mineralization, nitrification and denitrification across global terrestrial ecosystems. *Glob. Chang. Biol.* **2020**, *26*, 5267–5276. [CrossRef] [PubMed]
7. Wu, Z.T.; Dijkstra, P.; Koch, G.W.; Peñuelas, J.; Hungate, B.A. Responses of terrestrial ecosystems to temperature and precipitation change: A meta-analysis of experimental manipulation. *Glob. Chang. Biol.* **2011**, *17*, 927–942. [CrossRef]
8. Ollinger, S.V.; Richardson, A.D.; Martin, M.E.; Hollinger, D.Y.; Frolking, S.E.; Reich, P.B.; Plourde, L.C.; Katul, G.G.; Munger, J.W.; Oren, R.; et al. Canopy nitrogen, carbon assimilation, and albedo in temperate and boreal forests: Functional relations and potential climate feedbacks. *Proc. Natl. Acad. Sci. USA* **2008**, *105*, 19336–19341. [CrossRef]
9. Wang, C.H.; Han, X.G.; Xing, X.R. Effects of grazing exclusion on soil net nitrogen mineralization and nitrogen availability in a temperate steppe in northern China. *J. Arid Environm.* **2010**, *74*, 1287–1293. [CrossRef]
10. Fanin, N.; Mooshammer, M.; Sauvadet, M.; Meng, C.; Alvarez, G.; Bernard, L.; Bertrand, I.; Blagodatskaya, E.; Bon, L.; Fontaine, S.; et al. Soil enzymes in response to climate warming: Mechanisms and feedbacks. *Functi. Ecol.* **2022**, *36*, 1378–1395. [CrossRef]
11. Zuccarini, P.; Asensio, D.; Ogaya, R.; Sardans, J.; Peuelas, J. Effects of seasonal and decadal warming on soil enzymatic activity in a P-deficient Mediterranean shrubland. *Glob. Chang. Biol.* **2020**, *26*, 3698–3714. [CrossRef]
12. Yin, H.J.; Chen, Z.; Liu, Q. Effects of experimental warming on soil N transformations of two coniferous species, Eastern Tibetan Plateau, China. *Soil Biol. Biochem.* **2012**, *50*, 77–84. [CrossRef]
13. Butler, S.M.; Melillo, J.M.; Johnson, J.E.; Mohan, J.; Steudler, P.A.; Lux, H.; Burrows, E.; Smith, R.M.; Vario, C.L.; Scott, L. Soil warming alters nitrogen cycling in a New England forest: Implications for ecosystem function and structure. *Oecologia* **2011**, *168*, 819–828. [CrossRef]
14. Li, Y.Q.; Zhang, X.P.; Ma, X.Z.; Li, C.S.; Hao, C.Y.; Tian, H.; Wu, H.; Liang, Z. Effects of short-term warming on soil nitrogen mineralization rate of *Pinus tabulaeformis* plantation in Daqingshan, Inner Mongolia. *J. Anhui Agric. Sci.* **2023**, *51*, 126–131+175. (In Chinese)
15. Wang, X.N.; Xiong, D.C.; Zhang, Y.H.; Xi, Y.Q.; Huang, J.X.; Chen, S.D.; Liu, X.F.; Yang, Z.J. Effects of warming and nitrogen Addition on Soil Nitrogen Mineralization and N_2O Emission in a Mid-subtropical *Cunninghamia lanceolata* Plantation. *For. Res.* **2023**, *36*, 22–31. (In Chinese)
16. Bai, E.; Li, S.L.; Xu, W.H.; Li, W.; Dai, W.W.; Jiang, P. A meta-analysis of experimental warming effects on terrestrial nitrogen pools and dynamics. *New Phytol.* **2013**, *199*, 441–451. [CrossRef]
17. Bai, T.S.; Wang, P.; Qiu, Y.P.; Zhang, Y.; Hu, S.J. Nitrogen availability mediates soil carbon cycling response to climate warming: A meta-analysis. *Glob. Chang. Biol.* **2023**, *29*, 2608–2626. [CrossRef]
18. Wang, H.; Liu, S.R.; Wang, J.X.; Li, D.J.; Shi, Z.M.; Liu, Y.C.; Xu, J.; Hong, P.Z.; Yu, H.L.; Zhao, Z.; et al. Contrasting responses of heterotrophic and root-dependent respiration to soil warming in a subtropical plantation. *Agric. For. Meteorol.* **2017**, *247*, 221–228. [CrossRef]
19. Wang, H.; Liu, S.R.; Schindlbacher, A.; Wang, J.X.; Yang, Y.J.; Song, Z.C.; You, Y.M.; Shi, Z.M.; Li, Z.Y.; Chen, L.; et al. Experimental warming reduced topsoil carbon content and increased soil bacterial diversity in a subtropical planted forest. *Soil Biol. Biochem.* **2019**, *133*, 155–164. [CrossRef]
20. Liu, Y.C.; Liu, S.R.; Wan, S.Q.; Wang, J.X.; Luan, J.W.; Wang, H. Differential responses of soil respiration to soil warming and experimental throughfall reduction in a transitional oak forest in central China. *Agric. Meteorol.* **2016**, *226*, 186–198. [CrossRef]
21. Wang, Y.; Liu, S.R.; Wang, J.X.; Chang, S.X.; Luan, J.W.; Liu, Y.C.; Lu, H.B.; Liu, X.J. Microbe-mediated attenuation of soil respiration in response to soil warming in a temperate oak forest—Science Direct. *Sci. Total Environ.* **2020**, *711*, 134563. [CrossRef]
22. Tsui, C.C.; Chen, Z.S. Net nitrogen mineralization and nitrification of different landscape positions in a lowland subtropical rainforest in Taiwan. *Soil Sci. Plant Nutr.* **2010**, *56*, 319–331. [CrossRef]
23. Huang, C.Y.; Vance, E.D.; Brookes, P.C.; Jenkinson, D.S. An extraction method for measuring soil microbial biomass C. *Soil Biol. Biochem.* **1987**, *19*, 703–707.
24. Ruiz, R.; Nickel, B.; Koch, N.; Feldman, L.C.; Haglund, R.F.; Kahn, A.; Family, F.; Scoles, G. Dynamic scaling, island size distribution, and morphology in the aggregation regime of submonolayer pentacene films. *Phys. Rev. Lett.* **2003**, *91*, 136102. [CrossRef]
25. Yang, J.J.; Wu, A.C.; Li, J.H.; Wei, H.H.; Qin, J.; Tian, H.D.; Fan, D.H.; Wu, W.D.; Chen, S.; Tong, X.; et al. Structured and unstructured intraspecific propagule trait variation across environmental gradients in a widespread mangrove. *Ecol. Evol.* **2024**, *14*, e10835. [CrossRef]
26. Markesteijn, L.; Poorter, L. Seedling root morphology and biomass allocation of 62 tropical tree species in relation to drought- and shade-tolerance. *J. Ecol.* **2009**, *97*, 311–325. [CrossRef]

27. Wang, J.S.; Defrenne, C.; McCormack, M.L.; Yang, L.; Tian, D.S.; Luo, Y.Q.; Hou, E.Q.; Yan, T.; Li, Z.L.; Bu, W.S.; et al. Fine-root functional trait responses to experimental warming: A global meta-analysis. *New Phytol.* **2021**, *230*, 1856–1867. [CrossRef] [PubMed]
28. Zhao, X.X.; Tian, Q.X.; Michelsen, A.; Lu, M.Z.; Ren, B.S.; Huang, L.; Zhao, R.D. The effect of experimental warming on fine root functional traits of woody plants: Data synthesis. *Sci. Total Environ.* **2023**, *894*, 165003. [CrossRef]
29. Leppälammi-Kujansuu, J.; Ostonen, I.; Strömgren, M.; Nilsson, L.O.; Kleja, D.B.; Sah, S.P.; Helmisaari, H.S. Effects of long-term temperature and nutrient manipulation on Norway spruce fine roots and mycelia production. *Plant Soil* **2013**, *366*, 287–303. [CrossRef]
30. Parts, K.; Tedersoo, L.; Schindlbacher, A.; Sigurdsson, B.D.; Leblans, N.I.W.; Oddsdóttir, E.S.; Borken, W.; Ostonen, I. Acclimation of Fine Root Systems to Soil Warming: Comparison of an Experimental Setup and a Natural Soil Temperature Gradient. *Ecosystems* **2019**, *22*, 457–472. [CrossRef]
31. Xu, Z.F.; Hu, R.; Xiong, P.; Wan, C.A.; Cao, G.; Liu, Q. Initial soil responses to experimental warming in two contrasting forest ecosystems, Eastern Tibetan Plateau, China: Nutrient availabilities, microbial properties and enzyme activities. *Appl. Soil Ecol.* **2010**, *46*, 291–299. [CrossRef]
32. Biasi, C.; Meyer, H.; Rusalimova, O.; Hämmerle, R.; Kaiser, C.; Baranyi, C.; Daims, H.; Lashchinsky, N.; Barsukov, P.; Richter, A. Initial effects of experimental warming on carbon exchange rates, plant growth and microbial dynamics of a lichen-rich dwarf shrub tundra in Siberia. *Plant Soil* **2008**, *307*, 191–205. [CrossRef]
33. Zhang, W.; Parker, K.M.; Luo, Y.; Wan, S.; Wallace, L.L.; Hu, S. Soil microbial responses to experimental warming and clipping in a tallgrass prairie. *Glob. Change Biol.* **2005**, *11*, 266–277. [CrossRef]
34. Xu, W.F.; Yuan, W. Responses of microbial biomass carbon and nitrogen to experimental warming: A meta-analysis. *Soil Biol. Biochem.* **2017**, *15*, 265–274. [CrossRef]
35. Fu, W.; Wang, X.; Wei, X.R. No response of soil N mineralization to experimental warming in a northern middle-high latitude agro-ecosystem. *Sci. Total Environ.* **2019**, *659*, 240–248. [CrossRef] [PubMed]
36. Chantigny, M.H.; Curtin, D.; Beare, M.H.; Greenfield, L.G. Influence of temperature on water-extractable organic matter and ammonium production in mineral soils. *Soil Sci. Soc. Am. J.* **2010**, *74*, 517–524. [CrossRef]
37. Gao, Y.; Ding, J.J.; Yuan, M.T.; Chiariello, N.; Yang, Y.F. Long-term warming in a Mediterranean-type grassland affects soil bacterial functional potential but not bacterial taxonomic composition. *NPJ Biofilms Microbi.* **2021**, *7*, 17. [CrossRef] [PubMed]
38. Heinzle, J.; Liu, X.; Tian, Y.; Kengdo, S.; Heinze, B.; Nirschi, A.; Borken, W.; Inselsbacher, E.; Wanek, W.; Schindlbacher, A. Increase in fine root biomass enhances root exudation by long-term soil warming in a temperate forest. *Front. For. Glob. Chang.* **2023**, *6*, 1152142. [CrossRef]
39. Dijkstra, F.A.; Cheng, W.X. Moisture modulates rhizosphere effects on C decomposition in two different soil types. *Soil Biol. Bioch.* **2007**, *39*, 2264–2274. [CrossRef]
40. Huang, Z.Q.; Wan, X.H.; He, Z.M.; Yu, Z.P.; Wang, M.H.; Hu, Z.H.; Yang, Y.S. Soil microbial biomass, community composition and soil nitrogen cycling in relation to tree species in subtropical China. *Soil Biol. Biochem.* **2013**, *62*, 68–75. [CrossRef]
41. Drake, J.E.; Oishi, A.C.; Giasson, M.A.; Oren, R.; Johnsen, K.H.; Finzi, A.C. Trenching reduces soil heterotrophic activity in a loblolly pine (*Pinus taeda*) forest exposed to elevated atmospheric [CO_2] and N fertilization. *Agric. For. Meteorol.* **2012**, *165*, 43–52. [CrossRef]
42. Wilson, L. Characterization of Soil Carbon and Nitrogen Pools Following a Decade of Detritus Manipulation in a Temperate Coniferous Forest. Bachelor's Thesis, Oregon State University, Eugene, OR, USA, 2008.
43. Matsushima, M.; Chang, S.X. Effects of understory removal, N fertilization, and litter layer removal on soil N cycling in a 13-year-old white spruce plantation infested with Canada bluejoint grass. *Plant Soil* **2007**, *292*, 243–258. [CrossRef]
44. Fang, X.M.; Wang, G.G.; Xu, Z.J.; Zong, Y.Y.; Zhang, X.L.; Li, J.J.; Wang, H.M.; Chen, F.S. Litter addition and understory removal influenced soil organic carbon quality and mineral nitrogen supply in a subtropical plantation forest. *Plant Soil* **2021**, *460*, 527–540. [CrossRef]
45. Salazar, A.; Rousk, K.; Jónsdóttir, I.S.; Bellenger, J.P.; Andrésson, O.S. Faster nitrogen cycling and more fungal and root biomass in cold ecosystems under experimental warming: A meta-analysis. *Ecology* **2020**, *101*, e02938. [CrossRef]
46. Zhao, J.X.; Li, R.C.; Li, X.; Tian, L.H. Environmental controls on soil respiration in alpine meadow along a large altitudinal gradient on the central Tibetan Plateau. *Catena* **2017**, *159*, 84–92. [CrossRef]
47. Smithwick, E.A.; Eissenstat, D.M.; Lovett, G.M.; Bowden, R.D.; Rustad, L.E.; Driscoll, C.T. Root stress and nitrogen deposition: Consequences and research priorities. *New Phytol.* **2013**, *197*, 712–719. [CrossRef] [PubMed]
48. Liu, Q.; Yin, H.; Chen, J.; Zhao, C.; Cheng, X.; Wei, Y.; Lin, B. Belowground responses of *Picea asperata* seedlings to warming and nitrogen fertilization in the eastern Tibetan Plateau. *Ecol. Res.* **2011**, *26*, 637–648. [CrossRef]
49. Allison, S.D.; Treseder, K.K. Warming and drying suppress microbial activity and carbon cycling in boreal forest soils. *Glob. Chang. Biol.* **2008**, *14*, 2898–2909. [CrossRef]
50. Pietri, J.C.A.; Brookes, P.C. Relationships between soil pH and microbial properties in a UK arable soil. *Soil Biol. Biochem.* **2008**, *40*, 1856–1861. [CrossRef]
51. Gao, W.L.; Yan, D.H. Warming suppresses microbial biomass but enhances N recycling. *Soil Biol. Biochem.* **2019**, *131*, 111–118. [CrossRef]

52. Li, Z.L.; Zeng, Z.Q.; Song, Z.P.; Wang, F.Q.; Tian, D.S.; Mi, W.H.; Huang, X.; Wang, J.S.; Song, L.; Yang, Z.K.; et al. Vital roles of soil microbes in driving terrestrial nitrogen immobilization. *Glob. Chang. Biol.* **2021**, *27*, 1848–1858. [CrossRef]
53. Tian, Y.; Shi, C.P.; Malo, C.U.; Kengdo, S.K.; Heinzle, J.; Inselsbacher, E.; Ottner, F.; Borken, W.; Michel, K.; Schindlbacher, A.; et al. Long-term soil warming decreases microbial phosphorus utilization by increasing abiotic phosphorus sorption and phosphorus losses. *Nat. Commun.* **2023**, *14*, 864. [CrossRef] [PubMed]
54. Jia, L.Q.; Jiang, Q.; Sun, J.; Robinson, D.; Yang, Z.J.; Yao, X.D.; Wang, X.H.; Dai, X.L.; Chen, T.T.; Wu, D.M.; et al. Contrasting depth-related fine root plastic responses to soil warming in a subtropical Chinese fir plantation. *J. Ecol.* **2024**, *112*, 1058–1073. [CrossRef]
55. Wang, Y.; Mao, C.; Lin, W.S.; Yang, Z.J.; Xiong, D.C.; Xu, C.; Liu, X.F.; Chen, S.D. Effects of warming on soil nitrogen transformations in subtropical Chinese-fir plantations. *Chin. J. Appl. Environ. Biol.* **2024**, *30*, 537–543. (In Chinese)
56. Msimbira, L.A.; Smith, D.L. The roles of plant growth promoting microbes in enhancing plant tolerance to acidity and alkalinity stresses. *Front. Sustain. Food Syst.* **2020**, *4*, 160. [CrossRef]
57. Saad, O.A.L.O.; Conrad, R. Temperature dependence of nitrification, denitrification, and turnover of nitric oxide in different soils. *Biol. Fertil. Soils* **1993**, *15*, 21–27. [CrossRef]

forests

MDPI

Article

Environmental Driving Mechanism and Response of Soil's Fungal Functional Structure to Near-Naturalization in a Warm Temperate Plantation

Zhenlu Qiu, Huan Liu, Chunli Chen, Congcong Liu and Jing Shu *

College of Forestry Engineering, Shandong Agriculture and Engineering University, Jinan 250100, China; z2023011@sdaeu.edu.cn (Z.Q.); z2022022@sdaeu.edu.cn (H.L.); z2013221@sdaeu.edu.cn (C.C.); z2019082@sdaeu.edu.cn (C.L.)
* Correspondence: z2013222@sdaeu.edu.cn

Abstract: In this study, the near-naturalization process of *Pinus tabulaeformis* plantations in Baxianshan National Nature Reserve was divided into three stages depending on the proportion of *P. tabulaeformis* present, resulting in the following categories: the *P. tabulaeformis* forest stage, the mixed forest stage, and the near-natural forest stage. Natural secondary forests were selected as a control. We assessed alterations in the soil's fungal functional structures from three aspects: functional mode, vegetative mode, and growth mode, and their responses to vegetation and soil factors were also explored. The results showed that ectomycorrhizal, saprophytic, and plant pathogen types were dominant in the functional mode, and plant pathogens were most abundant in the *P. tabulaeformis* forest stage. Meanwhile, ectomycorrhizal fungi were the least abundant in the near-natural forest stage. In the vegetative mode, saprophytic, pathophysiological, and symbiotic types were dominant, and pathophysiological types were the most abundant in the *P. tabulaeformis* forest stage. In the growth mode, microfungi dominated, and the abundance of clavarioid decreased with near-naturalization. The degree of variation in functional structure in the three dimensions increased with near-naturalization, but the structure of natural secondary forests converged. The species composition of tree layer obviously affected the abundance and functional structure of fungi in the three modes, among which *Quercus mongolia* and *Carpinus hornbeam* were the most significant. The soil's pH and nitrate content significantly affected the structure of the functional mode, and the soil's dry matter content and C/N ratio significantly affected the structure of the vegetative mode. In this study, we explored the interaction between the plant community and soil ecological system during the near-naturalization process of plantations in terms of soil fungi functions, further clarifying the role of soil functions in the succession of plant communities and providing a new perspective on the in-depth exploration of ecosystem interactions during the succession of plantations.

Keywords: near-naturalization; functional structure; plant diversity; vegetation composition; soil properties

1. Introduction

The forest ecosystem is the most complex and diverse terrestrial ecosystem; it has important ecological functions such as primary productivity, nutrient biogeochemistry, carbon fixation, oxygen release, and soil and water conservation [1]. Plantations are an important type of forest ecosystem, and their characteristic large-scale planting can satisfy demands for wood and effectively promote the carbon cycle [2]. However, due to plantations' single-tree species and simple structure, they can suffer from a lack of diversity, which gives them a fragile structure and causes limited functioning [3]. The goal of plantation management and care is to promote their development into a healthy and stable community, and the ideal state is one in which both vegetation and soil ecosystems have a structure and function close to, or even consistent with, those of regional original ecosystems. This includes active processes achieved through human participation in management and natural development processes achieved through the interaction of planted forests with natural tree species in the inner regions of the community [4].

Fungi are important components of soil ecosystems, and their composition and structure respond to changes in tree species during the development of the forest, and further influence the development of soil nutrients and the above-ground community structure [5,6]. During secondary succession, the composition and yield of vegetation change greatly, which leads to changes in microbial communities within a short period of time [7]. However, natural succession has more utility than secondary succession in predicting the long-term development of ecological succession [8,9]. The species and composition of soil fungi are usually strongly correlated with the type of plant community and plant species, which means that tree species act as important mediators in shaping the structure of a soil fungal community. Studies have also shown that the composition, content, and diversity of fungi are significantly correlated with habitat [10–12]. These previous studies were only based on the similarity of the taxonomic composition of fungi. The authors have previously discussed the characteristic responses exhibited by soil micro-organisms during forest succession, focusing on their habitat-specialization adaptations [13]. However, functional structures that positively respond to ecological factors have rarely been reported [14,15]. Different functional groups of fungi are closely related to plant nutrient absorption, decomposition of organic matter, and the alleviation of plant diseases [16]. For example, mycorrhizal fungi can form mycorrhizas with plant roots in exchange for C, N, P, and other resources obtained from the soil [17] to promote the absorption of nutrients into plants [18]. Studies have shown that mycorrhiza provides nearly half of the organic nitrogen of trees and most of the new carbon in soil [19], while ectomycorrhiza fungi and arbuscular mycorrhiza (mycorrhiza groups) input varying levels of nitrogen and phosphorus to their hosts [20]. They also vary in their contribution of carbon to soil [21]. Saprotrophic fungi play a vital role in decomposing soil organic matter [22], and their biochemical process affects the element cycling rate [23]. Previous studies have found that fungal-driven ecosystem processes differ among different stand types [24]. Changes in soil properties and tree species composition can lead to changes in the diversity and functional structure of soil fungal communities [10,11]. Therefore, exploring the effects of forest community succession on fungal community structure and diversity from a functional perspective is the aim of this study, and we anticipate that it will more comprehensively show the interaction between the forest plant community and soil ecosystem. The FUNGuild package in R programs [25–27] has the advantage of directly analyzing and manipulating taxon data instead of genetic loci, and can divide sequenced OTUs data into groups using three dimensions: functional mode, vegetative mode, and growth mode. This makes it easier to explore responses to ecological factors in terms of fungal functional structure.

Baxianshan National Nature Reserve is rich in natural species, and its diversity ranks among the highest in the world among areas at the same latitude. However, its forest ecosystem has historically been severely damaged, and only regional vegetation has been preserved in the core area. In the 1950s, a large-scale *Pinus tabulaeformis* plantation was established in the sparsely occupied area of the reserve, and the forest was closed. In interacting with regional deciduous broad-leaved tree species over the past 70 years, *P. tabulaeformis* plantations have gradually developed the characteristics of regional natural forests and have formed mixed forests of *P. tabulaeformis* and a variety of regional deciduous broad-leaved tree species [28]. This process fully reflects the ecological effects of the interaction between artificial forests and regional tree species during near-natural development. In this study, we hope to reveal changes in the functional abundance and functional structure of soil fungi (in three dimensions) during near-naturalization, and explain their driving mechanisms. We predict that (1) the degree of variation in functional structure will increase as near-naturalization progresses; (2) changes in the composition of dominant tree species in the plant community are the influences on the abundance and structure of functional fungal groups in the process of near-naturalization; and (3) changes in soil physicochemical properties also affect the abundance and structure of functional groups to a certain extent. Our study will provide a new perspective on the interaction between the plant community and soil ecosystem during the establishment of plantations; we will also provide a scientific basis for the healthy development and sustainable management of stands.

2. Materials and Methods

2.1. Description of Research Area and Setup of Quadrats

Baxianshan National Nature Reserve is located in the North China, at the eastern foot of the Yanshan Mountains. Its original deciduous broad-leaved natural forests have been well preserved since the Tertiary and Cenozoic periods. By the decline of the Qing Dynasty, large numbers of primary forests were severely damaged. At the core of the reserve, a large number of deciduous broad-leaved tree species were preserved, and the felled communities began secondary succession and recovery, forming the existing natural secondary forest. In 1954, a large-scale *P. tabulaeformis* plantation was established in a sparsely vegetated area, and the mountain was closed for cultivation. During its interaction with the original deciduous broad-leaved tree species, the *P. tabulaeformis* plantation's natural characteristics were restored, forming a sequence of naturalized plantations. Therefore, the succession stands in this study represent a successful restoration project; the naturalized plantation recovered after around 70 years, and natural secondary forests began to recover after 110 years. In this study, the restoration of *P. tabulaeformis* plantations was divided into three distinct stages according to the proportion of *P. tabulaefomis* occupying a tree layer within the community. The *P. tabulaeformis* forest stage was defined by the coverage of *P. tabulaeformis* within the tree layer exceeding 90%, whereas the mixed forest stage had a coverage ranging from 40% to 60%, and the near-natural forest stage featured less than 20% coverage. For comparison, natural secondary forests were also chosen. Due to the rapid growth of broad-leaved trees in the process of near-naturalization and their superiority in competition, the area still occupied by the *P. tabulaeformins* forest stage is limited. Therefore, we selected 3 quadrats from the *P. tabulaeformis* forest stage, 4 from the mixed forest stage, 6 from the near-natural forest stage, and 14 from the natural secondary forest stage for control. The area of a single quadrat was 600 m^2 with 20 $\times$ 30 m, and the aspect of the quadrats was mainly to the south (Table S1).

2.2. Analysis of Plant Community Structure and Diversity

The species name, DBH (diameter at beast height), and height of all trees with DBH $\geq$ 3 cm were recorded. DBH was measured using a tree gauge, and tree height was calculated using a laser compass combined with trigonometric function.

2.2.1. Calculation of Importance Values

The importance value (IV) of each species was calculated with the following equation [29].

$$IV = \frac{\frac{n_i}{\sum_{i=1}^{s} n_i} * 100 + \frac{a_i}{\sum_{i=1}^{s} a_i} * 100 + \frac{f_i}{\sum_{i=1}^{s} f_i} * 100}{3}$$

where *ni* represents the number of individuals of the *i*th species; *ai* represents DBH; *fi* represents number of quadrats in which the *i*th species existed; and *s* represents the total number of species.

2.2.2. Biomass Calculation

Felling is prohibited in the protected area, and obtaining an allometry equation of biomass by means of standard wood analysis is not advisable. Therefore, in this study, we carried out biomass estimation using the allometry equation of biomass outlined in other studies on nearby mountains (Table 1).

Table 1. Allometry equation for calculating the biomass of each tree species in the tree layer.

Species	Biomass Allometry Equation
Pinus tabulaeformis	$Y = e^{-1.41+6.92/T} \times D^{1.03} \times H^{1.08} + 13.41$ Y: (t), D: (m^2), H: (m)
Quercus variabilis	$Y = 0.022337662(D^2H)^{0.993056421} + 0.006221667(D^2H)^{1.008154429} + 0.001179057(D^2H)^{1.298105392} + 0.018493229(D^2H)^{0.671232912} + 0.014665102(D^2H)^{0.950577264}$ Y: (kg), D: (cm^2), H: (m)
Robinia pseudoacacia	$Y = 0.020(D^2H) + 1.974$ Y: (kg), D: (cm^2), H: (m)
Juglans mandshurica	$\ln Y = -2.471 + 2.667 \times \ln(D)$
Quercus mongolica	$\ln Y = -3.453 + 1.004 \times \ln(D^2 + H)$
Tilia amurensis	$\ln Y = -3.771 + 1.013 \times \ln(D^2 + H)$
Other species	$\ln Y = -2.560 + 2.308 \times \ln(DBH) + 0.341 \times \ln H$ Y: (g), D: (cm^2), H: (m)

In the above formula, D is the chest height area, H is the tree height, and Y is the biomass (see table).

2.2.3. α-Diversity Calculation

Based on the importance of each species in the quadrat, the richness index, Shannon–Winner diversity index [30], Simpson diversity index [31], Shannon evenness index, Simpson evenness index, and Pielou evenness index (J) were calculated [32]. The richness index indicates the number of species in the quadrat, and the remaining indicators are calculated as follows:

$$Shannon - Winner\ index = -\sum_{i=1}^{m} P_i \times lnP_i$$

$$Pielou\ \text{eveness}\ index = -\frac{-\sum_{i=1}^{m} P_i logP_i}{logm}$$

$$Simpson\ index = 1 - \sum_{i=1}^{m} \frac{P_i(P_i - 1)}{N(\mathrm{N} - 1)}$$

$$\text{Shannon eveness}\ index = \frac{\text{Shannon} - \text{Winner index}}{\log(\mathrm{m})}$$

$$Simpson\ \text{eveness}\ index = \frac{\text{Simpsonindex}}{\log(\mathrm{m})}$$

In the above formula, m signifies the count of species in the quadrat, N signifies the total number of individuals of all species, and Pi signifies the importance value of the ith species in the quadrat.

2.3. Determination of Soil Physicochemical Properties

2.3.1. Soil Sampling

Topsoil within a depth range of 0–10 cm was gathered using the five-point sampling method, with the sampling locations positioned at the center point and the four corners of each quadrat. The samples were gathered from the five locations within each quadrat, blended thoroughly, and subsequently partitioned into two separate portions. This resulted in a total of 54 soil samples. Half of the samples were preserved in liquid nitrogen for subsequent molecular biological extraction and analysis, and the other half were kept at ambient temperature and promptly transported back to the laboratory for physicochemical evaluations.

2.3.2. Determination of Soil Indices

Soil pH was measured via the KCl dissolution electrode method, which is based on the conversion of electromotive force values to obtain pH values; it is highly accurate and suitable for soils of different pH ranges. A CN analyzer was employed to measure total nitrogen and total carbon content (Perkin-Elmer Optima8300), and the organic carbon content

(SOC) was determined via the potassium dichromate oxidation method [33]. Various ionic forms of nitrogen—including ammonia nitrogen (AN) and nitrate nitrogen (NAN)—were determined using the KCl leaching spectrophotometric method [34,35]. Total phosphorus (TP) was determined via microwave digestion spectrophotometry and available phosphorus (AP) by NH4F-HCl leaching spectrophotometry [36]. The soil's dry matter content was measured via the water content–gravimetric method [37]. Metrics of soil enzymatic activity, including cellulase (CA), urease (UA), β-glucosidase (GA), dehydrogenase (DHA), and acid phosphatase (ACP) were measured using the method of Lin et al. [38].

2.4. Soil Microbial Sequencing and Annotation

DNA extraction, amplification and purification: Briefly, 0.5 g of each soil sample was used to extract genomic DNA with the reagent kits (Manufacturer: Tiagen Biochemical Technology Co., Ltd., (Beijing, China) model: DP812), and then fungal ITS sequences (primer ITS1F: 5′-CTTGGTCATTTAGAGGAAGTAA-3; ITS2: 5′-GCTGCGTTCTTCATCGATGC-3′) were amplified [39]. The PCR products were purified and tested for purity and quantification. A unique barcode was used to separate each sample to prevent cross-contamination. The PCR system was as follows: 25 μL of Taq enzyme, 1 μL each of double-ended primers (10 mM), and 3 μL of template DNA (20 ng · μL^{-1}) supplemented with ddH_2O to 50 μL. The reaction conditions were as follows: pre-denaturation at 95 °C for 5 min, denaturation at 95 °C for 30 s, annealing at 50 °C for 30 s (extended to 72 °C for 40 s), and a total of 25 cycles. Finally, the reaction was extended to 72 °C for 7 min.

Sequencing and data processing **(1) Filtering of raw data:** Trimmomatic [40] was applied to filter dual-end sequencing files. For example, for a 50 bp window, if the average mass within the window was less than 20, the back-end base was truncated from the window. **(2) Identification and removal of primer sequences:** Cutadapt (Version 1.9.1) was utilized to identify primer sequences based on the specified parameters [41], which tolerated a maximum error rate of 20% and required a minimum coverage of 80%. **(3) Splicing of double-ended reads:** USEARCH (Version 10) [42] was employed for stitching double-ended reads from samples, adhering to a minimum overlap length of 10 bp, allowing a minimum similarity of 90% within the overlapping region, and tolerating a maximum of 5 bp error bases. **(4) Removal of chimera:** The chimera criteria split the query sequence into non-overlapping chunks, and compared these with the database. The best match for each block in the database was selected, and finally the two best parental sequences were selected. The sequence to be detected was compared with two parents. If the similarity between the parent sequence and the query sequence was greater than 80%, the query was judged to be a chimera. Chimeras were removed using UCHIME [43]. Each sample was separated using a unique barcode to prevent cross-contamination.

OTU clustering and species annotation: Cluster analysis was executed utilizing UPARSE with a 97% similarity threshold, followed by the application of the usearch command to eliminate redundant sequences and standalone operational taxonomic units (OTUs) during the process. Each clustered OTU was annotated into seven taxonomic classes of boundary, phylum, order, family, genus, and species by utilizing the sine method, referencing either the 16S rRNA Silva database or the ITS database in Unite. Consequently, community abundance tables were generated for each taxonomic rank.

2.5. Prediction of Fungal Function

FUNGuild (Fungi Functional Guild) within the R program was used to conduct functional analysis of fungal communities with OUT abundance > 1%, and the running results were downloaded and screened. In order to improve the reliability of our predictions of fungal functional groups, only two grades of confidence ("highly probable" and "probable") were retained [25].

2.6. Data Processing and Analysis

An ANOVA test conducted via the Scheffe method was used to analyze the differences between the plant diversity index, soil physicochemical properties, and abundance of soil microbial functional groups in three dimensions among different stands. The significance level was defined as 0.05. This analysis was implemented in the multicomp package of the R program. The "corr.test" function within the vegan package in R language was used to calculate the correlation coefficient (i.e., the Pearson coefficient) between the species composition of the plant community, diversity index, soil physicochemical properties, soil enzyme activity, and functional abundance of each dimension. The "Pheatmap" function was used to draw heatmaps. The RDA redundancy function within the vegan package of R software 4.2.2. was used to interpret the effect of species composition, diversity index, soil physicochemical properties, and enzyme activity on the functional structure of the fungal community. The Monte Carlo permutation test was used to analyze the significance of factors influencing each explanatory variable in the RDA model, and it was performed by means of the "anova.cca" function with 999 permutations. The correlation degree R^2 and significance degree P of each factor's contribution to the model were calculated using the "envfit" function, and the graph was completed in the R language ggplot2 package.

3. Results and Analysis

3.1. Changes in Plant Community Structure and Soil Properties

The results showed that, during the naturalization of the *Pinus tabulaeformis* plantation, the composition of the dominant species in the community tended to be stable, and the plantation gradually developed into a mixed community of plantation and natural forest with *Quercus mongolica, Q. variabilis,* and *Juglans mandshurica* as the dominant species. The diversity of tree layer increased, exceeding that of the natural secondary forest in the near-natural forest stage (Figure 1a–d). Soil pH value, dry matter content, and cellulase and urease activities showed an increasing trend. The organic carbon, ammonia nitrogen content, and acid phosphatase activity decreased. Total carbon, total nitrogen, total phosphorus, available phosphorus content, β-glucosidase, and dehydrogenase activity decreased first and then increased. The total organic carbon content and carbon–nitrogen ratio increased first and then decreased (Table 2).

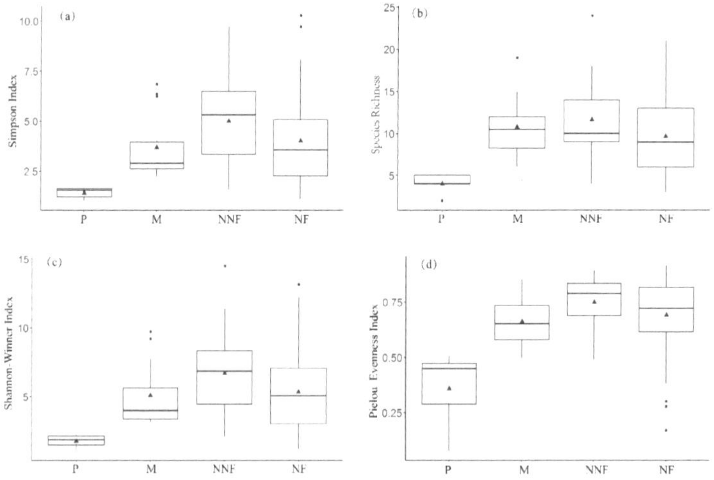

Figure 1. Characteristics of α-diversity indices in different forest stands. (**a**) Simpson index; (**b**) species richness; (**c**) Shannon–Winner index; (**d**) Pielou evenness index. Note: P represents pure *Pinus tabulaeformis* forests, M represents mixed forests, NNF represents near-natural forests, and NF represents natural forests.

Table 2. Soil chemical properties and enzyme activities of near-naturalized stand and natural secondary forest (SOC, soil organic carbon; AP, available phosphorus; TP, total phosphorus; TC, total carbon; TN, total nitrogen; C/N, carbon–nitrogen ratio; DW, dry matter content; CA, cellulase; UA, urease; GA, β-glucosidase; DHA, dehydrogenase; ACP, acid phosphatase).

Soil Indicator	P	M	NNF	NF
SOC (mg·kg^{-1})	79.40 ± 14.02	95.88 ± 13.18	79.94 ± 17.85	105.41 ± 21.03
AP (mg·kg^{-1})	38.58 ± 32.35	33.89 ± 43.06	45.40 ± 10.99	40.28 ± 33.67
TP (mg·kg^{-1})	545.80 ± 159.6	472.02 ± 88.10	563.10 ± 193.3	532.09 ± 159.4
NO_3^--N (mg·kg^{-1})	13.32 ± 2.26	9.97 ± 3.55	15.42 ± 3.73	21.17 ± 8.00
NH_4^+-N (mg·kg^{-1})	8.10 ± 0.72	8.72 ± 7.46	4.84 ± 3.05	10.50 ± 3.98
pH	4.79 ± 0.19	4.64 ± 0.34	5.85 ± 0.38	5.74 ± 0.88
DW (%)	95.71 ± 1.19	97.11 ± 0.97	96.73 ± 3.27	95.89 ± 1.90
TC (%)	7.18 ± 3.55	5.16 ± 0.37	7.15 ± 1.16	7.55 ± 2.78
TN (%)	0.52 ± 0.26	0.36 ± 0.04	0.56 ± 0.12	0.61 ± 0.06
C/N	13.87 ± 1.26	14.55 ± 0.71	12.97 ± 0.77	12.35 ± 0.86
CA (μg·g^{-1}·min^{-1})	0.10 ± 0.09	0.18 ± 0.05	0.20 ± 0.10	0.21 ± 0.06
UA (μg·g^{-1}·h^{-1})	5.23 ± 1.63	5.24 ± 1.81	9.61 ± 1.34	8.98 ± 2.78
GA (μg·g^{-1}·h^{-1})	0.83 ± 0.78	1.24 ± 0.98	1.01 ± 0.51	0.59 ± 0.54
DHA (μg·g^{-1}·h^{-1})	0.39 ± 0.35	0.61 ± 0.54	0.46 ± 0.37	0.33 ± 0.38
ACP (μg·g^{-1}·min^{-1})	5.83 ± 1.10	5.29 ± 1.29	4.36 ± 1.52	6.85 ± 3.73

3.2. Influence of Near-Naturalization on the Abundance of Fungal Functional Groups

The results showed that the abundance of ectomycorrhizal fungi, saprophytic fungi, and plant pathogens dominated in the functional mode groups. The abundance of plant pathogens in the *P. tabulaeformis* forest stage was significantly higher than in other stages, and the abundance of ectomycorrhizal fungi in the near-natural forest stage was significantly lower than in other stages (Figure 2). The abundance of saprophytic, pathophysiological, and symbiotic nutrition dominated in the vegetative mode group. The abundance of pathotrophic fungi in the *P. tabulaeformis* forest stage was significantly higher than that in other stages, which was consistent with the highest abundance in functional mode. The abundance of symbiotroph in the near-natural forest stage was significantly lower than in other stages, which was also consistent with the ectomycorrhizal fungi in the functional mode group (Figure 3). The abundance of microfungi accounted for 53.06% of the growth mode group, which was the dominant group. The abundance of clavarioid decreased with the naturalization process (Figure 4).

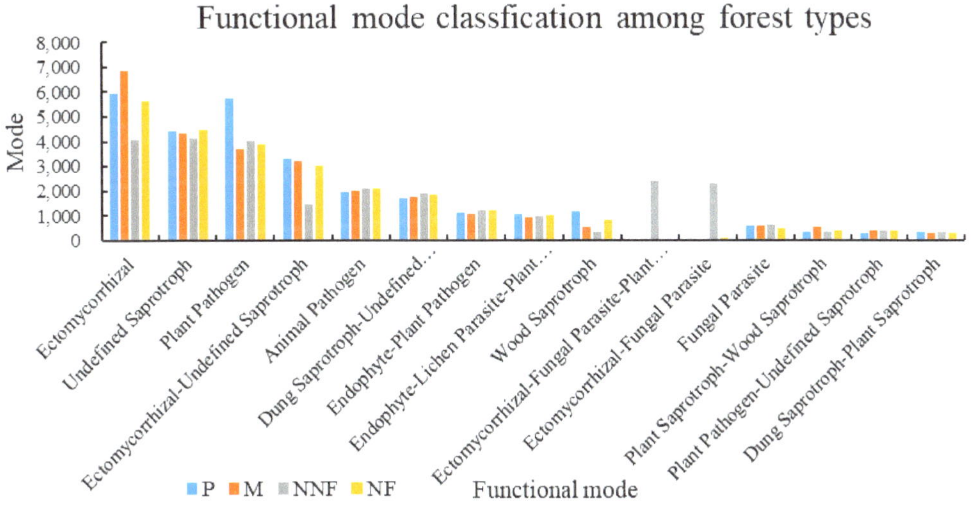

Figure 2. Changes in the abundance of functional mode of soil fungi during near-naturalization. Notes: P, *Pinus tabulaeformis* forest stage; M, mixed forest stage; NNF, near-natural forest stage; NF, natural secondary forest.

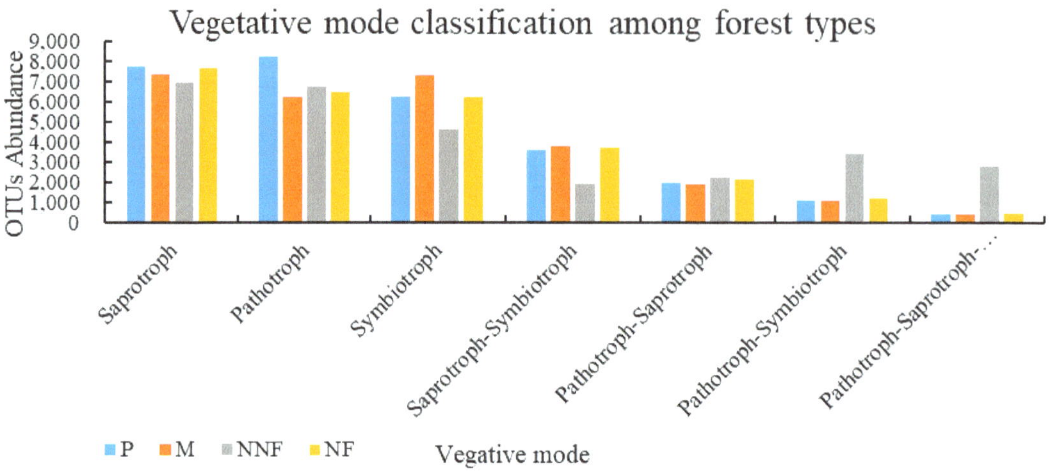

Figure 3. Changes in the abundance of soil fungal nutrient mode in near-naturalization P, *Pinus tabulaeformis* forest stage; M, mixed forest stage; NNF, near-natural forest stage; NF, natural secondary forest.

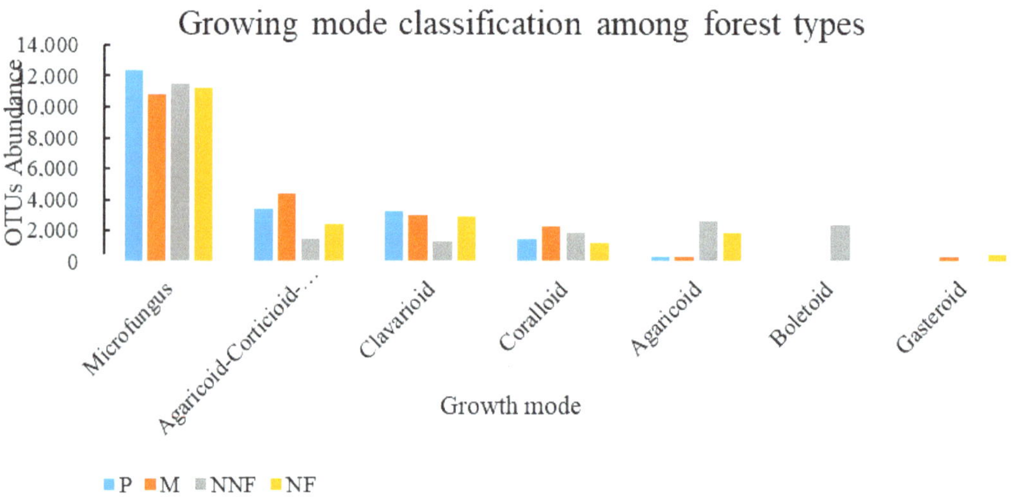

Figure 4. Changes in the abundance of soil fungal growth mode during near-naturalization. P, *P. tabulaeformis* forest stage; M, mixed forest stage; NNF, near-natural forest stage; NF, natural secondary forest.

3.3. Correlation between Vegetation and Soil Properties, and the Abundance of Soil Fungal Functional Groups

The diversity of the tree layer rarely correlated with the abundance of functional groups in the three groups. In the functional mode, the biomass of tree layer was significantly positively correlated with litter saprotroph and soil saprotroph, but significantly negatively correlated with the abundance of epiphyte–litter saprotroph (Figure 5). In the vegetative mode, the highest positive correlation coefficient was found in the biomass and the abundance pathotroph–saprotroph groups (Figure S1). In the growth mode, only the Shannon evenness index and Simpson evenness index showed a significant positive correlation with the abundance of yeast-type fungi (Figure S2).

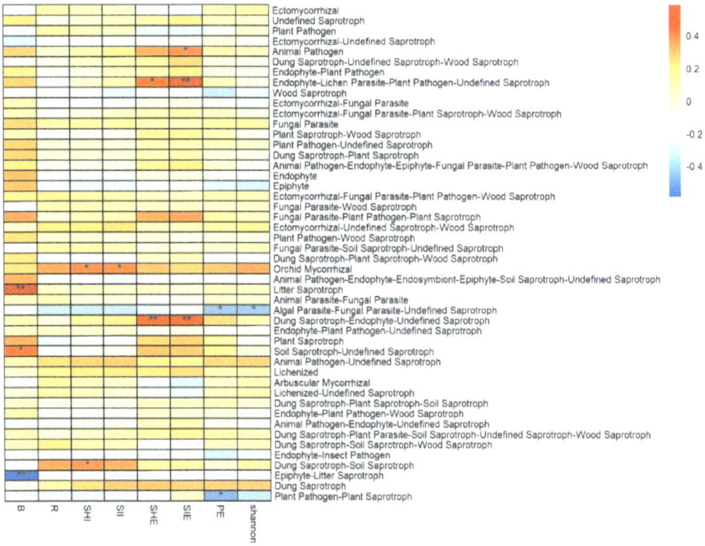

Figure 5. Correlation between tree layer diversity index and the abundance of soil fungi functional mode groups. B, biomass; R, richness index; SHI, Shannon—Wiener index; SII, Simpson index; SHE, Shannon—Wiener evenness index; SIE, Simpson evenness index; PE, Pielou evenness index. * means the significance level $p < 0.05$, ** means the significance level $p < 0.01$.

The composition of tree species has been shown to correlate significantly with the abundance of functional groups in the three dimensions. In the functional mode, the importance value of *Carpinus turczaninowii* showed a significant positive correlation with the ectomycorrhizal group, but a negative relationship with most other functional mode groups. The importance value of *Q. mongolica* was also significantly correlated with the abundance of various functional mode groups, and in some, the correlation coefficient exceeded 0.8. The importance value of *Tilia amurensis* significantly correlated with the ectomycorrhizal–unknown and saprophytic fungi (Figure 6). In the vegetative mode, the importance values of most tree species were negatively correlated with the most abundant types (e.g., the saprophytic and pathophysiological). The importance of *C. turczaninowii* was positively correlated with the abundance of symbiotroph, Tilia amurensis with saprotroph–symbiotroph, and Q. mongolica with pathotroph–symbiotroph and pathotroph–saprotroph–symbiotroph. (Figure S3). In the growth mode, no significant relationship could be found between the tree species composition and the most abundant types of growth mode (Figure S4).

Soil physicochemical properties showed a strong correlation with the abundance of functional and growth mode types, but a weak correlation with the abundance of vegetative mode. In the functional mode, ectomycorrhizal fungi with the highest abundance showed no significant correlation with soil physicochemical properties. Soil pH correlated positively with the abundance of orchid mycorrhizal and litter saprotroph. Soil organic carbon negatively correlated with plant pathogens and ectomycorrhizal–saprotroph–wood saprotroph, but correlated positively with wood saprotroph. Total nitrogen correlated positively with ectomycorrhizal–fungal–parasite–plant groups, pathogen–wood saprotroph bacteria, epiphytes, and plant pathogen–wood saprotroph. Soil ammonium nitrogen correlated positively with soil saprotroph–undefined saprotroph. All of the above relationships were significant (Figure 7). In the vegetative mode, the nitrate nitrogen content was significantly positively correlated with saprotroph, which was the most abundant type. Soil organic carbon content was negatively correlated with pathotroph, and a similar negative relationship was found between pathotroph–saprotroph and soil dry matter content (Figure S5). In the growth mode, available phosphorus content was significantly negatively correlated with the abundance of microfungus and agaricoid–corticioid–gasteroid–secotioid, but positively correlated with the abundance of clavarioid. In addition, the soil organic carbon, total phosphorus, total nitrogen content, pH, dry matter content, and carbon–nitrogen ratio were also strongly correlated with the abundance of the growth mode group (Figure S6).

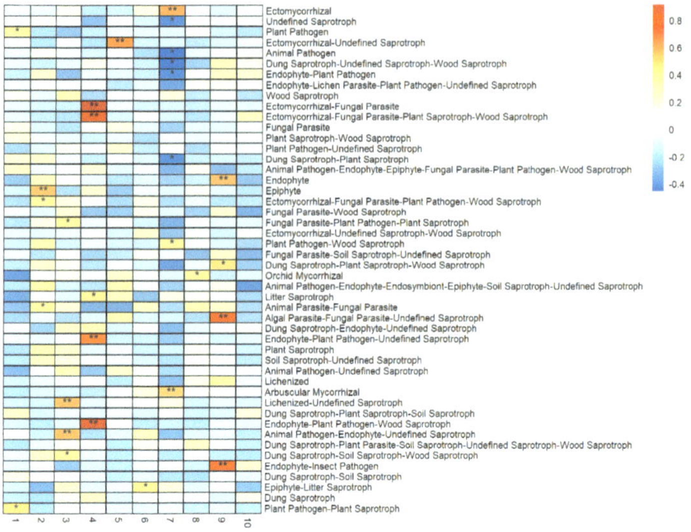

Figure 6. Correlation between species composition in the tree layer and the abundance of soil fungi functional mode groups. 1. *Pinus tabulaeformis;* 2. *Quercus variabilis;* 3. *Juglans mandshuruca;* 4. *Quercus mongolica;* 5. *Tilia amurensis;* 6. *Fraxinus chinensis;* 7. *Carpinus turczaninowii;* 8. *Quercus aliena;* 9. *Quercus dentata;* 10 *Tetradium daniellii.* * means the significance level $p < 0.05$, ** means the significance level $p < 0.01$.

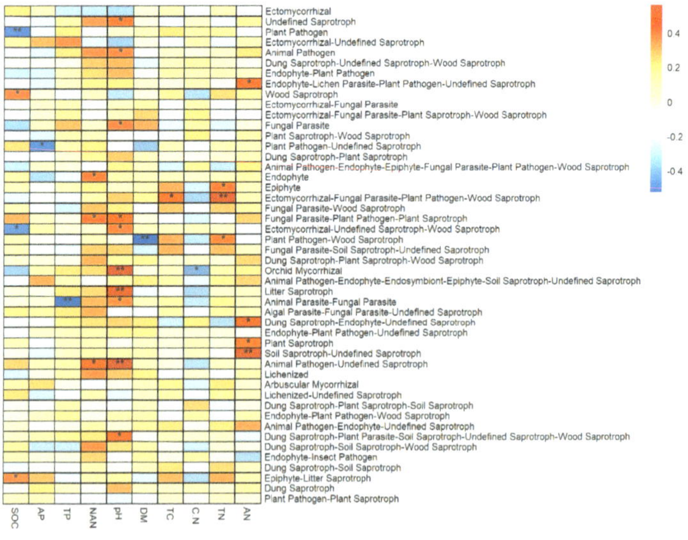

Figure 7. Correlation between soil properties and the abundance of fungal functional mode groups in soil. SOC, soil organic carbon; AP, available phosphorus; TP, total phosphorus; NAN, nitrate nitrogen; DM, dry matter content; TC, total carbon; C.N, carbon–nitrogen ratio; TN, total nitrogen; AN, ammonia nitrogen. * means the significance level $p < 0.05$, ** means the significance level $p < 0.01$.

Metrics of soil enzyme activity such as soil urease, acid phosphatase, and dehydrogenase were strongly correlated with functional mode groups, but cellulase showed no significant correlation. Dehydrogenase activity significantly positively correlated with the abundance of saprotroph (Figure 8). In the vegetative mode, no significant correlation was found between soil enzyme activity and highly abundant types (Figure S7). Among the growth mode groups, urease activity was significantly negatively correlated with the abundance of the agaricoid–corticioid–gasteroid–secotioid type (Figure S8).

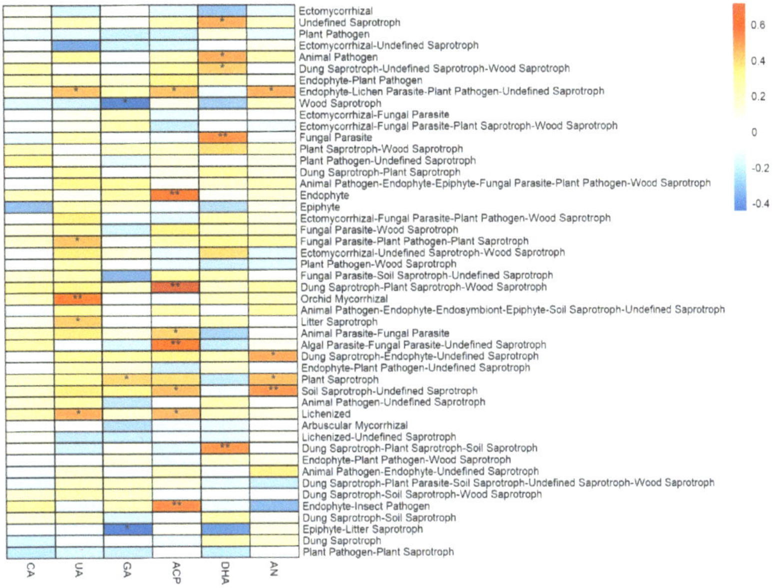

Figure 8. Correlation between soil enzyme activity and the abundance of soil fungi functional mode groups. CA, cellulase; UA, urease; GA, β-glucosidase; ACP, acid phosphatase; DHA, dehydrogenase. * means the significance level $p < 0.05$, ** means the significance level $p < 0.01$.

3.4. *Effects of Near-Naturalization on the Functional Structure of Fungal Community of Soil Fungi*

The results showed that the variation in community structure in the functional mode group and vegetative mode group in the first two axes of PcoA was 64.05% and 70.73%, respectively. With near-naturalization, the distribution of the co-ordinate points of the two modes on the PcoA axis 2 diverged, indicating that the degree of variation increased. On the contrary, the co-ordinate points of natural secondary forests were concentrated, indicating convergence of community structure. This indicated that the distribution of the functional mode and vegetative mode of soil fungi in areas without afforestation and restored near-naturalization was relatively stable. In the process of near-naturalization, the changes in the dominant plant species caused changes in soil nutrients and enzyme activity, which led to the recombination and distribution of soil fungi in functional and vegetative dimensions. The variation in the growth mode's group structure in the first two axes of PcoA reached 70.73%, and the community structure of the near-naturalized forest stages showed a tendency to diverge on both axes 1 and 2, while the growth mode group structure of the natural secondary forest still showed convergence (Figure 9).

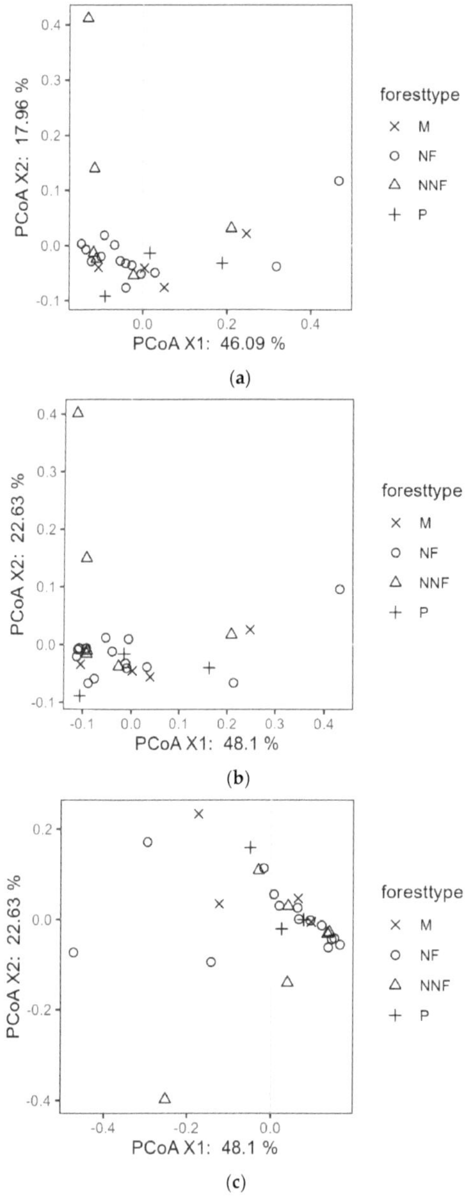

Figure 9. Differences in the functional structure of soil fungal communities at different stages of naturalization: (**a**) functional mode; (**b**) vegetative mode; (**c**) growth mode. P: *P. tabulaeformis* forest stage; M: mixed forest stage; NNF: near-natural forest stage; NF: natural secondary forest.

3.5. Factors Influencing the Near-Naturalization of the Soil Fungi Community's Functional Structure

The Monte Carlo test was used to explore the effects of the diversity and species composition of the tree layer, soil physicochemical properties, and enzyme activity on the functional structure of the fungal community. The results indicated a non-significant correlation (Table 3).

Table 3. Results of the Monte Carlo test used to evaluate the influence of plant and soil environmental factors on three dimensions of the fungal community's functional structure.

	Soil Physicochemical Properties	Plant Diversity	Plant Composition	Enzyme Activity
Functional mode	R = 0.021 p = 0.417	R = −0.130 p = 0.983	R = 0.098 p = 0.229	R = −0.099 p = 0.794
Vegetative mode	R = 0.048 p = 0.302	R = −0.128 p = 0.892	R = 0.055 p = 0.343	R = −0.107 p = 0.798
Growth mode	R = 0.053 p = 0.333	R = −0.151 p = 0.929	R = 0.111 p = 0.195	R = −0.114 p = 0.859

3.5.1. Influence of Tree Layer Diversity

The Monte Carlo test of RDA analysis showed that the indices of diversity had no significant effect on the functional structure of the three dimensions (Table 4). The two-dimensional RDA map showed that the diversity of the first two axes of the RDA model in functional mode, vegetative mode, and growth mode reached 50.89%, 79.33%, and 61.17%, respectively. For the functional mode, the biomass and evenness index correlated most strongly with structural changes in functional mode groups, especially the *P. tabulaeformis* forest stage and the mixed forest stage (Figure 10). For the vegetative mode, the evenness index also showed a strong correlation with structural changes, and the Shannon diversity index and biomass had a strong correlation with structural changes in natural secondary forests (Figure S9). For the growth mode, the community structure had the most significant response to the evenness index, which was consistent with the change between the *P. tabulaeformis* forest stage and the mixed forest stage. Meanwhile, the change in community structure from the mixed forest stage to the near-natural forest stage showed obvious consistency with both the Pielou evenness index and the Shannon diversity index (Figure S10).

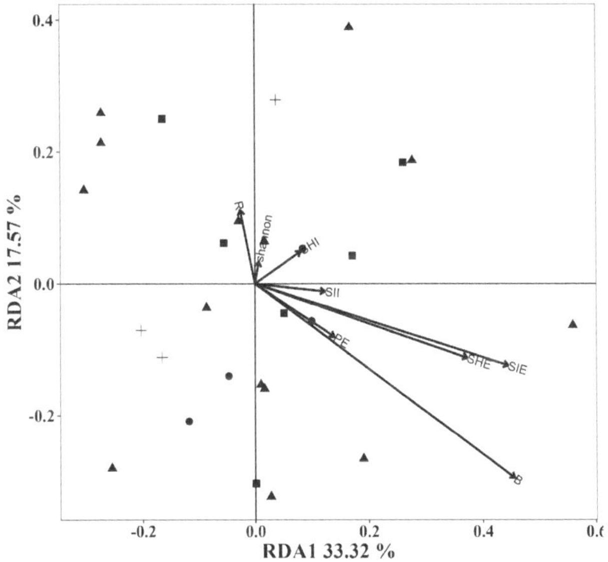

Figure 10. RDA analysis results of plant diversity on functional structure of soil fungal community in functional mode. Note: P: *P. tabulaeformis* forest stage; M: mixed forest stage; NNF: near-natural forest stage; NF: natural secondary forests; B: biomass; R: richness index; SHI: Shannon–Wiener index; SII: Simpson index; SHE: Shannon–Wiener evenness index; SIE: Simpson evenness index; PE Pielou evenness index.

Table 4. Monte Carlo test results of the RDA model and the influence of plant diversity factors on the functional structure of the soil fungal community in three dimensions. B, biomass; R, richness index; SHI, Shannon–Wiener index; SII, Simpson index; SHE, Shannon–Wiener evenness index; SIE, Simpson evenness index; PE, Pielou evenness index.

	Functional Mode		Vegetative Mode		Growth Mode	
	R^2	*p*	R^2	*p*	R^2	*p*
B	0.197	0.075	0.088	0.331	0.021	0.788
R	0.010	0.858	0.039	0.552	0.013	0.855
SHI	0.006	0.936	0.015	0.809	0.005	0.934
SII	0.009	0.905	0.006	0.938	0.017	0.801
SHE	0.095	0.291	0.031	0.710	0.050	0.548
SIE	0.134	0.176	0.075	0.387	0.113	0.265
PE	0.017	0.802	0.015	0.844	0.063	0.442
Shannon	0.001	0.999	0.042	0.588	0.031	0.662

3.5.2. Influence of Vegetation Composition

The Monte Carlo test results of RDA analysis showed that the changes in tree layer composition during near-naturalization had significant effects on soil fungal structure in three dimensions, among which the abundance of *Quercus mongolica* and *Carpinus turczaninowii* were the most significantly correlated (Table 5). RDA analysis revealed that the interpretation rates of species composition in the structure of functional, vegetative, and growth modes of soil fungi were 40.16%, 75.4%, and 48.25%, respectively. The changes in the functional and growth modes of the fungi community structure during near-naturalization responded positively to the increase in the importance of *Quercus mongolica* and *Carpinus turczaninowii* within the community (Figures 11 and S11). The changes in the structure of fungal vegetative mode mainly correlated positively with the changes in the importance value of *Carpinus turczaninowii* (Figure S12).

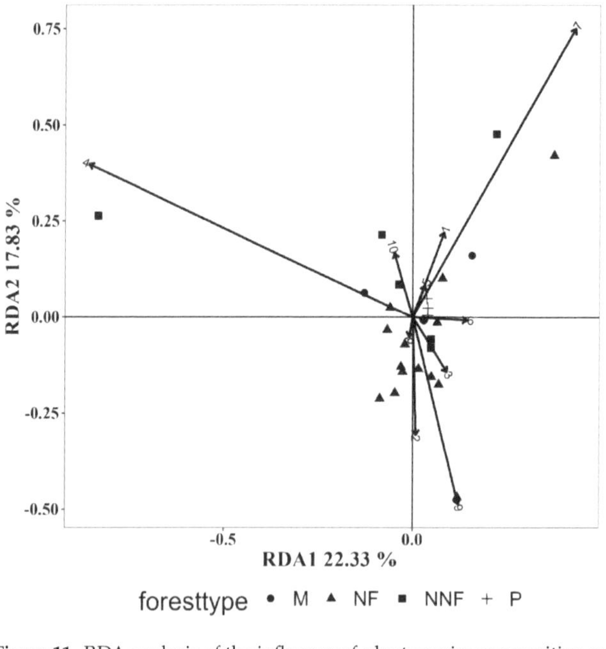

Figure 11. RDA analysis of the influence of plant species composition on the functional structure of the soil fungal community in functional mode. P, *P. tabulaeformis* forest stage; M, mixed forest stage; NNF, near-natural forest stage; NF, natural secondary forests; 1. *Pinus tabulaeformis*; 2. *Quercus variabilis*; 3. *Juglans mandshuruca*; 4. *Quercus mongolica*; 5. *Tilia amurensis*; 6. *Fraxinus chinensis*; 7. *Carpinus turczaninowii*; 8. *Quercus aliena*; 9. *Quercus dentata*; 10. *Tetradium daniellii*.

Table 5. Monte Carlo test of the RDA model and plant species composition factors influencing the functional structure of the soil fungal community in three dimensions. 1. *Pinus tabulaeformis*; 2. *Quercus variabilis*; 3. *Juglans mandshuruca*; 4. *Quercus mongolica*; 5. *Tilia amurensis*; 6. *Fraxinus chinensis*; 7. *Carpinus turczaninowii*; 8. *Quercus aliena*; 9. *Quercus dentata*; 10. *Tetradium daniellii*. Bold indicates a significance level ≤ 0.05.

	Functional Mode		Vegetative Mode		Growth Mode	
	R^2	p	R^2	p	R^2	p
1	0.025	0.696	0.050	0.473	0.052	0.487
2	0.046	0.529	0.011	0.899	0.110	0.241
3	0.021	0.739	0.058	0.414	0.019	0.781
4	0.832	**0.001**	0.867	**0.002**	0.754	**0.001**
5	0.004	0.887	0.116	0.134	0.034	0.530
6	0.019	0.688	0.050	0.378	0.026	0.676
7	0.349	**0.043**	0.410	**0.028**	0.303	**0.050**
8	0.001	0.984	0.015	0.839	0.000	0.998
9	0.145	0.125	0.000	0.984	0.020	0.574
10	0.018	0.761	0.016	0.834	0.087	0.316

3.5.3. Influence of Soil Physicochemical Properties

The Monte Carlo test results of RDA analysis showed that soil pH significantly affected the structure of the functional mode, while pH, organic carbon content, and soil water content significantly affected the structure of the vegetative mode. Soil organic carbon, total nitrogen content, and carbon to nitrogen ratio significantly affected the structure of the growth mode (Table 6). The RDA analysis showed that the model's interpretation of soil properties in the functional, vegetative, and growth modes returned values of 45.26%, 70.16%, and 51.52%, respectively. The functional mode of the community structure responded positively to the changes in soil pH during near-naturalization and showed a strong correlation with nitrate nitrogen content, however, it had an opposing relationship with changes in available phosphorus and organic carbon content (Figure 12). The variation in vegetative mode was most consistent with changes in dry matter content and C–N ratio during near-naturalization (Figure S13). Soil properties did not show significant effects on the structure of the growth mode during near-naturalization (Figure S14).

Table 6. Monte Carlo test of the RDA model and the influence of soil physicochemical properties on the functional structure of the soil fungal community in three dimensions. SOC, soil organic carbon; AP, available phosphorus; TP, total phosphorus; NAN, nitrate nitrogen; DM, dry matter content; TC, total carbon; CN, carbon–nitrogen ratio; TN, total nitrogen; AN, ammonia nitrogen. Bold indicates a significance level ≤ 0.05.

	Functional Mode		Vegetative Mode		Growth Mode	
	R^2	p	R^2	p	R^2	p
SOC	0.073	0.408	0.258	**0.037**	0.231	**0.037**
AP	0.020	0.815	0.128	0.203	0.100	0.291
TP	0.004	0.949	0.014	0.845	0.081	0.350
NAN	0.130	0.207	0.185	0.080	0.089	0.315
pH	0.394	**0.002**	0.232	**0.043**	0.180	0.092
DM	0.032	0.679	0.242	**0.031**	0.036	0.648
TC	0.074	0.390	0.027	0.739	0.138	0.161
CN	0.138	0.165	0.068	0.411	0.270	**0.019**
TN	0.135	0.171	0.042	0.599	0.249	**0.028**
AN	0.137	0.166	0.086	0.314	0.187	0.085

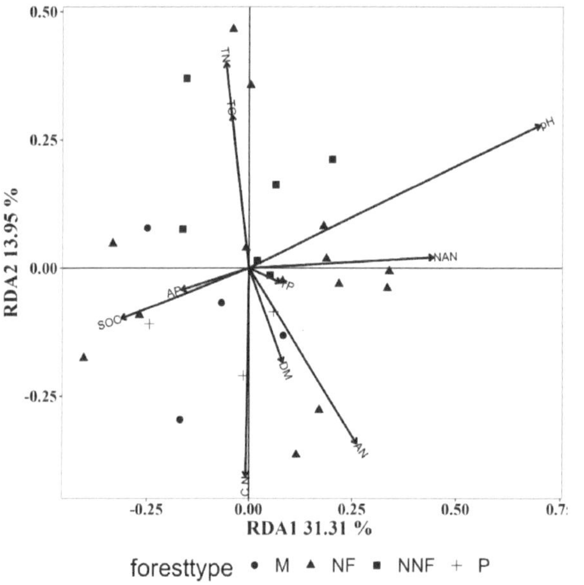

Figure 12. RDA analysis of the influence of soil physicochemical properties on the functional structure of the soil fungal community in functional mode. P, *P. tabulaeformis* forest stage; M, mixed forest stage; NNF, near-natural forest stage; NF, natural secondary forests; SOC, soil organic carbon; AP, available phosphorus; TP, total phosphorus; NAN, nitrate nitrogen; DM, dry matter content; TC, total carbon; CN, carbon nitrogen ratio; TN, total nitrogen; AN, ammonia nitrogen.

3.5.4. Influence of Soil Enzyme Activity

The Monte Carlo results of the RDA analysis showed that soil enzyme activities had significant effects on both the functional mode and growth mode, among which the activities of UA, ACP, and DHA significantly affected the former, and the activities of UA and GA significantly affected the latter (Table 7). The RDA analysis showed that the structural variation of functional mode was consistent with change in ACP activity during the progression from the *P. tabulaeformis* forest stage to the mixed forest stage. The community structure of the functional mode in the near-natural forest stage and the natural secondary forests was convergent on the RDA2 axis and was not significantly affected by soil enzyme activity (Figure 13). The structure of the soil fungal vegetative mode in the *P. tabulaeformis* forest stage and mixed forest stage was not significantly affected by soil enzyme activity but clearly changed in the near-natural forest stage, which was consistent with the change in soil GA activity (Figure S15). The community structure of the growth mode of soil fungi showed a tendency to converge gradually under the influence of enzyme activity during near-naturalization (Figure S16).

Table 7. Monte Carlo test of the RDA model and the influence of soil enzyme activity on the functional structure of the soil fungal community in three dimensions. Note: CA, cellulase; UA, urease; GA, β-glucosidase; ACP, acid phosphatase; DHA, dehydrogenase; AN, ammonia nitrogen. Bold indicates a significance level ≤ 0.05.

	Functional Mode		Vegetative Mode		Growth Mode	
	R^2	*p*	R^2	*p*	R^2	*p*
CA	0.062	0.427	0.022	0.768	0.044	0.610
UA	0.243	**0.027**	0.221	0.053	0.206	**0.047**
GA	0.127	0.185	0.107	0.269	0.371	**0.005**
ACP	0.520	**0.012**	0.154	0.149	0.123	0.183
DHA	0.438	**0.001**	0.159	0.126	0.163	0.111
AN	0.148	0.157	0.042	0.621	0.201	0.062

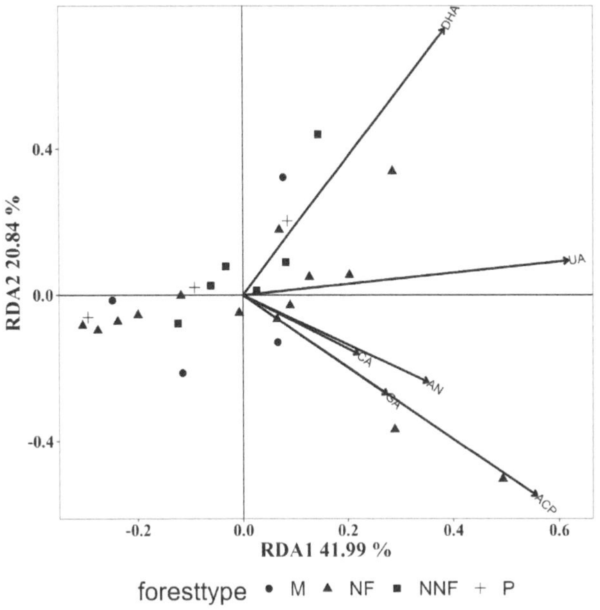

Figure 13. RDA analysis of the influence of soil enzyme activity on the functional structure of the soil fungal community in functional mode. P, *P. tabulaeformis* forest stage; M, mixed forest stage; NNF, near-natural forest stage; NF, natural secondary forests; CA, cellulase; UA, urease; GA, β-glucosidase; ACP, acid phosphatase; DHA, dehydrogenase.

4. Discussion

This study shows that the functional abundance and structure of soil fungal communities exhibit pronounced changes during the process of near-natural succession. Among them, the abundance of pathogenic fungi in the functional mode and that of pathophysiological fungi in the nutritional mode are highest in *P. tabuliformis* forests. The process of near-natural succession accelerates the variation in the functional structure of fungi, while natural secondary forests tend to be stable. This change is consistent with the increase in the complexity of plant communities during succession, and this study indicates that functional structure is significantly influenced by species composition and soil properties.

4.1. Effects of Near-Naturalization on the Abundance of Fungal Functional Groups

Our results revealed that ectomycorrhizal fungi, saprophytic fungi, and plant pathogens exhibited the highest abundance ratio in the functional mode. Similarly, for the vegetative mode, saprophytic, pathologic, and symbiotic nutrients accounted for the highest abundance ratio, and such results are consistent with other research findings on boreal coniferous forests, global forests, and forests in eastern China [20,44–46]. Our study demonstrated that the abundance of ectomycorrhizal fungi is significantly lower in natural secondary forests than in near-naturalization succession stages, which can be linked to a decrease in soil pH [47]. Previous studies have shown that the relative abundance of soil ectomycorrhiza increased during the progression of boreal coniferous forests [44], which aligns with the progression we observed from *P. tabulaeformis* forests to mixed forests. Additional studies have indicated that soil rich in ECM fungi tends to be more acidic, and a low soil pH value can prevent the decline of the ECM system. This could explain the decline in the relative abundance of EMC fungi from the mixed forest stage to the near-natural forest stage [48]. Soil with a high abundance of ectomycorrhizal fungi is conducive to enhancing overall ecological functions [49,50], particularly improving nutrients and water absorption by plants. Larger networks of EM fungi can directly degrade and acquire organic forms of nitrogen and are expected to dominate 'slow' nitrogen-cycling ecosystems with low levels of inorganic nutrients [51–53]. The fluctuation of EM fungi observed in succession in our study may indicate a change in the nitrogen cycle. This study showed that the abundance of saprophytic fungi in both functional and vegetative modes decreased with near-naturalization and was lower than that in natural secondary forests. Saprophytic fungi, as important decomposers in soil, play a crucial role in nutrient cycling [54], and also serve to enhance plant water absorption [55]. This may suggest that the function and nutrient structure of soil micro-organisms during near-naturalization are, to some extent, unfavorable for the decomposition of nutrients. The decline in the abundance of pathogenic fungi is consistent with Jiang's study [44] and may be attributed to nutrient competition between mycorrhizal fungi and free micro-organisms, during which EM fungi can secrete enzymes to degrade organic matter. Therefore, the nitrogen utilization path is more concise than the traditional nitrogen mineralization path [21]. Such concision intensifies the nitrogen restriction of free micro-organisms and inhibits the process of decomposing organic matter that they undertake [56]. Our research is also supported by findings indicating that increases in species diversity and richness are negatively correlated with the abundance of soil pathogens [57]. The decrease in the abundance of clavarioid during near-naturalization in the growth mode may be related to changes in soil factors and competition among species.

4.2. *Effects of Environmental Factors on the Abundance of Functional Groups of Fungi*

The results of this study showed that the biomass of the tree layer was significantly correlated with the abundance of saprophytic fungi groups during near-naturalization. This may due to increased biomass and biodiversity, which enhance litter input to the soil, providing suitable resources for fungi. Relevant studies confirm that plant diversity significantly drives changes in fungal composition [44]. Our study also revealed that species composition is an influential factor, with *Carpinus turczaninowii*'s importance positively correlating with the abundance of ectomycorrhizal fungi, as supported by their tree-host specificity [58] and role in symbiotic relationships [59]. Soil pH, organic carbon, and soil nitrogen contents significantly affect functional dimensions, as soil quality is closely linked to litter changes caused by species shifts, leading to microbial community recombination [60]. Thus, changes in tree species, rather than soil nutrients, at different successional stages may cause fungal changes. pH positively correlates with the abundance of litter saprophytes, which have been confirmed as a main factor affecting fungal functional modes during forest conversion in Southeast Asia [47]. Soil enzyme activities such as urease, acid phosphatase, and dehydrogenase correlate with functional mode groups, with fungi being more sensitive to these changes than other microbial groups [61]. Some studies have shown that saprophytic fungi are most closely related to higher soil nutrient levels. This study shows that variation in dehydrogenase activity aligns with the abundance of saprophytic fungi, possibly due to higher nutrient levels in certain soils [12].

4.3. *Effects of Near-Naturalization on the Functional Structure of the Fungal Community*

This study demonstrated that, during near-naturalization, the co-ordinate points representing the fungal community's functional, vegetative, and growth mode structures exhibited a tendency to diverge. This divergence likely stems from the functional structure of the fungal community adapting to the transformations occurring within these three dimensions during the near-naturalization process [62]. The functional characteristics of the fungal community remained relatively stable in the *Pinus tabulaeformis* forest stage and the natural secondary forests that had not undergone near-naturalization. This stability can be attributed to the consistent quality and quantity of litter returned to the soil by plants, mirroring the long-term adaptability of species composition and soil environmental stability. No significant differences were observed in the functional structure of fungal community across the three stages of near-naturalization, as assessed along these three dimensions, suggesting that changes in soil fungal functional structure are not solely driven by alterations in above-ground species, but are also influenced by multiple complex factors.

4.4. Effects of Environmental Factors on the Functional Structure of the Soil Fungal Community

With respect to the influence of environmental factors on fungal community functional structure, our analysis indicates that, while vegetative composition, plant diversity, soil properties, and soil enzyme activity do not show an overarching significant effect on the structure of fungal functional groups, notable effects can be identified within each of these categories. Prior research underscores the ability of tree species composition to modulate soil's abiotic attributes, such as pH and nutrient availability [63], which serve as pivotal links between plant vitality and soil microbial community dynamics [64,65]. *Quercus Mongolica* and *Carpinus turczaninowii* emerged as highly influential across multiple functional dimensions. This observation may stem from the distinct role played by *Quercu mongolica* in altering community composition during near-naturalization, and the rapid regeneration of *Carpinus turczaninowii* under forest conditions. In terms of soil nutrients' effects on the fungal community's functional structure, soil pH, total nitrogen content, and organic carbon are well-established and critical determinants of fungal community composition and function [66,67].

5. Conclusions

This study focused on variation in the functional structure of the soil fungal community and the role of environmental drivers during the near-naturalization of coniferous plantations in the North Warm Temperate Zone. Our results showed that variation in fungal functional structure intensifies as near-naturalization progresses, and is significantly influenced by changes in species composition, particularly *Quercus mongolica* and *Carpinus turczaninowii*. Additionally, soil pH, dry matter content, carbon-to-nitrogen ratio, and nitrate content also significantly affect the functional structure of fungi. This study provides a new perspective that will help to develop our collective understanding of the functional adaptation of soil fungi to forest succession, and the interaction of above- and below-ground mechanisms in artificial forest ecosystems.

Supplementary Materials: The following supporting information can be downloaded at: https://www.mdpi.com/article/10.3390/f15091540/s1, Figure S1: Correlation between tree layer diversity index and the abundance of soil fungal vegetative mode. Note: B Biomass; R Richness index; SHI Shannon-Winner index; SII Simpson index; SHE Shannon-Winner evenness index; SIE: Simpson evenness index; PE Pielou evenness index; Figure S2: Correlation between tree layer diversity index and the abundance of soil fungal growth mode Note: B Biomass; R Richness index; SHI Shannon-Winner index; SII Simpson index; SHE Shannon-Winner evenness index; SIE: Simpson evenness index; PE Pielou evenness index; Figure S3: Correlation between tree composition in the tree layer and the abundance of soil fungal vegetative mode Note: 1 *Pinus tabulaeformis;* 2 *Quercus variabilis;* 3. *Juglans mandshuruca;* 4. *Quercus mogolica;* 5. *Tilia amurensis;* 6. *Fraxinus chinensis;* 7. *Carpinus turczaninowii;* 8. *Quercus aliena;* 9. *Quercus dentata;* 10. *Tetradium daniellii;* Figure S4: Correlation between species composition in the tree layer and the abundance of soil fungal growth mode Note: 1 *Pinus tabulaeformis;* 2 *Quercus variabilis;* 3. *Juglans mandshuruca;* 4. *Quercus mogolica;* 5. *Tilia amurensis;* 6. *Fraxinus chinensis;* 7. *Carpinus turczaninowii;* 8. *Quercus aliena;* 9. *Quercus dentata;* 10 *Tetradium daniellii;* Figure S5: Correlation between soil properties and the abundance of soil fungal vegetative mode Notes: SOC Soil organic carbon; AP Available phosphorus; TP Total phosphorus; NAN Nitrate nitrogen; DM Dry matter content; 89TC Total carban; C.N Carbon nitrogen ratio; TN Total nitrogen; AN Ammonia nitrogen; Figure S6: Correlation between soil properties and the abundance of soil fungal growth mode Notes: SOC: Soil organic carbon; AP Available phosphorus; TP Total phosphorus; NAN Nitrate nitrogen; DM Dry matter content; TC Total carbon; C.N Carbon nitrogen ratio; TN Total nitrogen; AN Ammonia nitrogen; Figure S7: Correlation between soil enzyme activity and the abundance of soil fungal trophic mode groups Notes: CA Cellulase; UA Urease; GA β-Glucosidase; ACP Acid phosphatase; DHA Dehydrogenase; Figure S8:Correlation between soil enzyme activity and the abundance of soil fungal growth mode Notes: CA Cellulase; UA Urease; GA β-Glucosidase;

ACP Acid phosphatase; DHA Dehydrogenase; Figure S9: RDA analysis results of plant diversity on functional structure of soil fungal community in vegetative mode; Note: P *P. tabulaeformis* forest stage, M mixed forest stage, NNF near-natural forest stage; NF natural secondary forests; B Biomass; R Richness index; SHI Shannon-Winner index; SII Simpson index; SHE Shannon-Winner evenness index; SIE Simpson evenness index; PE Pielou evenness index; Figure S10: RDA analysis results of plant diversity on functional structure of soil fungal community in growth mode; Note: P *P. tabulaeformis* forest stage, M mixed forest stage, NNF near-natural forest stage; NF natural secondary forests; B Biomass; R Richness index; SHI Shannon-Winner index; SII Simpson index; SHE Shannon-Winner evenness index; SIE Simpson evenness index; PE Pielou evenness index; Figure S11: RDA analysis results of plant species composition on functional structure of soil fungal community in growth mode; Note P: *P. tabulaeformis* forest stage; M mixed forest stage; NNF near-natural forest stage; NF: natural secondary forests; 1 *Pinus tabu*-laeformis; 2 *Quercus variabilis*; 3. *Juglans mandshuruca*; 4. *Quercus mogolica*; 5. *Tilia amurensis*; 6. *Fraxinus chinensis*; 7. *Carpinus turcza-ninowii*; 8. *Quercus aliena*; 9. *Quercus dentata*; 10 *Tetradium daniellii*; Figure S12: RDA analysis results of plant species composition on functional structure of soil fungal community in vegetative mode; Note P: *P. tabulaeformis* forest stage; M mixed forest stage; NNF near-natural forest stage; NF: natural secondary forests; 1 *Pinus tabulaeformis*; 2 *Quercus variabilis*; 3. *Juglans mandshuruca*; 4. *Quercus mogolica*; 5. *Tilia amurensis*; 6. *Fraxinus chinensis*; 7. *Carpinus turczaninowii*; 8. *Quercus aliena*; 9. *Quercus dentata*; 10 *Tetradium daniellii*; Figure S13: RDA analysis results of soil physicochemical properties on functional structure of soil fungal community in vegetative mode; Note: P *P. tabulaeformis* forest stage, M mixed forest stage, NNF near-natural forest stage; NF natural secondary forests; SOC Soil organic carbon AP Available phosphorus TP Total phosphorus; NAN Nitrate nitrogen; DM Dry matter content; TC Total carban; C.N Carbon nitrogen ratio; TN Total nitrogen; AN Ammonia nitrogen; Figure S14: RDA analysis results of soil physicochemical properties on functional structure of soil fungal community in growth mode; Note: P *P. tabulaeformis* forest stage, M mixed forest stage, NNF near-natural forest stage; NF natural secondary forests; SOC Soil organic carbon AP Available phosphorus TP Total phosphorus; NAN Nitrate nitrogen; DM Dry matter content; TC Total carban; C.N Carbon nitrogen ratio; TN Total nitrogen; AN Ammonia nitrogen; Figure S15: RDA analysis results of soil enzyme activity on functional structure of soil fungal community in vegetative mode; Note: P *P. tabulaeformis* forest stage; M mixed forest stage; NNF near-natural forest stage; NF natural secondary forests; CA Cellulase; UA Urease; GA β-Glucosidase; ACP Acid phosphatase; DHA Dehydrogenase; Figure S16: RDA analysis results of soil enzyme activity on functional structure of soil fungal community in growth mode; Note: P *P. tabulaeformis* forest stage; M mixed forest stage; NNF near-natural forest stage; NF natural secondary forests; CA Cellulase; UA Urease; GA β-Glucosidase; ACP Acid phosphatase; DHA Dehydrogenase; Table S1. Basic information of the community quadrats.

Author Contributions: Methodology, C.C.; Investigation, Z.Q., H.L., C.C. and C.L.; Data curation, Z.Q. and H.L.; Writing—original draft, Z.Q.; Writing—review & editing, Z.Q., C.C., C.L. and J.S.; Funding acquisition, J.S. All authors have read and agreed to the published version of the manuscript.

Funding: This research received no external funding.

Data Availability Statement: The raw data supporting the conclusions of this article will be made available by the authors on request.

Conflicts of Interest: The authors declare no conflict of interest.

References

1. Sasmito, S.-D.; Kuzyakov, Y.; Lubis, A.-A.; Murdiyarso, D.; Hutley, L.-B.; BAchri, S.; Friess, D.-A.; Martius, C.; Borchard, N. Organic carbon burial and sources in soils of coastal mudflat and mangrove ecosystems. *Catena* **2020**, *187*, 104414. [CrossRef]
2. Li, X.-J.; Li, X.-R.; Wang, X.-P.; Yang, H.-T. Changes in soil organic carbon fractions after afforestation with xerophytic shrubs in the Tengger Desert, northern China. *Eur. J. Soil. Sci.* **2016**, *67*, 184–195. [CrossRef]
3. Zhou, Q.-Q.; Li, F.; Cai, X.-A.; Rao, X.-Q.; Zhou, L.-X.; Liu, Z.-F.; Lin, Y.-B.; Fu, S.-L. Survivorship of plant species from soil seedbank after translocation from subtropical natural forests to plantation forests. *For. Ecol. Manag.* **2019**, *432*, 741–747. [CrossRef]
4. Chen, G.-P.; Gao, Z.-Y.; Zu, L.-H.; Tang, L.-L.; Yang, T.; Feng, X.-M.; Zhao, T.-J.; Shi, F.-C. Soil aggregate characteristics and stability of soil carbon stocks in a *Pinus tabulaeformis* plantation. *New For.* **2017**, *48*, 837–853. [CrossRef]
5. Peay, K.-G.; Baraloto, C.; Fine, P.-V.-A. Strong coupling of plant and fungal community structure across western Amazonian rainforests. *ISME J.* **2013**, *7*, 1852–1861. [CrossRef]

6. Bahram, M.; Plme, S.; Kljalg, U.; Zarre, S.; Tedersoo, L. Regional and local patterns of ectomycorrhizal fungal diversity and community structure along an altitudinal gradient in the Hyrcanian forests of northern Iran. *New Phytol.* **2015**, *193*, 465–473. [CrossRef] [PubMed]
7. Holtkamp, R.; van der Wal, A.; Kardol, P.; van der Putten, W.-H.; de Ruiter, P.-C.; Dekker, S.-C. Modelling C and N mineralisation in soil food webs during secondary succession on ex-arable land. *Soil Biol. Biochem.* **2021**, *43*, 251–260. [CrossRef]
8. Kardol, P.; Todd, D.-E.; Hanson, P.-J.; Mulholland, P.-J. Long-term successional forest dynamics: Species and community responses to climatic variability. *J. Veg. Sci.* **2021**, *21*, 627–642. [CrossRef]
9. Sun, C.; Liu, G.; Xue, S. Natural succession of grassland on the Loess Plateau of China affects multifractal characteristics of soil particle-size distribution and soil nutrients. *Ecol. Res.* **2016**, *31*, 891–902. [CrossRef]
10. Jaroslav, Š.; Petra, D.; Michaela, U.; Mirka, P.; Tomáš, C.; Jan, F.; Petr, B. Dominant trees affect microbial community composition and activity in post-mining afforested soils. *Soil Biol. Biochem.* **2013**, *56*, 105–115. [CrossRef]
11. Tedersoo, L.; Bahram, M.; Polme, S.; Kõljalg, U.; Yorou, N.S.; Wijesundera, R.; Ruiz, L.V.; Vasco-Palacios, A.M.; Thu, P.Q.; Suija, A.; et al. Global diversity and geography of soil fungi. *Science* **2014**, *346*, 1078. [CrossRef]
12. Wu, D.; Zhang, M.-M.; Peng, M.; Sui, X.-H.; Li, W.; Sun, G.-Y. Variations in soil functional fungal community structure associated with pure and mixed plantations in Typical Temperate forests of China. *Front. Microbiol.* **2019**, *10*, 1636. [CrossRef] [PubMed]
13. Qiu, Z.-L.; Shi, C.; Zhang, M.; Shi, F.-C. Effects of close-to-nature management of plantation on the structure and ecological functions of soil microorganisms with different habitat specialization. *Plant Soil* **2022**, *482*, 347–367. [CrossRef]
14. Flores-Rentería, D.; Rincón, A.; Valladares, F.; Yuste, J. Agricultural matrix affects differently the alpha and beta structural and functional diversity of soil microbial communities in a fragmented mediterranean holm oak forest. *Soil Biol. Biochem.* **2016**, *92*, 79–90. [CrossRef]
15. Žifčáková, L.; Vetrovsky, T.; Howe, A.; Baldrian, P. Microbial activity in forest soil reflects the changes in ecosystem properties between summer and winter. *Environ. Microbiol.* **2016**, *18*, 288–301. [CrossRef] [PubMed]
16. Frac, M.; Hannula, S.-E.; Bełka, M.; Jędryczka, M. Fungal biodiversity and their role in soil health. *Front. Microbiol.* **2018**, *9*, 707. [CrossRef] [PubMed]
17. Genre, A.; Lanfranco, L.; Perotto, S.; Bonfante, P. Unique and common traits in mycorrhizal symbioses. *Nat. Rev. Microbiol.* **2020**, *18*, 649–660. [CrossRef]
18. Tomer, A.; Singh, R.; Singh, S.-K.; Dwivedi, S.-A.; Reddy, C.-U.; Ram, M.; Keloth, A.; Rachel, R. Role of fungi in bioremediation and environmental sustainability. In *Fungal Biology*; Springer: Cham, Switzerland, 2021; pp. 187–200. [CrossRef]
19. Zhang, Z.; Yuan, Y.; Liu, Q.; Yin, H. Plant nitrogen acquisition from inorganic and organic sources via root and mycelia pathways in ecto mycorrhizal alpine forests. *Soil Biol. Biochem.* **2019**, *136*, 107517. [CrossRef]
20. Teste, F.-P.; Jones, M.-D.; Dickie, I.-A. Dual-mycorrhizal plants: Their ecology and relevance. *New Phytol.* **2020**, *225*, 1835–1851. [CrossRef]
21. Averill, C.; Turner, B.L.; Finzi, A.C. Mycorrhiza-mediated competition between plants and decomposers drives soil carbon storage. *Nature* **2014**, *505*, 543–545. [CrossRef]
22. Li, X.-L.; Qu, Z.-L.; Zhang, Y.-M.; Ge, Y.; Sun, H. Soil fungal community and potential function in different forest ecosystems. *Diversity* **2022**, *14*, 520. [CrossRef]
23. Frey, S.-D. Mycorrhizal Fungi as Mediators of Soil Organic Matter Dynamics. *Annu. Rev. Ecol. Evol. Syst.* **2019**, *50*, 237–259. [CrossRef]
24. Chen, L.; Xiang, W.-H.; Wu, H.-L.; Ouyang, S.; Lei, P.-F.; Hu, Y.-J.; Ge, T.-D.; Ye, J.; Kuzyakov, Y. Contrasting patterns and drivers of soil fungal communities in subtropical deciduous and evergreen broadleaved forests. *Appl. Microbiol. Biotechnol.* **2019**, *103*, 5421–5433. [CrossRef]
25. Nguyen, N.-H.; Song, Z.-W.; Bates, S.-T.; Branco, S.; Tedersoo, L.; Menke, J.; Schilling, J.-S.; Kennedy, P.-G. FUNGuild: An open annotation tool for parsing fungal community datasets by ecological guild. *Fungal Ecol.* **2016**, *20*, 241–248. [CrossRef]
26. Bello, A.; Wang, B.; Zhao, Y.; Yang, W.; Ogundeji, A.; Deng, L.; Egbeagu, U.-U.; Yu, S.; Zhao, L.-Y.; Li, D.-T.; et al. Composted biochar affects structural dynamics, function and co-occurrence network patterns of fungi community. *Sci. Total Environ.* **2021**, *775*, 145672. [CrossRef]
27. Põlme, S.; Abarenkov, K.; Nisson, R.-H.; Lindahl, B.D.; Clemmensen, K.E.; Kauserud, H.; Nguyen, N.; Kjøller, R.; Bates, S.T.; Baldrian, P.; et al. FungalTraits: A user-friendly traits database of fungi and fungus-like stramenopiles. *Fungal Divers.* **2020**, *105*, 1–16. [CrossRef]
28. Qiu, Z.-L.; Zhang, M.; Wang, K.-F.; Shi, F.-C. Vegetation community dynamics during naturalized developmental restoration of *Pinus tabulaeformis* plantation in North warm temperate zone. *J. Plant Ecol.* **2023**, *16*, rtac102. [CrossRef]
29. Feroz, S.-M.; Hagihara, Y.-A.; Hagihara, A. Stand stratification and woody species diversity of a subtropical forest in limestone habitat in the northern part of Okinawa Island. *J. Plant Res.* **2008**, *121*, 329–337. [CrossRef]
30. Shannon, C.-E.; Weaver, W. *The Mathematical Theory of Communication*; University of Illinois Press: Urbana, IL, USA, 1949.
31. Simpson, E.-H. Measurement of diversity. *Nature* **1949**, *163*, 688. [CrossRef]
32. Pielou, E.-C. An Introduction to mathematical ecology. *BioScience* **2011**, *24*, 7–12.
33. *HJ615–2011b*; Soil Determination of Orgaic Carbon-Potassium Dichromate Oxidation Spectrophotometric Method. Ministry of Environmental Protection, PRC: Beijing, China, 2011. (In Chinese)

34. *HJ634–2012*; Soil-Determination of Ammonium, Nitrite and Nitrate by Extraction with Potassium Chloride Solution-Spectrophotometric Methods. Ministry of Environmental Protection, PRC: Beijing, China, 2012. (In Chinese)
35. *GB/T 32737–2016*; Determination of Nitrate Nitrogen in Soil-Ultraviolet Spectrophotometry Method. Standardization Administration of China: Beijing, China, 2016. (In Chinese)
36. *NY/T 1121.7–2014*; Soil Testing-Method for Determination of Available Phosphorus in Soil. Ministry of Agriculture, PRC: Beijing, China, 2012. (In Chinese)
37. *HJ613–2011a*; Soil-Determination of Dry Matter and Water Content-Gravimetric Method. Ministry of Environmental Protection, PRC: Beijing, China, 2011. (In Chinese)
38. Lin, X.-G. *Principles and Methods of Soil Microbiology Research*; High Education Press: Beijing, China, 2010. (In Chinese)
39. White, T.-J.; Bruns, T.; Lee, S.; Taylor, J. Amplification and direct sequencing of fungal ribosomal RNA genes for phylogenetics. In *PCR Protocols: A Guide to Methods and Applications*; Academic Press: Cambridge, MA, USA, 1990; Volume 18, pp. 315–322.
40. Bolger, A.-M.; Lohse, M.; Usadel, B. Trimmomatic: A flexible trimmer for Illumina sequence data. *Bioinformatics* **2014**, *30*, 2114–2120. [CrossRef] [PubMed]
41. Martin, M. Cutadapt removes adapter sequences from high-throughput sequencing reads. *EMBnet J.* **2011**, *17*, 10–12. [CrossRef]
42. Edgar, R.-C. UPARSE: Highly accurate OTU sequences from microbial amplicon reads. *Nat. Methods* **2013**, *10*, 996–998. [CrossRef]
43. Edgar, R.-C.; Haas, B.-J.; Clemente, J.-C.; Quince, C.; Knight, R. UCHIME improves sensitivity and speed of chimera detection. *Bioinformatic* **2011**, *27*, 2194–2200. [CrossRef] [PubMed]
44. Jiang, S.; Xing, Y.-J.; Liu, G.-C.; Hu, C.-Y.; Wang, X.-C.; Yan, G.-Y.; Wang, Q.-G. Changes in soil bacterial and fungal community composition and functional groups during the succession of boreal forests. *Soil Biol. Biochem.* **2021**, *161*, 108393. [CrossRef]
45. Liu, S.-E.; Wang, H.; Tian, P.; Yao, X.; Sun, H.; Wang, Q.-K.; Baqueizo, M.-D. Decoupled diversity patterns in bacteria and fungi across continental forest ecosystems. *Soil Biol. Biochem.* **2020**, *144*, 107763. [CrossRef]
46. Sheng, Y.-Y.; Cong, W.; Yang, L.-S.; Liu, Q.; Zhang, Y.-G. Forest soil fungal community elevational distribution pattern and their ecological assembly processes. *Front. Microbiol.* **2019**, *10*, 02226. [CrossRef] [PubMed]
47. Monkai, J.; Purahong, W.; Nawaz, A.; Wubet, T.; Hyde, K.-D.; Goldberg, S.-D.; Mortimer, P.-E.; Xu, J.-C.; Harrison, R.-D. Conversion of rainforest to rubber plantations impacts the rhizosphere soil mycobiome and alters soil biological activity. *Land Degrad. Dev.* **2022**, *33*, 3411–3426. [CrossRef]
48. Ceci, A.; Pinzari, F.; Russo, F.; Persiani, A.-M.; Gadd, G.-M. Roles of saprotrophic fungi in biodegradation or transformation of organic and inorganic pollutants in co-contaminated sites. *Appl. Microbiol. Biotechnol.* **2019**, *103*, 53–68. [CrossRef]
49. Al-Yahya'ei, M.N.; Oehl, F.; Vallino, M.; Lumini, E.; Redecker, D.; Wiemken, A.; Bonfante, P. Unique arbuscular mycorrhizal fungal communities uncovered in date palm plantations and surrounding desert habitats of Southern Arabia. *Mycorrhiza* **2021**, *21*, 195–209. [CrossRef]
50. Wilson, H.; Johnson, R.-R.; Bohannan, B.; Pfeifer-Meister, L.; Mueller, R.; Bridgham, S.-D. Experimental warming decreases arbuscular mycorrhizal fungal colonization in prairie plants along a Mediterranean climate gradient. *PeerJ* **2016**, *4*, e2083. [CrossRef] [PubMed]
51. Phillips, R.-P.; Brzostek, E.; Midgley, M.-G. The mycorrhizal-associated nutrient economy: A new framework for predicting carbon-nutrient couplings in temperate forests. *New Phytol.* **2013**, *199*, 41–51. [CrossRef] [PubMed]
52. Lindahl, B.-D.; Tunlid, A. Ectomycorrhizal fungi-potential organic matter decomposers, yet not saprotrophs. *New Phytol.* **2015**, *205*, 1443–1447. [CrossRef] [PubMed]
53. Crowther, T.-W.; van den Hoogen, J.; Wan, J.; Mayes, M.-A.; Keiser, A.-D.; Mo, L.; Averill, C.; Maynard, D.-S. The global soil community and its influence on biogeochemistry. *Science* **2019**, *365*, eaav0550. [CrossRef]
54. Nie, S.-A.; Lei, S.-M.; Zhao, L.-X.; Brookes, P.-C.; Wang, F.; Chen, C.-R.; Yang, W.-H.; Xing, S.-H. Fungal communities and functions response to long-term fertilization in paddy soils. *Appl. Soil Ecol.* **2018**, *130*, 251–258. [CrossRef]
55. Gilmartin, E.-C.; Jusino, M.-A.; Pyne, E.-J.; Banik, M.-T.; Lindner, D.-L.; Boddy, L. Fungal endophytes and origins of decay in beech (*Fagus sylvatica*) sapwood. *Fungal Ecol.* **2022**, *59*, 101161. [CrossRef]
56. Averill, C. Slowed decomposition in ectomycorrhizal ecosystems is independent of plant chemistry. *Soil Biol. Biochem.* **2016**, *102*, 52–54. [CrossRef]
57. Liang, M.-X.; Liu, X.-B.; Parker, I.-M.; Johnson, D.; Zheng, Y.; Luo, S.; Gilbert, G.-S.; Yu, S.-X. Soil microbes drive phylogenetic diversity-productivity relationships in a subtropical forest. *Sci. Adv.* **2019**, *5*, eaax5088. [CrossRef]
58. Lang, C.; Seven, J.; Polle, A. Host preferences and differential contributions of deciduous tree species shape mycorrhizal species richness in a mixed Central European forest. *Mycorrhiza* **2011**, *21*, 297–308. [CrossRef]
59. Uroz, S.; Buee, M.; Deveau, A.; Mieszkin, S.; Martin, F. Ecology of the forest microbiome: Highlights of temperate and boreal ecosystems. *Soil Biol. Biochem.* **2016**, *103*, 471–488. [CrossRef]
60. Shao, P.-S.; Liang, C.; Rubert-Nason, K.; Li, X.-Z.; Xie, H.-T.; Bao, X.L. Secondary successional forests undergo tightly-coupled changes in soil microbial community structure and soil organic matter. *Soil Biol. Biochem.* **2019**, *128*, 56–65. [CrossRef]
61. Sturosova, M.; Barta, J.; Santruckova, H.; Baldrian, P. Small-scale spatial heterogeneity of ecosystem properties, microbial community composition and microbial activities in a temperate mountain forest soil. *Fems Microbiol. Ecol.* **2016**, *92*, fiw185. [CrossRef] [PubMed]

62. Francioli, D.; van Rijssel, S.-Q.; van Ruijven, J.; Termorshuizen, A.-J.; Cotton, T.; Dumbrell, A.-J.; Raaijmakers, J.-M.; Weigelt, A.; Mommer, L. Plant functional group drives the community structure of saprophytic fungi in a grassland biodiversity experiment. *Plant Soil* **2021**, *461*, 91–105. [CrossRef]
63. Xu, J.; Liu, B.; Qu, Z.-L.; Ma, Y.; Sun, H. Age and species of Eucalyptus plantations affect soil microbial biomass and enzymatic activities. *Microorgianisms* **2020**, *8*, 811. [CrossRef]
64. Bannert, A.; Kleineidam, K.; Wissing, L.; Mueller-Niggemann, C.; Vogelsang, V.; Welzl, G.; Cao, Z.; Schloter, M. Changes in diversity and functional gene abundances of microbial communities involved in nitrogen fixation, nitrification, and denitrification in a tidal wetland versus paddy soils cultivated for different time periods. *Appl. Environ. Microbiol.* **2011**, *77*, 6109–6116. [CrossRef] [PubMed]
65. Prescott, C.-E.; Grayston, S.-J. Tree species influence on microbial com munities in litter and soil: Current knowledge and research needs. *Forest Ecol. Manag.* **2013**, *309*, 19–27. [CrossRef]
66. Ding, J.; Jiang, X.; Guan, D.; Zhao, B.; Ma, M.; Zhou, B.; Cao, F.; Yang, X.; Li, L.; Li, J. Influence of inorganic fertilizer and organic manure application on fungal communities in a long-term field experiment of Chinese Mollisols. *Appl. Soil. Ecol.* **2017**, *111*, 114–122. [CrossRef]
67. Liu, J.; Sui, Y.; Yu, Z.; Shi, Y.; Chu, H.; Jin, J.; Liu, X.; Wang, G. Soil carbon content drives the biogeographical distribution of fungal communities in the black soil zone of northeast China. *Soil Biol. Biochem.* **2015**, *83*, 29–39. [CrossRef]

forests

MDPI

Article

A Survey of Organic Carbon Stocks in Mineral Soils of *Eucalyptus globulus* Labill. Plantations under Mediterranean Climate Conditions

Ana Quintela [1,*], Daniela Ferreira [1], Sérgio Fabres [1] and João Coutinho [2]

1 RAIZ—Forest and Paper Research Institute, Quinta de S. Francisco, Rua José Estevão, 3800-783 Aveiro, Portugal; daniela.ferreira@thenavigatorcompany.com (D.F.); sergio.fabres@thenavigatorcompany.com (S.F.)

2 Centro de Química—Vila Real, Universidade de Trás-os-Montes e Alto Douro, 5001-911 Vila Real, Portugal; j_coutin@utad.pt

* Correspondence: ana.quintela@thenavigatorcompany.com; Tel.: +351-234-920-130

Abstract: The main aim of this study was to assess the amount of carbon (C) stored in the upper 30 cm layer of mineral soils in eucalypt plantations in Portugal, with a Mediterranean-type climate. Soil sampling data (2468 samples), field evaluations (soil profile description) and relevant information on the particle size distribution, climate, bedrock and reference soil group were accomplished. Bulk density per sample was assessed using pedo-transfer functions and soil C stock was estimated. The results showed an average of 41.2 t C ha^{-1} stored in the soil. In the northern regions of Portugal, the coldest and wettest areas of the country with better stand productivity, a higher soil organic carbon (SOC) is achieved (median SOC of 39.2 g kg^{-1} and soil C stock of 55 t ha^{-1}) than in southern and inland regions, with a warmer and drier climate (median SOC of 15.2 g kg^{-1} and soil C stock of 28 t ha^{-1}). The assessment of mean soil C stock per bedrock type revealed higher C stored in granites followed by conglomerates, coal shales and clay shales. Regarding soil type, the results showed a higher C stock in *Cambisols*, *Leptosols* and *Fluvisols* (>50 t C ha^{-1}), whereas *Regosols* and *Luvisols* stored less, following the same trend presented for reference soil groups in Europe. Comparing the geographic distribution of the C stock in the upper layer of the mineral soils with the amount of C in eucalyptus stands (root and aboveground biomass—data from national forest inventory), the mineral soil pool can represent more than two-thirds of the total C stored in eucalyptus plantations in Portugal. Further studies should focus on the evolution of C stocks in eucalypt plantations during different stages of stand growth and under different management practices.

Keywords: carbon cycle; carbon sequestration; mineral topsoil; Köppen climate classification; Mediterranean planted forests

1. Introduction

Forests account for around 31% of the planet's continental surface [1] and play an important role in regulating hydrological and carbon (C) cycles at a global level [2–5]. Forests and their derived products are key elements in mitigating global warming, acting as reservoirs and, in some cases, also as C sinks [6–8]. A net C sink of −7.6 ± 49 $PgCO_2e$ yr^{-1} was estimated for global forests, considering a balance between gross C removals (−15.6 ± 49 $PgCO_2e$ yr^{-1}) and gross emissions from deforestation and other disturbances (8.1 ± 2.5 $PgCO_2e$ yr^{-1}) [9]. In fact, the noticeable potential of forests for C sequestration has been extensively addressed [10–13]. Carbon stocks are dynamic and influenced by continuous processes, such as tree growth, and by discrete events, namely fires and disturbances caused by pests and diseases or land-use change [14]. Carbon pools may be gathered in five main components as described by the Intergovernmental Panel on Climate Change (IPCC): aboveground biomass (wood and leaves), belowground biomass (roots), soil C in mineral layers, soil organic layer (forest floor, dead biomass in decomposition above soil surface) and harvested wood products. These pools of the C cycle, input and output fluxes from the system, and transfers, are presented in Figure 1.

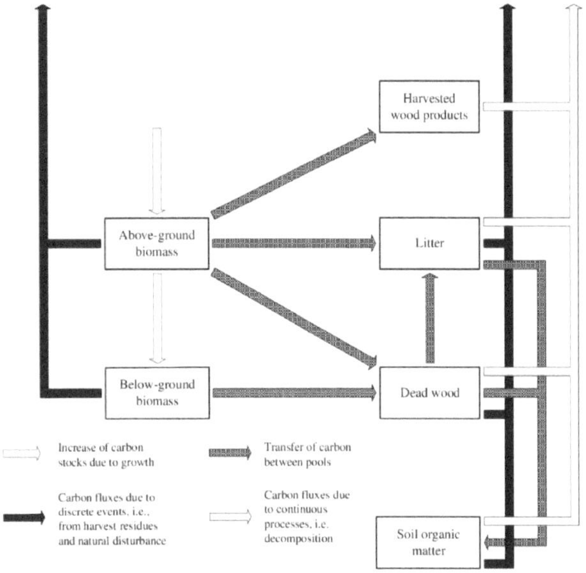

Figure 1. Generalized carbon cycle showing the flows of carbon between the five C pools for agriculture, forestry and other land use ecosystems [15].

Around two-thirds of global soil C is held as soil organic C (SOC), as a result of decaying vegetation, fungi and bacteria growth and the metabolic activities of living organisms, the remainder being inorganic C [16,17]. The average SOC stocks in Europe account for 22.1 t C ha^{-1} in forest floors, 108 t C ha^{-1} in mineral soils and 578 t C ha^{-1} in peat soils to a 1 m depth [18]. Considering the total area covered by forest at the European scale (163 M ha), a total stock of 3.50–3.94 Pg C was estimated in forest floors and 21.4–22.7 Pg C in mineral and peat soils up to a 1 m depth.

About 36% of the Portuguese mainland is forested, with more than 800 thousand hectares occupied with eucalypt, mostly *Eucalyptus globulus* L. [19]. Around 50% of the total C in forest areas is stored in the soil, even if it is quite variable with region, forest type and trees' age [20]. Some studies suggest even a higher proportion in the soil of *E. globulus* stands, with around 200 t C ha^{-1} in the soil, up to 1 m deep [21]. Also, in a six-year-old stand of *E. globulus* established in *Arenosols*, on the central coast of Portugal, the C storage in the soil and in the organic layer was estimated to be 42 to 63 t C ha^{-1}, respectively, depending on the water and nutrient availability [22]. In Chile, the total C stocks in the aboveground and belowground biomass, forest floor and soil in *E. globulus* plantations were around 254 Mg ha^{-1} (at the end of rotation), with 30% to 50% being stored in the soil [23]. Despite the studies carried out thus far, the knowledge about the amount of C stored in soils on eucalypt stands in Portugal and its spatial distribution is still very limited. Thus, the main aim of this study is to assess the amount of C stored in Portuguese mineral topsoil (0–30 cm) in eucalypt managed forests.

2. Materials and Methods

2.1. Study Area and Soil Sampling

The study area covers *E. globulus* plantations on Portugal's mainland. Sampling covered different eucalypt stand ages and has been carried out since 1996 comprising 2468 soil samples (Figure 2). About 70% of the samples are under a Mediterranean-type climate with mild winters and dry warm summers and 30% under a Mediterranean-type climate with an Atlantic influence (Csa and Csb Köppen classification, respectively) [24].

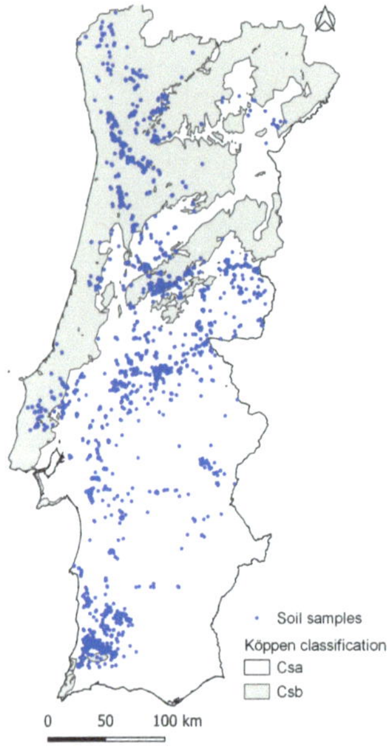

Figure 2. Soil sampling sites in eucalyptus plantations, Portugal (Csa—Mediterranean-type climate with mild winters and dry warm summers; Csb—Mediterranean-type climate with an Atlantic influence).

In each sampling site, a composite sample was collected inside forest plantation avoiding areas close to roads, firebreaks, plantation borders or any other disturbances. Samples were composed of 15 subsamples randomly distributed in the area. All samples were taken from 0 to 30 cm of soil after removing superficial dead organic material following the IPCC guidelines [15]. For samples with soil depth of less than 30 cm, the real depth was registered and considered for the C stock estimation.

Soil profile evaluation per sampling site allowed us to estimate the relative percentage of coarse fragments of soil volume following the FSCC guidelines [25]. Additionally, according to local climate characteristics and soil profile features, a site productivity index—indicating the potential for the aboveground biomass production of *E. globulus*—was assigned per site. This index was derived as described by RAIZ [26] for 70% of the total soil samples collected.

Prior to C analysis, soil samples were air-dried and passed through a 2 mm sieve. SOC concentration was then determined by dry combustion following ISO 10694 [27] at accredited national laboratories.

2.2. SOC Assessment

Particle size distribution data were not available for the soil samples. Therefore, this information was extracted from the Infossolo database [28] complemented with analytical data from 102 soil samples collected in eucalypt plantations by RAIZ. The assessment of the particle size distribution of the fraction <2 mm followed the Gee and Or method [29].

From Infossolo, 400 records related to the particle size distribution of soil samples collected under forest occupation in Portugal were considered. This information of 502 records was grouped according to the bedrock type (map provided by the Portuguese Environment Agency) and was used to fill the data on coarse sand (2–0.2 mm), fine sand (0.2–0.02 mm), silt (0.02–0.002 mm) and clay (<0.002 mm) content per soil sampled.

For C stock estimation, the bulk density (BD, g cm^{-3}) of the fine earth fraction (<2 mm) of the soil samples was determined using a pedo-transfer function [30] (Equation (1)):

$$BD = 0.69794 + 0.750636^{(-0.230355 \times SOC)} + (0.0008687 \times \text{sand}) - (0.0005164 \times \text{clay}) \quad (1)$$

where SOC is the organic C concentration (g kg^{-1}) of soil samples (fraction < 2 mm), sand (%) is the content in particles between 2 and 0.02 mm and clay (%) is the content in particles below 0.002 mm.

Carbon stock (CS, t ha^{-1}) per sample was determined according to Equation (2) [31,32]:

$$CS = SOC \times depth \times BD \times \left(1 - \frac{Cf}{100}\right) \times 10 \quad (2)$$

where SOC is the SOC concentration (g kg^{-1}) of soil samples (fraction < 2 mm), depth (m) is the thickness of the soil layer sampled (0–30 cm), BD is the bulk density (g cm^{-3}) of the fraction < 2 mm and Cf is the coarse fraction (>2 mm) of the soil (%).

Finally, data from the soil sampling (2468 samples) and soil profile information (coarse fragments) were organized in a layer and processed stepwise (Figure 3): (i) layers provided by the Portuguese Environment Agency on lithology/bedrock (map 1:1,000,000), soil type according to the reference soil group [33], and Köppen classification [24] completed the information per soil sample; (ii) information on the particle size distribution of the fraction < 2 mm was completed using the mean values obtained from the Infossolo and RAIZ database per lithology/bedrock group; (iii) the BD per sample was computed using pedo-transfer functions [30]; and (iv) the C stock was estimated [32]. The main physical features of the soil samples are presented in Table 1.

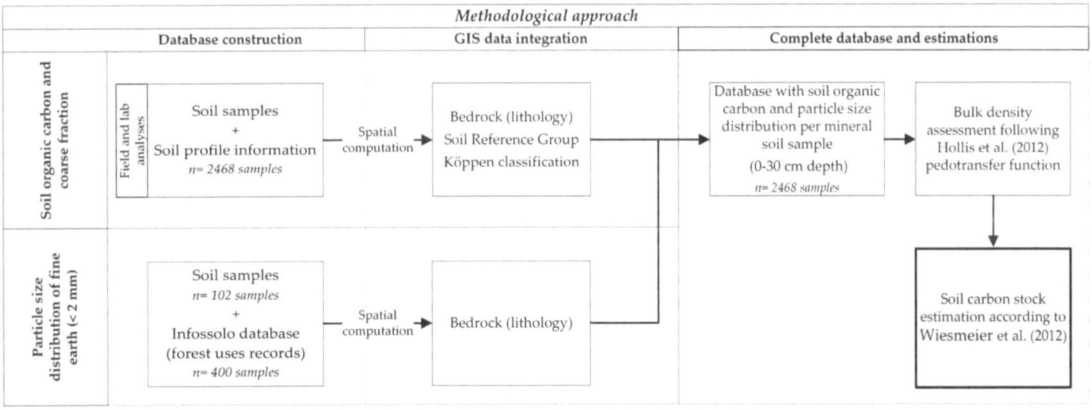

Figure 3. Methodological approach applied for soil carbon stock estimation: soil collection and analysis, relevant information completion (particle size distribution, bulk density, climate, bedrock and reference soil group) and C stock calculation [30,32].

Table 1. Number of samples (N) and mean physical characterization of soil samples by lithology group (bedrock).

Bedrock ID	Bedrock	N	Coarse Fraction (>2 mm)	Fine Fraction (<2 mm)	Particle Size Distribution (%) Fine Fraction (<2 mm) Classes				Bulk Density (g cm^{-3})
					Coarse Sand	Fine Sand	Silt	Clay	
1	Alluvium	29	35.9	64.1	29.7	36.8	24.2	9.4	1.28
2	Sands and gravel	17	34.0	66.0	30.4	41.6	15.8	12.2	1.36
3	Sands, pebbles, poorly consolidated sandstones, clays	298	25.2	74.8	50.4	27.5	13.0	9.1	1.32
4	Arcose sandstones and sandstones	9	35.0	65.0	55.5	26.6	8.0	9.9	1.18
5	Sandstones, more or less marly limestones, sands, gravel, clays	278	31.9	68.1	55.5	26.6	8.0	9.9	1.35
6	Sandstones, conglomerates, limestones, dolomitic limestones, marly limestones, marls	34	30.9	69.1	63.3	22.1	6.2	8.3	1.26
7	Basalts	2	50.3	49.7	43.5	25.0	11.7	19.8	1.15
8	Limestones, dolomitic limestones, marly limestones, marls	11	29.0	71.0	86.1	11.2	0.4	2.3	1.38
9	Plateau gravel pits, Beira Baixa arches, sandstones, limestones	24	29.4	70.6	43.1	32.3	13.5	11.0	1.24
10	Conglomerates, sandstones, white limestones, reddish marls	5	17.0	83.0	31.7	32.6	13.2	22.5	1.34
11	Conglomerates, sandstones, limestones, dolomitic limestones, marly limestones, marls	36	20.1	79.9	31.7	32.6	13.2	22.5	1.29
12	Conglomerates, coal shales, clay shales	9	43.8	56.2	25.1	23.4	30.9	20.6	1.07
13	Dolerites	4	44.5	55.5	43.5	25.0	11.7	19.8	1.36
14	Dunes and aeolian sands	1	10.0	90.0	50.0	47.4	0.9	1.7	1.45
15	Granites and similar rocks	225	27.7	72.3	46.8	31.1	12.4	9.7	1.15
16	Red sandstones (from Silves), conglomerates, marls, limestones generally dolomitic	3	30.7	69.3	31.7	32.6	13.2	22.5	1.27
17	Metavolcanites	39	33.2	66.8	33.2	25.3	25.7	15.9	1.28
18	Granite porphyries	1	61.8	38.2	46.8	31.1	12.4	9.7	1.35
19	Quartziferous porphyries	2	31.0	69.0	48.2	31.6	12.1	8.1	1.34
20	Quartzites	64	51.2	48.8	22.8	35.4	24.6	17.1	1.14
21	Quartzdiorites	4	40.5	59.5	59.6	24.0	8.5	7.9	1.44
22	Nepheline syenites	2	13.0	87.0	46.8	31.1	12.4	9.7	1.28
23	Clay shale, greywacke, sandstone	438	46.4	53.6	27.2	26.1	22.8	24.0	1.22
24	Schists, amphibolites, mica schists, greywacke quartzites, carbonate rocks, gneisses	19	41.7	58.3	23.5	40.2	22.5	13.9	1.23
25	Shale, greywacke	228	50.3	49.7	28.9	38.1	19.4	13.6	1.14
26	Shale, greywacke (shale–greywacke complex)	682	46.5	53.5	22.7	32.2	30.9	14.3	1.10
27	Shales, quartzites, amphibolites	4	48.0	52.0	47.3	24.3	12.8	15.7	1.39

2.3. Statistical Analysis

Statistical analyses were performed using Statistica, TIBCO software Inc. (Santa Clara, CA, USA), version 13.5.0.17. The medians of SOC and C stock by climate stratification (Csa and Csb) were computed and presented with percentiles 25%–75% in a boxplot. This information was complemented with the Tukey test (HSD) to check the significance of the results ($p \leq 0.05$). Principal component analysis (PCA) was applied considering the following as active variables: Köppen group, SOC, soil C stock, productivity index (aboveground biomass production potential), coarse fraction (>2 mm), BD, particle size distribution, and as supplementary variables: altitude and slope. As composition data, for the particle size distribution of the fine fraction (<2 mm) an additive log-ratio transformation (alr) was applied and for the coarse fraction a logarithmic transformation [34]. Slope and altitude data were provided by ASTER-GDEM [35]. The steps in the PCA included computing the univariate statistics, covariance matrix, correlation matrix, eigenvalues and eigenvectors, degree of correlation, and projection in principal components (PCs). Prior to analysis, the database was standardized, and qualitative variables were transformed into numerical variables.

3. Results

3.1. SOC and C Stock by Lithology Group (Bedrock)

The SOC ranged between 1.6 and 104.4 g kg^{-1}. The mean SOC concentration ranged from 4.8 to 37.0 g kg^{-1} (Figure 4a), and the highest mean values were achieved for soil samples derived from conglomerates, coal shales and clay shales (ID12). Also, soils derived from shale, greywacke and shale–greywacke complexes (IDs 25 and 26), granites (ID15) and arcose sandstones and sandstones (ID4) presented ≥30 g kg^{-1}. The lowest mean values were obtained for soil samples of dunes and aeolian sands (ID14), followed by quartzdiorites (ID21) and shales, quartzites and amphibolites (ID27).

The distribution of C stock by lithology group is shown in Figure 4b and ranged in average between 9.0 and 70 t ha^{-1}. Granites and similar rocks (ID15) revealed the highest mean value, followed by conglomerates, coal shales and clay shales (ID12), arcose sandstones and sandstones (ID4) and nepheline syenites (ID22) which also presented C stocks higher than 50 t ha^{-1}. The lowest values (<25 t ha^{-1}) were found for sands and gravel (ID2), dolerites (ID13), dunes and aeolian sands (ID14), shales, quartzites and amphibolites (ID27), quartzdiorites (ID21) and granite porphyries (ID18). The remaining groups displayed stocks between 25 and 50 t C ha^{-1}.

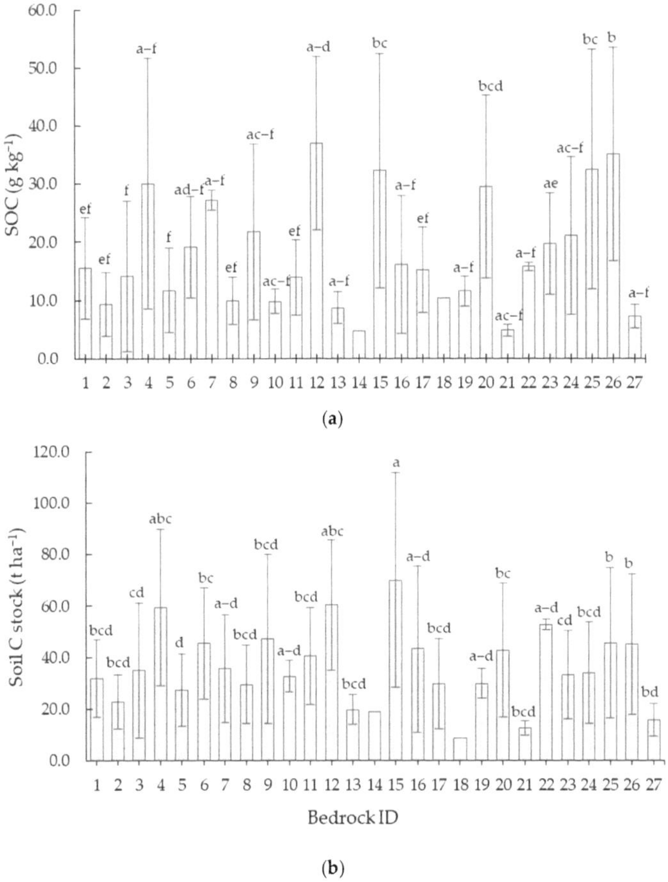

Figure 4. Mean SOC content (**a**) and C stock (**b**), and standard deviation in soil samples (0–30 cm depth) by lithology group (bedrock ID presented in Table 1). Significant differences are denoted by different letters (Tukey's test, $p \leq 0.05$).

3.2. SOC and C Stock by Reference Soil Group (Soil Type)

The bulk densities (BDs), particle size distribution (fine and clay fractions), SOC in mineral 30 cm topsoil and C stocks by soil reference group are provided in Table 2. The most representative sampled groups were *Leptic Cambisols* (Humic), *Eutric Leptosols*, *Umbric Podzols* and *Leptic Luvisols*. The average BD of the sampled soils was 1.20 g cm^{-3} ranging between 1.01 and 1.43 g cm^{-3}, a mean fine fraction $\geq$37% and a clay proportion within 9.9 and 24.0% of the total soil volume. The average SOC was 24.7 g kg^{-1} and allowed a mean C stock of 41.2 t ha^{-1}. *Eutric Fluvisols, Leptosols, Leptic Cambisols (Humic)* and *Cambic Calcisols* revealed a C stock higher than 50 t ha^{-1}; the lowest stock belonged to *Leptic Calcaric Regosols, Albic Gleyic Luvisols, Chromic Leptic Luvisols, Eutric Leptic Regosols* and *Vertic Luvisols* (C stock $\leq$ 25 t ha^{-1}).

Table 2. Number of samples (N), mean and quartiles (25% and 75%) of physical properties, SOC and C stocks (0–30 cm depth) according to the reference soil group ([36], * previous classification reference [33]).

Reference Soil Group	N	Bulk Density (g cm^{-3})			Fine Fraction (<2 mm) (%)			Clay Fraction (<0.002 mm) (%)			SOC (g kg^{-1})			C Stock (t ha^{-1})		
		Mean	Q25	Q75	Mean	Q25	Q75	Mean	Q25	Q75	Mean	Q25	Q75	Mean	Q25	Q75
Cambic Calcisols	22	1.22	1.16	1.29	73.8	69.0	85.0	13.9	8.3	22.5	21.7	14.3	28.4	53.7	39.5	67.2
Chromic Cambisols	30	1.29	1.23	1.36	70.7	58.4	83.4	16.3	8.3	22.5	14.8	9.8	18.4	36.9	24.5	49.0
Leptic Calcaric Regosols/Calcaric Chromic Cambisols *	7	1.36	1.27	1.43	57.7	56.0	56.0	9.9	9.9	9.9	10.8	5.5	17.4	22.7	13.2	36.9
Dystric Cambisols	36	1.34	1.31	1.41	69.5	62.0	85.0	10.1	9.7	9.7	11.4	6.0	13.7	28.5	13.4	35.4
Eutric Cambisols	99	1.36	1.31	1.41	69.8	62.0	81.7	10.8	9.1	9.9	10.4	6.0	13.0	26.3	15.3	34.3
Leptic Cambisols (Humic)	751	1.06	0.97	1.12	57.2	40.0	72.3	13.1	9.7	14.3	41.6	30.2	52.3	63.8	39.6	84.0
Eutric Fluvisols	3	1.01	0.93	1.18	56.0	56.0	56.0	9.9	9.9	9.9	52.5	25.6	65.9	90.7	50.7	110.7
Eutric Leptosols/Eutric Lithosols *	664	1.21	1.14	1.31	53.6	41.6	63.3	17.5	14.3	24.0	22.1	12.0	27.0	30.6	17.0	41.0
Leptic Ferric Luvisols/Ferric Luvisols *	28	1.26	1.17	1.35	50.9	51.7	52.0	19.1	13.6	24.0	17.0	9.3	22.9	28.4	18.8	37.9
Albic Gleyic Luvisols	10	1.38	1.36	1.40	77.1	69.0	90.0	11.7	9.1	13.6	8.4	6.4	9.8	21.3	14.3	28.6
Leptic Luvisols/Orthic Luvisols *	244	1.22	1.17	1.28	57.6	51.2	66.7	21.6	24.0	24.0	19.8	14.1	23.7	36.4	23.7	46.5
Chromic Leptic Luvisols /Rhodochromic Luvisols *	9	1.26	1.20	1.30	37.6	30.0	40.0	15.6	13.6	17.1	16.7	13.1	22.4	19.1	11.7	30.4
Calcic Chromic Luvisols	15	1.29	1.22	1.36	69.5	60.0	78.4	15.1	9.4	22.5	15.3	8.2	20.2	37.7	23.8	59.8
Vertic Luvisols	3	1.43	1.41	1.44	52.0	52.0	52.0	11.7	7.9	13.6	4.8	4.4	5.2	10.6	9.8	11.5
Umbric Podzols/Orthic Podzols *	542	1.34	1.28	1.42	71.6	62.0	85.0	9.9	9.1	9.9	12.1	5.8	15.5	29.8	17.0	36.4
Leptosols/Rankers *	2	1.02	1.01	1.02	61.0	49.7	72.3	11.7	9.7	13.6	45.8	45.8	45.8	83.9	68.4	99.4
Plintic Luvisols	1	1.32	-	-	60.0	-	-	24.0	-	-	10.8	-	-	25.7	-	-
Eutric Leptic Regosols	1	1.41	-	-	90.0	-	-	9.9	-	-	6.4	-	-	17.4	-	-
Endosalic Gleysols/Gleyic Solonchaks *	1	1.35	-	-	95.0	-	-	9.9	-	-	10.5	-	-	40.5	-	-

3.3. Major Drivers for Soil C Stock

To further identify the major drivers of SOC and C stock, attributes were extracted by PCA. Two principal components with eigenvalues >1 were extracted with a cumulative contribution of 66.3%, which could explain most of the results. The first principal component (PC1) had the highest contribution, accounting for 46.8%, followed by PC2 with 19.5%, respectively.

As presented in Figure 5, the SOC concentration and soil C stock were loaded in PC1 with a positive relation with Köppen classification and site productivity index (biomass production), suggesting that higher SOC concentrations and C stocks are related with better climate conditions and higher biomass production. A higher BD of soil is in opposition to the increment of SOC. The coarse fragments and the particle size distribution of fine fraction were accounted for by PC2, showing the opposition of the coarse fraction and the higher proportion of sand in the fine elements (<2 mm), suggesting that the soils studied have a higher component in fractions between 2 mm and 0.02 mm.

A box-plot with the distribution of SOC concentration and C stock according to the Köppen classification is displayed in Figure 6, following the PCA results suggesting a positive correlation between these parameters. As suspected, Csb revealed a significantly higher SOC and soil C stock than Csa. For Csb, the SOC values ranged between 27.4 (Q25) and 51.6 g kg^{-1} (Q75), whilst for Csa results from 9.1 (Q25) to 23.5 g kg^{-1} (Q75) were recorded. Concerning soil C stocks, Csb presented a median of 55 t ha^{-1} (38.8–80.3 t ha^{-1}, lower and upper quartile) and Csa showed a median of around 28 t ha^{-1} (17.8–40.8 t ha^{-1}, lower and upper quartile).

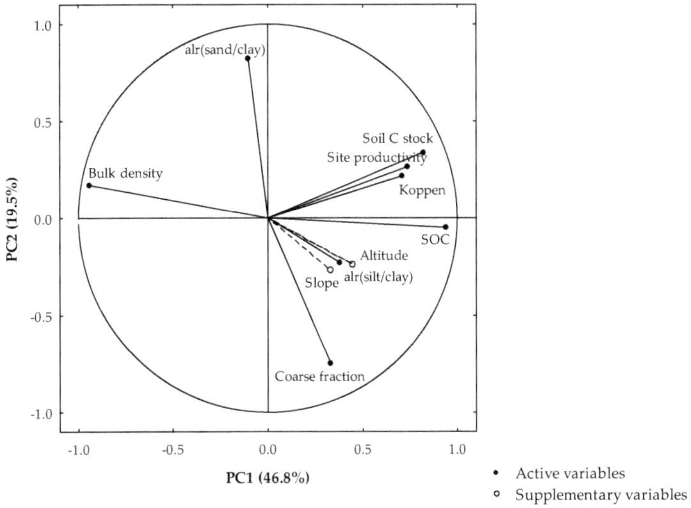

Figure 5. Spatial ordination resulting from the principal components analysis (PCA) for soil samples features and climate attributes (PC1: 3.74, variance explained: 46.8%; PC2: 1.56, variance explained: 19.5%).

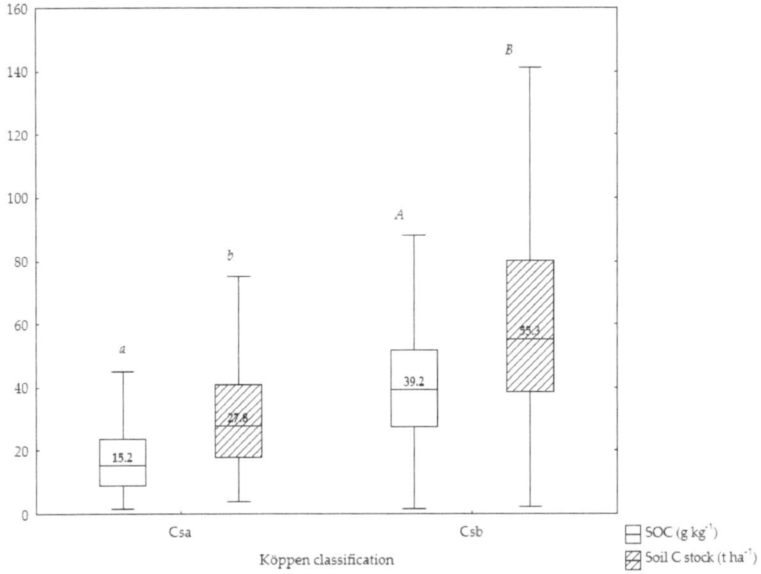

Figure 6. Median, upper and lower quartiles of soil organic carbon content (g kg^{-1}) and carbon stock (t ha^{-1}) in mineral soil samples (0–30 cm depth) according to the climate Köppen classification (Csa—Mediterranean-type climate with mild winters and dry warm summers; Csb—Mediterranean-type climate with an Atlantic influence). Within each variable, significant differences among climate classification are denoted by small and capital letters (Tukey's test, $p \leq 0.05$).

Apart for arcose sandstones and sandstones (ID4), the remaining groups of bedrock lithology agree with the previous box-plot, revealing that within the same bedrock type, mineral soil samples with the Csb climate type have higher C stocks than Csa (Figure 7). Focusing in the lithologies with the highest frequency of samples distributed by different climate types such as granites and similar rocks (ID 15) and shale and greywacke groups (ID 25 and 26), this is reinforced.

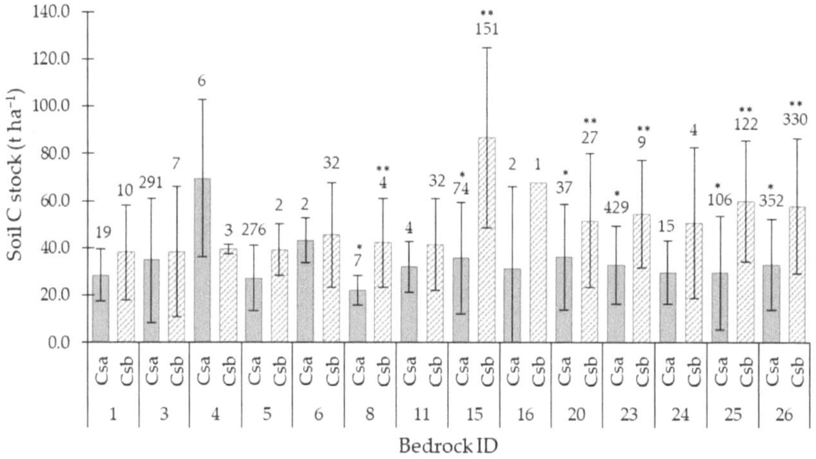

Figure 7. Mean and standard deviation of carbon stock (t ha^{-1}) in mineral soil samples (0–30 cm depth) by lithology group and climate classification of Köppen (Csa—Mediterranean-type climate with mild winters and dry warm summers; Csb—Mediterranean-type climate with an Atlantic influence) (Lithology ID in Table 1 and number of soil samples in italics above graph bars). Within each bedrock ID, significant differences among climate classification are denoted by * and ** (Tukey's test, $p \leq 0.05$).

4. Discussion

4.1. C Stored in Eucalypt Stands and the Influence of Mediterranean Climate Conditions

Data from the last Portuguese National Forestry Inventory—2015 [19] show that the total plant biomass (above and belowground) accounts for around 19.5 t C ha^{-1} in pure eucalypt plantations (the majority of plantations until 12-year-old stands). This inventory is used to assess the land use land cover (LULC) and forecast wood availability and the C stored in the biomass. The data support that eucalypt plantations are one of the forestry species in Portugal that most contributes to C stored in biomass. Northern regions and the central part of Portugal are the greatest contributors to the C in total plant biomass (25.4 and 20.4 t ha^{-1}, respectively), followed by Alentejo (15.4 t ha^{-1}) and finally Algarve in the south of the country (11.0 t ha^{-1}). The amount of biomass produced and the ability to store C in the aboveground biomass in eucalypt plantations has been well documented and evaluated [6,19,22,37]. Furthermore, the results also show that around 50 to 55% of plant C is stored in wood, suggesting that if management practices consider leaving leaves, branches, bark and roots in the area after harvesting, this may contribute to the conservation and increase of C in the soil. The forest floor was found to be significantly lower (less than 4% of the total) when compared with other pools and showed a large spatial variability [19,23].

Stand productivity is highly influenced by climatic and topographic variability, with temperature, precipitation and soils (physical and chemical features) being major drivers for wood biomass production [37]. Concerning the soil pool, studies found SOC concentrations in forest soils between 3 and 115 g kg^{-1} in the upper 20 cm of the mineral soil [18,38], in line with this study. The present results are also supported by Gómez-Rey et al. [39]

who reported SOC values from 9.8 to 64.4 g kg^{-1} for five eucalypt plantations in Portugal across different mean annual precipitation values (500 to 2000 mm). The accumulation of C is a function of the prevailing edaphoclimatic characteristics at each site. In Portugal, two distinct climate regions are defined by the Köppen classification [24]. In the northern regions of Portugal, colder and wetter areas of the country (Csb), a higher amount of C is in stored in the soil (median SOC of 39.2 g kg^{-1} and soil C stock of 55 t ha^{-1}), whereas in southern and inland regions, the climate is warmer, drier and with a very pronounced dry season (usually March to September) (Csa). As a result, productivity is lower in Csa as stated by Pereira et al. (2007), and consequently litter inputs to soils decrease and SOC also reduces (median SOC of 15.2 g kg^{-1} and soil C stock of 28 t ha^{-1}). These values are in line with (but slightly lower than) IPCC references which have reported for warm temperate climate regions an average of 38 t C ha^{-1} (dry) and 88 t C ha^{-1} (moist) for mineral soils under native vegetation at a 30 cm depth [15].

4.2. SOC Stock and Soil Variability

Our study shows that eucalypt plantations in Portugal can store a substantial amount of C in soils, which is in line with other studies showing that a large amount of the total C stock is stored in the soil [40]. The present study revealed that mineral topsoil (0–30 cm) in eucalypt plantations is able to store an average of 41.2 t C ha^{-1}, which is not out of line with the data presented in other studies that found average SOC stocks of 108 t C ha^{-1} in mineral soils in European forests to a 1 m depth [18] and between 11.3 and 126.3 t C ha^{-1} for a 0–20 cm soil depth [41].

Considering the analysis of soil C stock by the northern, central, Alentejo, and Algarve regions of the country, an average amount of C is found to be stored in the upper 30 cm of the mineral soil of 65.8, 43.4, 28.2 and 38.0 t ha^{-1}, respectively. As expected, sites with the most favorable climatic and soil conditions for tree growth have a higher biomass production and soil C storage potential. This is in accordance with data from forest plantations including *E. globulus* in Chile, which revealed that 30 to 50% of the total C stock is stored in the soil (0–30 cm) [23]. Also, Scharlemann et al. [17] highlight the relevance of C stored in the topsoil (0–30 cm), which represents between 46% and 65% of the total SOC. This last study also identified differences related to climatic conditions.

Mediterranean soils are known for their variability, reduced water holding capacity, shallow horizons and high stoniness on the soil surface [42]. Therefore, storing C in the soil is essential to promote an increase in organic matter and soil quality, also mitigating the effect of climate change. Differences were found in SOC by the soil parent material (bedrock) and reference soil group. According to the FAO database, in the Mediterranean region, *Cambisols* prevail but *Fluvisols*, *Luvisols* and *Leptosols* are also common, which agrees with the main reference soil group sampled. The results showed a higher C stock in *Cambisols*, *Leptosols* and *Fluvisols* (>50 t C ha^{-1}), whereas *Regosols* and *Luvisols* stored less—following the same trend presented for reference soil groups in Europe [18,43]. Similar results were found for soil samples according to the soil group reference in eucalypt plantations in northwestern Spain [10].

4.3. Methodological Considerations

This study contributes to the characterization of C stocks in soil in eucalypt managed forests. However, some limitations must be acknowledged to contextualize the findings and guide future research. Firstly, our sampling, although extensive, was limited to Portugal and may not fully capture the variability of C content across Mediterranean conditions in planted forests. Furthermore, the C stock per sampling site reflects a measurement at a single point in time and the time span of soil sampling may not provide a comprehensive understanding of soil C stocks throughout the stand growth cycle. Future research should consider implementing long-term monitoring per location and different soil layers to capture soil C changes over multiple years, thereby better elucidating temporal dynamics and the influence of management practices. Additionally, it is advisable to assess the aboveground and root biomass of eucalypt stands per location in order to establish more reliable comparisons with soil C stocks.

We also faced some analytical constraints. Ideally, BD, coarse fragments and particle size distribution should be assessed by analysis per soil sample. In this study, we used a pedo-transfer function for predicting soil BD that is applicable across all soil types within Europe with a mean percentage error of ± 11%, which introduces some uncertainty in the data [30]. However, we found values of BD around 1.20 g cm^{-3}, being higher for *Vertic Luvisols* and *Eutric Leptic Regosols* (around 1.40 g cm^{-3}) and lower to *Eutric Fluvisols* and *Leptosols* (around 1.02 g cm^{-3}) which agrees with the results found for soil BD in Europe [44]. Concerning coarse fragments, we considered the data from the soil profile evaluation in the assessment of C stocks. As expected from field observations, usually, forest soils present a high stoniness as soils are pedogenetically underdeveloped [40] which is also supported by the lower amount of clay fraction in the soils. For the assessment of particle size distribution, we used laboratory results from soil samples and Infossolo data to fill the main database of the 2468 soil samples with this information. Although the results are consistent with the literature, the reliance on other databases may have introduced some uncertainty in the estimation of C stocks.

New approaches using remote sensing, machine learning, and other technologies are being tested and optimized to accurately quantify the SOC, which could represent a promising solution for C stock estimation at the management unit level, rather than a point-based analysis [45], eventually helping in the future to overcome some of the data gaps faced in this study.

4.4. The Effect of Management Practices on SOC Stock and Further Studies

The C held in the mineral topsoil (upper 30 cm layer) is often the most chemically decomposable, and the most directly exposed to natural and anthropogenic disturbances [15]; this is the reason why this study focused on this layer of mineral soil. Nevertheless, C stored in deeper soil layers can be relevant for the soil C sequestration potential in a changing forest ecosystem (e.g., afforestation and forest management). The amount and rate of organic C accumulation associated with afforestation depend on several factors such as climate, topography, soil characteristics, species and forestry management practices [38,46]. Forest management can improve C sequestration amounts, and understanding the effects of forest management is even more important in the Mediterranean area, given the current high climatic variability and disturbance events [47]. Harvesting operation and soil tillage, particularly, can lead to a reduction in soil C stocks if mitigation measures are not considered. Applying conservation practices in site preparation and leaving harvesting residues on the soil (roots, leaves, stumps and bark), for instance, and the proper management of stand growth may outweigh soil C losses over a rotation [38,47]. Management practices pursuing a canopy cover of the soil and minimizing the disturbances in soil are likely to improve the C storage [38]. Ruiz-Peinado et al. [47] reviewed the implications of different forest management practices on C sequestration (e.g., stocking, thinning, coppice, rotation period and harvesting operations) in the Mediterranean region and highlighted the need to carry out further integrated studies to investigate forest management alternatives.

This study provided estimations that can be valuable for the validation and parameterization of C models, also providing relevant information for management practices. Further studies should be conducted to assess the evolution of C stocks in eucalypt plantations during different stages of stand growth and under different management practices.

5. Conclusions

Across the country, we assessed the ability of forests soils to store C under a wide range of bedrock and reference soil groups. The SOC assessed in the first 30 cm of mineral soils in eucalypt plantations ranged between 1.6 and 104.4 g kg^{-1}. The total soil stocks averaged 41.2 t C ha^{-1} and were clearly influenced by climate conditions. Better conditions for tree growth, warm temperatures and higher annual precipitation (Csb) lead to a higher stand productivity as supported by the national forest inventory. This results in a higher input of litter in the soil and higher C stocks. Comparing different C pools and considering the amount of C stored in plant biomass according to the most recent national inventory [19], the relevance of the C stock in the upper 30 cm of the mineral soil is recognized, which may represent more than 65% the total C retention (considering soil and plant biomass pools).

While acknowledging that this study presented some methodological limitations, the data compiled, and the methodological approach used in this study contribute to the establishment of a forest soil C baseline and highlights the importance of soil and its conservation for C sequestration. Given the relevance of soil to the total C stock in eucalyptus plantations, special attention should be addressed to the management practices that may affect this organic C pool, and further studies should include long-term monitoring and time-dynamic assessment to provide new insights on C sequestration strategies.

Author Contributions: Conceptualization, A.Q., S.F., D.F. and J.C.; methodology, A.Q., S.F. and J.C.; software, A.Q.; validation, J.C.; formal analysis, A.Q.; investigation, A.Q., D.F. and S.F.; writing—original draft preparation, A.Q.; writing—review and editing, A.Q., D.F., S.F. and J.C.; visualization, A.Q.; supervision, J.C.; project administration, A.Q.; funding acquisition, S.F. All authors have read and agreed to the published version of the manuscript.

Funding: This research received no external funding.

Data Availability Statement: The raw data supporting the conclusions of this article may be made available by the authors upon request.

Acknowledgments: The authors would like to thank João Rocha and André Duarte for their assistance in spatial layers completion. We also thank the RAIZ team that has being performing soil profile evaluation, currently coordinated by Cláudio Teixeira, for the information provided; and finally, thanks are due to all the RAIZ colleagues that helped in building the soil database by carrying out soil sampling over time.

Conflicts of Interest: The authors declare no conflicts of interest.

References

1. FAO. *The State of the World's Forests 2020. Forests, Biodiversity and People*; UNEP: Rome, Italy, 2020. [CrossRef]
2. Sheil, D.; Murdiyarso, D. How forests attract rain: An examination of a new hypothesis. *BioScience* **2009**, *59*, 341–347. [CrossRef]
3. Ellison, D.; Futter, M.N.; Bishop, K. On the forest cover-water yield debate: From demand- to supply-side thinking. *Glob. Chang. Biol.* **2012**, *18*, 806–820. [CrossRef]
4. Ellison, D.; Morris, C.E.; Locatelli, B.; Sheil, D.; Cohen, J.; Murdiyarso, D.; Gutierrez, V.; van Noordwijk, M.; Creed, I.F.; Pokorny, J.; et al. Trees, forests and water: Cool insights for a hot world. *Glob. Environ. Chang.* **2017**, *43*, 51–61. [CrossRef]
5. Sheil, D. Forests, atmospheric water and an uncertain future: The new biology of the global water cycle. *For. Ecosyst.* **2018**, *5*, 19. [CrossRef]
6. Arroja, L.; Dias, A.C.; Capela, I. The role of *Eucalyptus globulus* forest and products in carbon sequestration. *Climat. Chang.* **2006**, *74*, 123–140. [CrossRef]
7. Strandberg, G.; Chen, J.; Fyfe, R.; Kjellström, E.; Lindström, J.; Poska, A.; Zhang, Q.; Gaillard, M.-J. Did the Bronze Age deforestation of Europe affect its climate? A regional climate model study using pollen-based land cover reconstructions. *Clim. Past* **2023**, *19*, 1507–1530. [CrossRef]
8. Barnes, M.L.; Zhang, Q.; Robeson, S.M.; Young, L.; Burakowski, E.A.; Oishi, A.C.; Stoy, P.C.; Katul, G.; Novick, K.A. A century of reforestation reduced anthropogenic warming in the Eastern United States. *Earth's Future* **2024**, *12*, e2023EF003663. [CrossRef]
9. Harris, N.L.; Gibbs, D.A.; Baccini, A.; Birdsey, R.A.; de Bruin, S.; Farina, M.; Fatoyinbo, L.; Hansen, M.C.; Herold, M.; Houghton, R.A.; et al. Global maps of twenty-first century forest carbon fluxes. *Nat. Clim. Chang.* **2021**, *11*, 234–240. [CrossRef]
10. Gómez-García, E.; Biging, G.; García-Villabrille, J.D.; Crecente-Campo, F.; Castedo-Dorado, F.; Rojo-Alboreca, A. Cumulative continuous predictions for bole and aboveground woody biomass in *Eucalyptus globulus* plantations in northwestern Spain. *Biomass Bioenergy* **2015**, *77*, 155–164. [CrossRef]
11. Scalenghe, R.; Celi, L.; Costa, G.; Laudicina, V.A.; Santoni, S.; Vespertino, D.; La Mantia, T. Carbon stocks in a 50 year old *Eucalyptus camaldulensis* stand in Sicily, Italy. *South. For. J. For. Sci.* **2015**, *77*, 263–267. [CrossRef]
12. Infante-Amate, J.; Iriarte-Goñi, I.; Aguilera, E. Historical changes in biomass carbon stocks in the Mediterranean (Spain, 1860–2010). *Anthropocene* **2023**, *44*, 100416. [CrossRef]
13. Mo, L.; Zohner, C.M.; Reich, P.B.; Liang, J.; de Miguel, S.; Nabuurs, G.; Renner, S.S.; Hoogen, J.V.D.; Araza, A.; Herold, M.; et al. Integrated global assessment of the natural forest carbon potential. *Nature* **2023**, *624*, 92–101. [CrossRef]

14. Aalde, H.; Gonzalez, P.; Gytarsky, M.; Krug, T.; Kurz, W.A.; Ogle, S.; Raison, J.; Schoene, D.; Ravindranath, N.H.; Elhassan, N.G.; et al. Forest land. IPCC Guidel. *Natl. Greenh. Gas Invent.* **2006**, *4*, 1–83.
15. IPCC. *IPCC Guidelines for National Greenhouse Gas Inventories*; Institute for Global Environmental Strategies (IGES): Hayama, Japan, 2006.
16. Batjes, N.H. Total carbon and nitrogen in the soils of the world. *Eur. J. Soil Sci.* **1996**, *47*, 151–163. [CrossRef]
17. Scharlemann, J.P.; Tanner, E.V.; Hiederer, R.; Kapos, V. Global soil carbon: Understanding and managing the largest terrestrial carbon pool. *Carbon Manag.* **2014**, *5*, 81–91. [CrossRef]
18. De Vos, B.; Cools, N.; Ilvesniemi, H.; Vesterdal, L.; Vanguelova, E.; Carnicelli, S. Benchmark values for forest soil carbon stocks in Europe: Results from a large scale forest soil survey. *Geoderma* **2015**, *251–252*, 33–46. [CrossRef]
19. ICNF. *IFN6—Principais Resultados—Relatório Sumário*; Instituto da Conservação da Natureza e das Florestas: Lisboa, Portugal, 2019.
20. Dias, A.C.; Martins, M.C.; Arroja, L.; Capela, I. *Management of Forest Resources in the Perspective of the Kyoto Protocol. CARBOCENTRO Project Report*; Universidade de Aveiro: Aveiro, Portugal, 2004.
21. Fabião, A.; Madeira, M.; Steen, E. Root mass in plantations of *Eucalyptus globulus* in Portugal in relation to soil characteristics. *Arid Land Res. Manag.* **1987**, *1*, 185–194.
22. Madeira, M.V.; Fabião, A.; Pereira, J.S.; Araújo, M.C.; Ribeiro, C. Changes in carbon stocks in *Eucalyptus globulus* Labill. Plantations induced by different water and nutrient availability. *For. Ecol. Manag.* **2002**, *171*, 75–85. [CrossRef]
23. Olmedo, G.F.; Guevara, M.; Gilabert, H.; Montes, C.R.; Arellano, E.C.; Barría-Knopf, B.; Gárate, F.; Mena-Quijada, P.; Acuña, E.; Bown, H.E.; et al. Baseline of Carbon Stocks in *Pinus radiata* and *Eucalyptus* spp. Plantations of Chile. *Forests* **2020**, *11*, 1063. [CrossRef]
24. Köppen, W. *Grundriss der Klimakunde*; Walter de Gruyter: Berlin, Germany, 1931.
25. FSCC. Manual IIIa: Sampling and Analysis of Soil. ICP Forests. In *Methods and Criteria for Harmonized Sampling, Assessment, Monitoring and Analysis of the Effects of air Pollution on Forests*; UN/ECE ICP Forests Programme Coordinating Centre: Hamburg, Germany, 2006.
26. RAIZ. Vinte Anos a Criar Conhecimento Para a Bioeconomia de Base Florestal 1996–2016. 2019. Available online: https://20-anos-de-i-d-do-raiz-1996-2016 (accessed on 19 January 2024).
27. *ISO 10694*; Soil Quality—Determination of Organic and Total Carbon After dry Combustion (Elementary Analysis). International Organization for Standardization: Geneva, Switzerland, 1995.
28. Ramos, T.B.; Horta, A.; Gonçalves, M.C.; Pires, F.P.; Duffy, D.; Martins, J.C. The INFOSOLO database as a first step towards the development of a soil information system in Portugal. *CATENA* **2017**, *158*, 390–412. [CrossRef]
29. Gee, G.W.; Or, D. Particle-size analysis. In *Methods of soil Analysis, Part 4. Physical Methods*; Dane, J.H., Topp, G.C., Eds.; SSSA Book Ser. 5; SSSA: Madison, WI, USA, 2002; pp. 255–294.
30. Hollis, J.M.; Hannam, J.; Bellamy, P.H. Empirically-derived pedotransfer functions for predicting bulk density in European soils. *Eur. J. Soil Sci.* **2012**, *63*, 96–109. [CrossRef]
31. Ogle, S.M.; Breidt, F.J.; Eve, M.D.; Paustian, K. Uncertainty in estimating land-use and management impacts on soil organic carbon storage for U.S. agricultural lands between 1982 and 1997. *Glob. Chang. Biol.* **2003**, *9*, 1521–1542. [CrossRef]
32. Wiesmeier, M.; Spörlein, P.; Geuß, U.; Hangen, E.; Haug, S.; Reischl, A.; Schilling, B.; von Lützow, M.; Kögel-Knabner, I. Soil organic carbon stocks in southeast Germany (Bavaria) as affected by land use, soil type and sampling depth. *Glob. Chang. Biol.* **2012**, *18*, 2233–2245. [CrossRef]
33. SROA. *Carta de Solos Reprodução da Carta Apresentada à FAO—1971*; Agência Portuguesa do Ambiente, I.P.: Amadora, Portugal, 2010.
34. Gonçalves, J.A.; Morgado, A. Use of the SRTM DEM as a geo-referencing tool by elevation matching. *Int. Arch. Photogramm. Remote Sens. Spat. Inf. Sci.* **2008**, *37*, 879–883.
35. Aitchison, J. The single principle of compositional data analysis, continuing fallacies, confusions and misunderstandings and some suggested remedies. In Proceedings of the CoDaWork'08, Girona, Spain, 27–30 May 2008; pp. 1–28.
36. IUSS Working Group WRB. *World Reference Base for Soil Resources 2014 International Soil Classification System for Naming Soils and Creating Legends for Soil Maps*; World Soil Resources Reports No. 106; FAO: Rome, Italy, 2015.
37. Pereira, J.; Mateus, J.; Aires, L.; Pita, G.; Pio, C.; David, J.S.; Andrade, V.; Banza, J.; David, T.S.; Paço, T.A.; et al. Net ecosystem carbon exchange in three contrasting Mediterranean ecosystems—The effect of drought. *Biogeosciences* **2007**, *4*, 791–802. [CrossRef]
38. Lal, R. Forest soils and carbon sequestration. *For. Ecol. Manag.* **2005**, *220*, 242–258. [CrossRef]
39. Gómez-Rey, M.; Madeira, M.; González-Prieto, S.; Coutinho, J. Soil C and N dynamics within a precipitation gradient in Mediterranean eucalypt plantations. *Plant Soil* **2010**, *336*, 157–171. [CrossRef]
40. Mayer, M.; Prescott, C.E.; Abaker, W.E.A.; Augusto, L.; Cécillon, L.; Ferreira, G.W.D.; James, J.; Jandl, R.; Katzensteiner, K.; Laclau, J.-P.; et al. Tamm Review: Influence of forest management activities on soil organic carbon stocks: A knowledge synthesis. *For. Ecol. Manag.* **2020**, *466*, 118–127. [CrossRef]
41. Baritz, R.; Seufert, G.; Montanarella, L.; Van Ranst, E. Carbon concentrations and stocks in forest soils of Europe. *For. Ecol. Manag.* **2010**, *260*, 262–277. [CrossRef]

42. Rodeghiero, M.; Rubio, A.; Díaz-Pinés, E.; Romanyà, J.; Marañón-Jiménez, S.; Levy, G.J.; Fernandez-Getino, A.P.; Sebastià, M.T.; Karyotis, T.; Chiti, T.; et al. Soil carbon in Mediterranean ecosystems and related management problems. In *Soil Carbon in Sensitive European Ecosystems*; Jandl, R., Rodeghiero, M., Olsson, M., Eds.; John Wiley & Sons Ltd.: Hoboken, NJ, USA, 2011; pp. 175–218. [CrossRef]
43. IUSS Working Group WRB. *World Reference Base for Soil Resources*; First Update; FAO: Rome, Italy, 2006.
44. Panagos, P.; De Rosa, D.; Liakos, L.; Labouyrie, M.; Borrelli, P.; Ballabio, C. Soil bulk density assessment in Europe. *Agric. Ecosyst. Environ.* **2024**, *364*, 108907. [CrossRef]
45. Fu, P.; Clanton, C.; Demuth, K.M.; Goodman, V.; Griffith, L.; Khim-Young, M.; Maddalena, J.; LaMarca, K.; Wright, L.A.; Schurman, D.W.; et al. Accurate Quantification of 0–30 cm Soil Organic Carbon in Croplands over the Continental United States Using Machine Learning. *Remote Sens.* **2024**, *16*, 2217. [CrossRef]
46. de Rigo, D.; Bosco, C.; San-Miguel-Ayanz, J.; Houston Durrant, T.; Barredo, J.I.; Strona, G.; Caudullo, G.; Di Leo, M.; Boca, R. Forest resources in Europe: An integrated perspective on ecosystem services, disturbances and threats. In *European Atlas of Forest Tree Species*; JRC103181; Publication Office of the European Union: Luxembourg, 2016; pp. 8–19.
47. Ruiz-Peinado, R.; Bravo-Oviedo, A.; López-Senespleda, E.; Bravo, F.; Río, M. Forest management and carbon sequestration in the Mediterranean region: A review. *For. Syst.* **2017**, *26*, eR04S. [CrossRef]

forests

MDPI

Article

Differential Responses of Bacterial and Fungal Community Structure in Soil to Nitrogen Deposition in Two Planted Forests in Southwest China in Relation to pH

Zheng Hou [1,2,3,4,†], Xiaohua Zhang [1,5,†], Wen Chen [1], Ziqi Liang [1], Keqin Wang [6,7], Ya Zhang [2,3,4,*] and Yali Song [6,7,*]

1 College of Ecology and Environment, Southwest Forestry University, Kunming 650224, China; hzheng@mail.cgs.gov.cn (Z.H.); zxh1570494195@163.com (X.Z.); chenwen0610@163.com (W.C.); liangziqi0621@163.com (Z.L.)
2 Kunming General Survey of Natural Resources Center, China Geological Survey, Kunming 650100, China
3 Technology Innovation Center for Natural Ecosystem Carbon Sink, Ministry of Natural Resources, Kunming 650100, China
4 Innovation Base for Eco-Geological Evolution, Protection and Restoration of Southwest Mountainous Areas, Geological Society of China, Kunming 650100, China
5 School of Soil and Water Conservation, Beijing Forestry University, Beijing 100083, China
6 School of Soil and Water Conservation, Southwest Forestry University, Kunming 650224, China; wangkeqin@swfu.edu.cn
7 Yuxi Forestry Ecosystem Research Station of National Forestry and Grassland Administration, Kunming 650224, China
* Correspondence: zhangya@mail.cgs.gov.cn (Y.Z.); songyali@swfu.edu.cn (Y.S.)
† These authors contributed equally to this work.

Abstract: Increased nitrogen deposition profoundly impacts ecosystem nutrient cycling and poses a significant ecological challenge. Soil microorganisms are vital for carbon and nutrient cycling in ecosystems; however, the response of soil microbial communities in subtropical planted coniferous forests to nitrogen deposition remains poorly understood. This study carried out a four-year nitrogen addition experiment in the subtropical montane forests of central Yunnan to explore the microbial community dynamics and the primary regulatory factors in two coniferous forests (*P. yunnanensis* Franch. and *P. armandii* Franch.) under prolonged nitrogen addition. We observed that nitrogen addition elicited different responses in soil bacterial and fungal communities between the two forest types. In *P. yunnanensis* Franch. plantations, nitrogen supplementation notably reduced soil bacterial α-diversity but increased fungal diversity. In contrast, *P. armandii* Franch. forests showed the opposite trends, indicating stand-specific differences. Nitrogen addition also led to significant changes in soil nutrient dynamics, increasing soil pH in *P. yunnanensis* Franch. forests and decreasing it in *P. armandii* Franch. forests. These changes in soil nutrients significantly affected the diversity, community structure, and network interactions of soil microbial communities, with distinct responses noted between stands. Specifically, nitrogen addition significantly influenced the β-diversity of fungal communities more than that of bacterial communities. It also reduced the complexity of bacterial interspecies interactions in *P. yunnanensis* Franch. forests while enhancing it in *P. armandii* Franch. forests. Conversely, low levels of nitrogen addition improved the stability of fungal networks in both forest types. Using random forest and structural equation modeling, soil pH, $NH4^+$-N, and total nitrogen (TN) were identified as key factors regulating bacterial and fungal communities after nitrogen addition. The varied soil nutrient conditions led to different responses in microbial diversity to nitrogen deposition, with nitrogen treatments primarily shaping microbial communities through changes in soil pH and nitrogen availability. This study provides essential insights into the scientific and sustainable management of subtropical plantation forest ecosystems.

Keywords: nitrogen addition; planted coniferous forests; bacteria; fungi

1. Introduction

Increased atmospheric nitrogen (N) deposition is a significant environmental issue that affects global change [1–3]. Historical data indicate that atmospheric nitrogen deposition has increased from 3.4×10^{13} g·N·a^{-1} to 1.0×10^{14} g·N·a^{-1} between 1860 and 1995, with projections suggesting a rise to 2.0×10^{14} g·N·a^{-1} by 2050, demonstrating a clear upward trend [4]. Soil microorganisms, particularly bacteria and fungi, serve as primary decomposers in forest ecosystems and are crucial for ecosystem functionality [5]. They facilitate the cycling of materials and the flow of energy within ecosystems [6]. An increase in nitrogen deposition can enhance net primary production, which subsequently alters the availability of microbial nutrients such as carbon and nitrogen in the soil, affecting their physiological activities [7,8]. These changes can impact the composition of microbial communities, potentially having profound implications on the global carbon and nitrogen cycles and influencing climate change [8].

The impact of atmospheric nitrogen deposition on soil microbiomes, particularly in nitrogen-rich tropical and subtropical forests, has become a pivotal area of study within the ecological sciences. This interest stems from the profound effects that nitrogen deposition has on forest soil microbial communities [9–12]. Recent meta-analyses have attempted to clarify the general response of soil microbial diversity and composition to nitrogen enrichment. These studies have consistently found that nitrogen addition typically diminishes both bacterial and fungal diversity [13,14]. However, the data from these meta-analyses are often heterogeneous, complicating the task of distinguishing responses between different ecosystems, such as natural versus plantation forests. This variability limits the precision and applicability of the findings. Notably, soil microorganisms in tropical nitrogen-rich ecosystems exhibit marked changes in nutrient content and soil properties following nitrogen addition. For instance, in coastal mangrove ecosystems, nitrogen supplementation has been linked to decreased bacterial abundance and increased fungal presence [15]. In subtropical forests in southeastern China, similar nitrogen additions have generally reduced both bacterial and fungal abundances, with fungi showing a more pronounced sensitivity to nitrogen changes [6]. In tropical forests, nitrogen inputs have curtailed microbial growth in acidic soils, with fungi demonstrating better adaptability to these acidic conditions compared to bacteria [16,17]. Despite these insights, the effects of nitrogen deposition on soil microbial communities in nitrogen-rich tropical and subtropical forests remain incompletely understood. Clarifying how soil bacterial and fungal communities and their diversity respond to nitrogen deposition is essential to assess the resilience of forest ecosystems to global changes induced by nitrogen enrichment.

Since the 20th century, the subtropical Central Yunnan Plateau has witnessed extensive vegetation restoration efforts, notably converting farmland back to forests and establishing planted afforestation projects [18,19]. These initiatives have predominantly led to the creation of pure coniferous forests primarily composed of *P. yunnanensis* Franch. and *P. armandii* Franch. However, these forests now face escalating pressures from increased nitrogen deposition, a consequence of rising reactive nitrogen emissions from expanding industrial and agricultural activities in the region [20]. Afforestation has markedly altered species richness, diversity, and soil physicochemical properties within these ecosystems, significantly affecting the diversity and composition of soil microbial communities [21,22]. Soil acidity is a key factor in both soil biogeochemical cycles and the growth environment of soil microorganisms [23]. Previous research indicates that nitrogen addition typically promotes soil acidification, thereby influencing microbial composition and diversity through reduced soil pH [24,25]. Yet, these studies have predominantly focused on naturally complex forests, with limited research directed towards plantation forests characterized by single-species stands and simpler structures. Different tree species exhibit unique nitrogen nutrient utilization strategies, which can lead to variations in soil pH [26,27], influencing the resilience of soil microorganisms to environmental changes [28]. For instance, soil microorganisms in monoculture fir forests are particularly vulnerable to changes in soil pH and nutrient dynamics resulting from nitrogen addition [9,29], whereas planted broadleaf forests demonstrate greater soil-buffering capacities and lower sensitivity to long-term nitrogen impacts compared to their natural counterparts [30]. Clearly, the impacts of nitrogen addition on nitrogen-rich planted forests and natural ecosystems vary and are not uniformly applicable across different tree species. Thus, understanding the responses of soil microbial communities to nitrogen addition in differentiated plantation forests of major tree species is crucial for assessing the impacts of global change on subtropical forest ecosystems.

To explore the effects of long-term nitrogen addition on soil microbial communities in the subtropical Central Yunnan region, we carried out extended simulated nitrogen addition experiments at the Yuxi Forest Ecosystem National Positioning Observation and Research Station. These experiments were concentrated on the subalpine planted coniferous forests, predominantly consisting of *P. yunnanensis* Franch. and *P. armandii* Franch. Our study aimed to fill the gap in understanding the microbial response to nitrogen addition in this specific subalpine area of Central Yunnan and to examine how soil microbial communities adapt to environmental changes and stress caused by prolonged nitrogen addition. Given that subtropical forests are nitrogen-rich ecosystems, nitrogen additions may lead to soil nitrogen enrichment, resulting in soil acidification, nutrient imbalances [31], and adverse effects on soil bacterial and fungal communities. Accordingly, we hypothesized that (1) soil microbial diversity is affected by nitrogen addition, with variations in plantation types driving distinct responses in soil microbial diversity to nitrogen enrichment; (2) nitrogen addition significantly impacts bacterial communities, while fungal communities are relatively less affected; and (3) nitrogen addition primarily influences soil microbial communities through changes in soil pH and nitrogen availability in both types of coniferous forests. The findings of this research are pivotal for the scientific management of subtropical plantation ecosystems in relation to nitrogen deposition and augment our understanding of soil microbial functions and ecosystem processes amidst global changes.

2. Materials and Methods

2.1. Site Description

The study area is located at the Yunnan Yuxi Forest Ecosystem National Positioning Observation and Research Station (23°46′18″~23°54′34″ N, 101°16′06″~101°16′12″ E) (Figure 1), with an altitude ranging from 1260.0 to 2614.4 m. This area features a subtropical low-latitude plateau monsoon climate characterized by typical mountain climate features. The seasons are distinctly split into dry (November to April) and wet (May to October) periods, with rain and heat coinciding and concentrated rainfall occurring from June to August [32]. The average annual precipitation is 1050 mm, and the average annual temperature is 15 °C. The highest recorded temperature is 33 °C in June, while the lowest drops to −2.2 °C in February. Forest coverage in the study area is approximately 86%, consisting primarily of primary and secondary primeval forests, which are predominantly semi-humid evergreen broad-leaved forests. The area's vegetation is diverse, encompassing 98 families, 137 genera, and 324 species, including *A. palisotii* (Desv.) Alston, *Reevesia pubescens* Mast., and *R. delavayi* Franch., among others. The distribution of forests exhibits a vertical arrangement corresponding to increasing altitude. The primary forest vegetation types include natural forests such as subtropical evergreen broad-leaved forests and high mountain dwarf forests, as well as planted forests like mixed coniferous and broad-leaved forests and pure coniferous forests. *P. yunnanensis* Franch. and *P. armandii* Franch. forests are the most abundant species of planted forests in the region, coexisting in a relatively small area with similar terrain and microclimate conditions. Both plantations have been established for about 40 years. The predominant soil type is Argi-Udic ferrosols, with localized areas of Hapli-Udic argosols (World Reference Base for Soil Resource).

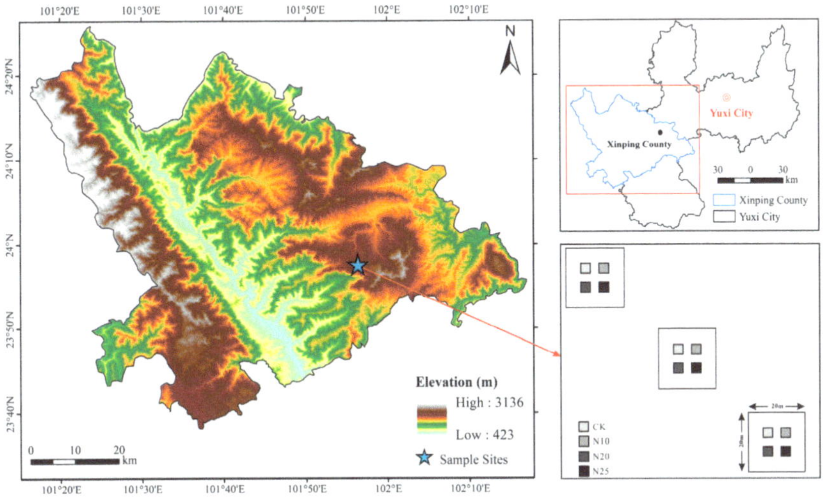

Figure 1. Overview map of the study area.

2.2. Experimental Design

This study focused on two types of planted forests, *P. yunnanensis* Franch. and *P. armandii* Franch., located within the research station, approximately 2.6 km apart. Nitrogen fertilization commenced in January 2019. After thorough on-site field investigations (Table S1), representative areas were selected for each forest type. Three 20 m × 20 m sample plots were established for each type, maintaining at least 10 m of separation between plots. A split-plot design was utilized within each plot, consisting of four smaller 3 m × 3 m quadrats. These quadrats were designated for the control group and three nitrogen fertilization gradients, creating a total of 12 small quadrats to provide three replicates for each treatment. Based on the atmospheric wet nitrogen deposition flux in Southwest China [33], the nitrogen deposition during dry and wet seasons (15 $g{\cdot}N{\cdot}m^{-2}{\cdot}a^{-1}$ and 10 $g{\cdot}N{\cdot}m^{-2}{\cdot}a^{-1}$, respectively) [34], and the annual increase in nitrogen deposition (0.05 $g{\cdot}N{\cdot}m^{-2}{\cdot}a^{-1}$) in Southwest China [32], we adjusted the nitrogen application to reflect the annual level of 3.84 $g{\cdot}N{\cdot}m^{-2}{\cdot}a^{-1}$. The nitrogen gradients were set as follows: 0 $g{\cdot}N{\cdot}m^{-2}{\cdot}a^{-1}$ (CK), 10 $g{\cdot}N{\cdot}m^{-2}{\cdot}a^{-1}$ (N10), 20 $g{\cdot}N{\cdot}m^{-2}{\cdot}a^{-1}$ (N20), and 25 $g{\cdot}N{\cdot}m^{-2}{\cdot}a^{-1}$ (N25). Experimental urea ($CO(NH_2)_2$ analytical grade, >=99.0%) was used as the nitrogen source. The annual nitrogen application was divided into 12 equal parts, with each portion dissolved in 1000 mL of water and applied monthly using a backpack sprayer. The control plots received an equivalent volume of water.

2.3. Soil Sampling

Soil samples were collected in early March 2023, four years and two months after the initiation of the simulated nitrogen deposition. Sampling began by removing the litter layer from the quadrat surface. Using the five-point sampling technique, soil from the top 20 cm was extracted with an auger. Soil from the three replicate quadrats under the same treatment was mixed thoroughly and sieved through a 2 mm nylon mesh to remove visible roots and stones. The sieved soil was then divided into two equal portions: one was immediately stored at −80 °C for subsequent DNA extraction and high-throughput sequencing analysis, and the other was air-dried for chemical property analysis.

2.4. Measurement of Soil Chemical Properties

The measurement of soil chemical properties was conducted using the methods outlined by Bao [35]. Soil pH was determined with a pH meter (m_{soil}:V_{water} = 1:5). Total nitrogen (TN) content was assessed via the semi-micro Kjeldahl method. Available phosphorus (AP) was quantified using inductively coupled plasma emission spectrometry. Potassium ions (K^+) were measured by extraction with 1 mol·L^{-1} neutral ammonium acetate, followed by flame photometer analysis. Soil organic matter (SOM) content was determined using the potassium dichromate volumetric method with external heating to ascertain organic carbon, which was then converted to organic matter using a conversion factor of 1.724. Ammonium nitrogen (NH_4^+-N) in the soil was quantified using the indophenol blue method, and soil nitrate nitrogen (NO_3^--N) was measured via ultraviolet spectrophotometry using the colorimetric method.

2.5. DNA Extraction and Illumina Sequencing

Total DNA from the samples was extracted using the EZNA™ Mag-Bind Soil DNA Kit from OMEGA Bio-tek. The extracted DNA samples were sent to Sangon Biotech (Shanghai) Co., Ltd. (Shanghai, China) for high-throughput sequencing. For soil bacteria, the V3–V4 variable region was PCR-amplified using primers 341F (5′-CCTACGGGNGGCWGCAG-3′) and 805R (5′-GACTACHVGGGTATCTAATCC-3′) [36]. For soil fungi, the ITS region was targeted using primers ITS1F (5′-CTTGGTCATTTAGAGGAAGTAA-3′) [37] and ITS2R (5′-GCTGGCTTTCTTCATCGATGC-3′) [38]. Both bacterial and fungal amplifications involved two rounds of PCR. In the first round, we used 16SV3-V4 primers for bacteria and ITS1-ITS2 primers for fungi, with a 30 μL reaction mix (15 μL 2 × Hieff® Robust PCR Master Mix, 1 μL of each primer, 10–20 ng of PCR products, and 9–12 μL of H_2O). The PCR conditions were an initial denaturation at 94 °C for 3 min, followed by 5 cycles at 94 °C for 20 s, annealing at 45 °C for 20 s, and extension at 65 °C for 30 s; then 20 cycles at 94 °C for 20 s, 55 °C for 20 s, and 72 °C for 30 s, with a final extension at 72 °C for 5 min and storage at 10 °C. In the second round, Illumina bridge PCR-compatible primers were used with a similar reaction mix and conditions: pre-denaturation at 95 °C for 3 min, followed by 5 cycles of denaturation at 94 °C for 20 s, annealing at 55 °C for 20 s, and extension at 72 °C for 30 s, concluding with a final extension at 72 °C for 5 min and storage at 10 °C. The PCR products were verified via 2% agarose gel electrophoresis, and library concentrations were measured using a Qubit 3.0 fluorometer. After confirming library quality, sequencing was conducted on the Illumina MiSeq platform.

2.6. Microbial Data Analysis and Co-Occurrence Network Construction

After quality control filtering of the sequencing data, Usearch 11.0.667 software [39] was used to cluster non-redundant sequences (excluding single sequences) at a 97% similarity threshold for operational taxonomic unit (OTU). Chimeras were removed during the clustering process, with similarity assessments conducted after comparing 0.1% of the sequences [40]. The sequences were preprocessed by trimming adapter sequences and removing low-quality reads using the FastQC and Trimmomatic tools. The SILVA database (version 132) was used for the identification of bacterial community composition, while the UNITE database (version 8.0) was used for the identification of fungal community composition. Microbial α-diversity and richness were calculated using Mothur 1.43.0. The Shannon index [41] was employed to quantify community diversity, while the Chao index [42] was used to estimate community richness. The relative abundance of bacterial and fungal communities was considered when constructing networks involving OTUs with a relative abundance above 0.1%. OTUs with a relative abundance of zero in two-thirds of the samples were excluded. Random matrix theory (RMT) was applied to determine the optimal similarity threshold, and the pairwise similarity matrix was calculated based on Spearman correlation. All analyses were performed using the Molecular Ecological Networks (MENs) analysis platform (iNAP platform, updated on 6 June 2023, https://inap.denglab.org.cn/, accessed on 20 June 2023) [43]. Nodes and edges were exported and processed using Gephi 0.9.2 software to generate co-occurrence networks for the bacterial and fungal communities.

2.7. Statistical Analysis

We investigated the effects of nitrogen addition on the soil chemical properties and the bacterial and fungal communities in the *P. yunnanensis* Franch. and *P. armandii* Franch. forests. Initially, homogeneity of variance tests were performed using SPSS 25.0 software (SPSS Inc., Chicago, IL, USA). Upon confirming homogeneity of variance, analysis of variance (ANOVA) was utilized to evaluate the impacts of different nitrogen addition treatments on soil chemical properties and microbial α-diversity. For cases of heterogeneous variance, the Welch correction method was applied prior to variance analysis. In the R software (v4.3.1; http://www.r-project.org/, accessed on 25 June 2023), the "vegan" [44] and "ggplot2" [45] packages were utilized for principal coordinate analysis (PCoA) and similarity analysis (ANOSIM) based on microbial genus-level classification to detect differences in microbial community composition using Bray–Curtis distance. Redundancy analysis (RDA) was conducted to ascertain whether soil properties were key factors influencing the composition of bacterial and fungal communities in these forests. Subsequently, the "randomForest" package was used for random forest analysis to assess the importance of individual soil chemical properties on the composition of microbial communities in *P. yunnanensis* Franch. and *P. armandii* Franch. forests, where community composition was represented by the first axis of PCoA. A structural equation model (SEM) was established using Amos 24.0 software to evaluate the effects of nitrogen addition and soil properties (TN, NH_4^+-N, and pH) on microbial communities. The goodness of fit of the model was assessed using the chi-square test (χ^2), *p*-value, and root mean square error of approximation (RMSEA).

3. Results

3.1. Soil Properties in Response to N Addition

As depicted in Table 1, significant differences were observed in soil nutrients between the two forest stands, primarily attributed to variations in tree species. Additionally, nitrogen input significantly influenced soil nutrients in both stands. Compared to CK, the soil SOM and K^+ content in the *P. yunnanensis* Franch. forest were significantly higher than in the *P. armandii* Franch. forest. Conversely, the levels of pH, NH_4^+-N, and NO_3^--N were notably lower in the *P. yunnanensis* Franch. forest. In the *P. yunnanensis* Franch. forest, an increase in nitrogen input was correlated with a gradual rise in soil pH and initial increases followed by decreases in NH_4^+-N and NO_3^--N. On the other hand, in the *P. armandii* Franch. forest, escalating nitrogen input resulted in a progressive decrease in soil pH and K^+ content, while NO_3^--N, SOM, and AP content showed an increase. The content of NH_4^+-N initially rose and then diminished with increased nitrogen input.

Table 1. Soil properties analysis.

Forest Types	Treatments	Soil Physicochemical Properties						
		pH	SOM ($g \cdot kg^{-1}$)	NH_4^+-N ($mg \cdot kg^{-1}$)	NO_3^--N ($mg \cdot kg^{-1}$)	AP ($mg \cdot kg^{-1}$)	K^+ ($mg \cdot kg^{-1}$)	TN ($g \cdot kg^{-1}$)
P. yunnanensis Franch. Forest	CK	4.17(0.04) ^{c}	64.67(4.13) ^{a}	11.62(0.34) ^{c}	1.09(0.18) ^{c}	0.19(0.01) ^{ab}	29.19(7.02) ^{b}	0.28(0.01) ^{b}
	N10	4.25(0.02) ^{b}	70.73(11.27) ^{a}	14.92(0.28) ^{b}	1.64(0.08) ^{a}	0.20(0.01) ^{ab}	19.81(1.23) ^{c}	0.30(0.00) ^{a}
	N20	4.32(0.04) ^{ab}	64.67(4.59) ^{a}	16.14(0.34) ^{a}	1.36(0.13) ^{b}	0.17(0.04) ^{b}	40.97(2.41) ^{a}	0.29(0.01) ^{ab}
	N25	4.35(0.04) ^{a}	64.27(3.69) ^{a}	15.08(0.20) ^{b}	1.04(0.06) ^{d}	0.22(0.02) ^{a}	21.42(1.23) ^{b}	0.31(0.02) ^{a}
P armandii Franch. Forest	CK	4.71(0.02) ^{a}	47.18(4.91) ^{b}	14.20(0.98) ^{c}	1.60(0.22) ^{d}	0.13(0.02) ^{b}	25.17(2.32) ^{a}	0.27(0.03) ^{b}
	N10	4.53(0.05) ^{b}	58.88(2.21) ^{a}	18.53(0.29) ^{a}	3.64(0.49) ^{c}	0.19(0.02) ^{a}	22.49(1.39) ^{ab}	0.32(0.02) ^{a}
	N20	4.42(0.04) ^{c}	48.16(2.44) ^{b}	15.39(0.16) ^{b}	8.80(0.24) ^{b}	0.21(0.03) ^{a}	19.28(1.61) ^{b}	0.26(0.03) ^{b}
	N25	4.18(0.04) ^{d}	62.00(3.84) ^{a}	11.40(0.57) ^{d}	9.71(0.21) ^{a}	0.18(0.02) ^{ab}	21.42(1.23) ^{b}	0.33(0.02) ^{a}
Total	N	0.000 ***	0.020 *	0.000 ***	0.000 ***	0.041 *	0.000 ***	0.001 **
	Forest types	0.000 ***	0.000 ***	0.035 *	0.000 ***	0.103	0.000 ***	0.864
	N× Forest types	0.000 ***	0.093	0.000 ***	0.000 ***	0.009 **	0.000 ***	0.17

(1) Data are presented as means (standard errors), and the different letters within columns indicate significant differences among treatments in the same stand ($p < 0.05$). (2) * indicates $p < 0.05$, ** indicates $p < 0.01$, *** indicates $p < 0.001$. (3) Abbreviations: CK, 0 $g \cdot N \cdot m^{-2} \cdot a^{-1}$; N10, 10 $g \cdot N \cdot m^{-2} \cdot a^{-1}$; N20, 20 $g \cdot N \cdot m^{-2} \cdot a^{-1}$; and N25, 25 $g \cdot N \cdot m^{-2} \cdot a^{-1}$. SOM, soil organic matter; TN, total nitrogen; AP, available phosphorus; K^+, potassium ion.

3.2. Soil Microbial Diversity in Response to N Addition

Utilizing α-diversity indices to evaluate soil microbial diversity and richness under control conditions (CK), the *P. yunnanensis* Franch. forest exhibited significantly higher numbers of OTUs, Shannon index, and Chao index for soil bacterial communities compared to the *P. armandii* Franch. forest. However, the trend was reversed for fungal communities. In the *P. yunnanensis* Franch. forest, nitrogen addition notably decreased the OTUs, Shannon index, and Chao index of bacterial communities in a dose-dependent manner (CK > N10 > N20 > N25). Conversely, in the *P. armandii* Franch. forest, nitrogen supplementation led to increases in these indices for bacterial communities, with the most pronounced effects observed at the lowest level of nitrogen addition (N10). Regarding soil fungal communities in the *P. yunnanensis* Franch. forest, nitrogen addition had a stimulatory effect on the OTUs, Shannon index, and Chao index, with the N20 treatment showing more significant enhancement than the N10 and N25 treatments. However, in the *P. armandii* Franch. forest, nitrogen addition resulted in decreased OTUs and Chao index for fungal communities, while the Shannon index increased. Overall, the varying levels of nitrogen addition and the distinct characteristics of the two forest types significantly influenced the α-diversity of both soil bacterial and fungal communities in the subalpine forests of central Yunnan. These effects were found to be statistically significant ($p < 0.001$; Table 2).

Table 2. Soil bacterial and fungal alpha diversity analysis.

Forest Types	Treatments	Bacteria			Fungi		
		OTUs	Shannon	Chao	OTUs	Shannon	Chao
P. yunnanensis Franch. Forest	CK	3416(189) ^{a}	5.92(0.05) ^{a}	4068(201) ^{a}	481(28) ^{c}	2.69(0.11) ^{c}	487(34) ^{c}
	N10	2787(17) ^{bc}	5.85(0.04) ^{ab}	3524(112) ^{b}	884(4) ^{a}	3.21(0.07) ^{b}	1087(34) ^{ab}
	N20	2578(95) ^{c}	5.82(0.05) ^{b}	3200(190) ^{b}	886(33) ^{a}	3.83(0.22) ^{a}	1128(58) ^{a}
	N25	2819(118) ^{b}	5.78(0.03) ^{b}	3322(231) ^{b}	793(16) ^{b}	3.07(0.02) ^{b}	1028(10) ^{b}
P. armandii Franch. Forest	CK	2388(153) ^{c}	5.63(0.06) ^{c}	2778(59) ^{c}	741(37) ^{a}	2.76(0.06) ^{a}	973(57) ^{a}
	N10	2979(103) ^{a}	6.00(0.05) ^{a}	3758(87) ^{a}	287(51) ^{c}	3.20(0.07) ^{b}	305(34) ^{c}
	N20	2689(65) ^{b}	5.99(0.02) ^{a}	3161(176) ^{b}	363(15) ^{bc}	2.80(0.02) ^{a}	456(44) ^{b}
	N25	2217(148) ^{cd}	5.83(0.10) ^{b}	2683(65) ^{c}	434(57) ^{b}	3.93(0.14) ^{c}	444(60) ^{b}
Total	N	0.000 ***	0.000 ***	0.000 ***	0.291	0.000 ***	0.014 *
	Forest types	0.000 ***	0.307	0.000 ***	0.000 ***	0.51	0.000 ***
	N×Forest types	0.000 ***	0.000 ***	0.000 ***	0.000 ***	0.000 ***	0.000 ***

(1) Data are presented as means (standard errors), and the different letters within columns indicate significant differences among treatments in the same stand ($p < 0.05$). (2) * indicates $p < 0.05$, *** indicates $p < 0.001$. (3) Abbreviations: CK, 0 $g \cdot N \cdot m^{-2} \cdot a^{-1}$; N10, 10 $g \cdot N \cdot m^{-2} \cdot a^{-1}$; N20, 20 $g \cdot N \cdot m^{-2} \cdot a^{-1}$; and N25, 25 $g \cdot N \cdot m^{-2} \cdot a^{-1}$.

Notably, significant alterations were observed in the fungal communities of both forest types ($p < 0.05$), whereas the bacterial communities exhibited no substantial changes, revealing that the composition of fungal species is more responsive to nitrogen addition compared to most bacterial species (Figure 2).

3.3. Effects of N Addition on Soil Microbial Community Composition

The predominant bacterial taxa in both the *P. yunnanensis* Franch. and *P. armandii* Franch. forest soils were *Acidobacteria* and *Proteobacteria*, constituting approximately 58.82% to 68.39% of the total community. In a similar pattern, the dominant fungal taxa were *Basidiomycota* and *Ascomycota*, comprising about 79.17% to 91.79% of the community. The *Acidobacteria* phylum is about twice as abundant in *P. yunnanensis* Franch. forests compared to *P. armandii* Franch. forests. Conversely, the *Proteobacteria* phylum was significantly more abundant in *P. armandii* Franch. forests ($p < 0.05$; Figure 3a). However, the difference in the relative abundance of fungal taxa between the two forests was not substantial (Figure 3b).

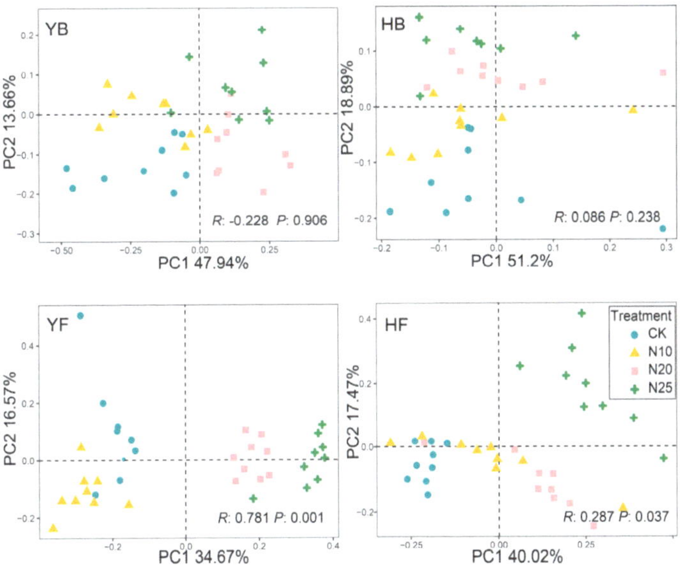

Figure 2. Principal coordinate analysis (PCoA) was performed to analyze the bacterial and fungal communities in different forest stands under varying levels of nitrogen addition using an OTU table. Abbreviations: CK, 0 g·N·m^{-2}·a^{-1}; N10, 10 g·N·m^{-2}·a^{-1}; N20, 20 g·N·m^{-2}·a^{-1}; and N25, 25 g·N·m^{-2}·a^{-1}. YB, the bacterial community of *P. yunnanensis* Franch. forest; YF, the fungal community of *P. yunnanensis* Franch. forest; HB, the bacterial community of *P. armandii* Franch. forest; HF, the fungal community of *P. armandii* Franch. forest.

In *P. yunnanensis* Franch. forests, the relative abundance of the *Acidobacteria* phylum within the soil bacterial community significantly increased at low nitrogen levels and decreased markedly at higher nitrogen levels (N10 > N20 > N25 > CK), while other bacterial phyla showed no notable response to nitrogen addition (Figure 3a; Table S2). For soil fungal taxa, the *Rozellomycota* phylum significantly declined with increasing nitrogen addition. *Ascomycota* showed a notable increase only at the N10 level, and *Mucoromycota* increased significantly at N25. Correlation analysis indicated a significant positive correlation between the *Acidobacteria* phylum and NO_3^--N ($p < 0.001$) and a significant negative correlation between the *Rozellomycota* phylum and both NH_4^+-N and pH ($p < 0.05$) (Table S3). In *P. armandii* Franch. forests (Table S4), nitrogen addition significantly affected the relative abundance of the *Acidobacteria* and *Chloroflexi* phyla in soil bacteria, with an increase observed at low nitrogen levels and a decrease at high levels. The relative abundance of *Proteobacteria* was substantially reduced, showing no significant differences among the various nitrogen levels. *Acidobacteria* were positively correlated with NH_4^+-N, whereas *Proteobacteria* showed a significant negative correlation with NO_3^--N and AP and a positive correlation with K^+. The *Actinobacteria* phylum was positively correlated with AP and negatively with K^+ (Table S4). Compared to CK, nitrogen addition significantly increased the relative abundance of *Ascomycota* and decreased that of *Basidiomycota* in fungal taxa. The relative abundance of *Mortierellomycota* was highest at the lowest nitrogen input level (N10) but was suppressed at N20 and N25. Correlation analyses revealed a significant positive correlation of *Basidiomycota* with pH and negative correlations with SOM and TN. *Ascomycota* showed significant negative correlations with pH and positive correlations with SOM and TN; *Mortierellomycota* were significantly positively correlated with NH_4^+-N (Table S5).

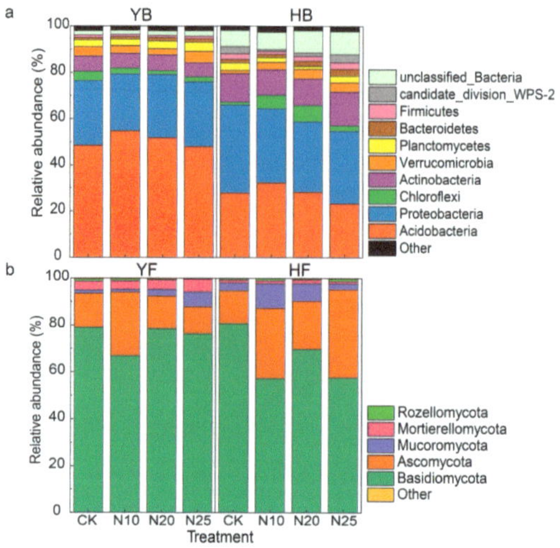

Figure 3. The relative abundance of bacteria and fungi at the phylum level. Relative abundance at the bacterial phylum level (**a**); relative abundance at the fungal phylum level (**b**). Abbreviations: CK, 0 g·N·m^{-2}·a^{-1}; N10, 10 g·N·m^{-2}·a^{-1}; N20, 20 g·N·m^{-2}·a^{-1}; and N25, 25 g·N·m^{-2}·a^{-1}. YB, the bacterial community of *P. yunnanensis* Franch. forest; YF, the fungal community of *P. yunnanensis* Franch. forest; HB, the bacterial community of *P. armandii* Franch. forest; HF, the fungal community of *P. armandii* Franch. forest.

3.4. Soil Microbial Co-Occurrence Networks

In the *P. yunnanensis* Franch. and *P. armandii* Franch. forests, the total number of nodes in the bacterial co-occurrence networks was comparable to those of their respective controls (CKs). However, the network of the *P. yunnanensis* Franch. forest exhibited a 39.5% increase in connecting edges compared to that of the *P. armandii* Franch. forest (Figure 4). This difference underscores the greater complexity of bacterial interspecies interactions within the *P. yunnanensis* Franch. forest. With increased nitrogen application in the *P. yunnanensis* Franch. forest, there was a notable decrease in the total number of nodes, the number of connecting edges, and the average degree of the soil bacterial co-occurrence network, accompanied by an increase in modularity value. Conversely, in the *P. armandii* Franch. forest, these parameters initially decreased and then increased as nitrogen levels rose, reaching their peak in the N20 treatment. The ratios of modularity values and positively correlated edges exhibited an inverse trend, suggesting that the complexity of bacterial interspecies relationships was reduced with increased nitrogen in the *P. yunnanensis* Franch. forest, whereas it was augmented in the *P. armandii* Franch. forest, reaching significant enhancement at the N20 treatment level and indicating a threshold for nitrogen-induced promotion.

The properties of the fungal co-occurrence networks in both forests demonstrated trends opposite to those observed in their bacterial counterparts (Figure 5). In both forest types, the number of total nodes under control conditions (CKs) was comparable, yet the total number of connected edges in the fungal network of the *P. yunnanensis* Franch. forest was significantly lower than that in the *P. armandii* Franch. forest, indicating lesser complexity in fungal interspecies relationships in the former. For both forests, the total number of nodes, the number of connecting edges, and the average degree of the soil fungal co-occurrence network initially increased and then decreased following nitrogen addition treatments. Conversely, the modularity values exhibited a decrease followed by an increase, peaking in the N20 treatments. These patterns suggest that lower levels of nitrogen addition promote stabilization and complexity of fungal networks in both forests, whereas higher levels of nitrogen do not sustain this positive effect and may even inhibit it.

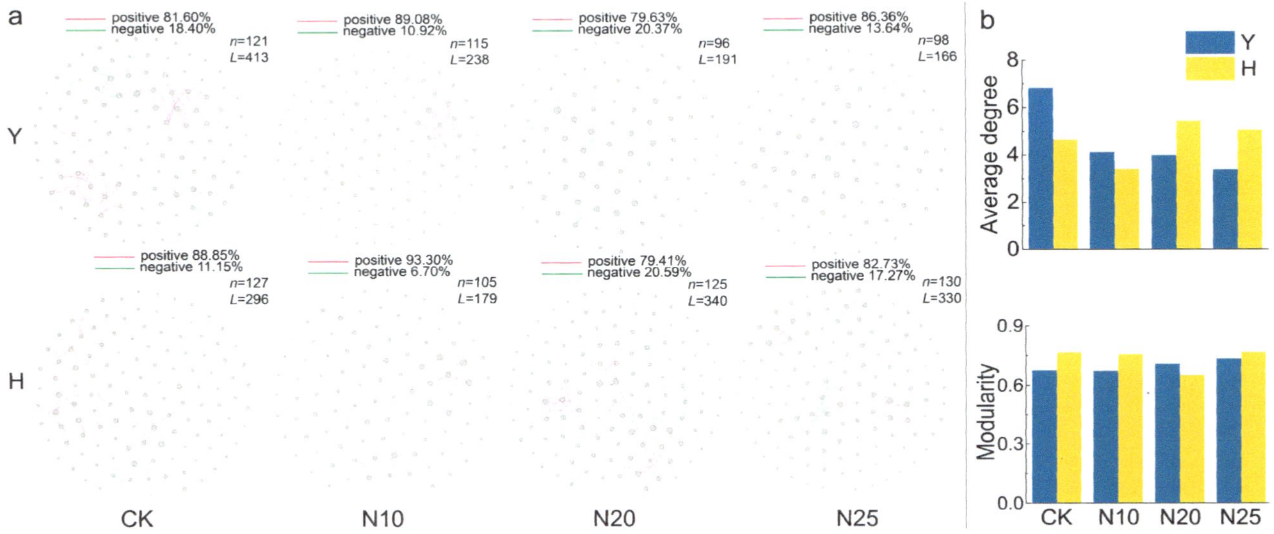

Figure 4. Soil bacterial co-occurrence network-based OTU profile (**a**). Characterization of the properties of bacterial co-occurrence networks (**b**). Abbreviation: n, total nodes; L, total linking edges. CK, 0 $g \cdot N \cdot m^{-2} \cdot a^{-1}$; N10, 10 $g \cdot N \cdot m^{-2} \cdot a^{-1}$; N20, 20 $g \cdot N \cdot m^{-2} \cdot a^{-1}$; and N25, 25 $g \cdot N \cdot m^{-2} \cdot a^{-1}$. Y: *P. yunnanensis* Franch. forest; H: *P. armandii* Franch. forest. Node size indicates the connection size of the module, red connecting lines indicate cooperative relationships between species, and green connecting lines indicate competitive relationships.

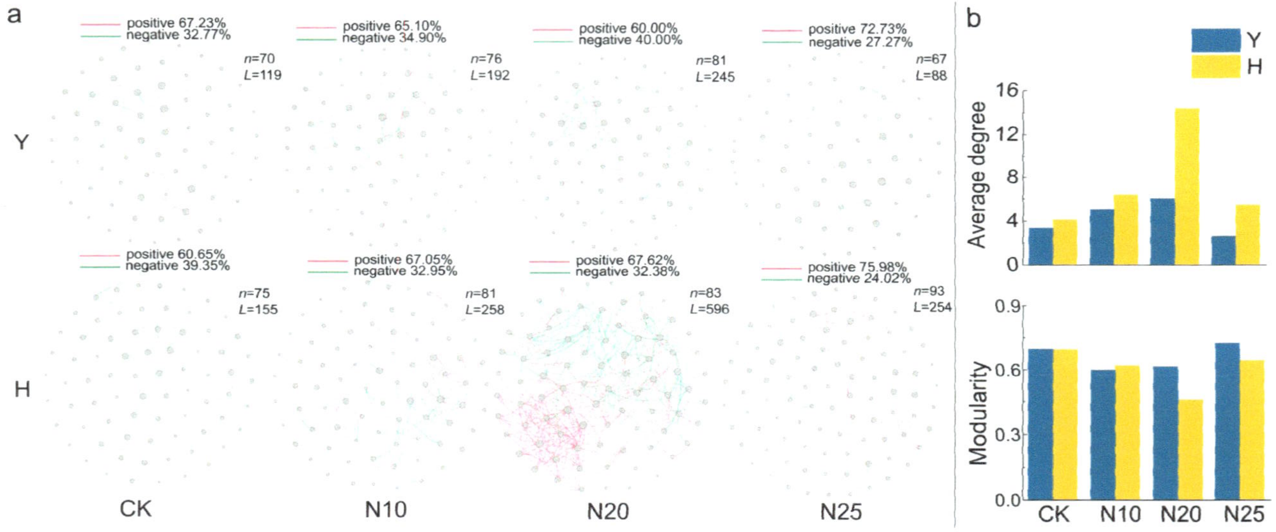

Figure 5. Soil fungal co-occurrence network-based OTU profile (**a**). Characterization of the properties of fungal co-occurrence networks (**b**). Abbreviations: n, total nodes; L, total linking edges. CK, 0 $g \cdot N \cdot m^{-2} \cdot a^{-1}$; N10, 10 $g \cdot N \cdot m^{-2} \cdot a^{-1}$; N20, 20 $g \cdot N \cdot m^{-2} \cdot a^{-1}$; and N25, 25 $g \cdot N \cdot m^{-2} \cdot a^{-1}$. Y: *P. yunnanensis* Franch. forest; H: *P. armandii* Franch. forest. Node size indicates the connection size of the module, red connecting lines indicate cooperative relationships between species, and green connecting lines indicate competitive relationships.

3.5. Relationships between Microbial Community and Soil Chemical Properties

Soil properties emerged as key determinants of the differences in bacterial and fungal community compositions across the forest types. RDA supported this observation, revealing that environmental factors accounted for 63.67% and 85.89% of the variations in soil bacterial communities and 81.03% and 77.04% in the fungal communities of *P. yunnanensis* Franch. and *P. armandii* Franch. forests, respectively (Figure S1). Random forest analysis identified NH_4^+-N as a crucial factor influencing the composition and abundance of soil bacterial and fungal communities in the *P. yunnanensis* Franch. forest and the bacterial communities in the *P. armandii* Franch. forest. Additionally, AP was pinpointed as a primary variable impacting the soil bacterial communities in the *P. armandii* Franch. forest. Soil pH also played a significant role in these community dynamics (Figure 6a). We further explored the effects of nitrogen addition on soil-microbe diversity in these subalpine planted forests through a structural equation model (Figure 6b). The model exhibited a good fit for the *P. yunnanensis* Franch. forest ($\chi^2 = 4.515$, $df = 7$, $p = 0.719$, RMSEA < 0.001) and for the *P. armandii* Franch. forest ($\chi^2 = 11.496$, $df = 6$, $p = 0.074$, RMSEA < 0.05). In the *P. yunnanensis* Franch. forest model, soil pH negatively influenced bacterial α-diversity, while NH_4^+-N content positively affected fungal α-diversity. TN content had a negative impact. In the *P. armandii* Franch. forest model, soil pH negatively affected both bacterial and fungal α-diversity. NH_4^+-N content positively influenced bacterial diversity, and TN content positively affected fungal diversity. A significant negative interaction between bacteria and fungi was also observed in this model (Figure 6b). In conclusion, NH_4^+-N and pH were identified as the primary factors influencing the soil bacterial and fungal communities in both forest types.

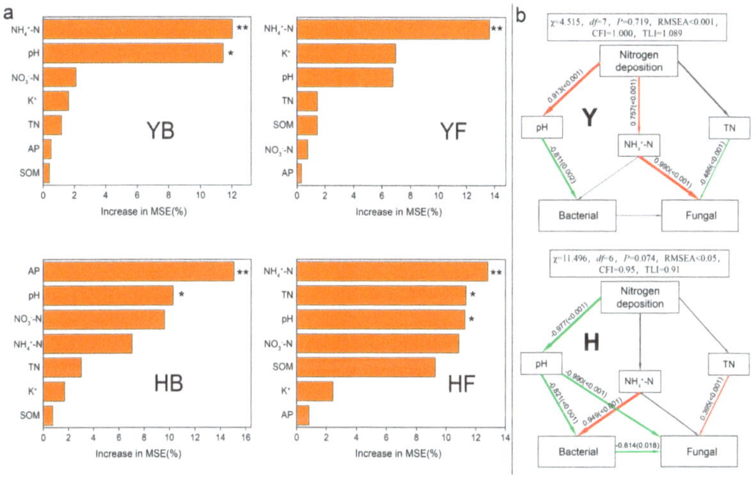

Figure 6. Soil properties drive the variation of soil bacterial/fungal community structure and richness. Random forests were used to determine the importance of the variable (**a**). Structural equation modeling (SEM) in soil-microbial community systems in response to N addition (**b**). Abbreviations: * indicates $p < 0.05$, ** indicates $p < 0.01$; pH, pondus hydrogenii; SOM, soil organic matter; NH_4^+-N, soil ammonium N; NO_3^--N, soil nitrate N; TN, total nitrogen; AP, available phosphorus; K^+, potassium ion; B. α index: bacterial α index; F. α index: fungal α index; YB: the bacterial community of *P. yunnanensis* Franch. forest; YF: the fungal community of *P. yunnanensis* Franch. forest; HB: the bacterial community of *P. armandii* Franch. forest; HF: the fungal community of *P. armandii* Franch. forest. Red arrows indicate significant positive correlations ($p < 0.05$), green arrows indicate significant negative correlations ($p < 0.05$), and gray arrows indicate non-significant relationships ($p > 0.05$).

4. Discussion

With the influence of global environmental changes and human activities, atmospheric nitrogen deposition in forest ecosystems in subtropical regions of China continues to increase. The study of soil microbial community response to nitrogen deposition in subtropical artificial coniferous forests provides an important basis for predicting the ecological consequences of the future increase in atmospheric nitrogen deposition as well as the scientific and sustainable management of the ecosystem. In this study, four nitrogen application gradients were established for long-term simulated nitrogen deposition experiments, and the response of soil microbial diversity and community composition of subtropical plantation forests to long-term nitrogen deposition in different stand types was systematically investigated. Our research found that different afforestation tree species have varying effects on soil pH, nutrient content, and microbial communities. Additionally, sustained high levels of nitrogen addition negatively impact microbial community diversity, with fungal communities being more sensitive to nitrogen addition than bacterial communities. Soil pH and nitrogen availability jointly regulate the diversity of soil bacterial and fungal communities.

4.1. Afforestation Tree Species Differences Lead to Distinct Responses of Soil Microbial Diversity to Nitrogen Addition

Consistent with our hypothesis, the continuous four-year nitrogen addition significantly altered soil microbial diversity in two planted forests, each exhibiting distinct responses to the nitrogen input. The divergence among ecosystem types and background pH values contributes to variations in soil resistance and stability in response to nitrogen deposition [31]. It is generally accepted that metal cations in soils act as the primary mechanism for buffering against soil acidification, which is typically balanced by the interaction between potassium (K^+) and acidic (hydrogen, H^+) cations [46,47]. Despite both forests being coniferous and sharing acidic soil characteristics within the same research area, their responses to nitrogen addition varied markedly. In the *P. yunnanensis* Franch. forest, nitrogen addition led to an increase in soil pH (from 4.17 to 4.35) and K^+ content. Conversely, in the *P. armandii* Franch. forest, both soil pH (from 4.71 to 4.18) and K^+ content gradually decreased, highlighting a notable difference in their buffering capacities against nitrogen disturbances. The well-known process of soil acidification in the nitrogen cycle occurs as bacteria oxidize ammonium to nitrate, producing hydrogen ions during nitrification [48], which was evident from the decrease in pH observed in the *P. armandii* Franch. forest, where ammonium and nitrate nitrogen showed opposing trends. Additionally, a significant increase in SOM content was observed in the *P. armandii* Franch. forest following nitrogen addition. From a microbial perspective, nitrogen addition typically reduces SOC mineralization, which is interpreted as a reduction in microbial nitrogen mining, thereby retaining more carbon in the soil [49]. Furthermore, the nitrogen-induced increase in SOM may also enhance soil cation exchange capacity (CEC) and contribute to a reduction in soil pH levels [47].

It is widely accepted that nitrogen addition generally has a negligible effect on the diversity of fungal communities [50]. Fungi, compared to bacteria, typically require less nitrogen [51] and are capable of degrading complex organic compounds more slowly [25]. This resistance to changes is primarily because fungi are not sensitive to alterations in soil pH, which are significantly influenced by nitrogen addition [47]. Contrary to our second hypothesis, our research revealed that the diversity of fungal communities was impacted by nitrogen addition in a manner similar to that of bacterial communities. This unexpected outcome could be attributed to the presence of polymerized phenolic compounds, such as lignin and tannins, in pine forest soils, which represent complex, low-quality carbon sources that are difficult to decompose. Interestingly, fungi do not necessarily favor these tough-to-degrade sources of organic matter [51]. Furthermore, a microculture experiment demonstrated that both fungi and bacteria were inefficient at utilizing these low-quality carbon sources during the later stages of apoplastic decomposition [52]. This suggests that additional nitrogen might have comparable effects on both bacterial and fungal communities. After nitrogen addition, the β-diversity of soil fungal communities in both *P. yunnanensis* Franch. and *P. armandii* Franch. forests exhibited more pronounced differences compared to bacterial communities. This highlights the heightened sensitivity of fungal communities, which play a crucial role in soil carbon cycling in organically rich soils, to nitrogen addition [53].

4.2. Nitrogen Addition Alters the Relative Abundance Composition of Microbial Communities

In the *P. yunnanensis* Franch. planted forests, the OTUs, Chao index, and Shannon index of soil bacteria significantly declined following nitrogen addition. Conversely, in the *P. armandii* Franch. forests, these indices for soil bacteria notably increased post-nitrogen addition compared to CK, although values were lower at N20-N25 levels than at N10. This indicates a threshold effect in nitrogen-induced microbial growth stimulation, potentially due to pH reductions at higher nitrogen levels (N20-N25) leading to magnesium/calcium deficiencies and aluminum toxicity in the soil, adversely affecting soil bacteria [54,55]. While tree species differences caused variations in soil bacterial responses to nitrogen addition, the observed inhibition at high nitrogen levels in the *P. armandii* Franch. forest suggests that prolonged high nitrogen addition could detrimentally affect the soil bacterial community in subtropical coniferous forests, albeit to varying extents [56]. The N10 and N20 treatments enhanced the α-diversity and richness of soil fungal communities in the *P. yunnanensis* Franch. forest. However, the α-diversity and richness in the N25 treatment were lower than those in the N10 and N20 treatments, signifying a nutrient limitation threshold. This could be due to increased nitrogen at N25 leading to limitations in other nutrients, such as phosphorus and potassium, thus reducing the diversity of root-associated fungi [57,58]. In contrast, nitrogen addition at all levels significantly reduced the OTU number and Chao index of soil fungal communities in the *P. armandii* Franch. forest while only increasing the Shannon index. Given the Chao index's sensitivity to changes in rare taxa [50], this suggests that nitrogen addition negatively impacts the soil fungal community in the *P. armandii* Franch. forest, although the impact is less pronounced than in the *P. yunnanensis* Franch. forest. Several studies have noted large differences in root exudation between species [59], which is a key mediator of plant–microbe–soil interactions, affecting soil organic matter decomposition, nutrient cycling, and microbial community composition [60,61]. Different patterns of root exudation in *P. yunnanensis* Franch. and *P. armandii* Franch. may differentially respond to additional nitrogen, leading to variations in changes in the relative abundance of soil microbes.

Notably, the relative abundance of *Acidobacteria* was about twice as high in the *P. yunnanensis* Franch. forest compared to the *P. armandii* Franch. forest, while *Proteobacteria* were significantly more abundant in the *P. armandii* Franch. forest than in the *P. yunnanensis* Franch. forest. This disparity can be linked to differences in soil pH and nitrogen availability between the two forest types. *Acidobacteria*, which are typically oligotrophic, favor acidic environments [62], whereas *Proteobacteria* often thrive in nutrient-rich conditions [63,64]. The levels of NO_3^--N and NH_4^+-N were substantially higher in the soil of the *P. armandii* Franch. forest, providing a nitrogen-rich environment conducive to *Proteobacteria* growth. Elevated nitrogen levels in the soil can increase competition for resources and alter soil nutrients, thereby affecting the ecological niches of certain microbial taxa and, thus, their relative abundance within the overall community [65]. Nitrogen addition significantly increases the relative abundance of *Chloroflexi*, which are involved in the second step of nitrification [66]. The higher soil NO_3^--N content resulting from nitrogen addition in the *P. armandii* Franch. forest provides ample nutrients for *Chloroflexi* growth, peaking in the N20 treatment [67]. However, nutrient enrichment from high-level nitrogen addition (N25) is not conducive to *Chloroflexi* reproduction, possibly due to close interactions between *Chloroflexi* and other microorganisms, with high levels of added nitrogen having a toxic effect on other microbial community members.

Unlike previous studies, although the fungal community composition of *P. yunnanensis* Franch. and *P. armandii* Franch. forests was similar, their responses to nitrogen addition differed significantly [68]. In the *P. yunnanensis* Franch. forest, most fungal phyla exhibited strong resilience against nitrogen-induced environmental stress, except for a significant decrease in *Rozellomycota* with increasing nitrogen levels. Simultaneously, a significant negative correlation was observed between the relative abundance of *Rozellomycota* and soil pH. This observation suggests that *Rozellomycota* may have higher competitive survival capabilities in acidic environments. However, due to the presence of numerous parasitic fungi within *Rozellomycota* [69], further investigation is warranted to elucidate its ecological adaptability and biological characteristics. In the *P. armandii* Franch. forest, nitrogen addition significantly reduced the relative abundance of *Basidiomycota* and increased that of *Ascomycota*. *Ascomycota*, known for their responsiveness to high nutrient levels, play a critical role in decomposing organic matter in the soil [65]. Here, a significant positive correlation was found between *Ascomycota*'s relative abundance and SOM, which increased with higher nitrogen content [70]. *Basidiomycota*, crucial for soil carbon cycling processes like litter and lignin decomposition in fertile forests [71], represents a slower-growing fungal type [72]. It competes with *Ascomycota* for resources in limited ecological niches. The decrease in *Basidiomycota*'s relative abundance might reduce competition for resources, thereby facilitating the increase in *Ascomycota*'s abundance [73].

Ecosystem stability hinges not only on community composition but also on the associations and co-occurrence networks among microbial communities [74,75]. In our study, we analyzed microbial community networks using correlations among OTUs, noting that positive and negative relationships among OTUs do not imply direct physical interactions but rather indicate co-occurrence [76]. Our results demonstrated that based on co-occurrence network attributes, bacterial interaction complexity was higher in *P. yunnanensis* Franch. forests than in *P. armandii* Franch. forests. Conversely, fungal interaction complexity was lower in *P. yunnanensis* Franch. forests. This disparity reflects the varying capacities of microbial communities to withstand nitrogen-induced disturbances influenced by differences in soil pH and nutrient availability. In *P. yunnanensis* Franch. forests, nitrogen addition led to reductions in the number of nodes, edge connections, and average degree in bacterial networks, paralleled by increased modularity and reduced network complexity. This underscores the detrimental impacts of nitrogen on bacterial communities in these forests, potentially due to constrained carbon (C) and phosphorus (P) resources following nitrogen addition, thereby diminishing their diversity and network complexity [77]. In *P. armandii* Franch. forests, bacterial interaction complexity was augmented with nitrogen addition, but a threshold was observed beyond continuous high-level nitrogen addition (N20), corroborating the α-diversity findings. This suggested that sustained high-level nitrogen addition could negatively impact soil bacterial communities in subtropical coniferous forests, with varying degrees of influence. In both forests, the fungal networks exhibited increases in nodes, edge connections, and average degree with nitrogen application, peaking at N20, revealing that low-level nitrogen addition fosters fungal community growth while continuous high-level nitrogen addition impedes this enhancement, adversely affecting fungal co-occurrence and network stability in both forest types.

4.3. Drivers of Microbial Community Responses to Nitrogen Addition

Consistent with our third hypothesis, nitrogen addition significantly influenced soil pH and nitrogen availability in two coniferous forests, consequently affecting soil microbial communities. The random forest model identified NH_4^+-N, pH, and AP as key factors shaping the composition and abundance of bacterial and fungal communities in both forest types. Further, the structural equation model demonstrated that nitrogen addition altered soil chemical properties (pH, NH_4^+-N, and TN), significantly impacting the alpha diversity of bacterial and fungal communities in both forests. Thus, $NH4^+$-N and pH emerged as the primary factors regulating these soil communities. Nitrogen inputs into ecosystems contribute to soil acidification mechanisms [31]. The conversion of urea to $NH4^+$-N in the soil, either fixed or volatilized by microorganisms or plants [78], generates H^+, contributing to acidity. The oxidation of NH_4^+ to NO_3^- also increases net H^+ production, further promoting soil acidification. Additionally, the rise in organic nitrogen can enhance root biomass, potentially increasing the uptake of alkaline cations (e.g., Ca^{2+}, Mg^{2+}, and K^+), thus reducing soil buffering capacity and inducing acidification [79], which detrimentally affects microbial communities. This is also the underlying cause of pH reduction in *P. armandii* Franch. forests. Conversely, nitrogen addition can accelerate the release of alkaline cations from litter and roots by promoting decomposition [80], enhancing soil buffering capacity, and mitigating acidification. The increase in soil pH in *P. yunnanensis* Franch. forests may be related to this process. NH_4^+-N positively affected fungal community abundance in *P. yunnanensis* Franch. forests and bacterial communities in *P. armandii* Franch. forests, likely due to excess nitrogen leading to $NH4^+$ accumulation, to which many microorganisms are sensitive [81], thus alleviating resource limitations and enabling specific species to thrive. TN exhibited contrasting effects on fungal α-diversity in the two forest types, possibly due to differing nitrogen utilization abilities among fungi. For example, in *P. yunnanensis* Franch. forests, the abundance of *Ascomycota* was positively correlated with TN, while the adverse impact of TN on fungal abundance in these forests may be linked to such differences [82]. AP, as a crucial factor in regulating the bacterial communities in *P. armandii* Franch. forests, could be attributed to low nitrogen addition, which enhances microbial phosphorus acquisition and stimulates microbial growth. However, continuous nitrogen increases may lead to phosphorus limitations in the ecosystem, causing an N and P nutrient imbalance and thus affecting the bacterial communities and abundance in *P. armandii* Franch. forests.

5. Conclusions

In conclusion, our study demonstrates that after four years of continuous nitrogen addition, the soil microbial communities in subtropical planted coniferous forests exhibit varied responses that are dependent on the afforestation tree species. This variability is reflected in changes to soil pH, nutrient content, and the dynamics of microbial communities. Notably, prolonged high-level nitrogen addition has consistently led to detrimental effects on microbial community diversity. Fungal communities, in particular, show greater sensitivity compared to bacterial communities, with significant changes observed in their diversity, community structure, and network properties in response to nitrogen addition. This study underscores that soil pH and nitrogen availability are crucial factors influencing the diversity of soil bacterial and fungal communities in these forests. These findings offer valuable theoretical insights into the long-term effects of nitrogen deposition on soil microbes in subtropical planted coniferous forests and elucidate the complex relationships between soil microbial communities and soil properties.

Supplementary Materials: The following supporting information can be downloaded at: https://www.mdpi.com/article/10.3390/f15071112/s1, Table S1: Site characteristics of two forest types.; Table S2: Response to N addition at the bacterial/fungal phylum level in *P. yunnanensis* Franch. forest soil. Table S3: Correlation of soil bacterial/fungal phylum levels and soil chemical properties in *P. yunnanensis* Franch. forests. Table S4: Response to N addition at the bacterial/fungal phylum level in *P. armandii* Franch. forest soil. Table S5: Correlation of soil bacterial/fungal phylum levels and soil chemical properties in *P. armandii* Franch. forests. Figure S1: RDA analysis of bacterial and fungal community structure in relation to environmental factors.

Author Contributions: Investigation, Methodology, Formal analysis, Writing—Original draft, Writing—Review and Editing, Z.H. Conceptualization, Visualization, Methodology, Data curation, Writing—Original draft, X.Z. Investigation, Methodology, Data curation, W.C. Investigation, Validation, Software, Z.L. Resources, Funding acquisition, Supervision, Project administration, K.W. Funding acquisition, Supervision, Y.Z. Project administration, Supervision, Writing—Review and Editing, Y.S. All authors have read and agreed to the published version of the manuscript.

Funding: This work was supported by the Agricultural Joint Special Project of Yunnan Province (202301BD070001-059), the project of China Geological Survey (No. DD20230482), the Scientific Research Foundation of Education Department of Yunnan Province (2022J0510), First-class Discipline Construction Project of Yunnan Province ([2022] No. 73), Natural Ecology Monitoring Network Project Operation Project of Yuxi Forest Ecological Station in Yunnan Province (2022-YN-13), Long-term Scientific Research Base of Yuxi Forest Ecosystem National in Yunnan Province (2020132550), the National Natural Science Foundation of China (No. 42271094), funded by Science and Technology Innovation Foundation of Command Center of Integrated Natural Resources Survey Center (No. KC20230019, KC20230021).

Data Availability Statement: The data are available upon request from the corresponding author.

Acknowledgments: We are particularly grateful to Yali Song (School of Soil and Water Conservation, Southwest Forestry University), who provided great support to this study and provided us with writing assistance, which enabled this study to be carried out smoothly.

Conflicts of Interest: The authors declare no conflicts of interest.

References

1. Ackerman, D.; Millet, D.B.; Chen, X. Global estimates of inorganic nitrogen deposition across four decades. *Glob. Biogeochem. Cycles* **2019**, *33*, 100–107. [CrossRef]
2. Feng, H.; Guo, J.; Peng, C.; Kneeshaw, D.; Roberge, G.; Pan, C.; Ma, X.; Zhou, D.; Wang, W. Nitrogen addition promotes terrestrial plants to allocate more biomass to aboveground organs: A global meta-analysis. *Glob. Chang. Biol.* **2023**, *29*, 3970–3989. [CrossRef]
3. Chen, T.; Cheng, R.; Xiao, W.; Shen, Y.; Wang, L.; Sun, P.; Zhang, M.; Li, J. Nitrogen Addition Enhances Soil Nitrogen Mineralization through an Increase in Mineralizable Organic Nitrogen and the Abundance of Functional Genes. *J. Soil Sci. Plant Nutr.* **2024**, *24*, 975–987. [CrossRef]
4. Galloway, J.N.; Townsend, A.R.; Erisman, J.W.; Bekunda, M.; Cai, Z.; Freney, J.R.; Martinelli, L.A.; Seitzinger, S.P.; Sutton, M.A. Transformation of the nitrogen cycle: Recent trends, questions, and potential solutions. *Science* **2008**, *320*, 889–892. [CrossRef]
5. Widdig, M.; Heintz-Buschart, A.; Schleuss, P.-M.; Guhr, A.; Borer, E.T.; Seabloom, E.W.; Spohn, M. Effects of nitrogen and phosphorus addition on microbial community composition and element cycling in a grassland soil. *Soil Biol. Biochem.* **2020**, *151*, 108041. [CrossRef]
6. Wang, J.; Shi, X.; Zheng, C.; Suter, H.; Huang, Z. Different responses of soil bacterial and fungal communities to nitrogen deposition in a subtropical forest. *Sci. Total Environ.* **2021**, *755*, 142449. [CrossRef] [PubMed]
7. Ke, Y.; Yu, Q.; Wang, H.; Zhao, Y.; Jia, X.; Yang, Y.; Zhang, Y.; Zhou, W.; Wu, H.; Xu, C. The potential bias of nitrogen deposition effects on primary productivity and biodiversity. *Glob. Chang. Biol.* **2023**, *29*, 1054–1061. [CrossRef]
8. Hu, J.; Huang, C.; Zhou, S.; Kuzyakov, Y. Nitrogen addition to soil affects microbial carbon use efficiency: Meta-analysis of similarities and differences in 13C and 18O approaches. *Glob. Chang. Biol.* **2022**, *28*, 4977–4988. [CrossRef]
9. Cao, M.; Zheng, X.; Cui, L.; Wu, F.; Gao, H.; Jiang, J. Soil bacterial communities are more sensitive to short-term nitrogen deposition than fungal communities in subtropical Chinese fir forests. *For. Ecol. Manag.* **2023**, *549*, 121490. [CrossRef]
10. Chen, W.; Su, F.; Nie, Y.; Zhong, B.; Zheng, Y.; Mo, J.; Xiong, B.; Lu, X. Divergent responses of soil microbial functional groups to long-term high nitrogen presence in the tropical forests. *Sci. Total Environ.* **2022**, *821*, 153251. [CrossRef]
11. He, J.; Jiao, S.; Tan, X.; Wei, H.; Ma, X.; Nie, Y.; Liu, J.; Lu, X.; Mo, J.; Shen, W. Adaptation of soil fungal community structure and assembly to long-versus short-term nitrogen addition in a tropical forest. *Front. Microbiol.* **2021**, *12*, 689674. [CrossRef]
12. Ma, S.; Verheyen, K.; Props, R.; Wasof, S.; Vanhellemont, M.; Boeckx, P.; Boon, N.; De Frenne, P. Plant and soil microbe responses to light, warming and nitrogen addition in a temperate forest. *Funct. Ecol.* **2018**, *32*, 1293–1303. [CrossRef]
13. Wang, C.; Liu, D.; Bai, E. Decreasing soil microbial diversity is associated with decreasing microbial biomass under nitrogen addition. *Soil Biol. Biochem.* **2018**, *120*, 126–133. [CrossRef]
14. Zhou, Z.; Wang, C.; Zheng, M.; Jiang, L.; Luo, Y. Patterns and mechanisms of responses by soil microbial communities to nitrogen addition. *Soil Biol. Biochem.* **2017**, *115*, 433–441. [CrossRef]
15. Luo, L.; Meng, H.; Wu, R.-N.; Gu, J.-D. Impact of nitrogen pollution/deposition on extracellular enzyme activity, microbial abundance and carbon storage in coastal mangrove sediment. *Chemosphere* **2017**, *177*, 275–283. [CrossRef]
16. Rousk, J.; Bååth, E.; Brookes, P.C.; Lauber, C.L.; Lozupone, C.; Caporaso, J.G.; Knight, R.; Fierer, N. Soil bacterial and fungal communities across a pH gradient in an arable soil. *ISME J.* **2010**, *4*, 1340–1351. [CrossRef]
17. Mao, Q.; Lu, X.; Zhou, K.; Chen, H.; Zhu, X.; Mori, T.; Mo, J. Effects of long-term nitrogen and phosphorus additions on soil acidification in an N-rich tropical forest. *Geoderma* **2017**, *285*, 57–63. [CrossRef]
18. Lu, F.; Hu, H.; Sun, W.; Zhu, J.; Liu, G.; Zhou, W.; Zhang, Q.; Shi, P.; Liu, X.; Wu, X. Effects of national ecological restoration projects on carbon sequestration in China from 2001 to 2010. *Proc. Natl. Acad. Sci. USA* **2018**, *115*, 4039–4044. [CrossRef]
19. Zhang, L.; Shen, Y.; Hu, Y.; Li, J.; Liu, Y.; Chen, S.; Wang, L.; Liu, S.; Li, H.; You, C. Response of soil phosphorus fractions to litter removal in subalpine coniferous forest. *Sci. Total Environ.* **2023**, *898*, 166383. [CrossRef]

20. Tian, Y.; Lu, H.; Wang, J.; Lin, Y.; Campbell, D.E.; Jian, S. Effects of canopy and understory nitrogen addition on the structure and eco-exergy of a subtropical forest community. *Ecol. Indic.* **2019**, *106*, 105459. [CrossRef]
21. Liu, J.; Le, T.H.; Zhu, H.; Yao, Y.; Zhu, H.; Cao, Y.; Zhao, Z. Afforestation of cropland fundamentally alters the soil fungal community. *Plant Soil* **2020**, *457*, 279–292. [CrossRef]
22. Liu, J.; Wang, Q.; Ku, Y.; Zhang, W.; Zhu, H.; Zhao, Z. Precipitation and soil pH drive the soil microbial spatial patterns in the *Robinia pseudoacacia* forests at the regional scale. *Catena* **2022**, *212*, 106120. [CrossRef]
23. Kang, E.; Li, Y.; Zhang, X.; Yan, Z.; Wu, H.; Li, M.; Yan, L.; Zhang, K.; Wang, J.; Kang, X. Soil pH and nutrients shape the vertical distribution of microbial communities in an alpine wetland. *Sci. Total Environ.* **2021**, *774*, 145780. [CrossRef]
24. Wang, H.; Liu, S.; Zhang, X.; Mao, Q.; Li, X.; You, Y.; Wang, J.; Zheng, M.; Zhang, W.; Lu, X. Nitrogen addition reduces soil bacterial richness, while phosphorus addition alters community composition in an old-growth N-rich tropical forest in southern China. *Soil Biol. Biochem.* **2018**, *127*, 22–30. [CrossRef]
25. Lauber, C.L.; Strickland, M.S.; Bradford, M.A.; Fierer, N. The influence of soil properties on the structure of bacterial and fungal communities across land-use types. *Soil Biol. Biochem.* **2008**, *40*, 2407–2415. [CrossRef]
26. Wang, L.; Macko, S.A. Constrained preferences in nitrogen uptake across plant species and environments. *Plant Cell Environ.* **2011**, *34*, 525–534. [CrossRef]
27. Huang, J.; Zhang, W.; Li, Y.; Wang, S.; Mao, J.; Mo, J.; Zheng, M. Long-term nitrogen deposition does not exacerbate soil acidification in tropical broadleaf plantations. *Environ. Res. Lett.* **2021**, *16*, 114042. [CrossRef]
28. Urbanová, M.; Šnajdr, J.; Baldrian, P. Composition of fungal and bacterial communities in forest litter and soil is largely determined by dominant trees. *Soil Biol. Biochem.* **2015**, *84*, 53–64. [CrossRef]
29. Zhou, Q.; Li, F.; Cai, X.-a.; Rao, X.; Zhou, L.; Liu, Z.; Lin, Y.; Fu, S. Survivorship of plant species from soil seedbank after translocation from subtropical natural forests to plantation forests. *For. Ecol. Manag.* **2019**, *432*, 741–747. [CrossRef]
30. Lu, X.; Mao, Q.; Mo, J.; Gilliam, F.S.; Zhou, G.; Luo, Y.; Zhang, W.; Huang, J. Divergent responses of soil buffering capacity to long-term N deposition in three typical tropical forests with different land-use history. *Environ. Sci. Technol.* **2015**, *49*, 4072–4080. [CrossRef]
31. Tian, D.; Niu, S. A global analysis of soil acidification caused by nitrogen addition. *Environ. Res. Lett.* **2015**, *10*, 024019. [CrossRef]
32. Xing, J.; Hu, C.; Song, C.; Wang, K.; Song, Y. Nitrogen Deposition Modulates Litter Decomposition and Enhances Water Retention in Subtropical Forests. *Forests* **2024**, *15*, 522. [CrossRef]
33. Leng, Q.; Cui, J.; Zhou, F.; Du, K.; Zhang, L.; Fu, C.; Liu, Y.; Wang, H.; Shi, G.; Gao, M. Wet-only deposition of atmospheric inorganic nitrogen and associated isotopic characteristics in a typical mountain area, southwestern China. *Sci. Total Environ.* **2018**, *616*, 55–63. [CrossRef] [PubMed]
34. Song, Y.; Xing, J.; Hu, C.; Song, C.; Wang, Q.; Wang, S. Decomposition and Carbon and Nitrogen Releases of Twig and Leaf Litter Were Inhibited by Increased Level of Nitrogen Deposition in a Subtropical Evergreen Broad-Leaved Forest in Southwest China. *Forests* **2024**, *15*, 492. [CrossRef]
35. Bao, S.D. *Soil and Agricultural Chemistry Analysis*; China Agricultural Press: Beijing, China, 2000.
36. Logue, J.B.; Stedmon, C.A.; Kellerman, A.M.; Nielsen, N.J.; Andersson, A.F.; Laudon, H.; Lindström, E.S.; Kritzberg, E.S. Experimental insights into the importance of aquatic bacterial community composition to the degradation of dissolved organic matter. *ISME J.* **2016**, *10*, 533–545. [CrossRef] [PubMed]
37. Gardes, M.; Bruns, T.D. ITS primers with enhanced specificity for basidiomycetes-application to the identification of mycorrhizae and rusts. *Mol. Ecol.* **1993**, *2*, 113–118. [CrossRef] [PubMed]
38. White, T.J.; Bruns, T.; Lee, S.; Taylor, J. Amplification and direct sequencing of fungal ribosomal RNA genes for phylogenetics. In *PCR Protocols: A Guide to Methods and Applications*; Academic Press: London, UK, 1990; pp. 315–322.
39. Edgar, R.C. UPARSE: Highly accurate OTU sequences from microbial amplicon reads. *Nat. Methods* **2013**, *10*, 996–998. [CrossRef] [PubMed]
40. Edgar, R.C.; Haas, B.J.; Clemente, J.C.; Quince, C.; Knight, R. UCHIME improves sensitivity and speed of chimera detection. *Bioinformatics* **2011**, *27*, 2194–2200. [CrossRef] [PubMed]
41. Shannon, C.E. A mathematical theory of communication. *Bell Syst. Tech. J.* **1948**, *27*, 379–423. [CrossRef]
42. Chao, A. Nonparametric estimation of the number of classes in a population. *Scand. J. Stat.* **1984**, *11*, 265–270.
43. Feng, K.; Peng, X.; Zhang, Z.; Gu, S.; He, Q.; Shen, W.; Wang, Z.; Wang, D.; Hu, Q.; Li, Y. iNAP: An Integrated Network Analysis Pipeline for Microbiome Studies. *iMeta* **2022**, *1*, e13. [CrossRef]
44. Oksanen, J.; Blanchet, F.G.; Kindt, R.; Legendre, P.; Minchin, P.R.; O'hara, R.; Simpson, G.L.; Solymos, P.; Stevens, M.H.H.; Wagner, H. R Package Version 2.2-0. Package 'Vegan'. *Community Ecology Package*. 2013. Available online: https://cran.r-project.org/web/packages/vegan/index.html (accessed on 22 May 2024).
45. Wickham, H.; Chang, W.; Wickham, M.H. Version 3.5.1. Package 'ggplot2'. *Create Elegant Data Visualisations Using the Grammar of Graphics*. 2016. Available online: https://search.r-project.org/CRAN/refmans/ggplot2/html/ggplot2-package.html (accessed on 22 May 2024).
46. Berthrong, S.T.; Jobbágy, E.G.; Jackson, R.B. A global meta-analysis of soil exchangeable cations, pH, carbon, and nitrogen with afforestation. *Ecol. Appl.* **2009**, *19*, 2228–2241. [CrossRef] [PubMed]
47. Lu, X.; Mao, Q.; Gilliam, F.S.; Luo, Y.; Mo, J. Nitrogen deposition contributes to soil acidification in tropical ecosystems. *Glob. Chang. Biol.* **2014**, *20*, 3790–3801. [CrossRef] [PubMed]

48. Philippot, L.; Chenu, C.; Kappler, A.; Rillig, M.C.; Fierer, N. The interplay between microbial communities and soil properties. *Nat. Rev. Microbiol.* **2024**, *22*, 226–239. [CrossRef]
49. Wild, B.; Alaei, S.; Bengtson, P.; Bodé, S.; Boeckx, P.; Schnecker, J.; Mayerhofer, W.; Rütting, T. Short-term carbon input increases microbial nitrogen demand, but not microbial nitrogen mining, in a set of boreal forest soils. *Biogeochemistry* **2017**, *136*, 261–278. [CrossRef]
50. Wang, X.; Feng, J.; Ao, G.; Qin, W.; Han, M.; Shen, Y.; Liu, M.; Chen, Y.; Zhu, B. Globally nitrogen addition alters soil microbial community structure, but has minor effects on soil microbial diversity and richness. *Soil Biol. Biochem.* **2023**, *179*, 108982. [CrossRef]
51. Strickland, M.S.; Rousk, J. Considering fungal: Bacterial dominance in soils–methods, controls, and ecosystem implications. *Soil Biol. Biochem.* **2010**, *42*, 1385–1395. [CrossRef]
52. Müller, K.; Marhan, S.; Kandeler, E.; Poll, C. Carbon flow from litter through soil microorganisms: From incorporation rates to mean residence times in bacteria and fungi. *Soil Biol. Biochem.* **2017**, *115*, 187–196. [CrossRef]
53. Lu, X.; Vitousek, P.M.; Mao, Q.; Gilliam, F.S.; Luo, Y.; Turner, B.L.; Zhou, G.; Mo, J. Nitrogen deposition accelerates soil carbon sequestration in tropical forests. *Proc. Natl. Acad. Sci. USA* **2021**, *118*, e2020790118. [CrossRef]
54. Cao, J.; Wang, H.; Holden, N.M.; Adamowski, J.F.; Biswas, A.; Zhang, X.; Feng, Q. Soil properties and microbiome of annual and perennial cultivated grasslands on the Qinghai–Tibetan Plateau. *Land Degrad. Dev.* **2021**, *32*, 5306–5321. [CrossRef]
55. Kaspari, M.; Bujan, J.; Weiser, M.D.; Ning, D.; Michaletz, S.T.; Zhili, H.; Enquist, B.J.; Waide, R.B.; Zhou, J.; Turner, B.L. Biogeochemistry drives diversity in the prokaryotes, fungi, and invertebrates of a Panama forest. *Ecology* **2017**, *98*, 2019–2028. [CrossRef] [PubMed]
56. Wang, Y.; Li, X.; Quan, X.; Liang, H.; Wang, L.; Yan, X. Effects of nitrogen stress and nitrogen form ratios on the bacterial community and diversity in the root surface and rhizosphere of *Cunninghamia lanceolata* and *Schima superba*. *Front. Plant Sci.* **2023**, *14*, 1240675. [CrossRef]
57. Chen, X.; Chen, H.Y.; Searle, E.B.; Chen, C.; Reich, P.B. Negative to positive shifts in diversity effects on soil nitrogen over time. *Nat. Sustain.* **2021**, *4*, 225–232. [CrossRef]
58. Sun, Y.; Wang, C.; Chen, X.; Liu, S.; Lu, X.; Chen, H.Y.; Ruan, H. Phosphorus additions imbalance terrestrial ecosystem C: N: P stoichiometry. *Glob. Chang. Biol.* **2022**, *28*, 7353–7365. [CrossRef]
59. Yang, L.; Wang, X.; Mao, Z.; Jiang, Z.; Gao, Y.; Chen, X.; Aubrey, D.P. Root exudation rates decrease with increasing latitude in some tree species. *Forests* **2020**, *11*, 1045. [CrossRef]
60. Zhang, Y.; Gu, M.; Xia, X.; Shi, K.; Zhou, Y.; Yu, J. Effects of phenylcarboxylic acids on mitosis, endoreduplication and expression of cell cycle-related genes in roots of cucumber (*Cucumis sativus* L.). *J. Chem. Ecol.* **2009**, *35*, 679–688. [CrossRef] [PubMed]
61. Haichar, F.e.Z.; Marol, C.; Berge, O.; Rangel-Castro, J.I.; Prosser, J.I.; Balesdent, J.; Heulin, T.; Achouak, W. Plant host habitat and root exudates shape soil bacterial community structure. *ISME J.* **2008**, *2*, 1221–1230. [CrossRef] [PubMed]
62. Fierer, N.; Bradford, M.A.; Jackson, R.B. Toward an ecological classification of soil bacteria. *Ecology* **2007**, *88*, 1354–1364. [CrossRef] [PubMed]
63. Gao, Y.; Wang, J.; Ge, Y.; Lei, Y.; Wei, X.; Xu, Y.; Zheng, X. Partial substitution of nitrogen fertilizers by organic products of rural waste co-composting impacts on farmland soil quality. *Environ. Technol. Innov.* **2023**, *33*, 103470. [CrossRef]
64. Hao, C.; Du, P.; Ren, J.; Hu, L.; Zhang, Z. Halophyte *Elymus dahuricus* colonization regulates microbial community succession by mediating saline-alkaline and biogenic organic matter in bauxite residue. *Sci. Total Environ.* **2023**, *905*, 167140. [CrossRef]
65. Yan, G.; Han, S.; Wang, Q.; Wang, X.; Hu, C.; Xing, Y. Variations of the effects of reduced precipitation and N addition on microbial diversity among different seasons in a temperate forest. *Appl. Soil Ecol.* **2021**, *166*, 103995. [CrossRef]
66. Sorokin, D.Y.; Vejmelkova, D.; Lücker, S.; Streshinskaya, G.M.; Rijpstra, W.I.C.; Sinninghe Damsté, J.S.; Kleerbezem, R.; van Loosdrecht, M.; Muyzer, G.; Daims, H. *Nitrolancea hollandica* gen. nov., sp. nov., a chemolithoautotrophic nitrite-oxidizing bacterium isolated from a bioreactor belonging to the phylum Chloroflexi. *Int. J. Syst. Evol. Microbiol.* **2014**, *64*, 1859–1865. [CrossRef] [PubMed]
67. Wendong, X.; Xiaotong, Z.; Wenjun, L. Research status and prospect on bacterial phylum Chloroflexi. *Acta Microbiol. Sin.* **2020**, *60*, 1801–1820.
68. Choma, M.; Tahovska, K.; Kaštovská, E.; Bárta, J.; Růžek, M.; Oulehle, F. Bacteria but not fungi respond to soil acidification rapidly and consistently in both a spruce and beech forest. *FEMS Microbiol. Ecol.* **2020**, *96*, fiaa174. [CrossRef] [PubMed]
69. Corsaro, D.; Walochnik, J.; Venditti, D.; Hauröder, B.; Michel, R. Solving an old enigma: *Morellospora saccamoebae* gen. nov., sp. nov. (Rozellomycota), a Sphaerita-like parasite of free-living amoebae. *Parasitol. Res.* **2020**, *119*, 925–934. [CrossRef] [PubMed]
70. Jiang, M.; Tian, Y.; Guo, R.; Li, S.; Guo, J.; Zhang, T. Effects of warming and nitrogen addition on soil fungal and bacterial community structures in a temperate meadow. *Front. Microbiol.* **2023**, *14*, 1231442. [CrossRef] [PubMed]
71. Liu, S.; Yang, H.; Zhou, L.; Jin, S.-S.; Xie, L.; Lin, C.; He, J.-Z.; Zheng, Y. Dynamic response of root-associated fungal community structure to nitrogen and phosphorus additions in a subtropical forest. *Pedobiologia* **2023**, *101*, 150909. [CrossRef]
72. Allison, S.D.; LeBauer, D.S.; Ofrecio, M.R.; Reyes, R.; Ta, A.-M.; Tran, T.M. Low levels of nitrogen addition stimulate decomposition by boreal forest fungi. *Soil Biol. Biochem.* **2009**, *41*, 293–302. [CrossRef]
73. Weber, C.F.; Vilgalys, R.; Kuske, C.R. Changes in fungal community composition in response to elevated atmospheric CO_2 and nitrogen fertilization varies with soil horizon. *Front. Microbiol.* **2013**, *4*, 78. [CrossRef]

74. de Vries, F.T.; Griffiths, R.I.; Bailey, M.; Craig, H.; Girlanda, M.; Gweon, H.S.; Hallin, S.; Kaisermann, A.; Keith, A.M.; Kretzschmar, M. Soil bacterial networks are less stable under drought than fungal networks. *Nat. Commun.* **2018**, *9*, 3033. [CrossRef]
75. Gao, C.; Xu, L.; Montoya, L.; Madera, M.; Hollingsworth, J.; Chen, L.; Purdom, E.; Singan, V.; Vogel, J.; Hutmacher, R.B. Co-occurrence networks reveal more complexity than community composition in resistance and resilience of microbial communities. *Nat. Commun.* **2022**, *13*, 3867. [CrossRef] [PubMed]
76. He, D.; Shen, W.; Eberwein, J.; Zhao, Q.; Ren, L.; Wu, Q.L. Diversity and co-occurrence network of soil fungi are more responsive than those of bacteria to shifts in precipitation seasonality in a subtropical forest. *Soil Biol. Biochem.* **2017**, *115*, 499–510. [CrossRef]
77. Ma, X.; Wang, T.; Shi, Z.; Chiariello, N.R.; Docherty, K.; Field, C.B.; Gutknecht, J.; Gao, Q.; Gu, Y.; Guo, X. Long-term nitrogen deposition enhances microbial capacities in soil carbon stabilization but reduces network complexity. *Microbiome* **2022**, *10*, 112.
78. Puri, G.; Ashman, M. Microbial immobilization of 15N-labelled ammonium and nitrate in a temperate woodland soil. *Soil Biol. Biochem.* **1999**, *31*, 929–932. [CrossRef]
79. Bai, T.; Wang, P.; Ye, C.; Hu, S. Form of nitrogen input dominates N effects on root growth and soil aggregation: A meta-analysis. *Soil Biol. Biochem.* **2021**, *157*, 108251. [CrossRef]
80. Dong, L.; Berg, B.; Sun, T.; Wang, Z.; Han, X. Response of fine root decomposition to different forms of N deposition in a temperate grassland. *Soil Biol. Biochem.* **2020**, *147*, 107845. [CrossRef]
81. Sun, K.; Jiang, H.-J.; Pan, Y.-T.; Lu, F.; Zhu, Q.; Ma, C.-Y.; Zhang, A.-Y.; Zhou, J.-Y.; Zhang, W.; Dai, C.-C. Hyphosphere microorganisms facilitate hyphal spreading and root colonization of plant symbiotic fungus in ammonium-enriched soil. *ISME J.* **2023**, *17*, 1626–1638. [CrossRef] [PubMed]
82. Xiang, X.; Liu, J.; Zhang, J.; Li, D.; Xu, C.; Kuzyakov, Y. Divergence in fungal abundance and community structure between soils under long-term mineral and organic fertilization. *Soil Tillage Res.* **2020**, *196*, 104491. [CrossRef]

forests

MDPI

Article

Production of Seedlings of *Corymbia citriodora* Inoculated with Endophytic Bacteria

Augusto Matias de Oliveira [1,*,†], Caique Menezes de Abreu [1], Paulo Henrique Grazziotti [1], Gabriel Faria Parreiras de Andrade [1], Jaqueline Vieira Gomes [1], Natanielly Rodrigues Avelino [1], June Faria Scherrer Menezes [2], Gabriela Madureira Barroso [1], José Barbosa dos Santos [1] and Márcia Regina da Costa [1]

[1] Department of Agronomy, Federal University of the Jequitinhonha and Mucuri Valleys, Diamantina 39100-000, Minas Gerais, Brazil; menezes.abreu@ufvjm.edu.br (C.M.d.A.); paulo.grazziotti@ufvjm.edu.br (P.H.G.); gabriel.parreiras@ufvjm.edu.br (G.F.P.d.A.); jaqueline.gomes@ufvjm.edu.br (J.V.G.); natanielly.avelino@ufvjm.edu.br (N.R.A.); gabriela.madureira@ufvjm.edu.br (G.M.B.); jbarbosa@ufvjm.edu.br (J.B.d.S.); marcia.costa@ufvjm.edu.br (M.R.d.C.)

[2] Department of Agronomy, University of Rio Verde, Rio Verde 75901-970, Goiás, Brazil; june@unirv.edu.br

* Correspondence: augusto.matias@ufvjm.edu.br

† This manuscript is part of a Ph.D. Thesis by the first author, Available online at link: https://www.sgppg.com.br/ppg/ppgpv-programa-de-pos-graduacao-em-producao-vegetal/1/dissertacoes-teses/2023/ (accessed on 22 May 2024).

Abstract: This study aimed to evaluate the effect of inoculants of endophytic bacteria producing indoleacetic acid (IAA) on the physiological quality of seeds and the production of seedlings of *Corymbia citriodora* (Hook.) KD Hill & LAS Johnson. In the physiological quality test of the seeds, the treatments used were individual inoculation with *Priestia megaterium*, *Exiguobacterium sibiricum*, *Pantoea vagans* strain 45URP4-1, and *Bacillus* sp.; joint effect of the four strains (mix); inoculation only with the carrier (cassava starch and activated charcoal); carrier with 1.0 μg mL^{-1} of IAA; and non-inoculated control without IAA and without a carrier. In the production of seedlings in a greenhouse, the treatments were the same, except for the mix, which was replaced by *P. vagans* strain 7URP1-6 (*Pvs*7), as inoculation with the mix increased the number of abnormal seedlings. In the physiological quality test of seeds, seeds inoculated with the bacteria individually did not have the physiological quality impaired and the carrier created a microenvironment around the seeds, benefiting germination percentage, germination speed index, average germination time, and average germination speed. In the greenhouse, seedlings inoculated with *Pvs*7, *P. megaterium* and *E. sibiricum* were taller, with a larger stem diameter and dry mass of shoot, roots, and total. Seeds inoculated with *E. sibiricum* had higher averages for height, chlorophyll *b* content, and shoot and total dry mass, as well as a greater ability to colonize the rhizosphere and roots of *C. citriodora*, resulting in the production of higher-quality seedlings. Inoculation of seeds of *C. citriodora* with endophytic bacteria proved to be a promising alternative for plant development.

Keywords: bacterial inoculant; microenvironment; IAA; colonization

1. Introduction

Plant growth-promoting bacteria (PGPB) can be divided into two groups regarding their colonization: rhizospherical and endophytic [1]. The rhizosphere bacteria colonize the roots externally, while the endophytic ones colonize internally. Endophytes, by colonizing plant tissues internally, may be more likely to associate with plants, as they will be protected from the adverse effects of the environment, such as edaphoclimatic variations (rainfall, drought, and pH changes, among others) and competition with other microorganisms [2,3]. PGPB benefits plants through the synthesis of phytohormones, facilitating the absorption of nutrients in the soil by promoting root growth and the synthesis of enzymes such as nitrogenase, nitrate reductase, urease, and catalase; siderophores production; fixation of atmospheric nitrogen; antagonizing plant pathogens; and inducing resistance to biotic and abiotic factors [4–6]. In return, the plant provides protection and photoassimilates for bacterial growth and propagation.

Inoculating seeds with bacteria is a common practice in oilseed crops, such as soybeans, as it improves plant development [2]. According to Paravar et al. [6], coating seeds with microorganisms is a method that can improve the physiological and physical properties of seeds, as well as plant growth rates, and alleviate biotic and abiotic stresses. Despite the benefits, this technique has yet to be widespread in forest species, as is the case with species of the genus *Corymbia*. *Corymbia* species are difficult to propagate vegetatively due to recalcitrance to rooting. For this reason, many plantations end up being established via seminal means, as occurs with the species *Corymbia citriodora* [7,8]. *C. citriodora* is a species of Australian origin, adapted to different soil and climate conditions; wood is used in the production of charcoal, firewood, posts, and sawmills, among other uses [7,9]. The essential oil extracted from its leaves is used as a raw material for the perfumery, cleaning, and pharmaceutical industries [7].

Although bacteria inoculation of seeds benefits plant development, bacteria can produce compounds that, in large quantities, are harmful to the physiological quality of seeds, such as indoleacetic acid (IAA). IAA is a hormone belonging to the auxin group, which acts mainly in the formation of lateral roots and root hair. Excess IAA becomes harmful due to auxins, in general, having herbicidal action [10]. The production of 171.1 $\mu g\ mL^{-1}$ of IAA by *Rhizobium leguminosarum* resulted in loss of seed vigor and the formation of abnormal lettuce seedlings [11]. The strains 7URP1-6 and 45URP4-1 of *Pantoea vagans*, 11URP4-2 of *Priestia megaterium*, 19RP3L2-7 of *Exiguobacterium sibiricum*, and 58CRP4-3 of *Bacillus* sp., used in the study, produce between 85 and 129 $\mu g\ mL^{-1}$ of IAA in vitro [12].

Therefore, physiological quality tests must be developed to verify whether the bacteria inhibit seed germination before inoculating them en masse. If bacteria do not inhibit the physiological quality of seeds, their potential in plant development can be explored. Thus, this study aimed to evaluate the effect of IAA-producing endophytic bacteria on the physiological quality of seeds and the production of C. citriodora seedlings.

2. Material and Methods

2.1. Physiological Quality of Seeds

2.1.1. Inoculant Production and Seed Inoculation

The experiment was implemented at the Soil Microbiology Laboratory, Department of Forest Engineering, Federal University of the Jequitinhonha and Mucuri Valleys (UFVJM), Diamantina, Minas Gerais State of Brazil, in April 2021. The seeds of *C. citriodora* were obtained from the company BENTEC-seeds®, inputs and technology, with a control record CITRINCO8NV and germination power of 80%. The seeds were disinfested in a laminar flow chamber (Filterflux, São Paulo, Brazil). For this, the seeds were washed six times with distilled and sterilized water and immersed in 70% alcohol solution for 30 s and 5% sodium hypochlorite for 15 min and then rinsed six times with distilled and sterilized water [13].

The bacteria were multiplied in 50 mL of Luria-Bertani liquid medium, composed of 10 g of tryptone, 5 g of yeast extract, and 5 g of sodium chloride per liter and incubated (Shaker Incubator Cienlab, São Paulo, Brazil) at 28 °C for 72 h under orbital agitation of 120 rpm. Then, the culture was centrifuged (Kasvi K14-400) at 4000 rpm for 10 min to remove the supernatant and preserve the bacterial mass. The bacterial cells were resuspended in 0.85% saline solution with the aid of the vortex (Kasvi K45-2810). Subsequently, the cell density was diluted in the saline solution until the cell suspension reached the optical density of 1.0 on a digital spectrophotometer (850MI) at a wavelength of 540 nm, which corresponds to 10^9 CFU mL^{-1}, using the saline solution as a blank [14].

The inoculation carrier was prepared using 5 g of cassava starch dissolved in 100 mL of distilled and sterilized water (Autoclave Marte av 11), which was previously heated until the gum appearance was formed. Each 120 g of seeds was mixed with 5 mL of the meticulously prepared gum. The carrier was composed of activated charcoal (porosity 0.8–2.0 nm), which corresponded to 30%–50% of the volume of the bacterial mass resuspended in saline solution. The charcoal was added to 5 mL of the suspension of each bacterium, leaving for three minutes and, after this period, the mixture of charcoal and bacteria was added to the seeds pre-treated with the gum [14] and placed to grow in rolls of Germitest® papers.

2.1.2. Treatments and Experimental Design

The treatments used in this study were individual inoculation with *Priestia megaterium* strain 11URP4-2, *Exiguobacterium sibiricum* strain 19RP3L2-7, *Pantoea vagans* strain 45URP4-1, and *Bacillus* sp. strain 58CRP4-3; inoculation of seeds with four bacterial strains together (Mix); inoculation of seeds only with the carrier used in the composition of the inoculant; inoculation of seed with the carrier supplemented with 1.0 μg mL^{-1} of IAA previously dissolved in distilled and sterilized water [15]; and non-inoculated control, without IAA and without inoculation carrier. The inoculant mix was prepared using the volume of $\frac{1}{4}$ of each strain.

The bacteria were chosen from the collection of the Soil Microbiology Laboratory of the UFVJM for the greater production capacity of IAA in vitro by the cultivation method supplemented with 10% tryptophan and determined by the occurrence of oxidation by the Salkowski mixture [16]. The strains were isolated from eucalyptus roots previously superficially disinfested to obtain endophytic ones [12]; their characteristics are presented in Table 1.

Table 1. Characteristics of the bacteria used in the experiment.

Species—Strain (GenBank Code)	Isolated From	Fertilization of the Substrate of Occurrence of the Host Plant	Synthesis of IAA In Vitro (μg mL^{-1})	Enzymes Synthesized In Vitro	Gram Staining
Priestia megaterium—11URP4-2 (OQ983563)	Rooted cutting of *Eucalyptus grandis* × *Eucalyptus urophylla*	Yes	87	Catalase, urease	Negative
Exiguobacterium sibiricum—19RP3L2-7 (OQ983562)	Adult plant of *E. camaldulensis*	No	90	Catalase, nitrate reductase, and phytate solubilizing phosphatase	Positive
Pantoe vagans—45URP4-1 (OQ983564)	Rooted cutting of *E. grandis* × *E. urophylla*	Yes	129	Catalase, nitrogenase, urease, nitrate reductase, and calcium and phytate solubilizing phosphatases	Negative
Bacillus sp.—58CRP4-3 (OQ983565)	Seedlings of *E. cloeziana*	Yes	95	Catalase, urease, nitrate reductase, and producer of siderophores	Positive

Source: Ramires [12].

The experimental design was completely randomized, with five replicates of 50 seeds each per treatment. The seeds, which were inoculated with their respective treatments, were packed in rolls with two sheets of paper towels Germitest® type moistened with distilled and sterilized water in the proportion of 2.5 times the mass of the dry paper [17]. Then, the seeds were incubated in a BOD germination chamber (Solab Sb224) at 25 °C and a photoperiod of 12 h.

2.1.3. Evaluations

Daily and at the same time, the number of germinated seeds was counted and seeds with radicle length equal to or greater than 2 mm were considered germinated. The Germination Speed Index (GSI) was calculated by the equation GSI = $\sum(n_i/t_i)$, where n_i = the number of seeds that germinated at time "i", t_i = time after installation of the test; and i = 1 $\longrightarrow$ 12 days [18]. The counts were made until the germination was stabilized.

After germination was stabilized, the following were evaluated:

- Germination percentage (GP): Calculated by the formula GP = (NG/NP) $\times$ 100, where NG = number of seeds germinated at the end of the test and NP = number of seeds placed to germinate;
- Average germination speed (AGS): Calculated by the equation AGS = 1/t where t = average germination time;
- Average germination time (AGT): calculated by the equation AGT = $(\sum n_i t_i)/\sum n_i$, where n_i = number of germinated seeds per day; t_i = incubation time; and i = 1 $\longrightarrow$ 11 days;
- Percentage of dead seeds (PDS): PDS = (ND/NG) $\times$ 100, where ND = number of dead seeds and NG = number of seeds placed to germinate;
- Entropy: calculated according to the procedure adopted by Labouriau and Valadares [19]. $E = \sum_{i=1}^{k}$ fi.log2.fi where: E = informational entropy; fi = relative frequency of germination; and log2 = logarithm in base 2.

The number of normal and abnormal seedlings was also counted. The seedlings considered abnormal were those that did not have the potential to continue their development and give rise to normal plants, even growing in favorable conditions, and the following characteristics were analyzed: damage, deformations, and deterioration, both shoot and of the root system [17].

2.1.4. Statistical Analysis

The data were submitted to the Durbin–Watson test, to verify the independence of the errors, Shapiro–Wilk, to verify normality, and Bartlett, to verify the homogeneity of variances. Then, the analysis of variance was performed ($p < 0.01$ and $p < 0.05$) and, when significant, the means were grouped by the Scott-Knott test ($p < 0.05$). The statistical program R version 4.3.1 and the package ExpDes.pt were used [20].

2.2. Production of Seedlings in Greenhouse

2.2.1. Treatments, Substrate, Conduction, Design, and Evaluations

The experiment was set up at the Olericulture Sector of UFVJM in March 2022. The treatments were similar to those used in the previous experiment, with the difference that the mix was not used due to the increase in the number of abnormal seedlings in the germination test caused by it and the addition of the bacterium *Pantoea vagans* strain 7URP1-6 (GenBank code: OQ996385). This strain is gram-negative, N-fixative, and isolated from rooted cutting roots of the hybrid *E. grandis* $\times$ *E. urophylla* in a fertilized environment, with in vitro production of 85 $\mu g\ mL^{-1}$ of IAA, synthesizing enzymes such as catalase, urease, and nitrate reductase, besides being able to solubilize inorganic and organic sources of phosphate, such as calcium phosphate and sodium phytate [12]. The inoculant production and inoculation were the same as in the previous experiment.

Tubes of 55 cm^3 were filled with substrate composed of 70% coconut husk fiber and 30% carbonized rice husk and fertilized with 2 kg m^{-3} of osmocote (NPK: 10-06-10), 1 kg m^{-3} of Superphosphate Simple (NPK: 00-17-00) and 1 kg m^{-3} of Mono Ammonium Phosphate (NPK: 12-61-00) and then three seeds inoculated with their respective treatments were sown in the substrate of each tube. Twelve days after sowing, once the emergence was stabilized, thinning was carried out, leaving one plant per tube. Fifteen days after sowing, a reinoculation with 5 mL per plant with its respective treatment was performed. The reinoculation was performed with the aid of a 20 mL disposable syringe coupled to a breast probe. The inoculant was applied close to the roots. Until the 60th day, the plants were fertilized weekly with Clark's nutrient solution with $\frac{1}{2}$ of the original strength and from 60° to 120° the complete Clark solution was applied [21].

The experiment was conducted in a randomized block design with five blocks, with each treatment, consisting of 15 plants. At 120 days after sowing, 10 plants were randomly harvested from each treatment and the following were evaluated: plant height, measured with a millimeter ruler; stem diameter, with a digital caliper (Digimess 100.179N); and chlorophyll *a* and *b* content, quantified directly with a ClorofiLOG brand chlorophyllometer (Falker CFL1030), according to the manufacturer's instructions. Then, the shoot and roots were separated. These were dried in an oven (Solab SL100) at 65 °C until reaching constant weight and weighed on a precision analytical balance (Marte Ay220) to determine the shoot dry mass (SDM) and the root dry mass (RDM) and calculated the total dry mass (TDM), which was the sum of the SDM with the RDM.

2.2.2. Bacterial Quantification of Shoots, Roots, and Rhizosphere

Two plants from each treatment were harvested randomly and shoots and roots were separated and then weighed. The plant tissues were superficially disinfested by sequential immersion in 70% ethyl alcohol for one minute, in the mixture of 2% sodium hypochlorite +0.1 mL of *polysorbate* Tween® 80 for two minutes for shoots and three minutes for roots and, finally, in 70% ethyl alcohol for 30 s. Then, the plant tissues were immersed in distilled and sterilized water and shaken manually for one minute; finally, they were rinsed with distilled water and sterilized four more times, changing water with each new wash [22].

Before root disinfestation, the substrate was removed manually and the roots were placed in saline solution (9 mL of saline solution per 1 g of roots) and agitated orbitally for 30 min at 200 rpm to displace the bacteria from the rhizosphere. The saline solution was composed of one liter of distilled water plus 3.4 g of KH_2PO_4; 0.2 g of $MgSO_4$; 0.1 g of NaCl; and 0.02 g of $CaCl_2$; 2 mL of the micronutrient solution (0.04 g of $CuSO_4$; 1.20 g of $ZnSO_4$; 1.40 g of H_3BO_3; 1.00 g of Na_2MoO_4; and 1.175 g of $MnSO_4$ per liter of distilled water); and 4 mL of ferric EDTA (1.64% solution) and 4.5 g of KOH, with the final pH adjusted at 7.0 [22]. Then, one gram of shoot and root tissue was macerated in grail containing 9 mL of the saline above the solution.

The suspensions resulting from maceration were transferred to 90 mL vials and shaken (Shaker Incubator Cienlab, São Paulo, Brazil) for 30 min at 200 rpm. Then, 1 mL of the suspension was transferred to a Falcon polypropylene tube containing 9 mL of saline, corresponding to the 10^{-1} dilution, with serial dilution up to 10^{-4} for shoots and 10^{-5} for roots and rhizosphere. From the tubes of the last three dilutions of each part (shoot, roots, and rhizosphere), 0.1 mL of the suspension was removed and transferred to Petri plates with TSA medium (Tryptone Soy Agar). For each dilution, four replications were used and the plates were incubated in a BOD (Solab Sb224, São Paulo, Brazil) chamber for three days at 28 °C. After this period, the number of colonies was counted in the dilutions that allowed growth between 25 and 300 colonies. The bacterial density was calculated considering the dilution and the aliquot inoculated in the plate and expressed in the number of CFU per gram of tissue, with the average of the plates of each treatment whose number of colonies was between 25 and 300.

2.2.3. Statistical Analysis

The data of height, stem diameter, chlorophyll *a* and *b* content, and dry mass of roots, shoot, and total were subjected to multivariate normality analysis, using the Doornik and Hansen [23] test ($p < 0.05$). Then, multivariate analysis of variance was conducted using Wilks Lambda, Pillai Trait, Hotelling–Lawley Trait, and Roy maximum root tests ($p < 0.05$). Grouping analysis of treatments was carried out using the Ward method (formation of homogeneous groups by the lowest minimum internal variance), being used as a reference to the Euclidean distance and the *Pearson* coefficient. To discriminate the treatment groups according to the agronomic variables of *C. citriodora*, canonical discriminant analysis was performed, which was represented in a biplot graph constructed for the first two canonical variables. Then, 95% confidence ellipses were constructed to detect statistical differences ($p < 0.05$) between treatment groups. All analyses were performed with R software version 4.2.2 [24]. The canonical discriminant analysis was performed with the aid of the candisc package [25].

The number of CFU in the rhizosphere, roots, and shoots was submitted to the Durbin–Watson, Shapiro-Wilk, Bartlett, and Tukey.1df tests to verify the independence of errors, normality, homogeneity of variances, and additivity, respectively, and then the analysis of variance. When there were significant differences, the means were grouped by the Scott-Knott test ($p < 0.05$) using the statistical program R [24].

Finally, Pearson correlation was performed between the variables (threshold set at 0.60). A correlation network was set up to graphically illustrate Pearson correlation analyses, in which the proximity between nodes is proportional to the absolute correlation values between the variables. These analyses were carried out using the Rbio version 190 software [26].

3. Results

3.1. Physiological Quality of Seeds

Inoculation influenced the germination percentage and dead seeds, germination speed index, average germination time, average germination speed, and number of normal and abnormal seedlings. Entropy was the only variable that was not influenced by the treatments (Table 2).

Table 2. Analysis of variance for germination percentage (GP), percentage of dead seeds (PDS), germination speed index (GSI), average germination time (AGT), average germination speed (AGS), number of normal (NS) and abnormal (AS) seedlings, and entropy (ENT) of *Corymbia citriodora* seeds inoculated with *Pantoea vagans* strain 45URP4-1, *Priestia megaterium*, *Exiguobacterium sibiricum*, *Bacillus* sp., inoculation carrier, inoculation carrier with 1.0 µg mL^{-1} of IAA and control without inoculation.

Source of Variation	Mean Square							
	GP (%)	PDS (%)	GSI	AGT (Days)	AGS (Days^{-1})	NS	AS $^{+}$	ENT (Bits)
Treatment	44.57 **	44.57 **	35.62 **	0.49 **	0.005 **	13.43 **	0.24 **	0.05 ^{ns}
Residue	9.95	9.50	3.58	0.07	0.001	3.07	0.06	0.04
CV (%)	3.59	25.86	9.37	9.1	9.42	4.06	18.85	9.30

** significant at 1%; ^{ns} not significant; CV: coefficient of variation; $^{+}$ statistics of data transformed by $\sqrt{(x + 1)}$.

The highest percentage of germination, lowest percentage of dead seeds, and highest number of normal seedlings were observed in seeds treated only with the inoculation carrier and in those inoculated with the *P. vagans* strain 45URP4-1 (Table 3). The lowest germination speed index, highest average germination time, and lowest average germination speed were observed in the non-inoculated seeds, without carrier and without IAA, and the other treatments did not differ for these variables. The seeds inoculated with the mix generated a higher number of abnormal seedlings (Table 3).

Table 3. Germination percentage (GP), percentage of dead seeds (PDS), germination speed index (GSI), average germination time (AGT), average germination speed (AGS), number of abnormal (AS), and normal (NS) seedlings of *Corymbia citriodora* seeds inoculated or not with endophytic bacteria.

Treatments	GP (%)	PDS (%)	GSI	AGT (Days)	AGS (Days^{-1})	AS *	NS
Control	85.20 b	14.80 a	14.52 b	3.59 a	0.28 b	0.8 b	41.80 b
Carrier	92.40 a	7.60 b	21.07 a	2.77 b	0.36 a	0.4 b	45.80 a

Table 3. *Cont.*

Treatments	GP (%)	PDS (%)	GSI	AGT (Days)	AGS (Days^{-1})	AS *	NS
Carrier with 1.0 μg mL^{-1} of IAA	86.80 b	13.20 a	20.78 a	2.68 b	0.38 a	0.6 b	42.80 b
Pantoea vagans strain 45URP4-1	91.60 a	8,40 b	23.58 a	2.57 b	0.39 a	0.4 b	45.40 a
Bacillus sp.	86.80 b	13.20 a	21.61 a	2.69 b	0.38 a	0.4 b	43.00 b
Exiguobacterium sibiricum	86.60 b	16.40 a	18.97 a	2.84 b	0.36 a	0.6 b	41.20 b
Priestia megaterium	88.40 b	11.60 a	19.61 a	2.88 b	0.35 a	0.8 b	43.40 b
Mix	87.40 b	12.40 a	21.34 a	2.86 b	0.35 a	2.4 a	41.40 b

Means followed by the same letter in the column did not differ statistically by the Scott-Knott test ($p < 0.05$).
* Original means with statistics of the data transformed by $\sqrt{(x + 1)}$.

3.2. Production of Seedlings in Greenhouse

There was a significant difference between the treatment by the multivariate Wilks Lambda test, Pillai Trait, Hotelling–Lawley Trait, and Roy maximum root for the variables plant height (PH), stem diameter (SD), and chlorophyll *a* (*Ca*) and *b* (*Cb*) content and dry mass of roots (RDM), shoot (SDM), and total (TDM) of *C. citriodora* plants (Table 4).

Table 4. Multivariate analysis of variance for the vectors of treatment averages of *Corymbia citriodora* plants inoculated with *Pantoea vagans* strains 7URP1-6 and 45URP4-1, *Priestia megaterium*, *Exiguobacterium sibiricum*, *Bacillus* sp., inoculation carrier, and inoculation carrier with 1.0 μg mL^{-1} of IAA and control without inoculation.

Statistics	Value	[1]num Df	[2]den Df	Aprox. F	Pr > F
Wilks Lambda	0.05	49	136.42	2.26	0.0001 ***
Pillai trace	1.72	49	224	1.49	0.03 *
Hotelling-Lawley trace	7.78	49	170	3.85	3.77×10^{-11} ***
Roy maximum root	6.63	7	32	30.33	2.16×10^{-12} ***

[1]num Df: degree of freedom of the numerator; [2]den Df: degree of freedom of the denominator. '***' 0.001 '*' 0.05; Pr > F: F-test significance.

The treatments were grouped into four groups according to the characteristics PH, SD, *Ca*, *Cb*, SDM, RDM, and TDM, being Group I. Control, control with carrier and carrier with 1.0 μg mL^{-1} of IAA. Group II: *P. vagans* strain 45URP4-1 and *Bacillus* sp. Group III: *E. sibiricum*. Group IV: *P. vagans* strain 7URP1-6 and *P. megaterium* (Figure 1).

The plants inoculated with *E. sibiricum* (group III), *P. megaterium*, and *P. vagans* strain 7URP1-6 (group IV) had higher PH, SD, *Cb* content, RDM, SDM, and TDM, not differing significantly from each other but differing from the plants inoculated with the other treatments (Figure 2). The plants inoculated with *E. sibiricum* (group III) had PH = 16.97 cm, SD = 1.42 mm, *Cb* = 8.71, RDM = 1.04 g, SDM = 4.79 g, and TDM = 5.83 g; on average, the plants inoculated with *P. megaterium* and *P. vagans* strain 7URP1-6 (group IV) had PH = 16.04 cm, SD = 1.46 mm, *Cb* = 8.23, RDM = 1.05 g, SDM = 4.12 g, and TDM = 5.16 g; and while the plants that received the control treatments (group I) obtained the means of PH = 14.08 cm, SD = 1.20 mm, *Cb* = 8.43, RDM = 1.04 g, SDM = 3.78 g, and TDM = 4.82 g (Table 5). The highest *Ca* content was 29.4, obtained in the plants treated with the control treatments (Group I). The average *Ca* contents of plants inoculated with *E. sibiricum* (group III), *P. megaterium*, and *P. vagans* strain 7URP1-6 (group IV) were 28.75 and 28.09, respectively. Plants inoculated with *P. vagans* strain 45URP4-1 and *Bacillus* sp. (Group II) presented intermediate results, with means taller than Group I for PH (14.44 cm), SD (1.41 mm), SDM (4.31 g), and TDM (5.33 g) (Figure 2).

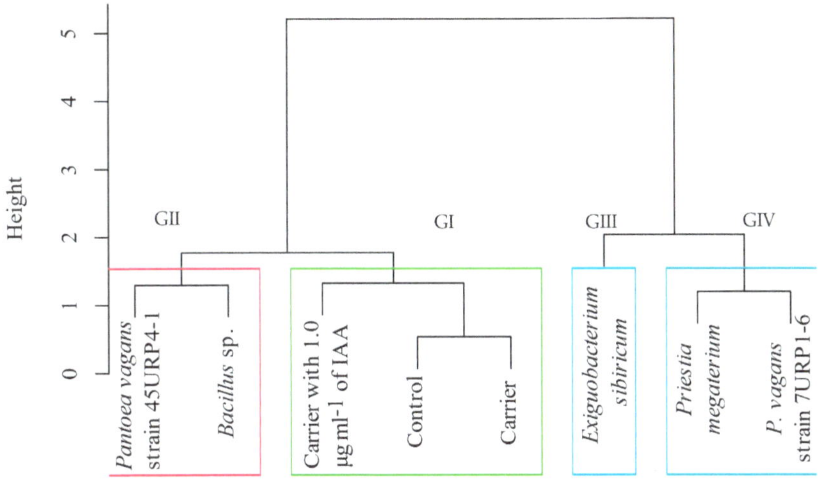

Figure 1. Grouping of dendrogram treatments with Euclidean distance considering the variables plant height, stem diameter, chlorophyll a and b content, and dry mass of roots, shoot, and total plants of *C. citriodora* inoculated with endophytic bacteria and treated with cassava starch and activated charcoal (carrier), carrier with 1.0 µg mL^{-1} of IAA and control without inoculation (Control).

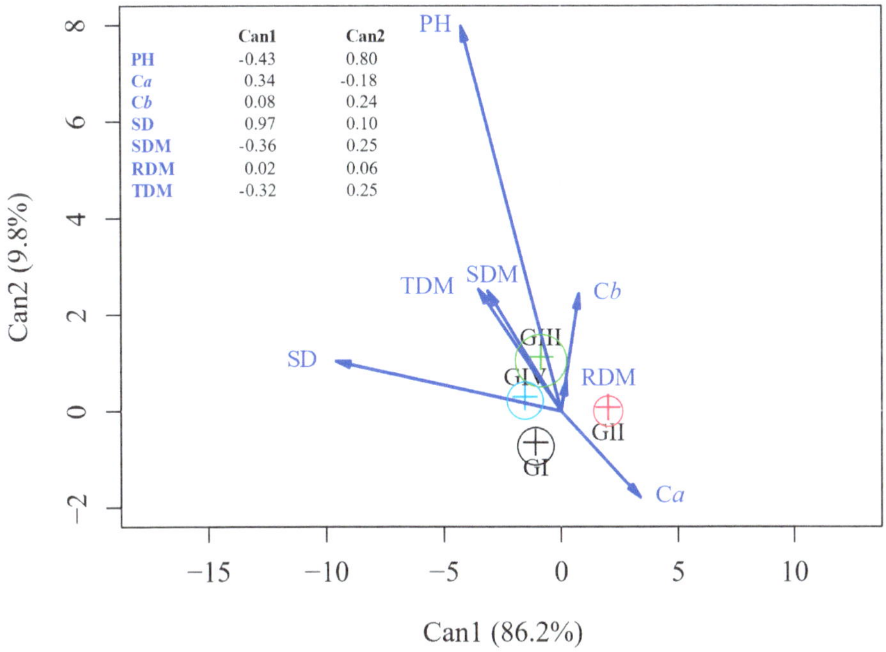

Figure 2. Canonical discriminant analysis of the variables plant height (PH); stem diameter (SD); shoot dry mass (SDM), root dry mass (RDM); total dry mass (TDM); chlorophyll *a* (C*a*) and *b* (C*b*) content of *Corymbia citriodora* plants treated with cassava starch and activated charcoal; the carrier with 1.0 µg mL^{-1} of IAA and control without inoculation (GI) and inoculated with *Pantoea vagans* strain 45URP4-1 and *Bacillus* sp. (GII); *Exiguobacterium sibiricum* (GIII); and *P. vagans* strain 7URP1-6 and *Priestia megaterium* (GIV).

Table 5. Means of plant height, stem diameter, chlorophyll *a* and *b* content, and dry mass of roots, shoot, and total of *Corymbia citriodora* plants treated with endophytic bacteria.

Variables	Group			
	I	II	III	IV
Plant height (cm)	14.08	14.44	16.97	16.04
Stem diameter (mm)	1.20	1.41	1.42	1.46
Chlorophyll *a*	29.4	28.89	28.75	28.09
Chlorophyll *b*	8.43	8.37	8.71	8.23
Roots dry mass (g)	1.04	1.02	1.04	1.05
Shoot dry mass (g)	3.78	4.31	4.79	4.12
Total dry mass (g)	4.82	5.33	5.83	5.16

GI: Cassava starch and activated charcoal, the carrier with 1.0 $\mu g\ mL^{-1}$ of IAA and control without inoculation; GII: *Pantoea vagans* strain 45URP4-1 and *Bacillus* sp.; GIII: *Exiguobacterium sibiricum*; GIV: *P. vagans* strain 7URP1-6 and *Priestia megaterium*.

The highest numbers of CFU in the roots and rhizosphere were observed in the plants inoculated with *E. sibiricum*, followed by the roots of the plants inoculated with *P. vagans* strain 7URP1-6 and in the rhizosphere of the plants inoculated with *P. vagans* strain 45URP4-1 (Table 6). In the shoot, the highest number of CFU was observed in plants inoculated with *P. vagans* strain 7URP1-6. In the non-inoculated plants (control without inoculation, only carrier and carrier with IAA), lower numbers of CFU in the roots and rhizosphere were observed (Table 6). Overall, the plants had lower numbers of CFU in the shoot and the CFU numbers of the rhizosphere and roots were close.

Table 6. Density of bacteria (CFU g of tissue^{-1}) in the rhizosphere, roots, and shoot of *Corymbia citriodora* plants inoculated or not with endophytic bacteria.

Treatments	CFU g of Tissue^{-1}		
	Rhizosphere	Root	Shoot
Control	4.0×10^7 e	4.2×10^7 e	6.4×10^4 d
Carrier	1.6×10^8 d	4.2×10^7 e	4.4×10^5 b
Carrier with 1.0 $\mu g\ mL^{-1}$ of IAA	2.2×10^7 e	1.8×10^8 c	3.0×10^5 c
Pantoea vagans strain 7URP1-6	7.1×10^7 e	2.9×10^8 b	6.1×10^5 a
Priestia megaterium	1.9×10^8 c	7.8×10^7 d	4.3×10^5 b
Exiguobacterium sibiricum	4.2×10^8 a	5.3×10^8 a	6.6×10^4 d
P. vagans strain 45URP4-1	2.9×10^8 b	1.5×10^8 c	2.0×10^5 c
Bacillus sp.	2.3×10^8 c	9.4×10^7 d	1.3×10^5 d
CV (%)	22.8	17.6	31.3

Means followed by the same letter in the column do not differ by the Scott-Knott test ($p < 0.05$). CV: Coefficient of variation.

The only significant positive correlations were between SDM and TDM (0.95) and between *Ca* and *Cb* content (0.65) (Figure 3).

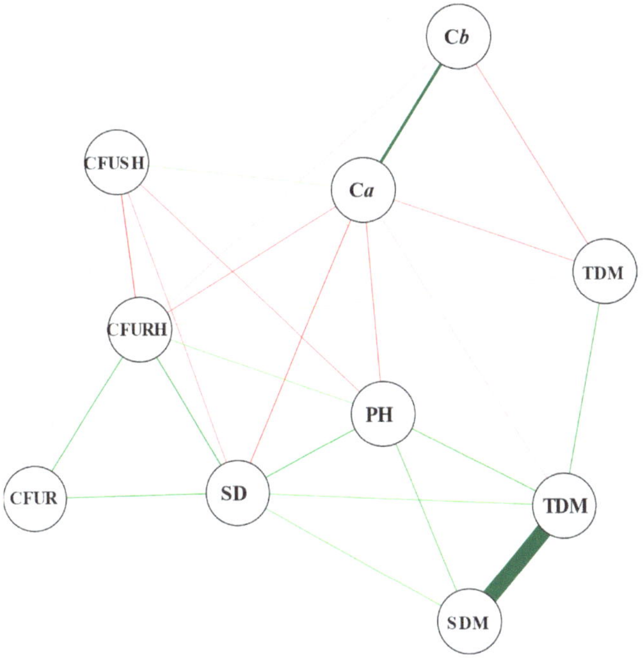

Figure 3. Correlation network illustrating Pearson's correlations between the variables chlorophyll *a* (C*a*) and *b* (C*b*); the number of colony-forming units of the shoot (CFUSH), roots (CFUR), and rhizosphere (CFURH); stem diameter (SD); plant height (PH); and dry mass of roots (RDM), shoot (SDM), and total (TDM) of *Corymbia citriodora* plants inoculated with *Pantoea vagans* strains 7URP1-6 and 45URP4-1, *Priestia megaterium*, *Exiguobacterium sibiricum*, *Bacillus* sp., inoculation carrier, and the carrier with 1.0 μg mL^{-1} of IAA and control without inoculation. The thicker greener lines represent the highest positive correlations. The red lines represent the negative correlations.

4. Discussion

The absence of s significant difference between the treatments for entropy (Table 2) shows that the bacteria and the carrier do not interfere in the organization and synchrony of seed germination. Entropy measures the organization of a system, so the lower the entropy, the greater the organization and synchrony of the system [27], in this case, represented by the population of seeds analyzed in the germination test. Therefore, the bacteria alone can be used in inoculant formulations for *C. citriodora* seeds as it does not interfere with entropy and other variables of the physiological quality of the seeds (Table 3).

The increase in the number of abnormal seedlings by mixed inoculation (Table 3) may have occurred due to competition between the strains for a niche [28], which may have resulted in the synthesis of compounds that impaired seedling development. Before establishing an association with plants, the bacterium–plant relationship can go through a parasitic phase, which is the period when bacteria require photosynthates from the plant to multiply and internally colonize the roots [29,30]. This may also have favored the increase in the number of abnormal seedlings due to the number of inoculated strains and the concentration of the inoculant. The success of co-inoculation is closely related to the genotype of the plant, with the proper selection of strains and the cellular concentration of each one [4]. In the present study, no previous tests were performed. Only the concentration of $\frac{1}{4}$ of each inoculant was used to formulate the mix. For further studies with co-inoculation, it is necessary first to select compatible strains and evaluate varying concentrations for each bacterium.

The benefits obtained with inoculations in the seed physiological quality test (Table 3) cannot be attributed to bacterial strains, as treatment with only the inoculation carrier also showed good results. Thus, it is inferred that the effects were not from the bacteria but from the carrier created a microenvironment around the seed, providing them with better germination conditions. One of the main characteristics of a good inoculant is its ability to retain moisture and avoid desiccation and, consequently, the death of microorganisms [31], which also stimulates the germination of the seeds because the humidity favors the germination process [32].

One of the main forms of application of inoculants is adhering the product to pre-inoculated seeds stored before sale or at sowing [31]. Bacterial survival in the seed is mainly diminished by three factors: desiccation, the toxic nature of the integument exudates, and high temperatures [33]. As the factors humidity and temperature were controlled in the physiological quality test of the seeds and the greenhouse experiment, the inoculation with *P. vagans* strain 7URP1-6, *P. megaterium*, and *E. sibiricum* benefited the growth of the plants (Figure 2); it is inferred that the benefits of the bacteria are not in the germination of the seeds but in the development of the seedlings.

In the greenhouse, the better development of plants after the seeds were inoculated with *E. sibiricum* (Figure 2) can be attributed to the colonization capacity of this species. Analyzing colonization, the numbers of CFU in the roots and rhizosphere were taller in the seeds inoculated with *E. sibiricum* (Table 6); one of the main points for the success of the inoculation is the ability of the bacteria to colonize the plants [3]. The *E. sibiricum* strain was the only one isolated from unfertilized environments and adult plants (Table 1). *Exiguobacterium* species can grow in extreme environments, with limited nutrients and temperatures ranging from -12 to 55 °C [34]. Thus, it is believed to have a greater ability to interact with *C. citriodora* plants; however, this effect may vary according to the genotype of the plant [35].

The in vitro IAA production of *P. vagans* strain 7URP1-6, *P. megaterium*, and *E. sibiricum* is close, ranging from 85 to 90 $\mu g\ mL^{-1}$ (Table 1). The best-known effect of auxins is the stimulation of rooting, increasing the area of soil exploitation by the roots and also the absorption of water and nutrients [36,37], consequently favoring plant development, as observed in the study, namely with respect to taller plants, with larger stem diameter and dry mass of the roots, shoot, and total (Figure 2). Sousa et al. [38] inoculated *E. urophylla* seeds with *Azospirillum amazonense* and *Stenotrophomonas maltophilia* and also observed better development of height, stem diameter, root length, and dry matter.

In addition to the synthesis of IAA, these strains synthesize catalase (Table 1), an enzyme responsible for remediating the toxic effects of hydrogen peroxide on plant metabolism. The *E. sibiricum* strain also synthesizes the enzyme nitrate reductase, mainly responsible for the assimilation of nitrogen by plants, and *P. megaterium* to urease, an enzyme responsible for the hydrolysis of urea into carbon dioxide and ammonia. The 7URP1-6 strain of *P. vagans* is N-fixing and synthesizes enzymes such as urease and nitrate reductase, which, in the long term, can improve plant development, as well as being able to solubilize inorganic and organic phosphate sources, such as calcium phosphate and sodium phytate. Inoculation of rice seeds with *P. vagans* strain LYY2b, also producing IAA, resulted in abundant root hair production, increased root and shoot length, and root hair formation [39].

Although the strains 7URP1-6 and 45URP4-1 of *P. vagans* are of the same species, their effects were different. 7URP1-6 improved seedling quality, while 45URP4-1 provided intermediate seedling quality (Figure 2). This can be explained by the genetic variability that exists within species, causing them to provide different results [35].

The lower number of CFU in the shoot of plants and the numbers of CFU in the rhizosphere and nearby roots (Table 6) are explained by the colonization process since endophytes usually colonize plants through the root system from sites with epidermal damage, which arise naturally, due to the development of lateral roots, or by means of root hairs. In the rhizosphere, exudates are also released in large quantities by the roots and both endophytic and rhizosphere bacteria grow. For this reason, it is believed that the highest bacterial densities are in the rhizosphere and progressively decrease from the roots through the stem and reach the leaves, where the density is lower [40].

The levels of chlorophyll *a* and *b* were positively and significantly correlated (Figure 3). Chlorophyll *a* and *b* are pigments are associated with photosynthesis. They are usually found in nature in a ratio of 3:1 [41], which may have balanced the effect of chlorophylls on plant development because the non-inoculated plants (group I) had the highest chlorophyll *a* content, while those inoculated with *E. sibiricum* (group III) had a taller chlorophyll *b* content. It acts in the production of organic substances, being used in the photochemical stage of photosynthesis since chlorophyll *b* is an accessory pigment and acts by increasing the range of light absorption that can be used in the photosynthetic process [41].

Given the importance of inoculation and not yet having an inoculant for eucalyptus seeds, the best development of plants after inoculation with *P. vagans* strain 7URP1-6, *P. megaterium*, and *E. sibiricum* is a promising result, given the importance of *C. citriodora* for the production of coal, cellulose, furniture, firewood, poles, sawmill, rails, manufacture of floors, and extraction of essential oils from its leaves, among other applications [42,43]. The production of seedlings with taller quality in the greenhouse increases the chances of survival in the field as it has greater chances of withstanding the adverse conditions of the environment. However, it is still necessary to expand the studies to field conditions and other tree species.

5. Conclusions

The inoculation of *C. citriodora* seeds with *P. megaterium*, *E. sibiricum*, *P. vagans* strain 45URP4-1, and *Bacillus* sp. does not interfere with the physiological quality of the seeds. The seeds of *C. citriodora* inoculated with *P. vagans* strain 7URP1-6, *P. megaterium*, and *E. sibiricum* present similar results in plant development. However, those inoculated with *E. sibiricum* had taller averages for height, chlorophyll *b* content, and shoot and total dry mass and greater ability to colonize the roots and rhizosphere of *C. citriodora*, resulting in the production of better-quality seedlings. Given the benefits of inoculating *C. citriodora* seeds with PGPB on plant development, future studies can evaluate the effect of PGPB on other forest species that are also propagated via seed, as inoculation of seeds of *C. citriodora* with endophytic bacteria proved to be a promising alternative in the plant development.

Author Contributions: Conceptualization, A.M.d.O., M.R.d.C. and P.H.G.; methodology, A.M.d.O., M.R.d.C. and C.M.d.A.; software, A.M.d.O.; validation, J.F.S.M., N.R.A., G.F.P.d.A., J.V.G. and P.H.G.; formal analysis, A.M.d.O., G.F.P.d.A., J.V.G., N.R.A. and C.M.d.A.; investigation, M.R.d.C., P.H.G., N.R.A. and C.M.d.A.; data curation, A.M.d.O., G.F.P.d.A. and J.V.G.; funding acquisition, J.B.d.S., G.M.B. and C.M.d.A.; writing—original draft preparation, J.V.G., G.F.P.d.A. and A.M.d.O.; writing—review and editing, A.M.d.O., C.M.d.A., M.R.d.C., G.M.B., J.B.d.S. and P.H.G.; visualization, G.M.B., J.F.S.M., N.R.A., J.B.d.S., P.H.G. and M.R.d.C.; supervision, M.R.d.C. and P.H.G.; project administration, A.M.d.O. All authors have read and agreed to the published version of the manuscript.

Funding: This research received no external funding.

Data Availability Statement: Data are contained within the article.

Acknowledgments: With acknowledgements to the Foundation for Research Support of the State of Minas Gerais (FAPEMIG), the National Council for Scientific and Technological Development (CNPq), and the Coordination for the Improvement of Higher Education Personnel (CAPES)—Financial Code 001.

Conflicts of Interest: The authors declare no conflicts of interest.

References

1. Hardoim, P.R.; Overbeek, L.S.V.; Berg, G.; Pirttila, A.M.; Compant, S.; Campisano, A.; Döring, M.; Sessitsch, A. The hidden world within plants: Ecological and evolutionary considerations for defining functioning of microbial endophytes. *Microbiol. Mol. Biol. Rev.* **2015**, *79*, 293–320. [CrossRef]

2. Oliveira, A.M.; Costa, M.R.; Grazziotti, P.H.; Abreu, C.M.; Bispo, N.S.; Roa, J.P.B.; Silva, D.M.; Miranda, J.M. Brazilian scenario of inoculant production: A look at patents. *Rev. Bras. Cienc. Solo* **2022**, *46*, e0210081. [CrossRef]
3. Souza, R.D.; Ambrosini, A.; Passaglia, L.M. Plant growth-promoting bacteria as inoculants in agricultural soils. *Genet. Mol. Biol.* **2015**, *38*, 401–419. [CrossRef] [PubMed]
4. Santos, M.S.; Nogueira, M.A.; Hungria, M. Microbial inoculants: Reviewing the past, discussing the present and previewing an outstanding future for the use of beneficial bacteria in agriculture. *AMB Express* **2019**, *9*, 1–22. [CrossRef]
5. Eid, A.M.; Fouda, A.; Abdel-Rahman, M.A.; Salem, S.S.; Elsaied, A.; Oelmüller, R.; Hijri, M.; Bhowmik, A.; Elkelish, A.; Hassan, S.E.D. Harnessing bacterial endophytes for promotion of plant growth and biotechnological applications: An overview. *Plants* **2021**, *10*, 935. [CrossRef] [PubMed]
6. Paravar, A.; Piri, R.; Balouchi, H.; Ma, Y. Microbial seed coating: An attractive tool for sustainable agriculture. *Biotechnol. Rep.* **2023**, *37*, e00781. [CrossRef]
7. Reis, C.A.F.; Assis, T.F.; Santos, A.M.; Paludzyszyn Filho, E. *Corymbia Citriodora: Estado da Arte de Pesquisas no Brasil*; Embrapa Florestas: Colombo, Brazil, 2013; 59p.
8. Lima, M.S.; Araujo, M.M.; Berghetti, Á.L.P.; Aimi, S.C.; Costella, C.; Griebeler, A.M.; Somavilla, L.M.; Santos, O.P.; Valente, B.M.R.T. Mini-cutting technique application in *Corymbia* and *Eucalyptus*: Effects of mini-tunnel use across seasons of the year. *New For.* **2022**, *53*, 161–179. [CrossRef]
9. Bonora, F.S.; Nahrung, H.F.; Hayes, R.A.; Scharaschkin, T.; Pegg, G.; Lee, D.J. Changes in leaf chemistry and anatomy of *Corymbia citriodora* subsp. *variegata* (Myrtaceae) in response to native and exotic pathogens. *Australas. Plant Pathol.* **2020**, *49*, 641–653. [CrossRef]
10. Todd, O.E.; Figueiredo, M.R.; Morran, S.; Soni, N.; Preston, C.; Kubeš, M.F.; Napier, R.; Gaines, T.A. Synthetic auxin herbicides: Finding the lock and key to weed resistance. *Plant Sci.* **2020**, *300*, 110631. [CrossRef] [PubMed]
11. Schlindwein, G.; Vargas, L.K.; Lisboa, B.B.; Azambuja, A.C.; Granada, C.E.; Gabiatti, N.C.; Prates, P.; Stumpf, R. Influência da inoculação de rizóbios sobre a germinação e o vigor de plântulas de alface. *Cienc. Rural* **2008**, *38*, 658–664. [CrossRef]
12. Ramires, R.V. Microrganismos Endofíticos em *Eucalyptus*. Ph.D. Thesis, Universidade Federal dos Vales do Jequitinhonha e Mucuri, Diamantina, Brazil, 2021. 110 f.
13. Titon, M.; Xavier, A.; Otoni, W.C.; Motoike, S.Y. Efeito dos reguladores de crescimento dicamba e picloram na embriogênese somática em *Eucalyptus grandis*. *Rev. Árvore* **2007**, *31*, 417–426. [CrossRef]
14. Oliveira, C.A.; Marriel, I.E.; Gomes, E.A.; Mattos, B.B.; Santos, F.C.; Oliveira, M.C.; Alves, V.M.C. *Metodologia de Aplicação de Microrgabnismos Solubilizadores de Fósforo em Sementes Visando Melhor Aproveitamento deste Nutriente Pelas Plantas*; Embrapa: Sete Lagoas, Brazil, 2013; 29p. Available online: https://ainfo.cnptia.embrapa.br/digital/bitstream/item/95211/1/bol-88.pdf (accessed on 5 April 2021).
15. Maravilha, L.F.; Titon, M.; Canguçu, V.S.; Rocha, F.M.; Oliveira, M.R. Enraizamento in vitro e aclimatação de plântulas de *Corymbia citriodora*. *Pesq. Flor. Bras.* **2023**, *43*, e202102232. [CrossRef]
16. Patten, C.L.; Glick, B.R. Role of *Pseudomonas putida* indoleacetic acid in development of the host plant root system. *Appl. Environ. Microbiol.* **2002**, *68*, 3795–3801. [CrossRef]
17. Mapa. Ministério da Agricultura, Pecuária e Abastecimento. Regras Para Análise de Sementes. 1a ed. Brasília, 2009. 398p. Available online: https://www.gov.br/agricultura/pt-br/assuntos/insumos-agropecuarios/arquivos-publicacoes-insumos/2946_regras_analise__sementes.pdf (accessed on 2 January 2022).
18. Maguire, J.D. Speed of germination-aid selection and evaluation for seedling emergence and vigor. *Crop Sci.* **1962**, *2*, 176–177. [CrossRef]
19. Labouriau, L.G.; Valadares, M.B. On the germination of seed of *Calotropis procera*. *An. Acad. Bras. Ciênc.* **1976**, *48*, 263–284.
20. Ferreira, E.B.; Cavalcanti, P.P.; Nogueira, D.A. ExpDes: An R package for ANOVA and experimental designs. *Appl. Math.* **2014**, *5*, 2952. [CrossRef]
21. Clark, R.B. Characterization of phosphatase of intact maize roots. *J. Agric. Food Chem.* **1975**, *23*, 458460. [CrossRef]
22. Döbereiner, J.; Baldani, V.L.D.; Baldani, J.I. *Como Isolar e Identificar Bactérias Diazotróficas de Plantas Não Leguminosas*; Embrapa-SPI: Itaguaí, Brazil, 1995; 60p.
23. Doornik, J.A.; Hansen, H. An omnibus test for univariate and multivariate normality. *Oxf. B Econ. Stat.* **2008**, *70*, 927–939. [CrossRef]
24. R Core Team. *R: A Language and Environment for Statistical Computing*; R Foundation for Statistical Computing: Vienna, Austria, 2022. Available online: https://www.R-project.org/ (accessed on 10 November 2022).
25. Friendly, M.; Fox, J. *Candisc: Visualizing Generalized Canonical Discriminant and Canonical Correlation Analysis*, R package version 0.8-0.2017; R Foundation for Statistical Computing: Vienna, Austria, 2015. Available online: https://CRAN.R-project.org/package=candisc(accessed on 9 November 2022).
26. Bhering, L.L. Rbio: A tool for biometric and statistical analysis using the R platform. *Crop Breed. Appl. Biotechnol.* **2017**, *17*, 187–190. [CrossRef]
27. Nassif, S.M.L.; Perez, S.C.J.G. Efeito da temperatura na germinação de sementes de amendoim-do-campo (*Pterogyne nitens* Tul.). *Rev. Bras. Sementes* **2000**, *22*, 1–6. [CrossRef]

28. Leppyanen, I.; Shtark, O.; Pavlova, O.; Bovin, A.; Ivanova, K.; Serova, T.; Dolgikh, E. Analysis of the effects of joint inoculation by arbuscular mycorrhizal fungi and rhizobia on the growth and development of pea plants *Pisum sativum* L. *Agric. Biol.* **2021**, *6*, 475–486. [CrossRef]
29. Toft, C.; Andersson, S. Evolutionary microbial genomics: Insights into bacterial host adaptation. *Nat. Rev. Genet.* **2010**, *11*, 465–475. [CrossRef]
30. Lyu, D.; Msimbira, L.A.; Nazari, M.; Antar, M.; Pagé, A.; Shah, A.; Monjezi, N.; Zajonc, J.; Tanney, C.A.S.; Backer, R.; et al. The coevolution of plants and microbes underpins sustainable agriculture. *Microorganisms* **2021**, *9*, 1036. [CrossRef]
31. Albareda, M.; Rodríguez-Navarro, D.N.; Camacho, M.; Temprano, F.J. Alternatives to peat as a carrier for rhizobia inoculants: Solid and liquid formulations. *Soil. Biol. Biochem.* **2008**, *40*, 2771–2779. [CrossRef]
32. Mbi, T.K.; Ntsefong, N.G.; Lenzemo, T.E. Seed Dormancy: Induction, maintenance and seed technology approaches to break dormancy. In *Seed Biology Updates*; Jimenez-Lopez, J.C., Ed.; IntechOpen: London, UK, 2022. [CrossRef]
33. Deaker, R.; Roughley, R.J.; Kennedy, I.R. Legume seed inoculation technology—A review. *Soil. Biol. Biochem.* **2004**, *36*, 1275–1288. [CrossRef]
34. Vishnivetskaya, T.A.; Kathariou, S.; Tiedje, J.M. The *Exiguobacterium* genus: Biodiversity and biogeography. *Extremophiles* **2009**, *13*, 541–555. [CrossRef]
35. Razgour, O.; Forester, B.; Taggart, J.B.; Bekaert, M.; Juste, J.; Ibáñez, C.; Puechmaille, S.J.; Novella-Fernandez, R.; Alberdi, A.; Manel, S. Considering adaptive genetic variation in climate change vulnerability assessment reduces species range loss projections. *Proc. Natl. Acad. Sci. USA* **2019**, *116*, 10418–10423. [CrossRef]
36. Pascale, S.; Rouphael, Y.; Colla, G. Plant biostimulants: Innovative tool for enhancing plant nutrition in organic farming. *Eur. J. Hortic. Sci.* **2017**, *82*, 277–285. [CrossRef]
37. Nakhforoosh, A.; Nagel, K.A.; Fiorani, F.; Bodner, G. Deep soil exploration vs. topsoil exploitation: Distinctive rooting strategies between wheat landraces and wild relatives. *Plant Soil.* **2021**, *459*, 397–421. [CrossRef]
38. Sousa, F.G.; Mielke, K.C.; Caldeira, D.R.M.; Baldani, V.L.D.; Baldani, J.I.; Silva, R.F.; Balbinot, E.; Klein, V.A.C. Genetic diversity and inoculation of plant-growth promoting diazotrophic bacteria for production of *Eucalyptus urophylla* seedlings. *Aust. J. Crop Sci.* **2022**, *16*, 35–44. [CrossRef]
39. Verma, S.K.; Kingsley, K.; Bergen, M.; English, C.; Elmore, M.; Kharwar, R.N.; White, J.F. Bacterial endophytes from rice cut grass (*Leersia oryzoides* L.) increase growth, promote root gravitropic response, stimulate root hair formation, and protect rice seedlings from disease. *Plant Soil* **2018**, *422*, 223–238. [CrossRef]
40. Souza, C.R.S.; Barbosa, A.C.O.; Ferreira, C.F.; Souza, F.V.D.; Rocha, L.S.; Souza, E.H.; Oliveira, S.A.S. Diversity of microorganisms associated to *Ananas* spp. from natural environment, cultivated and ex situ conservation areas. *Sci. Hortic.* **2019**, *243*, 544–551. [CrossRef]
41. Streit, N.M.; Canterle, L.P.; Canto, M.W.D.; Hecktheuer, L.H.H. The chlorophylls. *Cienc. Rural* **2005**, *35*, 748–755. [CrossRef]
42. Lin, L.; Chen, W.; Li, C.; Cui, H. Enhancing stability of *Eucalyptus citriodora* essential oil by solid nanoliposomes encapsulation. *Ind. Crops Prod.* **2019**, *140*, 111615. [CrossRef]
43. Souza, B.M.; Freitas, M.L.M.; Sebbenn, A.M.; Gezan, S.A.; Zanatto, B.; Zulian, D.F.; Lopes, M.T.G.; Longui, E.L.; Guerrini, I.A. Genotype-by-environment interaction in *Corymbia citriodora* (Hook.) KD Hill, & LAS Johnson progeny test in Luiz Antonio, Brazil. *For. Ecol. Manag.* **2020**, *460*, 117855. [CrossRef]

forests

MDPI

Article

The Changes in Soil Microbial Communities and Assembly Processes along Vegetation Succession in a Subtropical Forest

Jiusheng Ren [1,†], Kangxiang Huang [2,†], Fangfang Xu [1,2], Yuan Zhang [2], Bosen Yuan [2], Huimin Chen [3] and Fuxi Shi [2,4,*]

[1] School of Water Resources and Environmental Engineering/Jiangxi Province Key Laboratory of the Causes and Control of Atmospheric Pollution, East China University of Technology, Nanchang 330013, China; renjiusheng256@ecut.edu.cn (J.R.); xufangfang@stu.jxau.edu.cn (F.X.)
[2] Key Laboratory of National Forestry and Grassland Administration on Forest Ecosystem Protection and Restoration of Poyang Lake Watershed, College of Forestry, Jiangxi Agricultural University, Nanchang 330045, China; hkx2021@stu.jxau.edu.cn (K.H.); 15934791433@stu.jxau.edu.cn (Y.Z.); yuan476147253@jxau.edu.cn (B.Y.)
[3] Jiangxi Agricultural University Library, Jiangxi Agricultural University, Nanchang 330045, China; chenhm292@163.com
[4] Matoushan Observation and Research Station of Forest Ecosystem, Zixi, Fuzhou 335300, China
* Correspondence: shifuxi@jxau.edu.cn
† These authors contributed equally to this work.

Abstract: Soil microbes are the primary drivers of the material cycling of the forest ecosystem, and understanding how microbial structure and composition change across succession assists in clarifying the mechanisms behind succession dynamics. However, the response of soil microbial communities and assembly processes to succession is poorly understood in subtropical forests. Thus, through the "space instead of time" and high throughput sequencing method, the dynamics of the soil bacterial and fungal communities and assembly process along the succession were studied, where five succession stages, including Abandoned lands (AL), Deciduous broad-leaved forests (DB), Coniferous forests (CF), Coniferous broad-leaved mixed forests (CB), and Evergreen broad-leaved forests (EB), were selected in a subtropical forest on the western slope of Wuyi Mountain, southern China. The results demonstrated that succession significantly decreased soil bacterial α-diversity but had little effect on fungal α-diversity. The composition of soil bacterial and fungal communities shifted along with the succession stages. LEfSe analysis showed the transition from initial succession microbial communities dominated by *Firmicutes*, *Bacteroidota*, *Ascomycota*, and *Chytridiomycota* to terminal succession communities dominated by *Actinobacteriota* and *Basidiomycota*. Distance-based redundancy analysis (db-RDA) revealed that soil total organic carbon (TOC) was the main factor explaining variability in the structure of soil bacterial communities, and multiple soil environmental factors such as the TOC, soil total nitrogen (TN), C:N ratio, and pH co-regulated the structure of fungi. The null models illustrated that deterministic processes were dominant in the soil bacterial communities, while the stochastic processes contributed significantly to the soil fungal communities during succession. Collectively, our results suggest that different patterns are displayed by the soil bacterial and fungal communities during the succession. These findings enhance our comprehension of the processes that drive the formation and maintenance of soil microbial diversity throughout forest succession.

Keywords: soil microbes; microbial community; assembly process; high throughput sequencing; vegetation succession

1. Introduction

Forest succession is a continuous change in the species composition, structure, and function of a forest through time following disturbance. In general, vegetation succession could alter the functioning of forest ecosystems, affecting both above-ground ecological functions and below-ground material cycling processes [1–4]. Soil microorganisms are the main drivers of underground material cycling along the succession stages [5,6]. The diversity of soil microorganisms is an important foundation for driving material cycling, and it is mainly determined by multiple factors such as environmental factors, community structure, and community assembly [7,8]. Investigating how soil microbial community composition and assembly processes respond to forest succession helps to reveal the mechanisms driving below-ground material cycling.

Forest succession can affect soil microorganisms by changing vegetation and altering soil physicochemical properties [9,10]. In terms of soil physicochemical properties, substrates may be the most important factor regulating soil microbial communities [11,12]. The effects of substrates on soil microorganisms can change based on the species of trees present in a given forest succession [13]. Previous research has shown that soil pH tends to be more acidic with the progression of forest succession, and that soil pH also significantly affects microbial communities [14,15]. In the meanwhile, there are variations in soil background values in different regions, and whether succession series may cause soil acidity and alter the soil microbial communities is still up for debate. As the main distribution area of forests, the shift of soil microbial communities along with the forest succession through the impact on the soil's physical and chemical factors still needs further exploration in subtropical forests.

The establishment and maintenance of soil microbial communities are the core issues of microbial ecology [16]. Forest succession is typical environmental filtering, influencing the soil microbial communities through altering soil properties and vegetation [17,18]. However, there are currently two mainstream theories to explain microbial community structure, which are the niche theory and the neutral theory [19,20]. The niche theory emphasizes deterministic processes and holds that the succession of microbial communities is the consequence of selection and screening by biotic (e.g., competition, mutualism, and commensalism) and abiotic factors (e.g., pH and substrate) [21,22]. In contrast to the niche theory, the neutral theory emphasizes stochastic processes and asserts that community assembly is determined by stochastic processes (e.g., dispersal and drift) [22,23]. After a long debate, it is more widely acknowledged that microbial community assembly involves both stochastic and deterministic processes [23,24]. The majority of research on the influence of forest succession on soil microorganisms primarily examines the diversity and composition of the soil microbial communities [25,26]. Apart from the deterministic influence of forest succession, the neutral theory is also crucial for maintaining soil microbial diversity along the succession stages. However, the relative importance of the two processes during forest succession needs to be further studied, especially in subtropical forests.

Wuyi Mountain is one of the main distribution areas for subtropical forests in southern China [27]. It underwent deforestation during the early stages of its reform and opening up, and is currently in a period of forest vegetation recovery [28,29]. Therefore, this region will help us to understand microbial community succession patterns along forest succession. Previous studies have shown the direction of succession from conifer forests to evergreen broad-leaved forests in subtropical forests [30]. Evergreen broad-leaved forests are the most structurally derived, occurring at various elevations in subtropical forests [31]. According to the types of vegetation and leaves, we selected five succession stages: Abandoned lands (AL), Deciduous broad-leaved forests (DB), Coniferous forests (CF), Coniferous broad-leaved mixed forests (CB), and Evergreen broad-leaved forests (EB), to evaluate the impact of forest succession on soil microorganisms. The soil microbial communities can be impacted along forest succession by altering the input amount and types of substrates [17,32]. Increasing substrate input may increase microbial competition and select for specific microbial communities. Given this knowledge, we propose the following hypothesis: (i) the soil bacterial and fungal diversity and composition decreased along with the succession stages; (ii) the relative importance of deterministic and stochastic processes in soil bacterial and fungal assembly processes varies during the succession; and

(iii) the soil pH and total organic carbon (TOC) may be the key soil environmental factors regulating the structure of the soil bacterial and fungal communities.

2. Materials and Methods

2.1. Study Area and Soil Sampling

This study was conducted in a subtropical forest in Jiangxi Province, southern China (27°40′50″ N, 117°09′11″ E), within the Matoushan National Nature Reserve on the western slope of the Wuyi Mountains. The climate in the study area is classified as a humid subtropical monsoon climate, and average annual temperature was ranging from 16 to 18 °C. The temperature may drop as low as −5 °C in January and reach as high as 27.2 °C in July. The average annual precipitation is 1930 mm, with around 47% of it falling in the months of April and June. The main biome of this study site is primarily dominated by Evergreen broad-leaved forests and Coniferous broad-leaved mixed forests. The soil composition comprises mountain red, yellow-red, and yellow soil, which is a consequence of the weathering of granite [27].

In October 2020, we selected each stage of forest succession in Matoushan National Nature Reserve. The Abandoned lands (AL) are covered with herbaceous plants dominated by species such as *Euphorbia latiris*, *Echinochloa crus galli*, *Cyperus rotundis*, and *Equisetum ramosissimum*. *Liquidambar formosana* and *Cunninghamia lanceolata* are dominant tree species in Deciduous broad-leaved forests (DB) and Coniferous forests (CF), respectively. The Coniferous broad-leaved mixed forests (CB) are dominated by *C. lanceolata*, *Alniphyllum fortunei*, *Sassafras tzumu*, *Syzygium buxifolium*., etc. The Evergreen broad-leaved forests (EB) are dominated by that of evergreen and deciduous trees such as *Castanopsis eyrei*, *Castanopsis nigrescens*, *Cyclobalanopsis glauca*, *Syzygium buxifolium*., etc. [33]. Specific information about the sampling sites can be found in Table 1. A total of 30 composite soil samples were collected from five stages of succession, with three replicates and two depths (5 succession stages × 2 depths × 3 replicates). For each succession stage, three 40 m × 40 m plots were established, from which five soil cores (5 cm in diameter) were collected from each plot, encompassing the 0–10 cm and 10–30 cm layers. The soil samples, totaling 30, were transported to the laboratory packed with dry ice, and were subsequently sieved to eliminate rocks and visible roots. Soil samples were then stored at 4 °C for chemical analysis, or at −80 °C for DNA extraction.

Table 1. Detailed successional information on dominant tree species, soil types, and altitude in the subtropical forests on the western slope of the Wuyi Mountains. Abandoned lands (AL), Deciduous broad-leaved forests (DB), Coniferous forests (CF), Coniferous broad-leaved mixed forests (CB), and Evergreen broad-leaved forests (EB).

Successional Stages	Dominant Tree Species	Soil Types	Altitude (m)
AL	*Euphorbia latiris* and *Echinochloa crus galli*	yellow-red soil	250–318
DB	*Liquidambar formosana*	yellow-red soil	273–488
CF	*Cunninghamia lanceolata* and *Pinus massoniana*	yellow-red soil	452–501
CB	*Cunninghamia lanceolata*, *Alniphyllum fortunei*, *Sassafras tzumu*, *Syzygium buxifolium*	yellow-red soil	297–453
EB	*Castanopsis eyrei*, *Castanopsis nigrescens*, *Cyclobalanopsis glauca* and *Syzygium buxifolium*	yellow-red soil	287–454

2.2. Determination of Soil Physicochemical Properties

Soil pH was analyzed in a 1:5 (soil/water, *w*/*v*) ratio with a bench-top electrode pH meter. The oven-drying method was used to measure soil water content. In order to determine soil nutrient contents, air-dried soil samples were milled and passed through a 0.15 mm sieve. The concentrations of soil total organic C (TOC) and total N concentrations (TN) were measured using an elemental analyzer (Flash 2000 HT, Thermo Fisher Scientific, Bremen, Germany).

2.3. DNA Extraction, Amplicon Library Preparation, and Sequencing

Total microbial genomic DNA was extracted from 0.5 g of soil samples using the E.Z.N.A.® soil DNA Kit (Omega Bio-tek, Norcross, GA, USA) according to manufacturer instructions. The hypervariable region V3–V4 region for the bacterial 16S rRNA and the broad-spectrum primers for fungi were amplified with the primer pairs 338F (5′-ACTCCTACGGGAGGCAGCA-3′), 806R (5′-GGACTACHVGGGTWTCTAAT-3′) and ITS1 (5′-CTTGGTCATTTAGAGGAAGTAA-3′), ITS2 (5′-GCTGCGTTCTTCATCGATGC-3′), respectively [34]. All samples were amplified in triplicate. The amplified and purified amplicons were pooled in equimolar amounts and paired-end sequenced on an Illumina MiSeq PE300 platform (Illumina, San Diego, CA, USA) using the standard methods provided by Majorbio Bio-Pharm Technology Co., Ltd. (Shanghai, China).

2.4. Sequence Data Processing

Paired-end sequences were demultiplexed using QIIME 2 [35], followed by further processing of the raw sequences with DADA2. The default settings were used, with the exception of trimming the forward and reverse reads to 240 bases for the 16S rRNA gene. No length filtering was carried out for the ITS region. Following the removal of low-quality reads and the merging of paired reads, operational taxonomic units (OTUs) clustering was conducted based on a 97% similarity threshold using non-redundant sequences (excluding single sequences) based on the SILVA r138 database [36] and the UNITE v8.0 database [37] for prokaryotes and fungi, respectively, through naïve Bayes classifier [38]. Singletons and 16S rRNA sequences identified as belonging to the Eukaryota were discarded. In total, 1,831,407 high-quality sequences for prokaryotes and 2,569,298 high-quality sequences for fungi were obtained.

2.5. Statistical Analysis

All statistical analyses were conducted using the vegan package in R (Version 4.0.2), unless specifically mentioned. The Chao1 and Shannon index for both bacterial and fungal communities was computed to assess the changes in soil microbial species richness and diversity. To evaluate the significance of differences along successional stages at the $p < 0.05$ level, such as soil properties and microbial diversity indices, one-way analysis of variance (ANOVA) or the Kruskal–Wallis rank-sum test were performed employing the Pgirmess packages. The linear discriminant analysis (LDA) effect size (LEfSe) was used to detect significant variations in the phyla of bacteria and fungi (LDA score > 2.0, $p < 0.05$) [38]. The principal coordinate analysis (PCoA) based on Bray–Curtis dissimilarity was applied to examine the differences in community composition during the succession. The analysis of similarity (ANOSIM) and (Adonis) was performed based on the Bray–Curtis distance. Distance-based redundancy analysis (db-RDA) was performed to investigate the effect of soil physicochemical properties on soil bacterial and fungal community structure. The visualizations were finished using the ggplot2 package.

To understand the microbial community assembly, the Null model-based approach was applied to calculate the weighted β nearest taxon index (βNTI) and Bray–Curtis-based Raup-Crick (RC_{bray}) values [39,40]. Based on the βNTI and RC_{bray} indices, the relative importance of the deterministic and stochastic processes in soil microbial community assembly was determined [40]. The "iCAMP" package in R was used to calculate the βNTI by comparing the standard deviation of observed data with the null distribution of phylogenetic β-diversity metrics [20]. When the absolute value of βNTI is greater than 2, it means that the assembly of the community is dominated by deterministic processes. In a more specific categorization, the heterogeneous selection is represented by βNTI > 2, whereas the homogeneous selection is represented by βNTI < −2. In addition, the absolute value of βNTI is less than 2, indicating the dominance of stochastic processes. RC_{bray} was calculated by estimating the standard deviation between the empirical data and the null distribution of taxonomic β-diversity metrics [20]. Three types of random processes may be classified using βNTI and RC coupling. (1) $|\beta NTI| < 2$ and $RC_{bray} > 0.95$: dispersal limitation; (2) |βNTI| < 2 and $RC_{bray} < -0.95$: homogenizing dispersal; (3) |βNTI| < 2 and $RC_{bray} < -0.95$: drift [41].

3. Results

3.1. Soil Physicochemical Properties along the Succession Stages

Regarding soil characteristics, soil TOC ranged from 18.04 g kg^{-1} in the Abandoned lands (AL), 57.80 g kg^{-1} in the Coniferous forests (CF), and 79.14 g kg^{-1} in the Evergreen broad-leaved forests (EB) at 0–10 cm soil depths (Table S1). Soil TOC ranged from 16.33 g kg^{-1} in the Abandoned lands (AL) and 55.73 g kg^{-1} in the Evergreen broad-leaved forests (EB) at 10–30 cm soil depths (Table S1). Soil TOC showed a notable increasing trend from the initial to the terminal phases of succession ($p < 0.05$). pH decreased significantly from 5.33 in the AL to 4.96 in the EB at 10–30 cm soil depths ($p < 0.05$), and there were no dramatic changes in pH at 0–10 cm soil depths among succession stages (Table S1). However, there were no significant differences in TN, C:N ratio, and moisture among succession stages (Table S1).

3.2. The Diversity and Composition of Soil Microbial Communities along the Succession Stages

During the succession stages, 1,402,689 and 2,001,323 high-quality effective sequences of bacteria and fungi, respectively, were obtained. The sequences were resampled to 33,560 and 49,363 OTUs for bacteria and fungi, respectively. The six main phyla of bacteria along the succession stages were *Proteobacteria* (10.41%~51.41%), *Acidobacteria* (7.18%~34.22%), *Actinobacteria* (5.69%~28.71%), *Chloroflexi* (2.90%~20.85%), *Planctomycetota* (0.59%~9.88%), and *Firmicutes* (0.92%~10.28%) (Figure 1A). For fungi, *Ascomycota* (22.74%~90.28%) and *Basidiomycota* (2.99%~61.08%) were the dominant fungal phyla, whereas *Mortierellomycota* (0.37%~44.72%) and *Rozellomycota* (0.07%~13.10%) were the main fungal phyla (Figure 1B). There was similarity in the relative abundances of bacteria and fungi phyla at 0–10 cm and 10–30 cm soil depths.

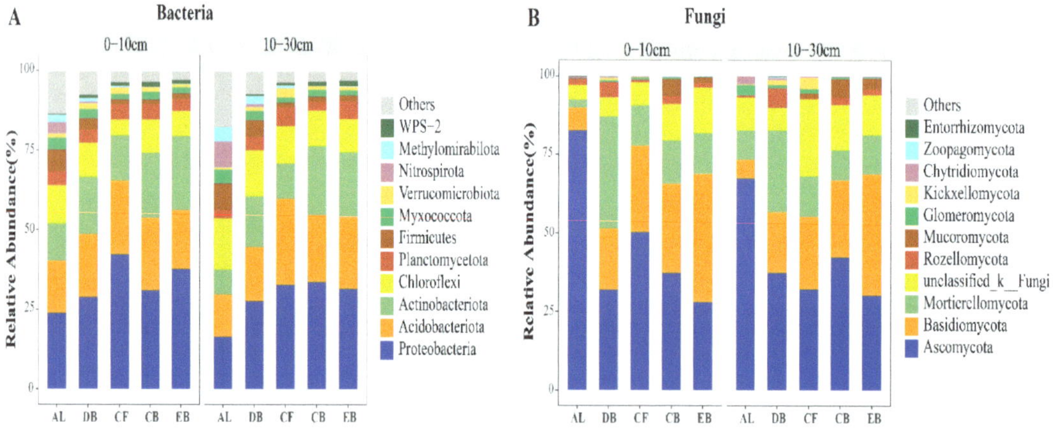

Figure 1. The relative abundances of the major soil bacterial (**A**) and fungal (**B**) phyla at different soil depths along the succession. The relative abundance of phyla less than 1% are merged into "Others". Abandoned lands (AL), Deciduous broad-leaved forests (DB), Coniferous forests (CF), Coniferous broad-leaved mixed forests (CB), and Evergreen broad-leaved forests (EB).

The diversity (Shannon index) and richness (Chao1 index) of soil bacteria and fungi changed inconsistently along the succession stages (Figure 2). The bacterial Chao1 and Shannon index showed a notable decreasing trend from Abandoned lands (AL) to Coniferous forests (CF) to Evergreen broad-leaved forests (EB) at 0–10 cm and 10–30 cm soil depths ($p < 0.05$) (Figure 2A,B). The change of fungal Chao1 at 10–30 cm soil depth was opposite that of bacteria, and it reached the maximum value in the EB forest (Figure 2C). The Chao1 index of fungi at 0–10 cm soil depth did not show an increasing trend (Figure 2C). The fungal Shannon indices showed no significant changes at 0–10 cm and 10–30 cm depths along the succession stages ($p > 0.05$; Figure 2D).

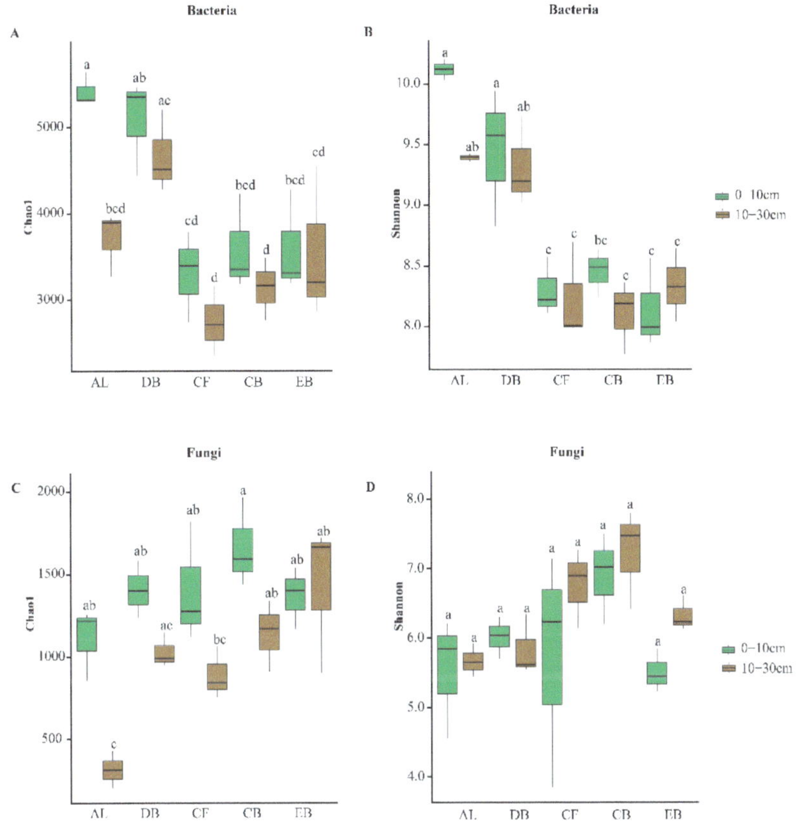

Figure 2. The changes in α-diversity index for both bacterial (**A**,**B**) and fungal communities (**C**,**D**) at different soil depths across the succession stages in subtropical forests. Significant changes are indicated by different letters ($p < 0.05$) among succession stages. Abandoned lands (AL), Deciduous broad-leaved forests (DB), Coniferous forests (CF), Coniferous broad-leaved mixed forests (CB), and Evergreen broad-leaved forests (EB).

3.3. The Beta Diversity of Soil Microbial Communities along the Succession Stages

The results of the principal coordinate analysis (PCoA) revealed that the structure of the soil bacterial and fungal communities showed the same change along the succession stages (Figure 3A,B). The results of Adonis tests (Bray–Curtis distance) showed no significant disparity between CB and EB ($p > 0.05$; Table S2). However, a notable divergence was observed in the bacterial and fungal community structure between AL and natural forests, including DB, CF, CB, and EB (Adonis test, $p < 0.01$; Adonis test, $p < 0.01$; Figure 3A,B).

According to LEfSe analysis, the soil bacterial and fungal communities significantly changed along the succession stages. Among them, 21 bacterial phyla were abundant differently along the succession stages (Figure 4A). The most phyla (*Chloroflexi*, *Firmicutes*, *Nitrospirota*, *Methylomirabilota*, and *Bacteroidota*) were mainly enriched in the Abandoned lands (AL). *Proteobacteria*, *Acidobacteriota*, and *Actinobacteriota* were enriched in Coniferous forests (CF) and Evergreen broad-leaved forests (EB), respectively. For fungi, seven fungal phyla were abundant differently along the succession stages (Figure 4B). *Ascomycota* and *Chytridiomycota* were mainly enriched in the Abandoned lands (AL), and *Basidiomycota* was enriched in the Evergreen broad-leaved forests (EB) (Figure 4B).

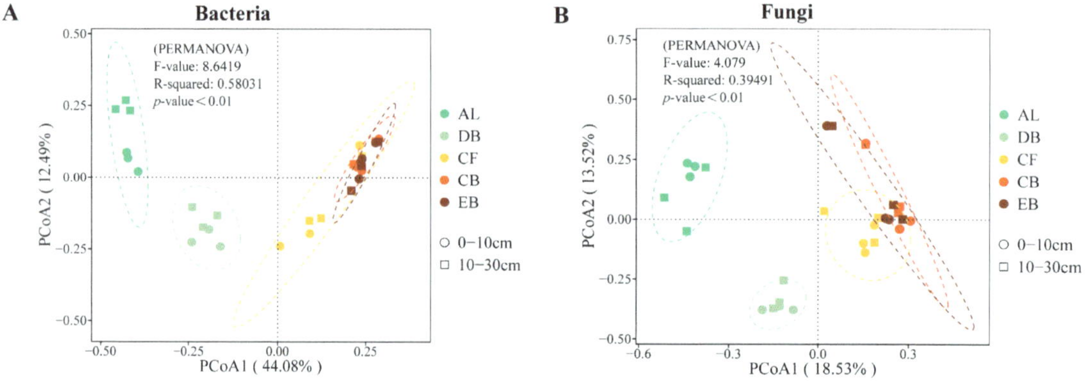

Figure 3. Principal co-ordinates analysis (PCoA) of soil bacterial (**A**) and fungal (**B**) communities at different depths along the succession stages. Abandoned lands (AL), Deciduous broad-leaved forests (DB), Coniferous forests (CF), Coniferous broad-leaved mixed forests (CB), and Evergreen broad-leaved forests (EB).

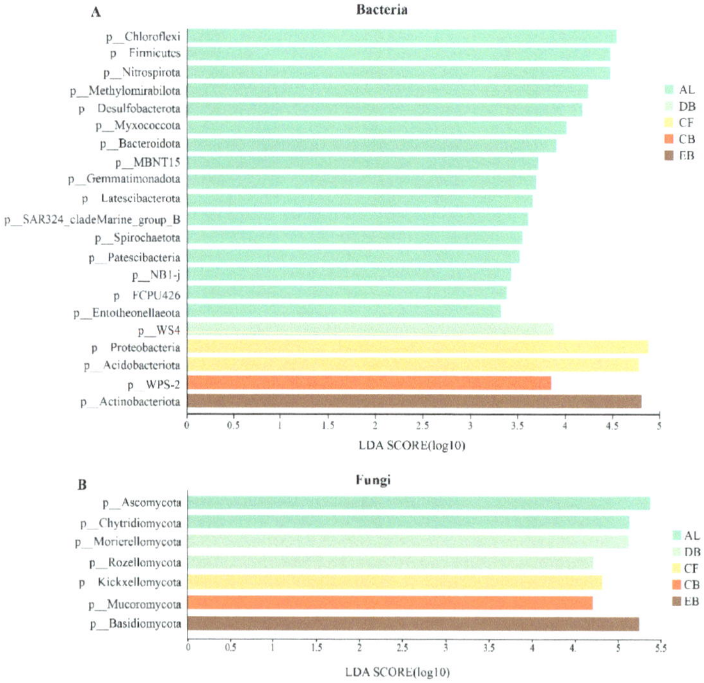

Figure 4. The linear discriminant analysis (LDA) effect size (LEfSe) illustrates the significant differences in bacterial (**A**) and fungal (**B**) phyla along the succession stages. The threshold value for distinguishing features based on the log LDA score was set at 2.0. Abandoned lands (AL), Deciduous broad-leaved forests (DB), Coniferous forests (CF), Coniferous broad-leaved mixed forests (CB), and Evergreen broad-leaved forests (EB).

3.4. The Changes in Microbial Assembly Processes along the Succession Stages

The β-NTI values and RCbray values were used to assess the ecological processes of soil microorganisms along the succession stages. The median βNTI values for the AL, DB, CB, and EB were greater than those of the CF ($p < 0.05$; Figure 5A). Meanwhile, the median βNTI values for the AL, DB, CB, and EB were more than two, whereas the median βNTI value for the CL was less than two (Figure 5A). This indicated that bacterial community assembly shifted from deterministic processes to stochastic processes to deterministic processes during the whole succession stages. For fungi, the community assembly processes were different from those of bacteria. The median βNTI values of the AL, DB, CB, and EB were higher than those of the CF ($p < 0.05$; Figure 5B). However, all values of the median βNTIs were less than two, indicating that stochastic processes mainly influenced fungal community assembly. Furthermore, the fungal community assembly remained unchanged across the whole succession stage (Figure 5B).

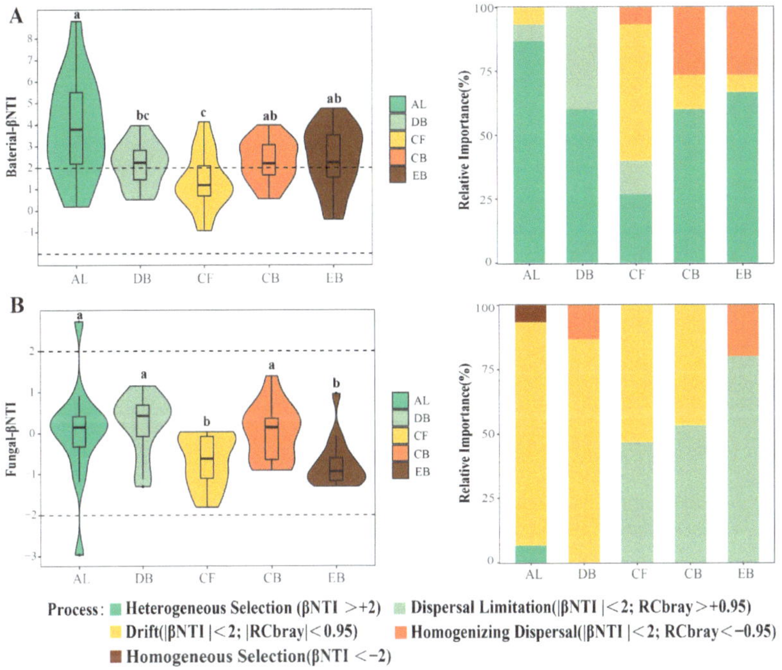

Figure 5. The βNTI values for pairwise community comparisons throughout the succession are distributed. Two dashed lines were added to distinguish between stochastic and deterministic processes according to βNTI values. Solid lines within violins represent the upper quartiles, median, and lower quartiles, respectively. The relative importance of assembly processes in bacterial (**A**) and fungal (**B**) communities during the succession. Significant changes are indicated by different letters ($p < 0.05$) among succession stages. Abandoned lands (AL), Deciduous broad-leaved forests (DB), Coniferous forests (CF), Coniferous broad-leaved mixed forests (CB), and Evergreen broad-leaved forests (EB).

3.5. The Correlations between Soil Microbial Communities and Soil Properties

Distance-based redundancy analysis (db-RDA) revealed the key soil environmental factors shaping the soil bacterial variation at the OTU level. (Figure 6A). The soil environmental factors accounted for a total of 58.89% of the variance in bacterial communities. Specifically, RDA1 and RDA2 explained 35.11% and 23.78% of the total variation, respectively (Figure 6A). The structure of the bacterial communities was mostly determined by TOC ($p < 0.01$; Figure 6A). The soil environmental factors accounted for 11.7% (CCA1 = 6.7%, CCA2 = 5.0%) of the variation in the fungal community structure (Figure 6B). The TOC, pH, C:N ratio, and TN significantly influenced the variability in soil fungal communities ($p < 0.01$; Figure 6B).

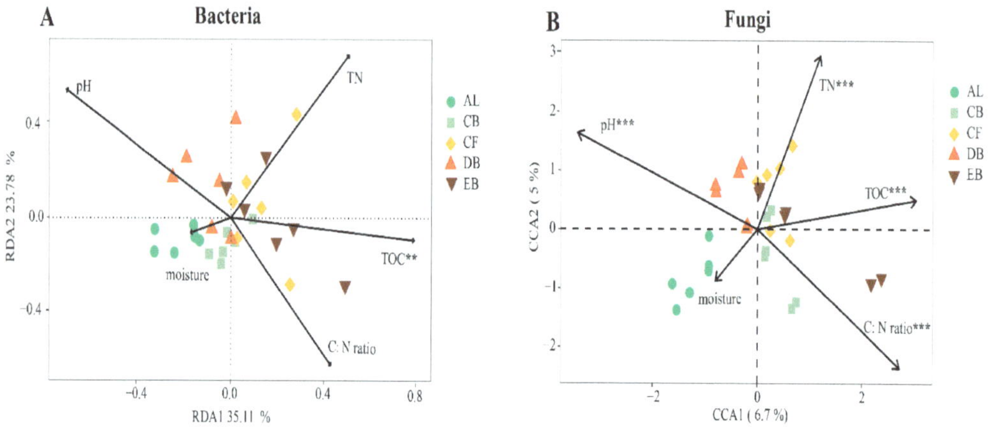

Figure 6. Distance-based redundancy analysis (db-RDA) of soil bacterial (**A**) and fungal (**B**) communities at different soil depths during the succession stages as related to soil environmental variables. TOC: total organic carbon; TN: total nitrogen; C: N ratio: the ratio of TOC and TN; pH: pondus Hydrogenii; Moisture: soil water content. **: $p < 0.01$; ***: $p < 0.001$. Abandoned lands (AL), Deciduous broad-leaved forests (DB), Coniferous forests (CF), Coniferous broad-leaved mixed forests (CB), and Evergreen broad-leaved forests (EB).

4. Discussion

4.1. The Structure and Diversity of Soil Microbial Communities during the Succession Stages

Soil bacteria and fungi are the main drivers of material cycling and play important roles along the forest succession [42]. The α-diversity of soil bacterial communities exhibited a notable decline from the initial to the terminal phases of succession, aligning with findings from previous research [15,26,43]. However, soil fungal α-diversity did not vary during the succession stages, which was supported by the previous studies [15,32].This does not completely align with our first hypothesis. The reason for the different responses of bacterial and fungal α-diversity to forest succession is that they have different survival strategies. Compared with the EB, more weeds like *Euphorbia latiris*, *Echinochloa crus galli*, and *Cyperus rotundis* were prevalent in the AL. Seasonal variation in the plant community in the AL may create more dynamic and unpredictable conditions [44]. To adapt to this relatively unfavorable environment, soil bacteria adopt an r-strategy survival mode, resulting in higher diversity. Beginning in the succession stage of CF, the environment becomes relatively stable, and bacteria are subject to selection pressure from soil organic carbon [10]. As a result, the abundance of rare bacterial communities decreased, whereas bacteria that can utilize the substrate were enriched [45]. As a K-strategy survival mode, the diversity of fungi was less sensitive to the succession of forests [46], and the null models also indicated that the response of fungi to succession is primarily driven by stochastic processes.

The PCoA and Adonis analyses revealed significant changes in the bacterial and fungal community structures at both depths throughout the succession. Many previous studies have obtained similar results [18]. LEfSe analysis indicated that there was a general shift from *Chloroflexi*, *Firmicutes*, *Nitrospirota*, and *Bacteoidota* in AL to *Proteobacteria* and *Acidobacteria* in the CF to *Actinobacteria* in the EB. There were two possible reasons for our results. Firstly, it is widely accepted that copiotrophic microorganisms are associated with the r-strategy, while oligotrophic microbes are regarded as being intimately linked to the K-strategy [31,47]. The relative abundance of *Actinobacteria* and *Acidobacteria*, which are indicative of k-strategy bacteria, has notably risen in comparison to the AL, whereas the relative abundance of *Bacteroides* belonging to the r-strategy bacteria showed an opposite trend. This suggests that the bacterial communities shifted from an r-strategy to a k-strategy in conjunction with the different phases of vegetal succession. Secondly, soil bacterial communities that were capable of sporulation and strong stress resistance were enriched in the AL. In hostile conditions, *Firmicutes* were able to produce resistant endospores [48]. For fungi, there was a general shift from *Ascomycota* enriched in the AL to *Basidiomycota* in the EB along the succession. This aligns with the findings of fungal community succession in the middle- and high-latitude regions [49,50]. Generally, members of the *Ascomycota* phylum demonstrate high resistance to disturbed environments. Meanwhile, *Basidiomycota* is crucial in breaking down stubborn organic matter because it contains several *ectomycorrhizal* and *saprophytic* fungi, which are vital for decomposing complex chemicals [51,52].

4.2. The Changes in Soil Microbial Community Assembly Processes during the Succession Stages

The establishment and maintenance of a vast variety of soil microorganisms has been a central issue in microbial ecology [16]. Our findings indicate that the median βNTI values decreased first and then increased along with the succession stages. The deterministic process was dominant in the four succession stages except for that in the CF. Although deterministic processes dominated both the initial and terminal succession stages, the factors of selection differed. At the initial stage of succession, soil bacteria are under multiple selective pressures due to the combination of harsh environments and substrates, whereas soil substrate was the main factor in selection at the terminal stage of succession [17]. The db-RDA findings indicated that TOC had a pivotal role in altering the bacterial community structures. Compared with the DB and EB, *Cunninghamia lanceolata* in the CF has lower nutrient content, higher lignin, and secondary metabolites, making it less susceptible to microbial utilization. Li et al. (2023) found that the decomposition rate of conifer litter was lower compared to that of broad-leaved species. Increased nutrient inputs may amplify the significance of stochastic processes in the relative abundance of taxonomic units [53].

For fungi, the median βNTI values significantly changed along with the succession stages. This is consistent with our second hypothesis. However, all of the median βNTI values were between 2 and −2, indicating that the stochastic process was dominant in the soil fungal community. These findings were consistent with the results of previous research [54,55]. This result was the consequence of multiple factors. As mentioned earlier, fungi are considered K-selected species and exhibit relatively low sensitivity to environmental changes [46]. The CCA findings revealed that soil fungal community structure was impacted by soil physicochemical properties beyond only TOC, suggesting that other soil environmental factors (pH and C:N ratio) play a role in regulating fungal community composition. Almost all tree species are ectomycorrhizal (ECM) tree species throughout the complete succession stage. Free-living soil fungi, especially saprotrophic fungi, compete with ECM fungi for resources [56].

4.3. Soil Physicochemical Properties Drive Soil Microbial Community Changes

Many studies have demonstrated that forest succession modifies soil characteristics, which in turn impacts the composition and diversity of the soil microbial communities [31,32,57]. According to our findings, TOC showed an increasing trend along with the succession stages (Table S1). It is worth pointing out that only TOC was the main factor in the shift of the soil bacterial communities. This partly supports our third hypothesis. Numerous earlier investigations revealed strong positive relationships between TOC and the soil microbial communities [10,58]. Elevated TOC along with the succession indicated that more substrate and nutrients became available for soil microbes. Furthermore, extensive research has unequivocally shown that the alterations in tree species resulting from secondary succession induce forest soil acidification. [15,59]. The soil pH exhibited a declining pattern in our investigation, while no notable disparity was observed (Table S1).

Unlike the findings for bacteria, the results of CCA analysis illustrated that soil pH, TN, and C:N ratio had an impact on the structure of the soil fungal community, in addition to TOC. The other related studies have also found that multiple environmental factors regulate the community structure of soil fungi [60,61]. Our research demonstrates that soil pH exerts a more significant influence on fungal communities in comparison to bacterial communities. Since fungi often have a greater pH tolerance range than bacteria, this is likely due to the intense interplay between pH and other variables [62]. Although multiple soil factors can affect the structure of fungal communities, the explanatory power of soil factors on fungi is lower than that of bacteria, which may be due to differences in the responses of bacterial and fungal community succession to forest succession.

5. Conclusions

From the Abandoned lands to the Evergreen broad-leaved forests, the diversity of soil bacteria decreased, while fungal diversity did not vary along with the succession stage, indicating that bacteria and fungi had distinct patterns along the forest succession. Moreover, soil bacterial and fungal communities shifted with the succession. More importantly, from Abandoned fields to Coniferous forests to Evergreen broad-leaved forests, the community assembly of soil bacteria shifts from a deterministic process to a stochastic process to a deterministic process. Throughout the whole succession, stochastic processes predominantly governed the community assembly of soil fungal communities. TOC was the influential soil environmental factor responsible for the change in the structure of the bacterial communities. More soil environmental factors such as, TOC, TN, C:N ratio, and pH co-regulate the changes in fungal community structure. Compared with soil fungi, the structure and composition of soil bacterial communities are more sensitive to substrates. In conclusion, our findings contribute to a better understanding of how the composition and assembly processes of the soil microbial community respond to plant succession in subtropical forests. For future work, co-occurrence networks, as well as abundant and rare taxa of soil microbial communities, will be further explored in subtropical forests.

Supplementary Materials: The following supporting information can be downloaded at: https://www.mdpi.com/article/10.3390/f15020242/s1, Table S1: Soil physical and chemical characteristics along the succession; Table S2: Effects of forest succession on the composition of fungal and bacterial communities.

Author Contributions: J.R.: investigation, writing—original draft; K.H.: investigation and data curation; F.X., Y.Z., B.Y. and H.C.: methodology and reviewing; F.S.: conceptualization, funding, writing—review and editing. All authors have read and agreed to the published version of the manuscript.

Funding: This study was supported by the National Natural Science Foundation of China (42067049), the Doctoral Scientific Research Foundation of East China University of Technology (DHBK2021001), the Science and Technology Project of the Jiangxi Provincial Department of Education (GJJ2200715), the Open Research Fund of Jiangxi Province Institute of Water Sciences (Grant Nos. 2022SKTR03 and 2022SKTR05), and the Double Thousand Plan of Jiangxi Province (jxsq2023102213).

Data Availability Statement: The data used to support the findings of this study are available from the corresponding author upon request.

Conflicts of Interest: The authors declare no conflicts of interest.

References

1. Chazdon, R.L. Beyond Deforestation: Restoring Forests and Ecosystem Services on Degraded Lands. *Science* **2008**, *320*, 1458–1460. [CrossRef] [PubMed]
2. Foley, J.A.; Asner, G.P.; Costa, M.H.; Coe, M.T.; Defries, R.; Gibbs, H.K.; Howard, E.A.; Olson, S.; Patz, J.; Ramankutty, N.; et al. Forest Degradation and Loss of Ecosystem Goods and Services in the Amazon Basin. *Front. Ecol. Environ.* **2007**, *5*, 25–32. [CrossRef]

3. Betts, M.G.; Wolf, C.; Ripple, W.J.; Phalan, B.; Millers, K.A.; Duarte, A.; Butchart, S.H.; Levi, T. Global Forest Loss Disproportionately Erodes Biodiversity in Intact Landscapes. *Nature* **2017**, *547*, 441–444. [CrossRef] [PubMed]
4. Strassburg, B.B.; Iribarrem, A.; Beyer, H.L.; Cordeiro, C.L.; Crouzeilles, R.; Jakovac, C.C.; Braga Junqueira, A.; Lacerda, E.; Latawiec, A.E.; Balmford, A.; et al. Global Priority Areas for Ecosystem Restoration. *Nature* **2020**, *586*, 724–729. [CrossRef] [PubMed]
5. Uroz, S.; Buee, M.; Deveau, A.; Mieszkin, S.; Martin, F. Ecology of the Forest Microbiome: Highlights of Temperate and Boreal Ecosystems. *Soil Biol. Biochem.* **2016**, *103*, 471–488. [CrossRef]
6. Lladó, S.; López-Mondéjar, R.; Baldrian, P. Forest Soil Bacteria: Diversity, Involvement in Ecosystem Processes, and Response to Global Change. *Microbiol. Mol. Biol. Rev.* **2017**, *81*, e00063-16. [CrossRef]
7. Bahram, M.; Kohout, P.; Anslan, S.; Harend, H.; Abarenkov, K.; Tedersoo, L. Stochastic Distribution of Small Soil Eukaryotes Resulting from High Dispersal and Drift in a Local Environment. *ISME J.* **2016**, *10*, 885–896. [CrossRef]
8. Meyer, K.M.; Memiaghe, H.; Korte, L.; Kenfack, D.; Alonso, A.; Bohannan, B.J. Why Do Microbes Exhibit Weak Biogeographic Patterns? *ISME J.* **2018**, *12*, 1404–1413. [CrossRef] [PubMed]
9. Wang, B.; Liu, G.B.; Xue, S.; Zhu, B. Changes in Soil Physico-Chemical and Microbiological Properties during Natural Succession on Abandoned Farmland in the Loess Plateau. *Environ. Earth Sci.* **2011**, *62*, 915–925. [CrossRef]
10. Jiang, S.; Xing, Y.; Liu, G.; Hu, C.; Wang, X.; Yan, G.; Wang, Q. Changes in Soil Bacterial and Fungal Community Composition and Functional Groups during the Succession of Boreal Forests. *Soil Biol. Biochem.* **2021**, *161*, 108393. [CrossRef]
11. Cline, L.C.; Zak, D.R. Soil Microbial Communities Are Shaped by Plant-driven Changes in Resource Availability during Secondary Succession. *Ecology* **2015**, *96*, 3374–3385. [CrossRef] [PubMed]
12. Zhao, F.Z.; Bai, L.; Wang, J.Y.; Deng, J.; Ren, C.J.; Han, X.H.; Yang, G.H.; Wang, J. Change in Soil Bacterial Community during Secondary Succession Depend on Plant and Soil Characteristics. *Catena* **2019**, *173*, 246–252. [CrossRef]
13. Prescott, C.E.; Grayston, S.J. Tree Species Influence on Microbial Communities in Litter and Soil: Current Knowledge and Research Needs. *For. Ecol. Manag.* **2013**, *309*, 19–27. [CrossRef]
14. Martins, K.G.; Marques, M.C.; dos Santos, E.; Marques, R. Effects of Soil Conditions on the Diversity of Tropical Forests across a Successional Gradient. *For. Ecol. Manag.* **2015**, *349*, 4–11. [CrossRef]
15. Delgado-Baquerizo, M.; Bardgett, R.D.; Vitousek, P.M.; Maestre, F.T.; Williams, M.A.; Eldridge, D.J.; Lambers, H.; Neuhauser, S.; Gallardo, A.; García-Velázquez, L.; et al. Changes in belowground biodiversity during ecosystem development. *Proc. Natl. Acad. Sci. USA* **2019**, *116*, 6891–6896. [CrossRef] [PubMed]
16. Zhou, J.; Ning, D. Stochastic Community Assembly: Does It Matter in Microbial Ecology? *Microbiol. Mol. Biol. Rev.* **2017**, *81*, e00002-17. [CrossRef]
17. Liu, L.; Zhu, K.; Krause, S.M.; Li, S.; Wang, X.; Zhang, Z.; Shen, M.; Yang, Q.; Lian, J.; Wang, X. Changes in Assembly Processes of Soil Microbial Communities during Secondary Succession in Two Subtropical Forests. *Soil Biol. Biochem.* **2021**, *154*, 108144. [CrossRef]
18. Qiang, W.; He, L.; Zhang, Y.; Liu, B.; Liu, Y.; Liu, Q.; Pang, X. Aboveground Vegetation and Soil Physicochemical Properties Jointly Drive the Shift of Soil Microbial Community during Subalpine Secondary Succession in Southwest China. *Catena* **2021**, *202*, 105251. [CrossRef]
19. Vandermeer, J.H. Niche Theory. *Annu. Rev. Ecol. Syst.* **1972**, *3*, 107–132. [CrossRef]
20. Ning, D.; Yuan, M.; Wu, L.; Zhang, Y.; Guo, X.; Zhou, X.; Yang, Y.; Arkin, A.P.; Firestone, M.K.; Zhou, J. A Quantitative Framework Reveals Ecological Drivers of Grassland Microbial Community Assembly in Response to Warming. *Nat. Commun.* **2020**, *11*, 4717. [CrossRef] [PubMed]
21. Vályi, K.; Mardhiah, U.; Rillig, M.C.; Hempel, S. Community Assembly and Coexistence in Communities of Arbuscular Mycorrhizal Fungi. *ISME J.* **2016**, *10*, 2341–2351. [CrossRef] [PubMed]
22. Chesson, P. Mechanisms of Maintenance of Species Diversity. *Annu. Rev. Ecol. Syst.* **2000**, *31*, 343–366. [CrossRef]
23. Chase, J.M.; Myers, J.A. Disentangling the Importance of Ecological Niches from Stochastic Processes across Scales. *Philos. Trans. R. Soc. B Biol. Sci.* **2011**, *366*, 2351–2363. [CrossRef]
24. Adler, P.B.; HilleRisLambers, J.; Levine, J.M. A Niche for Neutrality. *Ecol. Lett.* **2007**, *10*, 95–104. [CrossRef]
25. Ren, C.; Liu, W.; Zhao, F.; Zhong, Z.; Deng, J.; Han, X.; Yang, G.; Feng, Y.; Ren, G. Soil Bacterial and Fungal Diversity and Compositions Respond Differently to Forest Development. *Catena* **2019**, *181*, 104071. [CrossRef]
26. Cutler, N.A.; Chaput, D.L.; van der Gast, C.J. Long-Term Changes in Soil Microbial Communities during Primary Succession. *Soil Biol. Biochem.* **2014**, *69*, 359–370. [CrossRef]
27. Chen, H.-M.; Shi, F.-X.; Xu, J.-W.; Liu, X.-P.; Mao, R. Tree Mycorrhizal Type Controls over Soil Water-Extractable Organic Matter Quantity and Biodegradation in a Subtropical Forest of Southern China. *For. Ecol. Manag.* **2023**, *535*, 120900. [CrossRef]
28. Liu, Y.; Fang, F.; Li, Y. Key Issues of Land Use in China and Implications for Policy Making. *Land Use Policy* **2014**, *40*, 6–12. [CrossRef]
29. Tong, X.; Brandt, M.; Yue, Y.; Ciais, P.; Rudbeck Jepsen, M.; Penuelas, J.; Wigneron, J.-P.; Xiao, X.; Song, X.-P.; Horion, S. Forest Management in Southern China Generates Short Term Extensive Carbon Sequestration. *Nat. Commun.* **2020**, *11*, 129. [CrossRef]
30. Tang, C.Q.; He, L.-Y.; Su, W.-H.; Zhang, G.-F.; Wang, H.-C.; Peng, M.-C.; Wu, Z.-L.; Wang, C.-Y. Regeneration, Recovery and Succession of a Pinus Yunnanensis Community Five Years after a Mega-Fire in Central Yunnan, China. *For. Ecol. Manag.* **2013**, *294*, 188–196. [CrossRef]

31. Zhou, Z.; Wang, C.; Jiang, L.; Luo, Y. Trends in Soil Microbial Communities during Secondary Succession. *Soil Biol. Biochem.* **2017**, *115*, 92–99. [CrossRef]
32. Li, S.; Huang, X.; Shen, J.; Xu, F.; Su, J. Effects of Plant Diversity and Soil Properties on Soil Fungal Community Structure with Secondary Succession in the Pinus Yunnanensis Forest. *Geoderma* **2020**, *379*, 114646. [CrossRef]
33. Huang, K.-X.; Xue, Z.-J.; Wu, J.-C.; Wang, H.; Zhou, H.-Q.; Xiao, Z.-B.; Zhou, W.; Cai, J.-F.; Hu, L.-W.; Ren, J.-S. Water Use Efficiency of Five Tree Species and Its Relationships with Leaf Nutrients in a Subtropical Broad-Leaf Evergreen Forest of Southern China. *Forests* **2023**, *14*, 2298. [CrossRef]
34. Woese, C.R.; Kandler, O.; Wheelis, M.L. Towards a Natural System of Organisms: Proposal for the Domains Archaea, Bacteria, and Eucarya. *Proc. Natl. Acad. Sci. USA* **1990**, *87*, 4576–4579. [CrossRef] [PubMed]
35. Caporaso, J.G.; Kuczynski, J.; Stombaugh, J.; Bittinger, K.; Bushman, F.D.; Costello, E.K.; Fierer, N.; Peña, A.G.; Goodrich, J.K.; Gordon, J.I. QIIME Allows Analysis of High-Throughput Community Sequencing Data. *Nat. Methods* **2010**, *7*, 335–336. [CrossRef] [PubMed]
36. Quast, C.; Pruesse, E.; Yilmaz, P.; Gerken, J.; Schweer, T.; Yarza, P.; Peplies, J.; Glöckner, F.O. The SILVA Ribosomal RNA Gene Database Project: Improved Data Processing and Web-Based Tools. *Nucleic Acids Res.* **2012**, *41*, D590–D596. [CrossRef] [PubMed]
37. Nilsson, R.H.; Larsson, K.-H.; Taylor, A.F.S.; Bengtsson-Palme, J.; Jeppesen, T.S.; Schigel, D.; Kennedy, P.; Picard, K.; Glöckner, F.O.; Tedersoo, L. The UNITE Database for Molecular Identification of Fungi: Handling Dark Taxa and Parallel Taxonomic Classifications. *Nucleic Acids Res.* **2019**, *47*, D259–D264. [CrossRef] [PubMed]
38. Bokulich, N.A.; Kaehler, B.D.; Rideout, J.R.; Dillon, M.; Bolyen, E.; Knight, R.; Huttley, G.A.; Gregory Caporaso, J. Optimizing Taxonomic Classification of Marker-Gene Amplicon Sequences with QIIME 2's Q2-Feature-Classifier Plugin. *Microbiome* **2018**, *6*, 90. [CrossRef] [PubMed]
39. Paulson, J.N.; Stine, O.C.; Bravo, H.C.; Pop, M. Differential Abundance Analysis for Microbial Marker-Gene Surveys. *Nat. Methods* **2013**, *10*, 1200–1202. [CrossRef] [PubMed]
40. Chase, J.M.; Kraft, N.J.; Smith, K.G.; Vellend, M.; Inouye, B.D. Using Null Models to Disentangle Variation in Community Dissimilarity from Variation in A-diversity. *Ecosphere* **2011**, *2*, 1–11. [CrossRef]
41. Stegen, J.C.; Lin, X.; Fredrickson, J.K.; Chen, X.; Kennedy, D.W.; Murray, C.J.; Rockhold, M.L.; Konopka, A. Quantifying Community Assembly Processes and Identifying Features That Impose Them. *ISME J.* **2013**, *7*, 2069–2079. [CrossRef] [PubMed]
42. Stegen, J.C.; Lin, X.; Fredrickson, J.K.; Konopka, A.E. Estimating and Mapping Ecological Processes Influencing Microbial Community Assembly. *Front. Microbiol.* **2015**, *6*, 370. [CrossRef] [PubMed]
43. Zhong, Y.; Yan, W.; Wang, R.; Wang, W.; Shangguan, Z. Decreased Occurrence of Carbon Cycle Functions in Microbial Communities along with Long-Term Secondary Succession. *Soil Biol. Biochem.* **2018**, *123*, 207–217. [CrossRef]
44. Flores-Rentería, D.; Sánchez-Gallén, I.; Morales-Rojas, D.; Larsen, J.; Alvarez-Sanchez, J. Changes in the Abundance and Composition of a Microbial Community Associated with Land Use Change in a Mexican Tropical Rain Forest. *J. Soil Sci. Plant Nutr.* **2020**, *20*, 1144–1155. [CrossRef]
45. Eilers, K.G.; Lauber, C.L.; Knight, R.; Fierer, N. Shifts in Bacterial Community Structure Associated with Inputs of Low Molecular Weight Carbon Compounds to Soil. *Soil Biol. Biochem.* **2010**, *42*, 896–903. [CrossRef]
46. Ho, A.; Di Lonardo, D.P.; Bodelier, P.L. Revisiting Life Strategy Concepts in Environmental Microbial Ecology. *FEMS Microbiol. Ecol.* **2017**, *93*, fix006. [CrossRef]
47. Fierer, N.; Bradford, M.A.; Jackson, R.B. Toward an Ecological Classification of Soil Bacteria. *Ecology* **2007**, *88*, 1354–1364. [CrossRef] [PubMed]
48. Filippidou, S.; Wunderlin, T.; Junier, T.; Jeanneret, N.; Dorador, C.; Molina, V.; Johnson, D.R.; Junier, P. A Combination of Extreme Environmental Conditions Favor the Prevalence of Endospore-Forming Firmicutes. *Front. Microbiol.* **2016**, *7*, 1707. [CrossRef]
49. Chai, Y.; Cao, Y.; Yue, M.; Tian, T.; Yin, Q.; Dang, H.; Quan, J.; Zhang, R.; Wang, M. Soil Abiotic Properties and Plant Functional Traits Mediate Associations between Soil Microbial and Plant Communities during a Secondary Forest Succession on the Loess Plateau. *Front. Microbiol.* **2019**, *10*, 895. [CrossRef]
50. Dong, K.; Tripathi, B.; Moroenyane, I.; Kim, W.; Li, N.; Chu, H.; Adams, J. Soil Fungal Community Development in a High Arctic Glacier Foreland Follows a Directional Replacement Model, with a Mid-Successional Diversity Maximum. *Sci. Rep.* **2016**, *6*, 26360. [CrossRef]
51. Urbanová, M.; Šnajdr, J.; Baldrian, P. Composition of Fungal and Bacterial Communities in Forest Litter and Soil Is Largely Determined by Dominant Trees. *Soil Biol. Biochem.* **2015**, *84*, 53–64. [CrossRef]
52. Yan, B.; Sun, L.; Li, J.; Liang, C.; Wei, F.; Xue, S.; Wang, G. Change in Composition and Potential Functional Genes of Soil Bacterial and Fungal Communities with Secondary Succession in Quercus Liaotungensis Forests of the Loess Plateau, Western China. *Geoderma* **2020**, *364*, 114199. [CrossRef]
53. Liang, Y.; Ning, D.; Lu, Z.; Zhang, N.; Hale, L.; Wu, L.; Clark, I.M.; McGrath, S.P.; Storkey, J.; Hirsch, P.R. Century Long Fertilization Reduces Stochasticity Controlling Grassland Microbial Community Succession. *Soil Biol. Biochem.* **2020**, *151*, 108023. [CrossRef]
54. Liu, K.; Liu, Y.; Hu, A.; Wang, F.; Chen, Y.; Gu, Z.; Anslan, S.; Hou, J. Different Community Assembly Mechanisms Underlie Similar Biogeography of Bacteria and Microeukaryotes in Tibetan Lakes. *FEMS Microbiol. Ecol.* **2020**, *96*, fiaa071. [CrossRef]

55. Zhang, Q.; Goberna, M.; Liu, Y.; Cui, M.; Yang, H.; Sun, Q.; Insam, H.; Zhou, J. Competition and Habitat Filtering Jointly Explain Phylogenetic Structure of Soil Bacterial Communities across Elevational Gradients. *Environ. Microbiol.* **2018**, *20*, 2386–2396. [CrossRef]
56. Phillips, R.P.; Brzostek, E.; Midgley, M.G. The Mycorrhizal-associated Nutrient Economy: A New Framework for Predicting Carbon–Nutrient Couplings in Temperate Forests. *New Phytol.* **2013**, *199*, 41–51. [CrossRef]
57. Smith, A.P.; Marín-Spiotta, E.; Balser, T. Successional and Seasonal Variations in Soil and Litter Microbial Community Structure and Function during Tropical Postagricultural Forest Regeneration: A Multiyear Study. *Glob. Change Biol.* **2015**, *21*, 3532–3547. [CrossRef] [PubMed]
58. Shao, S.; Zhao, Y.; Zhang, W.; Hu, G.; Xie, H.; Yan, J.; Han, S.; He, H.; Zhang, X. Linkage of Microbial Residue Dynamics with Soil Organic Carbon Accumulation during Subtropical Forest Succession. *Soil Biol. Biochem.* **2017**, *114*, 114–120. [CrossRef]
59. Landesman, W.J.; Nelson, D.M.; Fitzpatrick, M.C. Soil Properties and Tree Species Drive SS-Diversity of Soil Bacterial Communities. *Soil Biol. Biochem.* **2014**, *76*, 201–209. [CrossRef]
60. Yan, G.; Luo, X.; Huang, B.; Wang, H.; Sun, X.; Gao, H.; Zhou, M.; Xing, Y.; Wang, Q. Assembly Processes, Driving Factors, and Shifts in Soil Microbial Communities across Secondary Forest Succession. *Land Degrad. Dev.* **2023**, *34*, 3130–3143. [CrossRef]
61. Goldmann, K.; Schöning, I.; Buscot, F.; Wubet, T. Forest Management Type Influences Diversity and Community Composition of Soil Fungi across Temperate Forest Ecosystems. *Front. Microbiol.* **2015**, *6*, 1300. [CrossRef] [PubMed]
62. Rousk, J.; Bååth, E.; Brookes, P.C.; Lauber, C.L.; Lozupone, C.; Caporaso, J.G.; Knight, R.; Fierer, N. Soil Bacterial and Fungal Communities across a pH Gradient in an Arable Soil. *ISME J.* **2010**, *4*, 1340–1351. [CrossRef] [PubMed]

forests

MDPI

Article

Soil Microbial Communities in *Pseudotsuga sinensis* Forests with Different Degrees of Rocky Desertification in the Karst Region, Southwest China

Wangjun Li [1], Bin He [1], Tu Feng [1], Xiaolong Bai [1], Shun Zou [1], Yang Chen [1], Yurong Yang [2,*] and Xuefeng Wu [3,*]

[1] Guizhou Province Key Laboratory of Ecological Protection and Restoration of Typical Plateau Wetlands, Guizhou University of Engineering Science, Bijie 551700, China; teesn470@gues.edu.cn (W.L.); hebin@gues.edu.cn (B.H.); fengtu@gues.edu.cn (T.F.); baixiaolong@gues.edu.cn (X.B.); zoushun@gues.edu.cn (S.Z.); chenyang@gues.edu.cn (Y.C.)

[2] Key Laboratory for Vegetation Ecology, Ministry of Education, State Environmental Protection Key Laboratory of Wetland Ecology and Vegetation Restoration, School of Environment, Northeast Normal University, Changchun 130117, China

[3] Chongqing Institute of Quality & Standardization, Chongqing 400023, China

* Correspondence: yangyr422@nenu.edu.cn (Y.Y.); wuxf112@nenu.edu.cn (X.W.); Fax: +86-0431-89165610 (Y.Y.); +86-023-89185796 (X.W.)

Abstract: Rocky desertification (RD), a natural and human-induced process of land degradation in karst areas, has become the primary ecological disaster and one of the obstacles to sustainable ecological development in southwest China. Nevertheless, the variation of soil physical and chemical properties, bacterial and fungal communities, and their relationships in RD forests remains limited. Therefore, soil samples were collected from forests under four degrees of RD (NRD, non-RD; LRD, light RD; MRD, moderate RD; and SRD, severe RD) and subjected to high-throughput sequencing of 16S rRNA and ITS1 genes. The results showed a significant reduction in bacterial richness and diversity, while fungal richness and diversity decreased markedly and then showed a balanced trend with the increase in RD degree, indicating that bacteria and fungi did not present the same dynamics in response to the process of RD. The bacterial communities were dominated by Proteobacteria, Actinobacteria, Acidobacteria, and Chloroflexi, while the fungal communities were dominated by Basidiomycota, Ascomycota, and Mortierellomycota. The PCoA and NMDS demonstrated significant differences in microbial communities in study sites, among which the fungal communities in non-RD forest and LRD forest clustered together, suggesting that fungal communities were more stable than bacteria in RD forest. The db-RDA, Mantel test, and random forest model confirmed the important role of soil BD, pH, SOC, AN, and AP in driving microbial diversity and communities. The IndVal analysis suggested that Chloroflexi, Patescibacteria, Atheliales, and Cantharellales with high indicator values were identified as potential bio-indicators for RD forests. This study could not only improve our understanding of bacterial and fungal community dynamics across RD gradients, but also could provide useful information for the further use of microorganisms as indicators to reflect the environmental changes and ecosystem status during forest RD.

Keywords: karst area; rocky desertification; soil properties; microbial community; bio-indicators

1. Introduction

Karst refers to a special landform formed by the dissolution and transformation of carbonate rocks by groundwater and surface water under special geological conditions, and is widely distributed globally [1]. Karst covers an area of approximately 5.1×10^7 km^2 globally, accounting for about 12% of the world's land area, mainly distributed in regions such as the Mediterranean, Eastern Europe, the Middle East, Southeast Asia, and the Caribbean [1]. China is one of the world's three major concentrated karst distribution areas, with a karst area of over 1.3 million km^2, accounting for 13.54% of the total land area [2]. Within this area, the southwestern region of China, mainly in Guizhou, has the largest and most concentrated contiguous karst ecological fragile area in Asia, with a karst area of 550,000 km^2, accounting for over 70% of the total land area of the province [3]. This region is not only the area with the widest and most developed distribution of tropical and subtropical karst in the world, but also the most typical and representative region for karst environmental and ecological issues. Due to the unique geological structure of karst and the long-term shaping by climatic and hydrological factors, the spatial distribution of water and soil resources is uneven, and the hydrothermal conditions exhibit high spatiotemporal heterogeneity. Moreover, population explosion, severe soil erosion, frequent natural disasters, and ecosystem vulnerability exacerbate the deterioration of the ecological environment, severely affecting the local people's living standards and hindering the social and economic development in the area [4].

Rocky desertification (RD) refers to a land degradation process occurring in the vulnerable karst environment of the subtropical regions, resulting from human-induced disturbances [5]. It leads to severe soil erosion, extensive exposure of bedrock, a significant decline in land productivity, and the appearance of desert-like landscapes on the surface. In recent years, with the rapid increase in population, the issue of RD has become increasingly severe and has emerged as a global environmental problem [6]. As one of the largest and most typical continuous karst distribution belts in the world, southwest China has the largest and most typical distribution of RD, with an annual growth rate of 0.91% [7]. The southwestern karst region, represented by Guizhou, is the central area of karst development in East Asia. Due to its lower environmental carrying capacity and significant human–land conflicts, a series of ecological degradation phenomena, such as soil erosion, vegetation destruction, and exposure of bedrock, are particularly common in this region, leading to the formation of RD landscapes. As the intensity of RD intensifies, soil erosion worsens, and the soil's capacity for nutrient retention, supply, and fertility rapidly diminishes, even leading to the loss of land productivity. RD has become one of the environmental challenges that restrict large-scale development and a significant ecological problem for agricultural production and economic development in China. Previous studies indicated that RD leads to changes in soil properties, a decline in plant diversity, and reductions in both above- and below-ground biomass [4]. Despite these findings, there is a significant gap in our understanding of how soil microbial composition and diversity are affected by RD in karst areas.

Soil microorganisms play a crucial role in the functioning and services of soil ecosystems, such as plant productivity, nutrient cycling, and organic matter decomposition [8]. They are essential components of the soil environment and regulate ecosystem nutrient cycling through their metabolic activities and interactions with other biotic and abiotic factors. Due to their short life cycles and sensitivity to environmental changes, soil microorganisms can respond rapidly to ecological and environmental perturbations, making them sensitive indicators of the soil environment [9]. The diversity of soil microorganisms is regulated by various factors, including both natural and anthropogenic influences. The importance of different factors in controlling soil microbial communities depends on the spatial scale. At the global and regional scales, microbial communities are more influenced by climatic factors, soil environment, and nutrients, such as soil pH and organic carbon content [10,11]. On the other hand, at local or smaller micro-scales, factors such as plant characteristics, soil microenvironment, interactions between trophic levels in the soil food web, and inter-species interactions among microorganisms might become more important in controlling microbial communities [12].

In the karst region, the ecological structure is fragile, and microorganisms actively participate in the weathering of karst rocks (limestone). They actively accelerate the dissolution of carbonate rocks. Studies have found the presence of autotrophic nitrogen-fixing microorganisms and autotrophic photosynthetic microorganisms on karst rocks undergoing autotrophic weathering [13]. Moreover, heterotrophic microorganisms are also found on autotrophically weathered rocks, and under favorable conditions, they can proliferate and spread rapidly, thereby accelerating the weathering of carbonate rocks to some extent [14]. The process of RD affects soil properties and vegetation community characteristics. As the degree of RD increases, the content of soil organic matter and plant biomass declines. RD has different influences on soil bacterial community composition, while the richness and diversity of bacterial communities do not show significant changes with the increase in RD [15]. In fact, the bacterial diversity slightly increases and the richness slightly decreases in severe RD areas. Moreover, in heavy RD areas, the soil fungal community differs significantly from other degrees of RD, with soil total nitrogen being the main factor correlated with fungal communities, followed by pH [16]. The phylum Ascomycota is the most abundant in non-degraded soils, while Basidiomycota dominates in severe RD soils, and the ratio of Ascomycota to Basidiomycota decreases significantly with the deepening of RD, which can serve as an indicator of the degree of RD [16]. In the degraded karst ecosystems, a study has shown that soil degradation in the karst region reduced soil bacterial genetic diversity and significantly changed community structure [17]. However, the dynamics of diversity and compositions of soil bacterial and fungal communities in response to RD, and their driving factors, are still unclear.

Furthermore, effective ecosystem management and restoration strategies require sensitive indicators to assess the state of the ecosystem. Soil microbes play a critical role in community reconstruction, serving as valuable indicators of ecosystem conditions [18]. Firstly, microorganisms represent the largest component of total soil biomass and are vital drivers of processes essential for ecosystem services, such as organic matter decomposition, biodiversity, and plant productivity [19]. Secondly, microorganisms respond rapidly to environmental conditions due to their high surface-to-volume ratio and intimate relationship with their surroundings. This rapid responsiveness makes them valuable indicators of environmental change. Thirdly, molecular methods offer a convenient way to track changes in microbial diversity and community composition, translating them into tangible parameters. Therefore, deriving parameters from microbial diversity, community composition, and sensitive taxa to degraded ecosystem offers a promising approach to evaluate ecosystem conditions. Thus, recognizing the importance of microbial indicators in evaluating habitat status, further exploration of microbial evaluation strategies utilizing high-throughput sequencing technologies is crucial.

In this study, soil physicochemical properties were determined and high-throughput sequencing technology was used to sequence the soil bacteria (16S rRNA) and fungi (ITS1) to assess the relationships between soil properties and microbial communities in forests under different degrees of RD in the karst region. The main purposes of this study are (i) to determine the diversity and taxonomic compositions of bacterial and fungal communities across the RD gradient, (ii) to uncover the key factors driving bacterial and fungal communities in karst area, and (iii) to identify the potential microbial taxa as bioindicators of the degrees of RD.

2. Materials and Methods

2.1. Study Site and Sampling

The study area is in Weining *Pseudotsuga sinensis* Nature Reserve of Bijie City, Guizhou province, southwest China (103.93–104.26° E, 26.54–26.76° N), with elevations ranging from 1800 to 2450 m, covering a total area of 951.86 hectares (Figure 1). Before the nature reserve was established, the study region faced intensifying anthropogenic disturbances, such as agricultural cultivation and logging, which inflicted destructive impacts on the already fragile forest, compromising its regenerative capacity and competitive survival. This led to severe soil erosion, exacerbating the process of rocky desertification. To sensibly protect biological resources and restore degraded karst habitats, the People's Government of Weining County officially approved the establishment of a county-level nature reserve in August 2000, aiming to prevent further degradation of the karst habitat. In order to further enhance the efforts in ecosystem conservation, the government undertook further optimization of the nature reserve in March 2022. The reserve experiences a warm and humid subtropical monsoon climate, with an average annual temperature of 10.5 °C and an annual average rainfall of 1000 mm, of which more than 70% occurs during the time from June to September. The predominant soil type in this area is Leptosols, which are soils limited by hard rock or have a low percentage of fine earth material according to World Reference Base for Soil Resources (WRB). The region boasts rich and well-preserved vegetation resources, primarily consisting of warm-temperate coniferous forests dominated by *P. sinensis* and *Pinus yunnanensis* Franch. The shrublands are mainly composed of *Cotoneaster franchetii* Bois., *Corylus yunnanensis*, *Viburnum dilatatum* Thunb., *Hypericum monogynum* L., and *Coriaria napalensis* Wall.

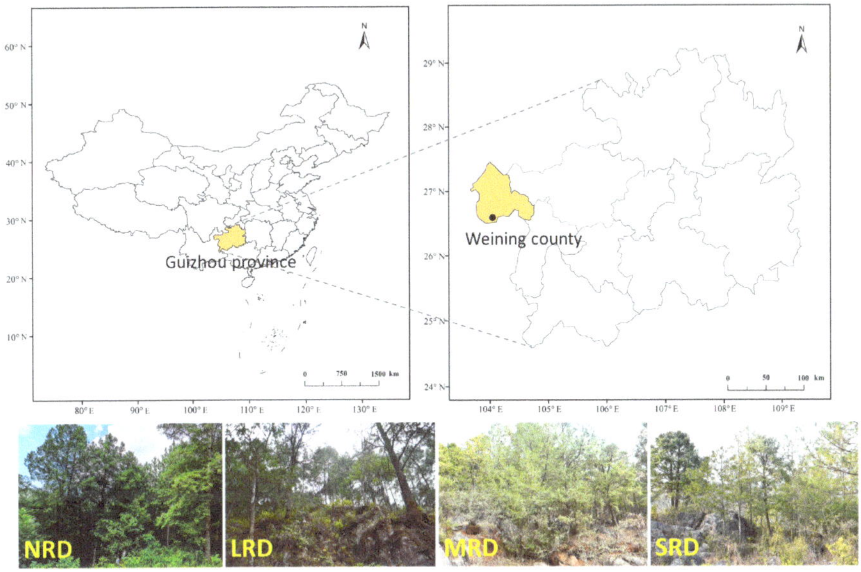

Figure 1. Location of the study area and the pictures of four degrees of rocky desertification in Guizhou Province. NRD, non-rocky desertification; LRD, light rocky desertification; MRD, moderate rocky desertification; and SRD, severe rocky desertification.

In June 2022, four regions with different degrees of rocky desertification (NRD: non-rocky desertification, LRD: light rocky desertification, MRD: moderate rocky desertification and SRD: severe rocky desertification) were selected according to the industry standard of the State Forestry Administration (LY/T 1840-2009) [20] and the classification standard of rocky desertification [21]. The soil depth in NRD, LRD, MRD, and SRD forests were approximately 50–60 cm, 30–40 cm, 20 cm, and 10 cm, respectively. Moreover, the bare rock rate in NRD, LRD, MRD, and SRD forests were approximately 20%–30%, 40%–50%, 70%, and 80%, respectively, based on visual assessment. Then, six plots of 50 × 50 m were set up within each region with different degrees of rocky desertification. The distance between each plot was at least 50 m. After removing surface litter, soil samples were collected from the 0–10 cm soil layer using a shovel and every sample was a mixture of three subsamples collected in an "S" shape. These soil samples were then passed through a 2 mm sieve to remove impurities such as stones and plant residues. The ring knife was inserted into the ground to obtain a soil core. The sieved soil samples were divided into two portions. The first potion (~5 g of soil) was preserved in a cooler and transported to the laboratory within 24 h. These samples were stored at −80 °C and used for the extraction of total soil DNA. The second potion (~200 g of soil) was stored in self-sealing bags, transported to the laboratory, and air dried naturally. After drying, the soil was ground and passed through a 100-mesh sieve. These samples were used for the determination of soil properties, including soil pH, soil organic carbon (SOC), total nitrogen (TN), available nitrogen (AN), total phosphorus (TP), available phosphorus (AP), total potassium (TK), and available potassium (AK). The third potion (soil core) was then placed in a sterile plastic bag for the measurement of soil bulk density (BD).

2.2. Soil Properties

Soil BD was measured via the ring knife method. Specifically, the metal ring was pressed into the soil (intact core), and then the soil samples were placed into a sterile plastic box to preserve for lab drying. The soil inside the metal ring was dried to constant weight at a high temperature of 105 °C (for at least 24 h) and weighed to calculate soil BD [22]. Weighed soil samples (20 g fresh weight) were made into a soil suspension via the addition of deionized water in a soil:water ratio of 1:5. Soil pH was measured using a FE28-Standard pH meter (Mettler-Toledo, CA, USA) [23]. SOC was determined via potassium dichromate oxidation with external heating [23]. Specifically, 0.5 g of the dry soil samples was weighed in a 250 mL Erlenmeyer flask containing 10 mL of K_2Cr_2O7 (0.8 M) and 10 mL of H_2SO_4, and then boiled at 170–180 °C for 5 min. TP was determined using the molybdate colorimetric method after perchloric acid digestion and ascorbic acid reduction [24]. AP was measured via extraction with 0.5 M $NaHCO_3$ (pH = 8.5) for 30 min and then assessed colorimetrically via the molybdate-ascorbic acid method using a UV spectrophotometer [25]. TN was determined via Kjeldahl digestion [26]. Specifically, 0.5 g soil was digested with 5 mL of concentrated H_2SO_4. The digestion was initially started at 50 °C and then raised gradually to 350 °C. The samples were analyzed using 40%NaOH solution, and 4% boric acid, 0.1 N standard HCl solutions. AN were qualified according to the dichromate oxidation [26]. TK in soil was determined via NaOH alkali fusion-flame photometry, while AK was extracted with 1 M ammonium acetate ($CH_3CO_2NH_4$) at pH 7.0 and subsequently measured using flame photometry [27].

2.3. DNA Extraction, Illumina Sequencing, and Sequencing Data Processing

The total microbial genomic DNA in the soil samples was extracted from 0.5 g of frozen soil using a PowerSoil DNA extraction kit (MoBio, Carlsbad, CA, USA) according to the manufacturer's recommendations. Detailed information on primer sets, PCR conditions, and reaction systems for bacterial 16S rRNA and fungal ITS1 genes are shown in Table S1. The high-throughput sequencing was performed by Shanghai Personal Biotechnology Co., Ltd. (Shanghai, China) based on an Illumina MiSeq platform (San Diego, CA, USA). Bioinformatic processing of sequencing data was conducted with QIIME 2 2019.4 with modifications according to the official tutorials (https://docs.qiime2.org/2019.4/tutorials/, accessed on 10 October 2022) (Supporting Information, Method 1). The Illumina sequencing raw read data deposited in the Sequence Read Archive (SRA) are available in the NCBI SRA portal with PRJNA797899, BioProject ID.

2.4. Statistical Analysis

After confirmation of the homogeneity of variances and normality of the data using the Levene and Shapiro–Wilk tests, respectively, one-way analysis of variance (ANOVA) was used to determine the differences in soil physicochemical properties, Bray–Curtis dissimilarities, richness (Sobs index and Chao1 index) and α-diversity (Shannon index) of microbial communities in different forests under different degrees of RD via SPSS (version 21.0, Chicago, IL, USA). The richness (Sobs index, and Chao1 index) and α-diversity (Shannon index) of bacterial and fungal communities were performed by the QIIME2 software (version 2019.4). Sobs was the number of species observed in the sample. The formulas used for the calculation of Chao1 and Shannon indices were as follows:

$$\mathrm{Chao1} = \mathrm{Sobs} + \frac{\mathrm{n}_1(\mathrm{n}_1 - 1)}{2(\mathrm{n}_2 + 1)} \tag{1}$$

$$\text{Shannon} = -\sum_{i=1}^{S} p_i(lnp_i) \quad (2)$$

where Sobs is the number of ASVs observed, n_1 is the number of OTUs with only one sequence, and n_2 is the number of ASVs with only two sequences. p_i is the proportion of individuals belonging to *i*-species in the sample.

The Venn and upset Venn diagrams were performed using the "VennDiagram" and "UpSetR" packages, respectively, to show the unique and shared amplicon sequencing variants (ASVs) in each study site. Principal coordinate analysis (PCoA) and non-metric multidimensional scaling (NMDS) were used to explore the structural variation of bacterial and fungal communities in forests under different degrees of RD based on Bray–Curtis distance using the "Vegan" package. Similarly, the β-diversity of bacterial and fungal communities in each study site was calculated based on the Bray–Curtis dissimilarity index using the "Vegan" package. The indicator value (IV) index was derived based on the relative abundance and relative frequency of occurrence to identify bio-indicators associated with each degree of RD using the "IndVal" package. Distance-based redundancy analysis (db-RDA) and Mantel tests were used to explore the correlations within the microbial communities in the context of soil properties based on relative abundances of ASVs using "vegan" and "LinKET" packages, respectively. Pearson's correlation coefficient was used to measure the relationships between the diversity of microbial communities and soil properties. Subsequently, random forest analysis was performed to determine the importance of each soil parameter for the α-diversity and β-diversity of bacterial and fungal communities based on MSE using the "randomforest" package (version 4.7-1.1) [28].

3. Results

3.1. Soil Properties in Different Study Sites

During the process of rocky desertification (RD), the soil's physical and chemical properties changed dramatically among different study sites (Figure 2). Compared with the control, i.e., NRD forest (1.01 ± 0.10 g cm^{-3}), soil bulk density (BD) significantly decreased in the MRD (1.14 ± 0.09 g cm^{-3}) and SRD forests (1.35 ± 0.06 g cm^{-3}), but did not change in the LRD forest (1.10 ± 0.12 g cm^{-3}). NRD and LRD forests had lower soil pH than MRD and SRD, but had no difference between them. The soil organic carbon (SOC) decreased significantly, from 53.19 ± 3.57 g kg^{-1} (NRD) to 49.79 ± 2.11 g kg^{-1} (MRD) and 40.39 ± 4.13 g kg^{-1} (SRD), but did not change in the LRD forest (55.65 ± 4.07 g kg^{-1}). Soil total nitrogen (TN) and total phosphorus (TP) did not change significantly among different study sites. Compared with NRD and MRD forests, soil total potassium (TK) content in LRD and SRD forests was significantly increased, and the change between LRD and SRD forests was not significant. The soil available nitrogen (AN) and available phosphorus (AP) showed significant variations among different study sites, with the AN content ranking, from high to low, as LRD > NRD > MRD > SRD, while the AP content ranking was LRD > NRD > SRD > MRD. Compared with the MRD forest, NRD, LRD, and SRD forests had significantly higher available nitrogen potassium content, while no difference in available potassium (AK) was found among the three study sites.

3.2. Sequencing Information and α-Diversity of Microbial Communities

A total of 1,960,415 and 2,114,212 reads were obtained for 16S V3-V4 and ITS1 datasets, respectively. The DADA2 pipeline allowed identification of 110,692 and 10,728 bacterial and fungal ASVs, respectively after removing singletons. Bacterial ASVs represented 36 phyla, 118 classes, 311 orders, 525 families, 1105 genera, and 1643 species, while fungal ASVs represented 16 phyla, 51 classes, 127 orders, 278 families, 601 genera, and 959 species. All rarefaction curves were saturated with increased sequencing amounts (Figure 3a,b) and the coverage exceeded 99%, which indicated that most soil bacteria and fungi were sequenced in all samples.

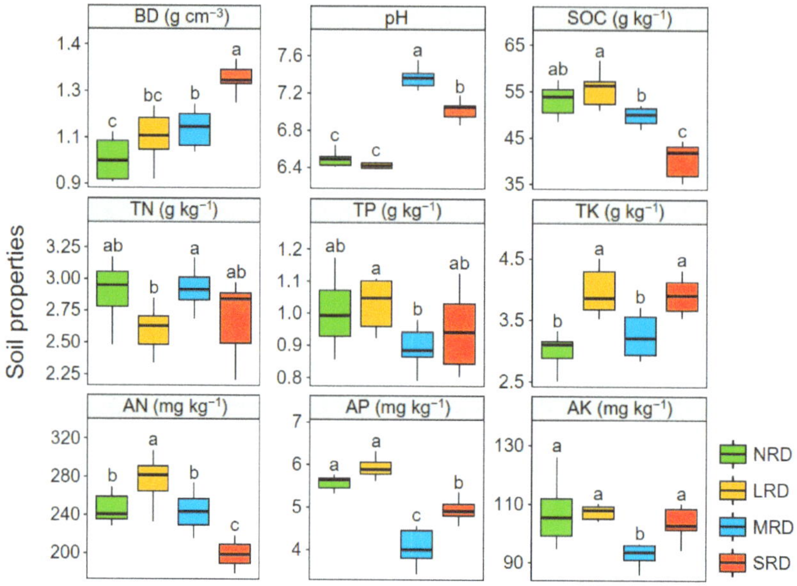

Figure 2. Variations in soil properties in study sites with different degrees of rocky desertification. Different letters indicate a significant difference at $p < 0.05$ according to Duncan's multiple range test. NRD, non-rocky desertification; LRD, light rocky desertification; MRD, moderate rocky desertification; and SRD, severe rocky desertification. BD, bulk density; SOC, soil organic carbon; TN, total nitrogen; TP, total phosphorus; TK, total potassium; AN, available nitrogen; AP, available phosphorus; and AK, available potassium.

In the current study, observed species (Sobs), Chao1, and Shannon indices were used to evaluate the alpha-diversity of soil microbial communities (Figure 3c–h). The results showed that RD had a strongly negative impact on the α-diversity of soil bacterial communities. NRD forest had the highest level of bacterial community richness. The Sobs, Chao1, and Shannon indices of bacterial communities in the SRD forest decreased significantly when compared with those in the NRD forest ($p < 0.05$). However, no difference was found in bacterial α-diversity between MRD and SRD forests. In terms of fungal α-diversity, the NRD forest showed the highest values of Sobs and Chao1 indices, while there was no difference among LRD, MRD, and SRD forests. The Shannon index of fungal communities in the NRD forest was higher compared with that in the MRD forest. Bacterial α-diversity was considerably higher than fungal diversity, e.g., there were approximately ten times as many bacterial ASVs as there were fungal ASVs.

3.3. Composition of Microbial Communities

As shown in Figure 4, the bacterial phyla Proteobacteria, Actinobacteria, Acidobacteria, and Chloroflexi accounted for the largest proportion in all soil samples, with percentages of 30.16%, 17.87%, and 16.58%, respectively (Figure 3a). By contrast, fungal phylum Basidiomycota, Ascomycota, and Mortierellomycota accounted for the largest proportion in all soil samples, with percentages of 64.33%, 27.84%, and 1.44%, respectively. Moreover, there were considerable differences in microbial community composition along the forests with different degrees of rocky desertification. For instance, Actinobacteria accounted for 34.23% of bacterial community composition in NRD forests but only 15.5% and 15.8% in MRD and SRD forests, respectively; moreover, Proteobacteria and Chloroflexi only accounted for 27.71% and 10.55%, respectively, of bacterial community composition in NRD forest but 34.15% and 23.72% in SRD forest. Soil bacterial and fungal phyla had different sensitivity and adaptability to RD. In this study, varying kinds of changes were found in the insensitive and sensitive phyla with the increase in RD gradient. For instance, Gemmatimonadetes, Verrucomicrobia, Rokubacteria, and Planctomycetes showed "arch" changes, whereas Ascomycota, Mortierellomycota, Rozellomycota, Zoopagomycota, and Glomeromycota showed "inverted arch" changes.

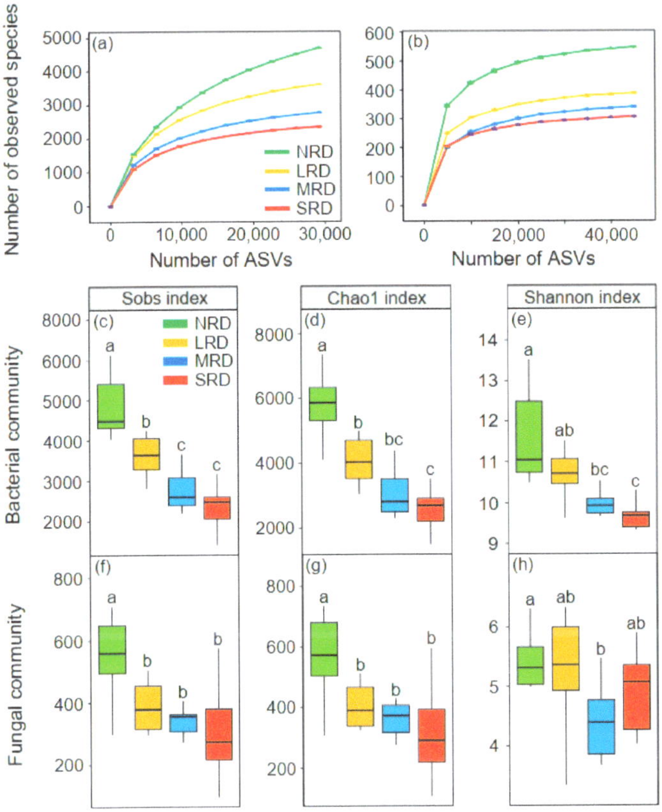

Figure 3. Rarefaction curves and alpha-diversity plots of soil microbial communities along four degrees of rocky desertification. Rarefaction curve constructed based on observed ASVs and species of bacterial (**a**) and fungal (**b**) communities. Alpha-diversity is assessed by richness (Sobs and Chao1 indices) and diversity (Shannon index) of bacterial (**c**–**e**) and fungal (**f**–**h**) communities. Different letters indicate a significant difference at $p < 0.05$ according to Duncan's multiple range test. NRD, non-rocky desertification; LRD, light rocky desertification; MRD, moderate rocky desertification; and SRD, severe rocky desertification.

The Venn diagrams showed the ASV level for the microbial community of soil samples (Figure 4c,d). The number of unique fungal ASVs decreased with increasing degrees of rocky desertification: the highest number was found in the NRD forest (1990), followed by the LRD forest (1148) and MRD forest (931), while the SRD forest had the lowest number (868). However, the distribution of unique bacterial ASVs was different compared with that of unique fungal ASVs. The number of unique bacterial ASVs increased firstly and then decreased: the highest number was found in the LRD forest (19,376), followed by the NRD forest (13,372) and MRD forest (8849), while the SRD forest had the lowest number (7158).

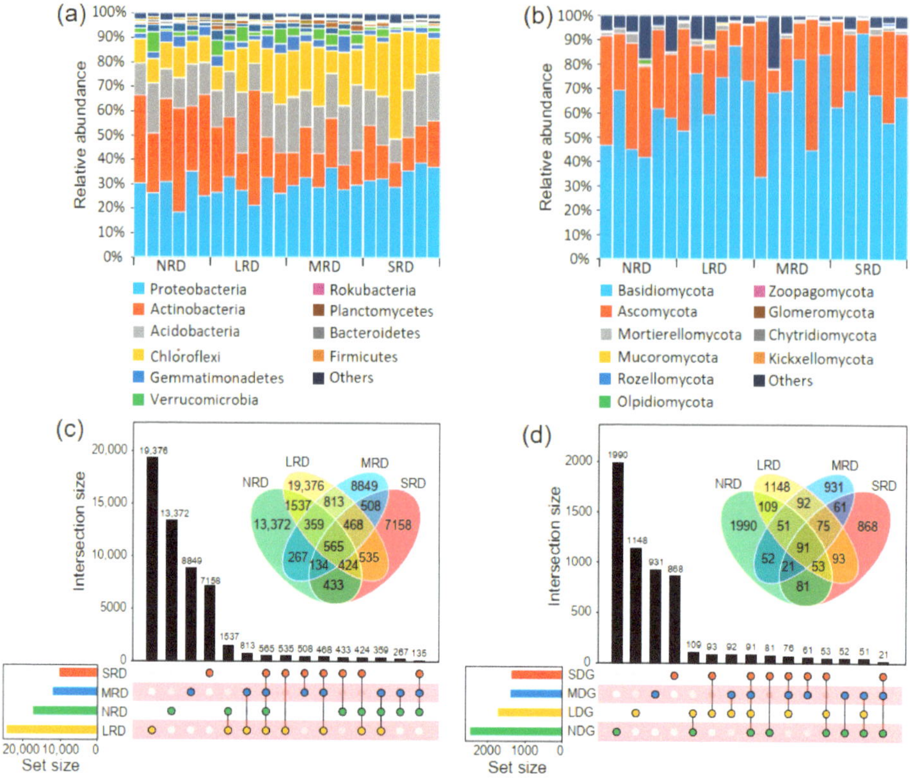

Figure 4. Percentage stacking diagrams show the community composition of bacteria (**a**) and fungi (**b**) at the phylum level. The top 10 phyla were selected for abundance analysis. Venn diagrams show the number of ASVs shared and unique in bacterial (**c**) and fungal (**d**) communities among study sites with four degrees of rocky desertification as shown in a Venn diagram. NRD, non-rocky desertification; LRD, light rocky desertification; MRD, moderate rocky desertification; and SRD, severe rocky desertification.

3.4. Composition and β-Diversity of Microbial Communities

To further compare the variations in the structure of bacterial and fungal communities in study sites with different degrees of rocky desertification, based on the Bray–Curtis algorithm, principal coordinate analysis (PCoA) and non-metric multidimensional scaling (NMDS) were employed (Figure 5). Study sites were separated from each other, indicating large differences in the structure of microbial communities at different degrees of rocky desertification. In addition, the degree of dispersion of fungal communities between NRD and LRD forests was smaller, suggesting smaller differences in the structure of fungal communities compared with bacterial communities between NRD and LRD forests. Furthermore, microbial β-diversity was calculated using Bray–Curtis dissimilarity based on the ASV table, and the results showed that the β-diversity of bacterial communities decreased with increasing rocky desertification (Figure 5e). The β-diversity of fungal communities in NRD forest was higher than that in SRD forest, while there was no difference between either NRD and LRD forests or MRD and SRD forests (Figure 5f).

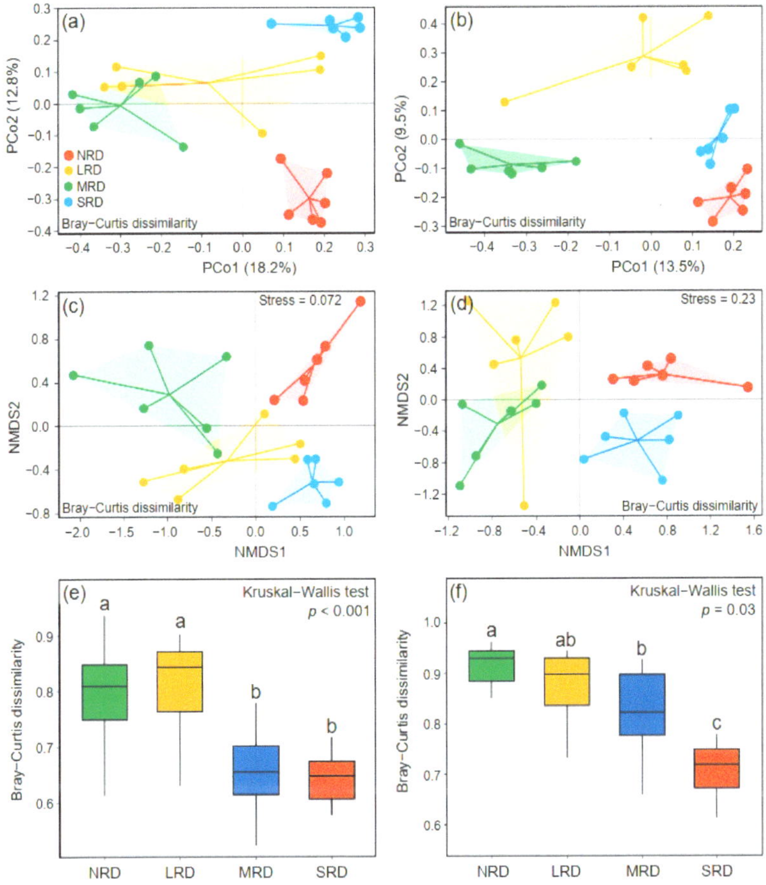

Figure 5. Principal coordinate analysis (PCoA), non-metric multidimensional scaling (NMDS) analysis, and Bray-Curtis dissimilarity of bacterial (**a**,**c**,**e**) and fungal (**b**,**d**,**f**) communities among study sites with four degrees of rocky desertification. Different letters indicate a significant difference at $p < 0.05$ according to Duncan's multiple range test. NRD, non-rocky desertification; LRD, light rocky desertification; MRD, moderate rocky desertification; and SRD, severe rocky desertification.

3.5. Indicator Taxa for Different Degrees of RD

Indicator taxa analysis was performed to explore whether bacterial or fungal taxa could be detected as being representative of specific degree of RD based on IndVal analysis (IndVal.g > 0.6, IndVal.$p < 0.05$), as shown in Figure 6. Six bacterial phyla were identified to be indicator taxa for different degrees of RD: Actinobacteria for NRD forest, Planctomycetes for LRD forest, Dependentiae for MRD forest, and Chloroflexi and Proteobacteria for SRD forest. However, there is no fungal phylum that could be considered as an indicator after calculation. For this reason, we computed the indicator analysis on the fungal order dataset. Finally, twelve fungal orders were selected and could be regarded as indicator taxa for different degrees of RD: Chaetosphaeriales, Diaporthales, Hysteriales, Pleosporales, Trechisporales, and Trichosphaeriales for NRD; Geminibasidiales, GS25, and Umbelopsidales for LRD; Leucosporidiales for MRD; and Atheliales and Cantharellales for SRD.

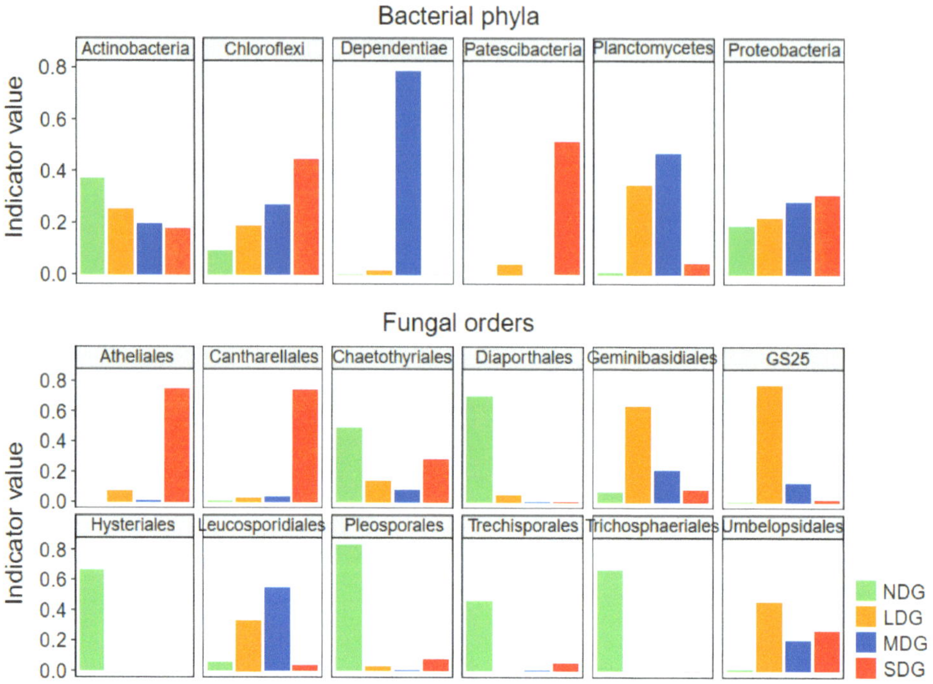

Figure 6. Indicator taxa identified for study sites with different degrees of RD. Values were obtained from the IndVal function, including IndVal.g (>0.6) and IndVal.*p* (<0.05). NRD, non-rocky desertification; LRD, light rocky desertification; MRD, moderate rocky desertification; and SRD, severe rocky desertification.

3.6. Main Factors Driving Soil Microbial Communities

As shown in Figure 7, the first and second ordination axes of distance-based redundancy analysis (db-RDA) totally explained 15.07 and 17.84% of the soil bacterial and fungal community changes, respectively. Soil microbes in NRD and LRD forests gathered in the second and third quadrants, which were obviously positively correlated with TP, and also have a certain correlation with SOC, AP, pH, and BD. The microbial communities in MRD forests were mainly distributed in the fourth quadrant, and were mainly restricted by the SOC and AN. The distribution of fungal communities in SRD was relatively scattered, and was mainly affected by pH, TP, and AP.

To explore the correlation between soil microbial community structures and environmental variables, Mantel test analysis was performed with community matrices at OTU level and environmental factors on four gradients of RD (Figure 7c). The results revealed that elevation, BD, pH, SOC, AP, and AN were the determinants of both soil bacterial and fungal community structures among study sites.

Based on the multiple regression model and random forest analysis, the interdependence between the soil microbial community and environmental factors was studied, and the influence of environmental factors on the soil microbial community in different RD forests was further elaborated (Figure 7d). We confirmed that BD, pH, SOC, and AN were the most important environmental factors, showing a strong correlation with microbial community richness, α-diversities, and β-diversities.

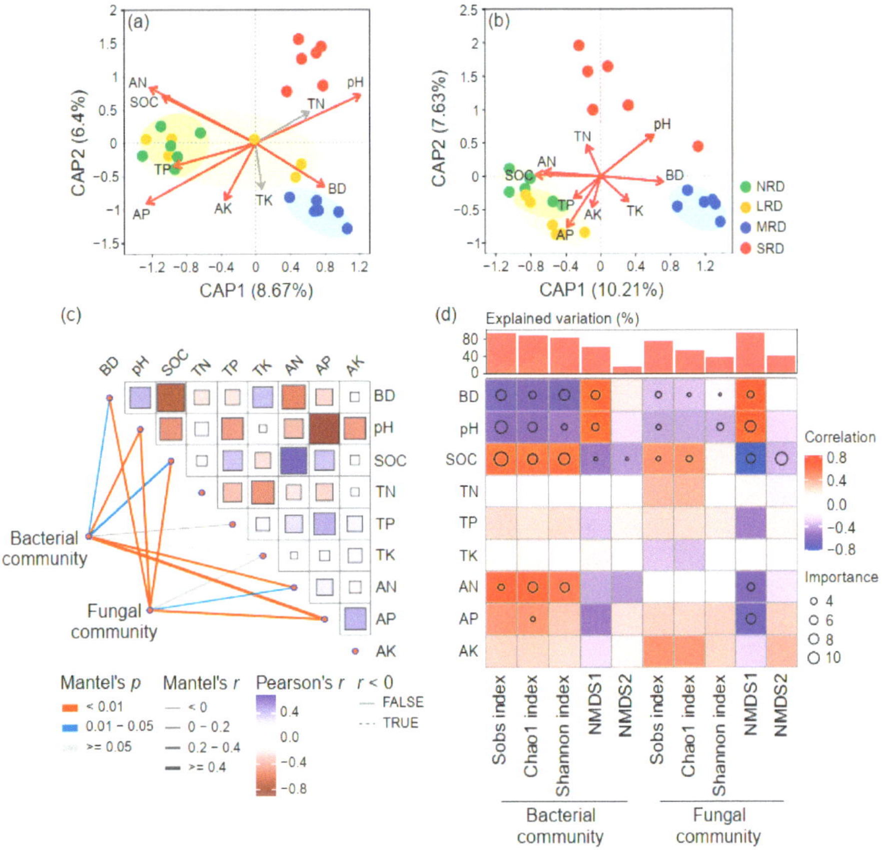

Figure 7. Relationship between soil physical/chemical properties and the diversity and compositions of soil bacterial and fungal communities in forests with different degrees of RD. Distance-based redundancy analysis (db-RDA) ordination plots of bacterial (**a**) and fungal communities (**b**) and soil properties. The environmental variables that significantly explained variability in microbial communities are shown in red color. (**c**) Mantel tests between the microbial communities and soil physical/chemical properties. The upper right corner presents the correlations (Pearson) among the nine soil properties. The lower light corner presents the correlations (Mantel test) between microbial communities (bacterial community and fungal community) and soil properties. Line width represents the significant correlation coefficients of the partial Mantel tests. The line color denotes the significance level. Line formatting indicates the correlation sign (a solid line means a positive correlation, while a dashed line means a negative correlation). (**d**) Correlation between microbial diversity and soil properties, and the main predictors of microbial diversity based on random forest model. The mean predictor importance (% of increase in MSE) indicates the importance of environmental drivers on microbial diversity. NRD, non−rocky desertification; LRD, light rocky desertification; MRD, moderate rocky desertification; and SRD, severe rocky desertification. BD, bulk density; SOC, soil organic carbon; TN, total nitrogen; TP, total phosphorus; TK, total potassium; AN, available nitrogen; AP, available phosphorus; and AK, available potassium.

4. Discussion

Rocky desertification is a common ecological problem in soil ecosystems worldwide. In this study, we revealed the effects of RD on soil properties, and soil bacterial and fungal communities, in the karst area of southwest China, uncovered the driving factors of the diversity and compositions of soil bacterial and fungal communities, and explored the indicator taxa for different degrees of RD.

4.1. Effects of RD on Soil Properties

Soil bulk density and porosity are crucial indicators of soil health, directly reflecting the soil's structure and its ability to hold water and resist erosion [29]. The present study showed that RD significantly altered soil physical properties in the karst area of southwest China. The high quality and quantity of plant litters, as well as low human disturbance in NRD forests, promoted more efficient accumulation of soil organic matter, which probably contributed to the relatively low BD in these study areas. With the increase in RD, soil bulk density (BD) increased continuously (Figure 2), which is consistent with other studies from various ecosystems [30,31]. In the alpine meadow of the Tibetan Plateau, grassland degradation significantly increased BD, from 0.99 g cm^{-3} of non-degraded grasslands to 1.38 g cm^{-3} of extremely degraded grasslands [32]. Additionally, in the forest of the Savan watershed, degraded forest (1.29 g cm^{-3}) exhibited higher BD than natural forest (1.03 g cm^{-3}) [33]. The high BD can be attributed to low soil moisture, loss of organic carbon, and weak soil structure. There was a negative correlation between the stock of soil organic carbon and soil bulk density in arid and semi-arid areas of North Africa [34]. In this study, RD caused the reduction in vegetation coverage and productivity, while increasing bare rock area. Subsequently, the soil water evaporation increased and soil water-retention capacity decreased. Additionally, RD exhibited negative effects on soil clay content while leading to an increase in soil sand content [35]. Moreover, wind erosion is one of the most crucial drivers in the RD process of the karst region [36]. Wind erosion breaks down aggregates and wind removes fine particles, mainly clay, increasing the soil's susceptibility to be further degraded [37]. The low vegetation coverage combined with the effect of wind erosion on the karst area destroyed soil structure and accelerated the loss of soil organic carbon. Greater loss of soil organic carbon, lower soil moisture, and poorer aggregation probably account for the higher BD in the RD area in comparison with the non-RD soils (Figure 2).

In addition, RD may affect soil chemistry via direct or indirect pathways, and significantly higher SOC, AN, and AP were found in NRD forests compared with SRD forests (Figure 2). Previous research also found that soil degradation can decrease soil nutrients in several ways. The loss of vegetation coverage caused by RD reduced the protection for soil nutrients, and the reduction in plant productivity decreased the input of soil nutrients from plant litter and root exudates [38]. Soil enzymes are crucial to catalyzing several important reactions necessary for the decomposition of organic matter; thus, their activities play a vital role in nutrient cycling [39]. RD inhibited the activity of soil enzymes, and thus reduced the available nutrients in soils, which explained the lower AN and AP in the SRD forest compared with in the NRD forest (Figure 2). Overall, our results indicated that RD would have adverse effects on soil physical and chemical attributes. Therefore, long-term monitoring of the variations in susceptible indices of soil properties in the early stage of RD helps understand the process of RD and to adjust the management strategies in time [40].

4.2. Effects of RD on the Diversity and Composition of Soil Microbial Communities

Soil microorganisms are a major engine of terrestrial biogeochemistry, playing an essential role in maintaining soil ecosystem functions, including plant growth stimulation, crop residue decomposition, and organic matter turnover [41]. In the current study, we performed high-throughput sequencing of 16S rRNA and ITS1 genes to assess the soil microbial communities of forests with different degrees of RD. RD can alter the diversity and composition of soil microbial communities through various biotic and abiotic factors. The bacterial richness (Sobs index and Chao1 index) and diversity (Shannon index) were lower in the RD forest than in the non-RD forest, which was consistent with a previous study that showed bacterial diversity decreased with the degrees of RD in karst areas [42]. During the RD process, soil nutrients were significantly reduced (Figure 2), which may inhibit bacterial growth and reproduction [43]. The lost vegetation coverage and productivity in the RD forest was another important reason for the decrease in bacterial and fungal richness [44]. However, our study found that RD did not pose a threat to the soil fungal diversity (Figure 3). This finding agrees with a previous study, which indicated a higher resistance of fungi compared with bacteria [45]. On the one hand, the fungal cell walls are composed of polymers of chitin and melanin, making them quite resistant to degradation. On the other hand, fungi are much more efficient at assimilating and storing nutrients than bacteria, which helps fungi to have a better adaptive ability to poor-resource RD soils. Furthermore, poor-resource RD areas probably provide higher soil heterogeneity and create more niche differentiation for fungal communities than rich-resource non-RD soils [46].

In this study, the bacterial and fungal community compositions in non-RD forests were different from those in RD forests based on both PCoA and NMDS (Figure 5). The dominant phyla in the bacterial microbial communities among different study sites were consistent, including Proteobacteria, Actinobacteria, Acidobacteria Chloroflexi, and Gemmatimonadetes. This result is in line with previous studies performed in the karst region [47,48]. The phylum Proteobacteria is one of the most abundantly distributed bacterial groups in various ecosystems based on numerous 16S rRNA gene surveys from forest, grassland, farmland, and polluted environments worldwide [49–51]. Actinobacteria are considered an indicator of harsh environment, and they play an important role in enhancing the weathering of carbonate rocks [52]. Both 16S rRNA gene-based approaches and shotgun metagenomic assessment have found that Acidobacteria was a highly diverse phylum across a wide range of habitats worldwide [53]. Chloroflexi and Verrucomicrobia are categorized as oligotrophic, having a strong resistance to poor nutrient conditions [54]. In this study, Proteobacteria and Chloroflexi were selected to be indicators for SRD based on IndVal analysis (Figure 6). Nevertheless, the relative abundances of these bacterial phyla were significantly different among study sites. In this study, the relative abundance of Proteobacteria (27.71% in NRD, 27.88% in LRD, 30.90% in MRD, and 34.15% in SRD) and Chloroflexi (10.55% in NRD, 14.06% in LRD, 18.00% in MRD, and 23.72% in SRD) increased with increasing RD, while the relative abundance of Actinobacteria (34.23% in NRD, 24.48% in LRD, 15.49% in MRD, and 15.80% in SRD) reduced gradually (Figure 4). Proteobacteria is one of the largest phyla of soil bacteria, having highly diverse metabolic capabilities, and many members in this phylum can be involved in N_2 fixation and the soil nitrogen cycle [55]. Moreover, Chloroflexi is found in various ecosystems such as soil and water, most of which have a very slow growth rate with an anti-stress strategy in disturbed and nutrient-limited environments [56]. Regarding the fungal communities, Basidiomycota and Ascomycota were the two dominant phyla in this study (Figure 4), and most members in the phyla were involved in carbon cycling through decomposing organic matters [57]. However, the relative abundance of Basidiomycota and Ascomycota presented the opposite trend (Figure 4). This result is consistent with a previous study which indicated that Basidiomycota was higher in degraded soils compared with non-degraded soils [58]. Overall, the results revealed that microbial groups have various adaptive strategies to soil conditions with different degrees of RD in karst habitats [41].

4.3. Factors Shaping Soil Bacterial and Fungal Communities

Extensive studies have indicated that soil variables are the main factors impacting the diversity and composition of soil microbial communities [59]. In the current study, soil pH was one of the most important environmental factors affecting both bacterial and fungal communities in karst areas based on distance-based redundancy analysis (db-RDA), the Mantel test, and the random forest model (Figure 7). Soil pH has been widely accepted as a key factor influencing microbial communities through directly or indirectly influencing microbial growth and soil nutrient availability [60]. Notably, pH impacts the activities of enzymes and each microbial taxa has a definitive pH growth range [61]. Microorganisms grow fast at their optimum growth pH, while growing slowly below the minimum growth pH and above the maximum growth pH. Furthermore, soil pH is regarded as a "master variable" that has a strong effect on the mobility of compounds in the soil and impacts many biogeochemical processes [62]. Generally, high soil pH can reduce the availability of soil micronutrients, while low soil pH will improve the availability of certain elements such as phosphorus (P), and thereby influence microbial abundance and composition. Moreover, soil pH exerts various effects on bacterial and fungal taxa. For instance, the relative abundances of Chloroflexi and Basidiomycota were positively correlated with pH, while the relative abundances of Actinobacteria, Zoopagomycota, and Kickxellomycota had a negative relationship with soil pH (Figure S1). This result indicated that microbial taxa have their habitat preferences and exhibit different responses to soil pH. It is reported that lower pH decreased bacterial growth and increased fungal growth [63]. On the contrary, bacterial and fungal richness and diversities had the same response patterns to pH in this study: all indices (Sobs index, Chao1 index, and Shannon index) decreased with increasing soil pH. These inconsistent results may be caused by the differences in climate condition, soil and vegetation properties, and the type of degradation among study sites. Furthermore, bacterial and fungal communities were strongly impacted by soil pH, and the bacteria communities ($r^2 = 0.947$, $p = 0.001$) were more strongly influenced by pH than the fungi ($r^2 = 0.699$, $p = 0.001$), based on db-RDA (Figure 7). This result might be due to the relatively wider optimal pH range for fungal growth than for bacteria growth [64]. The soil BD was also an important factor that impacted the diversity and composition of bacterial and fungal communities in all study sites (Figure 7). Similar results were found in a previous study, which showed that BD had a significant association with the communities of bacteria and fungi in karst areas [65]. It has been accepted that soil BD is an indicator of soil compaction and many key soil processes, such as infiltration, availability of nutrients, and activity of soil microorganisms [66]. High BD with soil compaction leads to the restriction of soil O_2 concentration, and thereby disturbs the microbial diversity and community. It was found that SOC, AN, and AP, but not pH or BD, were the significant variables that explained the majority of the variability in bacterial and fungal diversity and community (Figure 7). By changing microbial activities and plant characteristics, soil nutrients are considered to be an important factor that influences the growth and development of microbes in karst regions [41]. The results were consistent with a previous study that revealed that SOC, AN, AP, and AK were the main factors influencing the microbial community [67]. As a source of nutrients and energy to microorganisms, soil organic matter was reported to play a crucial role in regulating microbial communities [68]. In this study, the relative abundance of Proteobacteria decreased with an increase in SOC and AP, while the relative abundance of Actinobacteria showed a significantly positive correlation with AN and AP. Inconsistent with a previous study [65], there was no correlation between microbial diversity/community and TN and TP in the karst area. This result might be due to the fact that soil microbial taxa vary in their nutrient preferences and nutrient acquisition strategies, which influences the metabolism and species turnover rate of microorganisms. Thus, the current study demonstrated the potential of both soil physical and chemical properties in driving the diversity and composition of bacterial and fungal communities in karst areas, and soil available nutrients (i.e., AN, and AP) might play a more important role in shaping the microbial community than total nutrients (i.e., TN and TP). Notably, further studies are still needed to confirm the effects of environmental factors on shaping diversities and compositions of soil microbial communities in karst areas because the variations in some variables (especially pH and BD) were relatively low in the current study.

4.4. Linkage between Micorbial Indicators and Ecosystem Functions

The remarkable activity of the soil microbiome plays a critical role in numerous vital soil functions, including nutrient cycling, organic matter decomposition, and soil structure formation [19]. Consequently, the positive impacts of combined amendments on enhancing ecosystem health are intricately linked to the diversity and composition of soil microbial communities. In this study, we found that the bacterial phylum Actinobacteria can be considered a good indicator for NRD forests, while Planctomycetes, Dependentiae, Chloroflexi, and Proteobacteria had relatively high abundance in RD forests (Figure 6). Previous research has indicated that Actinobacteria members were clustered to be typical copiotrophic bacterial species, while Planctomycetes and Chloroflexi are typical oligotrophic bacterial species [69]. Here, we detected that the oligotrophic groups (Planctomycetes and Chloroflexi) decreased while copiotrophic groups (Actinobacteria) increased with the RD gradient, suggesting that there was a shift in microorganisms from oligotrophic to copiotrophic groups following forest RD. The oligotrophic taxa with a low growth rate play an essential role in organic matter decomposition [70]. Previous research has revealed that Actinobacteria comprise one of the most abundant and impactful groups of microorganisms within soil microbial communities [71]. They play an essential role in driving ecological nutrient cycling within the soil ecosystem. Notably, these Actinobacteria are renowned for their production of diverse biologically active substances, including enzymes, antibiotics, and vitamins, which contribute significantly to various ecological functions [71]. Furthermore, Dependentiae can be found in most soils with relatively low abundances, although information about their ecological functions remains scant [72]. Unlike the fast-growing copiotrophic taxa, the oligotrophic bacteria prioritize the production of extracellular enzymes during periods of resource scarcity. These enzymes act like molecular scissors, breaking down complex molecules into simpler forms that other organisms can readily utilize.

In addition, Chaetosphaeriales, Diaporthales, Hysteriales, Pleosporales, and Trichosphaeriales belong to the fungal phylum Ascomycota and were found to be enriched in NRD forests (Figure 6). Previous research has indicated that the Ascomycota phylum dominated the fungal communities in harsh environments [73]. These fungi are critical drivers of carbon and nitrogen cycling within these arid ecosystems, playing essential roles in soil stabilization, plant biomass decomposition, and endophytic interactions with plants. By contrast, Geminibasidiales, GS25, Leucosporidiales, Atheliales, and Cantharellales belong to Basidiomycota, and were mainly found in RD forests (Figure 6). Basidiomycota phylum fungi dominate the decomposition of deadwood, with white-rot and brown-rot species playing a pivotal role due to their potent lignin degradation capabilities. These fungal phyla achieve this feat through their impressive arsenal of extracellular lignocellulolytic enzymes, efficiently breaking down the complex lignin polymer [74]. However, further studies are required to reveal the ecological functions of these microbial indicators in forests with different degrees of RD in karst regions.

5. Conclusions

In conclusion, the responses of soil properties and microbial communities to rock desertification (RD) in karst area were systematically investigated. Based on our results, forest RD in this specific area resulted in the variations in the diversities and compositions of soil bacterial and fungal communities. The richness and diversities of microbial communities decreased with increasing degree of RD. The microbial community composition shifted from copiotrophic strategy groups dominating in the non-RD forest, to oligotrophic strategy groups dominating in the RD forest. The variations in soil bacterial and fungal community compositions were mainly explained by soil BD, pH, SOC, AN, and AP. The microbial taxa Chloroflexi, Patescibacteria, Atheliales, and Cantharellales could be regarded as the potential indicators for the specific degree of RD. Taken together, this study aids further understanding of the effects of RD on soil microbial communities and promotes the basic knowledge about the selection of biological indicators for RD forests. However, further studies are required to reveal the relationship between these bio-indicators and ecosystem functions at a large scale to improve activities aiming to restore karst systems via microbial technology.

Supplementary Materials: The following supporting information can be downloaded at: https://www.mdpi.com/article/10.3390/f15010047/s1, Method 1: DNA Extraction, pyrosequencing, and bioinformatic processing; Figure S1: The correlations between the relative abundance of microbial phyla and soil properties. BD, bulk density; SOC, soil organic carbon; TN, total nitrogen; TP, total phosphorus; TK, total potassium; AN, available nitrogen; AP, available phosphorus; and AK, available potassium;

Table S1: Primer sets and thermal profiles used in PCR amplification. References [75–80] are cited in the supplementary materials.

Author Contributions: W.L., Y.Y. and X.W. designed and conceived the experiment. W.L., X.B., S.Z. and Y.C. carried out the experiments and collected the empirical data. W.L., B.H. and Y.Y. performed the data analysis. W.L. wrote the original draft. T.F., Y.Y. and X.W. contributed to revising the paper. All authors have read and agreed to the published version of the manuscript.

Funding: This research was supported by the Project of Bijie Science and Technology Foundation (Bikelianhe [2023]10), the Jilin Scientific and Technological Development Program (20230101148JC), the Regional First-class Discipline of Ecology in Guizhou Province (XKTJ [2020]22), the Bijie Science and Technology Major Project (Bkhzdzx [2021]1), and the Bijie Talent Team of Biological Protection and Ecological Restoration in Liuchong River Basin (202112).

Data Availability Statement: Data will be made available on request.

Conflicts of Interest: The authors declare that they have no known competing financial interest or personal relationship that could have appeared to influence the work reported in this paper.

References

1. Ford, D.; Williams, P. *Karst Hydrogeology and Geomorphology*; John Wiley & Sons Ltd.: Chichester, UK, 2007.
2. Wang, X.; Lai, J.; Qiu, J.; Xu, W.; Wang, L.; Luo, Y. Geohazards, reflection and challenges in mountain tunnel construction of China: A data collection from 2002 to 2018. *Geomat. Nat. Hazards Risk* **2020**, *11*, 766–785. [CrossRef]
3. Li, Y.B.; Hou, J.J.; Xie, D.T. The recent development of research on karst ecology in Southwest China. *Sci. Geol. Sin.* **2002**, *22*, 365–370.
4. Jiang, Z.; Lian, Y.; Qin, X. Rocky desertification in Southwest China: Impacts, causes, and restoration. *Earth Sci. Rev.* **2014**, *132*, 1–12. [CrossRef]
5. Wang, S.J.; Liu, Q.M.; Zhang, D.F. Karst rocky desertification in southwestern China: Geomorphology, landuse, impact and rehabilitation. *Land Degrad. Dev.* **2004**, *15*, 115–121. [CrossRef]
6. Huang, Q.H.; Cai, Y.L. Spatial pattern of Karst rock desertification in the Middle of Guizhou Province, Southwestern China. *Environ. Geol.* **2007**, *52*, 1325–1330. [CrossRef]
7. Yang, Q.Q.; Wang, K.L.; Zhang, C.; Yue, Y.M.; Tian, R.C.; Fan, F.D. Spatio-temporal evolution of rocky desertification and its driving forces in karst areas of Northwestern Guangxi, China. *Environ. Earth Sci.* **2011**, *64*, 383–393. [CrossRef]
8. Singh, J.S.; Pandey, V.C.; Singh, D.P. Efficient soil microorganisms: A new dimension for sustainable agriculture and environmental development. *Agric. Ecosyst. Environ.* **2011**, *140*, 339–353. [CrossRef]
9. Nikitin, D.A.; Semenov, M.V.; Chernov, T.I.; Ksenofontova, N.A.; Zhelezova, A.D.; Ivanova, E.A.; Khitrov, N.B.; Stepanov, A.L. Microbiological indicators of soil ecological functions: A review. *Eurasian Soil Sci.* **2022**, *55*, 221–234. [CrossRef]
10. Fierer, N.; Jackson, R.B. The diversity and biogeography of soil bacterial communities. *Proc. Natl. Acad. Sci. USA* **2006**, *103*, 626–631. [CrossRef]
11. Delgado-Baquerizo, M.; Maestre, F.T.; Reich, P.B.; Trivedi, P.; Osanai, Y.; Liu, Y.R.; Hamonts, K.; Jeffries, T.C.; Singh, B.K. Carbon content and climate variability drive global soil bacterial diversity patterns. *Ecol. Monogr.* **2016**, *86*, 373–390. [CrossRef]
12. Hernández-Cáceres, D.; Stokes, A.; Angeles-Alvarez, G.; Abadie, J.; Anthelme, F.; Bounous, M.; Freschet, G.T.; Roumet, C.; Weemstra, M.; Merino-Martín, L.; et al. Vegetation creates microenvironments that influence soil microbial activity and functional diversity along an elevation gradient. *Soil Biol. Biochem.* **2022**, *165*, 108485. [CrossRef]
13. Liang, Y.; He, X.; Chen, X.; Su, Y.; Pan, F.; Hu, L. Low frequency of plants associated with symbiotic nitrogen-fixers exhibits high frequency of free-living nitrogen fixing bacteria: A study in karst shrub ecosystems of southwest China. *Forests* **2022**, *13*, 163. [CrossRef]
14. He, Q.; Xiao, Q.; Fan, J.; Zhao, H.; Cao, M.; Zhang, C.; Jiang, Y. The impact of heterotrophic bacteria on recalcitrant dissolved organic carbon formation in a typical karstic river. *Sci. Total Environ.* **2022**, *815*, 152576. [CrossRef] [PubMed]
15. Qi, D.; Wieneke, X.; Tao, J.; Zhou, X.; Desilva, U. Soil pH is the primary factor correlating with soil microbiome in karst rocky desertification regions in the Wushan County, Chongqing, China. *Front. Microbiol.* **2018**, *9*, 1027. [CrossRef] [PubMed]
16. Qi, D.; Wieneke, X.; Xue, P.; He, L.; DeSilva, U. Total nitrogen is the main soil property associated with soil fungal community in karst rocky desertification regions in southwest China. *Sci. Rep.* **2021**, *11*, 10809. [CrossRef] [PubMed]
17. Zhu, H.; He, X.; Wang, K.; Su, Y.; Wu, J. Interactions of vegetation succession, soil bio-chemical properties and microbial communities in a Karst ecosystem. *Eur. J. Soil Biol.* **2012**, *51*, 1–7. [CrossRef]
18. Singh, R.V.; Kaur, J.; Bhagwat, S.; Arora Pandit, M.; Dogra Rawat, C. Deploying microbes as drivers and indicators in ecological restoration. *Restor. Ecol.* **2023**, *31*, e13688. [CrossRef]
19. Delgado-Baquerizo, M.; Maestre, F.T.; Reich, P.B.; Jeffries, T.C.; Gaitan, J.J.; Encinar, D.; Berdugo, M.; Campbell, C.D.; Singh, B.K. Microbial diversity drives multifunctionality in terrestrial ecosystems. *Nat. Commun.* **2016**, *7*, 10541. [CrossRef]
20. LY/T 1840-2009*Technology Regulations of Vegetation Restoration in Karst Desertification Zone*; State Forestry Administration of the People's Republic of China: Beijing, China, 2011.

21. Ma, T.; Deng, X.; Chen, L.; Xiang, W. The soil properties and their effects on plant diversity in different degrees of rocky desertification. *Sci. Total Environ.* **2020**, *736*, 139667. [CrossRef]
22. Bao, S.D. *Soil Agrochemical Analysis*, 3rd ed.; China Agriculture Press: Beijing, China, 2000. (In Chinese)
23. Walkley, A.; Black, I.A. An examination of the Degtjareff method for determining soil organic matter, and a proposed modification of the chromic acid titration method. *Soil Sci.* **1934**, *37*, 29–38. [CrossRef]
24. Sommers, L.; Nelson, D. Determination of total phosphorus in soils: A rapid perchloric acid digestion procedure. *Soil Sci. Soc. Am. J.* **1972**, *36*, 902–904. [CrossRef]
25. Olsen, S.R.; Cole, C.V.; Watanabe, F.S.; Dean, L.A. *Estimation of Available Phosphorus in Soils by Extraction with Sodium Bicarbonate*; USDA Circular 939; U.S. Government Printing Office: Washington, DC, USA, 1954.
26. Keeney, D.R.; Nelson, D.W. Nitrogen inorganic forms. In *Methods of Soil Analysis*, 2nd ed; Agronomy Monograph 9 Part 2; Page, A.L., Miller, R.H., Keeney, D.R., Eds.; American Society of Agronomy: Madison, WI, USA, 1982.
27. Isaac, R.A.; Kerber, J.D. Atomic Absorption and Flame Photometry: Techniques and Uses in Soil, Plant, and Water Analysis. In *Instrumental Methods for Analysis of Soils and Plant Tissue*; Walsh, L.M., Ed.; Wiley Online Library: Hoboken, NJ, USA; The Soil Science Society of America: Madison, WI, USA, 1971.
28. Liaw, A.; Wiener, M. Classification and regression by randomForest. *R News* **2002**, *2*, 18–22.
29. Logsdon, S.D.; Karlen, D.L. Bulk density as a soil quality indicator during conversion to no-tillage. *Soil Tillage Res.* **2004**, *78*, 143–149. [CrossRef]
30. Shen, Y.; Yu, Y.; Lucas-Borja, M.E.; Chen, F.; Chen, Q.; Tang, Y. Change of soil K, N and P following forest restoration in rock outcrop rich karst area. *Catena* **2020**, *186*, 104395. [CrossRef]
31. Yang, X.; Zhang, Z.; Guo, X. Impact of soil structure and texture on occurrence of microplastics in agricultural soils of karst areas. *Sci. Total Environ.* **2023**, *902*, 166189. [CrossRef]
32. Zeng, C.; Zhang, F.; Wang, Q.; Chen, Y.; Joswiak, D.R. Impact of alpine meadow degradation on soil hydraulic properties over the Qinghai-Tibetan Plateau. *J. Hydrol.* **2013**, *478*, 148–156. [CrossRef]
33. Davari, M.; Gholami, L.; Nabiollahi, K.; Homaee, M.; Jafari, H.J. Deforestation and cultivation of sparse forest impacts on soil quality (case study: West Iran, Baneh). *Soil Tillage Res.* **2020**, *198*, 104504. [CrossRef]
34. Chenchouni, H.; Neffar, S. Soil organic carbon stock in arid and semi-arid steppe rangelands of North Africa. *Catena* **2022**, *211*, 106004. [CrossRef]
35. Peng, X.; Dai, Q.; Ding, G.; Shi, D.; Li, C. Impact of vegetation restoration on soil properties in near-surface fissures located in karst rocky desertification regions. *Soil Tillage Res.* **2020**, *200*, 104620. [CrossRef]
36. Li, S.; Wei, X.; Huang, J.; Wang, X.; Lu, G.; Li, H. The causes and processes responsible for rocky desertification in karst areas of southern China. Sci. *Cold Arid Reg.* **2009**, *1*, 80–90.
37. Colazo, J.C.; Buschiazzo, D. The impact of agriculture on soil texture due to wind erosion. *Land Degrad. Dev.* **2015**, *26*, 62–70. [CrossRef]
38. Han, R.; Zhang, Q.; Xu, Z. Responses of soil organic carbon cycle to land degradation by isotopically tracing in a typical karst area, southwest China. *PeerJ* **2023**, *11*, e15249. [CrossRef] [PubMed]
39. Zhou, Z.; Gao, T.; Van Zwieten, L.; Zhu, Q.; Yan, T.; Xue, J.; Wu, Y. Soil microbial community structure shifts induced by biochar and biochar-based fertilizer amendment to Karst calcareous soil. *Soil Sci. Soc. Am. J.* **2019**, *83*, 398–408. [CrossRef]
40. Jin, J.; Wang, L.; Müller, K.; Wu, J.; Wang, H.; Zhao, K.; Berninger, F.; Fu, W. A 10-year monitoring of soil properties dynamics and soil fertility evaluation in Chinese hickory plantation regions of southeastern China. *Sci. Rep.* **2021**, *11*, 23531. [CrossRef] [PubMed]
41. Sokol, N.W.; Slessarev, E.; Marschmann, G.L.; Nicolas, A.; Blazewicz, S.J.; Brodie, E.L.; Firestone, M.K.; Foley, M.M.; Hestrin, R.; Hungate, B.A.; et al. Life and death in the soil microbiome: How ecological processes influence biogeochemistry. *Nat. Rev. Microbiol.* **2022**, *20*, 415–430. [CrossRef]
42. Xiao, D.; He, X.; Zhang, W.; Hu, P.; Sun, M.; Wang, K. Comparison of bacterial and fungal diversity and network connectivity in karst and non-karst forests in southwest China. *Sci. Total Environ.* **2022**, *822*, 153179. [CrossRef]
43. Dai, T.; Wen, D.; Bates, C.T.; Wu, L.; Guo, X.; Liu, S.; Su, Y.; Lei, J.; Zhou, J.; Yang, Y. Nutrient supply controls the linkage between species abundance and ecological interactions in marine bacterial communities. *Nat. Commun.* **2022**, *13*, 175. [CrossRef]
44. Li, D.; Liu, J.; Chen, H.; Zheng, L.; Wang, K. Soil microbial community responses to forage grass cultivation in degraded karst soils, Southwest China. *Land Degrad. Dev.* **2018**, *29*, 4262–4270. [CrossRef]
45. Gao, C.; Xu, L.; Montoya, L.; Madera, M.; Hollingsworth, J.; Chen, L.; Purdom, E.; Singan, V.; Vogel, J.; Hutmacher, R.B.; et al. Co-occurrence networks reveal more complexity than community composition in resistance and resilience of microbial communities. *Nat. Commun.* **2022**, *13*, 3867. [CrossRef]
46. Schlatter, D.C.; Bakker, M.G.; Bradeen, J.M.; Kinkel, L.L. Plant community richness and microbial interactions structure bacterial communities in soil. *Ecology* **2015**, *96*, 134–142. [CrossRef]
47. Li, Q.; Qiu, J.; Liang, Y.; Lan, G. Soil bacterial community changes along elevation gradients in karst graben basin of Yunnan-Kweichow Plateau. *Front. Microbiol.* **2022**, *13*, 1054667. [CrossRef] [PubMed]
48. Lu, Z.X.; Wang, P.; Ou, H.B.; Wei, S.X.; Wu, L.C.; Jiang, Y.; Wang, R.J.; Liu, X.S.; Wang, Z.H.; Chen, L.J.; et al. Effects of different vegetation restoration on soil nutrients, enzyme activities, and microbial communities in degraded karst landscapes in southwest China. *For. Ecol. Manag.* **2022**, *508*, 120002. [CrossRef]

49. Janssen, P.H. Identifying the dominant soil bacterial taxa in libraries of 16S rRNA and 16S rRNA genes. *Appl. Environ. Microbiol.* **2006**, *72*, 1719–1728. [CrossRef] [PubMed]
50. Nagarajan, V.; Tsai, H.C.; Chen, J.S.; Hussain, B.; Koner, S.; Hseu, Z.Y.; Hsu, B.M. Comparison of bacterial communities and their functional profiling using 16S rRNA gene sequencing between the inherent serpentine-associated sites, hyper-accumulator, downgradient agricultural farmlands, and distal non-serpentine soils. *J. Hazard. Mater.* **2022**, *431*, 128557. [CrossRef] [PubMed]
51. Nicolas, A.M.; Sieradzki, E.T.; Pett-Ridge, J.; Banfield, J.F.; Taga, M.E.; Firestone, M.K.; Blazewicz, S.J. A subset of viruses thrives following microbial resuscitation during rewetting of a seasonally dry California grassland soil. *Nat. Commun.* **2023**, *14*, 5835. [CrossRef] [PubMed]
52. Borsodi, A.K.; Knáb, M.; Krett, G.; Makk, J.; Márialigeti, K.; Erőss, A.; Mádl-Szőnyi, J. Biofilm bacterial communities inhabiting the cave walls of the Buda Thermal Karst System, Hungary. *Geomicrobiol. J.* **2012**, *29*, 611–627. [CrossRef]
53. Kielak, A.M.; Barreto, C.C.; Kowalchuk, G.A.; Van Veen, J.A.; Kuramae, E.E. The ecology of Acidobacteria: Moving beyond genes and genomes. *Front. Microbiol.* **2016**, *7*, 744. [CrossRef] [PubMed]
54. Yang, L.; Barnard, R.; Kuzyakov, Y.; Tian, J. Bacterial communities drive the resistance of soil multifunctionality to land-use change in karst soils. *Eur. J. Soil Biol.* **2021**, *104*, 103313. [CrossRef]
55. Thajudeen, J.; Yousuf, J.; Veetil, V.P.; Varghese, S.; Singh, A.; Abdulla, M.H. Nitrogen fixing bacterial diversity in a tropical estuarine sediments. *World, J. Microbiol. Biotechnol.* **2017**, *33*, 41. [CrossRef]
56. Davis, K.E.; Sangwan, P.; Janssen, P.H. Acidobacteria, Rubrobacteridae and Chloroflexi are abundant among very slow-growing and mini-colony-forming soil bacteria. *Environ. Microbiol.* **2011**, *13*, 798–805. [CrossRef]
57. Luo, X.; Wen, D.; Hou, E.; Zhang, L.; Li, Y.; He, X. Changes in the composition of soil microbial communities and their carbon-cycle genes following the conversion of primary broadleaf forests to plantations and secondary forests. *Land Degrad. Dev.* **2022**, *33*, 974–985. [CrossRef]
58. Peng, X.; Liu, J.; Duan, X.; Yang, H.; Huang, Y. Key Soil Physicochemical Properties Regulating Microbial Community Structure under Vegetation Restoration in a Karst Region of China. *Ecosyst. Health Sust.* **2023**, *9*, 0031. [CrossRef]
59. Viruel, E.; Fontana, C.A.; Puglisi, E.; Nasca, J.A.; Banegas, N.R.; Cocconcelli, P.S. Land-use change affects the diversity and functionality of soil bacterial communities in semi-arid Chaco region, Argentina. *Appl. Soil Ecol.* **2022**, *172*, 104362. [CrossRef]
60. Wang, X.; Ren, Y.; Yu, Z.; Shen, G.; Cheng, H.; Tao, S. Effects of environmental factors on the distribution of microbial communities across soils and lake sediments in the Hoh Xil Nature Reserve of the Qinghai-Tibetan Plateau. *Sci. Total Environ.* **2022**, *838*, 156148. [CrossRef] [PubMed]
61. Rosso, L.; Lobry, J.R.; Bajard, S.; Flandrois, J.P. Convenient Model to Describe the Combined Effects of Temperature and pH on Microbial Growth. *Appl. Environ. Microbiol.* **1995**, *61*, 610–616. [CrossRef] [PubMed]
62. McCauley, A.; Jones, C.; Jacobsen, J. Soil pH and organic matter. *Nutrient Manag. Mod.* **2009**, *8*, 1–12.
63. Rousk, J.; Brookes, P.C.; Bååth, E. Fungal and bacterial growth responses to N fertilization and pH in the 150-year 'Park Grass' UK grassland experiment. *FEMS Microbiol. Ecol.* **2011**, *76*, 89–99. [CrossRef]
64. Zeng, Q.; Liu, D.; An, S. Decoupled diversity patterns in microbial geographic distributions on the arid area (the Loess Plateau). *Catena* **2021**, *196*, 104922. [CrossRef]
65. Jiang, C.; Sun, X.; Liu, Y.; Zhu, S.; Wu, K.; Li, H.; Shui, W. Karst tiankeng shapes the differential composition and structure of bacterial and fungal communities in karst land. *Environ. Sci. Pollut. Res.* **2023**, *30*, 32573–32584. [CrossRef]
66. Doran, J.W.; Parkin, T.B. Defining and assessing soil quality. In *Defining Soil Quality for a Sustainable Environment. SSSA Special Publication Number 35*; Doran, J.W., Coleman, D.C., Bezdicek, D.F., Stewart, B.A., Eds.; Soil Science Society of America: Madison, WI, USA, 1994.
67. Zheng, X.; Lin, C.; Guo, B.; Yu, J.; Ding, H.; Peng, S.; Sveen, T.R.; Zhang, Y. Effects of re-vegetation restoration on soil bacterial community structure in degraded land in subtropical China. *Eur. J. Soil Biol.* **2020**, *98*, 103184. [CrossRef]
68. Zheng, Q.; Hu, Y.; Zhang, S.; Noll, L.; Böckle, T.; Dietrich, M.; Herbold, C.W.; Eichorst, S.A.; Woebken, D.; Richter, A.; et al. Soil multifunctionality is affected by the soil environment and by microbial community composition and diversity. *Soil Biol. Biochem.* **2019**, *136*, 107521. [CrossRef] [PubMed]
69. Liang, C.; Schimel, J.P.; Jastrow, J.D. The importance of anabolism in microbial control over soil carbon storage. *Nat. Microbiol.* **2017**, *2*, 17105. [CrossRef] [PubMed]
70. Wu, X.; Liu, P.; Wegner, C.E.; Luo, Y.; Xiao, K.Q.; Cui, Z.; Zhang, F.; Liesack, W.; Peng, J. Deciphering microbial mechanisms underlying soil organic carbon storage in a wheat-maize rotation system. *Sci. Total Environ.* **2021**, *788*, 147798. [CrossRef] [PubMed]
71. van Bergeijk, D.A.; Terlouw, B.R.; Medema, M.H.; van Wezel, G.P. Ecology and genomics of Actinobacteria: New concepts for natural product discovery. *Nat. Rev. Microbiol.* **2020**, *18*, 546–558. [CrossRef]
72. Li, H.; Wang, J.; Liu, Q.; Zhou, Z.; Chen, F.; Xiang, D. Effects of consecutive monoculture of sweet potato on soil bacterial community as determined by pyrosequencing. *J. Basic Microbiol.* **2019**, *59*, 181–191. [CrossRef] [PubMed]
73. Challacombe, J.F.; Hesse, C.N.; Bramer, L.M.; McCue, L.A.; Lipton, M.; Purvine, S.; Nicora, C.; Gallegos-Graves, L.V.; Porras-Alfaro, A.; Kuske, C.R. Genomes and secretomes of Ascomycota fungi reveal diverse functions in plant biomass decomposition and pathogenesis. *BMC Genomics* **2019**, *20*, 1–27. [CrossRef]
74. Lundell, T.K.; Mäkelä, M.R.; Hildén, K. Lignin-modifying enzymes in filamentous basidiomycetes-ecological, functional and phylogenetic review. *J. Basic Microbiol.* **2010**, *50*, 5–20. [CrossRef]

75. Martin, M. Cutadapt removes adapter sequences from high-throughput sequencing reads. *EMBnet J.* **2011**, *17*, 10–12. [CrossRef]
76. Callahan, B.J.; McMurdie, P.J.; Rosen, M.J.; Han, A.W.; Johnson, A.J.A.; Holmes, S.P. DADA2: High-resolution sample inference from Illumina amplicon data. *Nat. Methods* **2016**, *13*, 581–583. [CrossRef]
77. Katoh, K.; Misawa, K.; Kuma, K.I.; Miyata, T. MAFFT: A novel method for rapid multiple sequence alignment based on fast Fourier transform. *Nucleic Acids Res.* **2002**, *30*, 3059–3066. [CrossRef]
78. Price, M.N.; Dehal, P.S.; Arkin, A.P. FastTree 2-approximately maximum-likelihood trees for large alignments. *PLoS ONE* **2010**, *5*, e9490. [CrossRef] [PubMed]
79. Bokulich, N.A.; Kaehler, B.D.; Rideout, J.R.; Dillon, M.; Bolyen, E.; Knight, R.; Huttley, G.A.; Caporaso, J.G. Optimizing taxonomic classification of marker-gene amplicon sequences with QIIME 2's q2-feature-classifier plugin. *Microbiome* **2018**, *6*, 90. [CrossRef] [PubMed]
80. McDonald, D.; Price, M.N.; Goodrich, J.; Nawrocki, E.P.; DeSantis, T.Z.; Probst, A.; Andersen, G.L.; Knight, R.; Hugenholtz, P. An improved Greengenes taxonomy with explicit ranks for ecological and evolutionary analyses of bacteria and archaea. *ISME J.* **2012**, *6*, 610–618. [CrossRef] [PubMed]

forests

MDPI

Article

Water Use Efficiency of Five Tree Species and Its Relationships with Leaf Nutrients in a Subtropical Broad-Leaf Evergreen Forest of Southern China

Kang-Xiang Huang [1], Zi-Jing Xue [1], Jian-Cheng Wu [1], Hong Wang [1], Hui-Qian Zhou [1], Ze-Bing Xiao [1], Wei Zhou [1,2,3], Jin-Feng Cai [1], Long-Wei Hu [2,3], Jiu-Sheng Ren [4], Yang Zhang [1,2], Sheng-Sheng Xiao [5] and Fu-Xi Shi [1,2,*]

1 Key Laboratory of National Forestry and Grassland Administration on Forest Ecosystem Protection and Restoration of Poyang Lake Watershed, College of Forestry, Jiangxi Agricultural University, Nanchang 330045, China; hkx2021@stu.jxau.edu.cn (K.-X.H.); xzj0202022354@stu.jxau.edu.cn (Z.-J.X.); wu2959440197@gmail.com (J.-C.W.); hongwang031217@gmail.com (H.W.); huiqianzhou24@gmail.com (H.-Q.Z.); zebingxiao@hotmail.com (Z.-B.X.); zhouwei980124@gmail.com (W.Z.); jinfengcai9@gmail.com (J.-F.C.); zhangyang0558@jxau.edu.cn (Y.Z.)
2 Matoushan Observation and Research Station of Forest Ecosystem, Zixi, Fuzhou 335300, China; lwhu999@gmail.com
3 Matoushan National Nature Reserve Administration of Jiangxi Province, Zixi, Fuzhou 335300, China
4 School of Water Resources and Environmental Engineering, East China University of Technology, Nanchang 330013, China; renjiusheng256@ecut.edu.cn
5 Jiangxi Provincial Academy of Water Resources Sciences, Nanchang 330029, China; xiaoshsh1213@hotmail.com
* Correspondence: shifuxi@jxau.edu.cn

Abstract: Water use efficiency (WUE) is key to linking the water, carbon, and nutrient cycles in terrestrial ecosystems. However, the coupling between WUE and leaf nutrients is still poorly understood in subtropical forests. Here, the stable carbon isotope technique was employed to estimate the leaf-scale WUE of five common tree species (*Castanopsis eyrei*, *Symplocos laurina*, *Machilus grijsii*, *Ternstroemia gymnanthera*, and *Rhododendron ovatum*) in different habitat types (i.e., hillside, near the top of the peak, and peak) in a subtropical broad-leaf evergreen forest on the western slope of Wuyi Mountain, southern China. In addition, leaf carbon (C), nitrogen (N), and phosphorus (P) contents were also measured to assess plant nutrient utilization and its relationship with WUE. From the hillside to the peak, soil water content showed a decreasing trend, whereas the soil total C, N, and P contents showed an increasing trend. Regardless of species, the leaf $\delta^{13}C_p$ value and WUE showed an increasing trend from the hillside to the peak, mainly due to an increase in soil water deficit and light. The leaf N and P contents showed an increasing trend from hillside to peak due to an increase in soil nutrients, while the leaf C: N ratio, C:P ratio, and N:P ratio showed a decreasing trend. The regression analysis showed that leaf-scale WUE was positively correlated with the leaf N and P contents but negatively correlated with the leaf N:P ratio, especially for the three species (*C. eyrei*, *S. laurina*, and *T. gymnanthera*). These results indicated that the differences in soil water availability, light, and soil development resulting from different habitats have a significant impact on leaf-scale WUE and nutrient status on Wuyi Mountain. Therefore, there may be a close relationship between WUE and leaf nutrients, which would help us to better understand the water-, carbon-, and nutrient-coupled relationships for the evergreen broad-leaved tree species in this region.

Keywords: subtropical forest; stable carbon isotope; C N P stoichiometry; habitat; Wuyi Mountain

1. Introduction

In recent years, global climate warming and corresponding environmental changes have significantly affected the plant community structure and function in terrestrial ecosystems [1]. Firstly, to cope with climate and environmental changes, plants will develop a series of physiological and ecological regulatory mechanisms, the critical aspect of which is to modulate plant water use efficiency (WUE) [2,3]. WUE is defined as the function between carbon assimilation rate and stomatal conductance [4]. Hence, WUE not only reflects the balance of water supply and consumption for plant growth [5] but also can indicate the plant's adaptation to environmental changes [2,6]. Secondly, as the essential nutrients for plant growth, carbon (C), nitrogen (N), and phosphorus (P) can substantially influence the physiological and ecological function of the plants [7–9]. As the most crucial structural element in plant tissues, C accounts for about 50% of plant dry matter and is a substrate and energy source for various physiological and biochemical processes in plants [10,11]. N and P are the two most important limiting elements for plant growth, and the C, N, and P contents, as well as their stoichiometric ratios (i.e., C:N ratio, C:P ratio, and N:P ratio), can effectively reflect the uptake assimilation capacity and nutrient limitation of plants [8,12,13]. There have been many studies on leaf-scale WUE [2,3,5,6,14–18] and C, N, and P nutrients [7,8,10,19,20]. However, the intrinsic correlation between the leaf-level WUE and leaf nutrients is not well studied [9,21], which limits our understanding of the coupled relationships between carbon–water cycling and nutrient cycling in terrestrial ecosystems.

In C_3 plants, leaf-scale 'intrinsic' WUE can be identified by the stable carbon isotope ratio in a leaf ($\delta^{13}C_p$), due to the fact that it is primarily related to the ratio of the intercellular (C_i) to the atmospheric (C_a) partial pressure of CO_2 (C_i/C_a; [4,22,23]). As a sensitive indicator in response to climate and environmental changes, leaf-scale WUE is generally influenced by external environmental conditions (such as water availability and light) and internal leaf nutrients. For example, Liu et al. (2021) found that the leaf $\delta^{13}C_p$ and WUE values of desert plants in northern China decreased with increasing annual precipitation, which was mainly because the increase in rainfall enhanced plant water availability and reduced light radiation [15]. Moreover, a soil water deficit can lead to stomatal closure and reduced stomatal conductance, resulting in a lower ratio of C_i/C_a and a higher leaf $\delta^{13}C_p$ and WUE [4,22–24]. Moreover, light differences can affect the leaf $\delta^{13}C_p$ and WUE by influencing chlorophyll content, leaf phototropism, carboxylase activity, and other biochemical processes that are related to photosynthesis in plants [11]. In addition to environmental factors, the plant's nutrient status also can influence leaf-scale WUE by modulating plant photosynthesis. For instance, the leaf P content tends to be positively correlated with the photosynthetic rate [25]. This is because, under a shortage of P, the influence of chloroplasts on the photosynthetic rate is greater than that of stomatal conductance [26]. Likewise, the increase in leaf N content could enhance photosynthesis mainly by promoting the activity of enzymes [27]. At the leaf level, WUE was positively correlated with leaf P content, but it was not significantly associated with leaf N content in a P-limited subtropical forest [9,28]. However, the WUE of broad-leaf trees was positively correlated both with the leaf N and P contents in a temperate forest [11]. In rubber forests, WUE was positively correlated with the leaf N and P content but negatively correlated with the leaf N:P ratio [29].

The subtropical evergreen broad-leaved forest is one of the most important vegetation types in the Wuyi Mountains of southern China [30]. Changes in intrinsic WUE have been reported for different life types (i.e., trees, shrubs, and herbs) along the altitudinal gradient of Wuyi Mountain [21]. However, to date, a systematic study on the leaf-scale WUE changes and their coupling relationship with leaf nutrients in this region under different habitat conditions is still unclear. In this study, we selected five common trees (i.e., *Castanopsis eyrei*, *Symplocos laurina*, *Machilus grijsii*, *Ternstroemia gymnanthera*, and *Rhododendron ovatum*) as the target plants under different habitats (i.e., hillside, near the top of the peak, and peak) in a subtropical broad-leaf evergreen forest on the western slope of Wuyi Mountain, southern China. By measuring the leaf $\delta^{13}C_p$ and the C, N, and P contents for these five tree species, our main aims are (1) to determine the variation patterns of leaf-scale WUE from hillside to peak; (2) to analyze the differences in leaf C, N, P contents and their stoichiometric ratios under different habitat conditions; and (3) to explore the relationship between intrinsic WUE and leaf nutrients for the five tree species. These results would help us to better understand the coupled relationships between WUE and nutrients for the evergreen broad-leaved tree species in this region.

2. Materials and Methods

2.1. Study Area, Experimental Design, and Sample Collection

We conducted the study in a subtropical evergreen broad-leaved forest (27°40′50″ N, 117°09′11″ E), which is located within Jiangxi Matou Mountain National Nature Reserve in Zixi County, Jiangxi Province, southern China (Figure 1). The study area lies on the western slope of Wuyi Mountain, where it preserves a larger area of native evergreen broad-leaved forests with a high forest coverage of 96.3%, making it an important gene pool of biological resources in the world. It has a typical subtropical humid monsoon climate with an annual average temperature of 17 °C, an annual average precipitation of 1930 mm, and a relative humidity of 83%. In this forest, the dominant species are mainly composed of evergreen and deciduous trees such as *C. eyrei*, *Lithocarpus litseifolius*, *Schima superba*, *S. laurina*, *M. grijsii*, *T. gymnanthera*, *Alniphyllum fortune.*, etc. The soil types belong to red soil and red–yellow soil according to the Genetic Soil Classification of China [31].

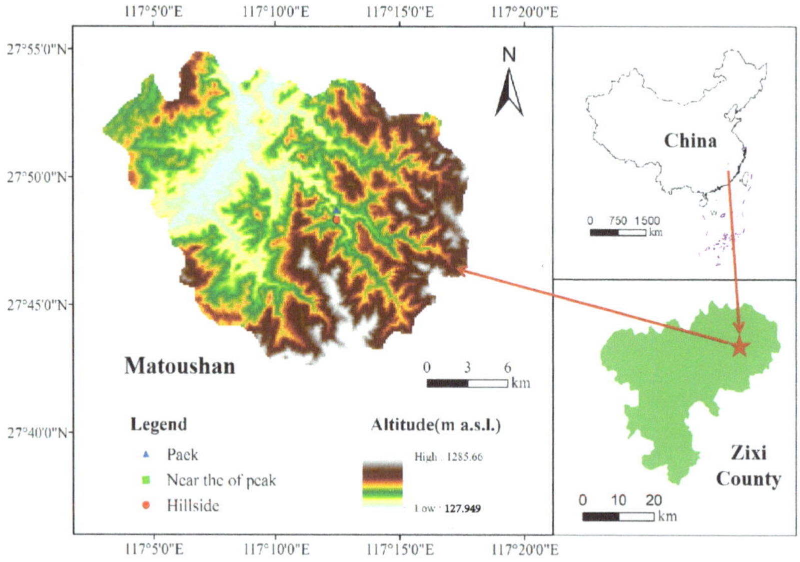

Figure 1. Location of the study area and sampling points.

In October 2020, three sample plots with different habitat types (i.e., hillside in the northeast slope (H–NE), near the top of the peak in the southwest slope (NP–SW), and peak in the southwest slope (P–SW)) were set up in this subtropical evergreen broad-leaved forest (Figure 1 and Table 1). Five common dominant tree species (i.e., *C. eyrei*, *S. laurina*, *M. grijsii*, *T. gymnanthera*, and *R. ovatum*) were selected as target plants in each sample plot. In each plot, four quadrats with a size of 20 m × 20 m were randomly set up to investigate the five target trees. For each target species, five plants with similar heights and diameters at breast height (DBH) in each quadrat were selected to collect 10–15 mature leaves. The collected leaf samples were washed with pure water to remove surface impurities and dust, brought back to the laboratory, and oven-dried at 80 °C for 48 h. Each dried sample was then crushed using an onyx grinder, sieved through a 100-mesh sieve, and used as a sample to be tested. Meanwhile, soil samples from 0 to 30 cm were collected with a soil core (5 cm in diameter) after removing the forest floor in each sample plot according to a 10 m × 10 m grid to determine the initial soil moisture and nutrient contents.

Table 1. Basic situation of plant growth environments of three sample plots in a subtropical broad-leaf evergreen forest on the western slope of Wuyi Mountain in southern China.

Habitat Type	Aspect	Slope Gradient (°)	Altitude (m)	Soil Water Content (%)	Soil Total C Content (mg g^{-1})	Soil Total N Content (mg g^{-1})	Soil Total P Content (mg g^{-1})
Hillside	NE	30	275	28.57 ± 0.51 ^{a}	41.02 ± 1.81 ^{c}	1.08 ± 0.07 ^{b}	0.16 ± 0.01 ^{c}
Near the top of the peak	SW	38	356	17.89 ± 0.79 ^{b}	62.17 ± 8.57 ^{b}	1.70 ± 0.24 ^{a}	0.35 ± 0.03 ^{b}
Peak	SW	40	370	19.96 ± 1.63 ^{b}	87.67 ± 12.83 ^{a}	1.81 ± 0.19 ^{a}	0.43 ± 0.01 ^{a}

Notes: NE: northeast slope; SW: southwest slope. Data are mean value ± standard error. The different lowercase letters in the same column indicate a significant difference in soil properties among stands at $p < 0.05$.

2.2. Sample Chemical Analysis

The soil water content was measured using the oven-drying method. Air-dried soil samples were milled to pass through a 0.15 mm sieve to analyze the soil nutrient contents. In this experiment, N and P contents were measured colorimetrically on an Autoanalyzer (AA3, Seal Analytical, Norderstedt, Germany) after acid digestion with 98% H_2SO_4 and 30% H_2O_2. Organic C content was measured with an elemental analyzer (Flash 2000 HT, Thermo Fisher Scientific, Bremen, Germany) [31]. In addition, 0.04~0.05 mg of dried leaf sample was weighed and placed in pre-burned tin capsules (3.5 × 9 mm) for stable carbon isotope measurement. The leaf $\delta^{13}C$ was determined using an isotope ratio mass spectrometer (IRMS) (Delta V Advantage, Thermo Fisher Scientific, Bremen, Germany) coupled with an elemental analyzer (Flash EA 2000 HT, Thermo Fisher Scientific, Bremen, Germany) at the IRMS laboratory of Jiangxi Agricultural University. The cellulose standard (IAEA-CH3; $\delta^{13}C = -24.724‰$), L-glutamic acid (USGS40; $\delta^{13}C = -26.39‰$), and urea ($CO(NH_2)_2$; $\delta^{13}C = -41.30‰$) were used to calibrate the $\delta^{13}C$ measurements, respectively. The standard deviation for the repeated analysis of an internal standard was <0.03‰.

2.3. Calculation of Water Use Efficiency (WUE)

The leaf-scale WUE is defined as the ratio of the rate of carbon assimilation (A, photosynthesis) and stomatal conductance (g_s) and can be estimated from the stable carbon isotope discrimination value in leaves ($\Delta^{13}C_p$) using the following equation [4,18,22,23]:

$$\mathrm{WUE} = A/g_s = (C_a - C_i)/1.6 = C_a\,(b - \Delta^{13}C_p)/[1.6\,(b - a)] \quad (1)$$

where $\Delta^{13}C_p$ is the stable carbon isotope discrimination value (Equation (2)), C_i is the intercellular CO_2 concentration, C_a is the atmospheric CO_2 concentration obtained from Equation (3), a (4.4‰) is the fractionation during the diffusion of CO_2, b (27‰) is the biochemical fractionation, and 1.6 is the ratio of the diffusivities of water vapor and carbon dioxide [22,23].

The stable carbon isotope discrimination value ($\Delta^{13}C_p$; Equation (2)) [4,22,23], atmospheric CO_2 concentration (C_a; Equation (3)) [5,17], and stable carbon isotopic ratio ($\delta^{13}C_{atm}$; Equation (4)) [5,17] were calculated from the following equations, respectively:

$$\Delta^{13}C_p = a + (b - a)\,C_i/C_a = (\delta^{13}C_{atm} - \delta^{13}C_p)/[(1 + \delta^{13}C_p)/1000] \quad (2)$$

$$C_a = 277.78 + 1.350\exp\,(0.01572\,(t - 1740)) \quad (3)$$

$$\delta^{13}C_{atm} = -6.429 - 0.006\exp\,(0.0217\,(t - 1740)) \quad (4)$$

In the formula, $\delta^{13}C_p$ is the stable carbon isotopic ratio of the leaves, and t is the sampling year. In this study, t = 2020, and substituting this into Equations (3) and (4), the C_a in 2020 was calculated to be 389.66 μmol mol^{-1}, and $\delta^{13}C_{atm}$ was −6.93‰.

2.4. Statistical Analyses

The data were statistically analyzed using R-studio 4.1 software. A one-way analysis of variance (ANOVA) was employed to compare the differences in the leaf $\delta^{13}C_p$ values, WUE, and nutrient contents of the same tree species under different habitat conditions. Additionally, the differences in soil water content, soil organic C content, soil total N content, and soil total P content were examined across different sampling plots. The significance level was set at 0.05. Linear fits were performed through a regression analysis of leaf-scale WUE to leaf N and P concentrations and the N:P ratio for each species, with 95% confidence intervals.

3. Results

3.1. Environmental Conditions under Different Habitat Types

There were significant differences in the environmental conditions under the three different habitat types (Table 1). Specifically, there was a smaller slope gradient and a lower altitude on the hillside on the northeast slope (H–NE), while there was the largest slope gradient and the highest altitude at the peak on the southwest slope (P–SW). For the soil properties, the soil water content decreased with an increasing altitudinal gradient from the hillside to the peak. However, the soil total C content, soil total N content, and soil total P content increased with an increasing altitudinal gradient from the hillside to the peak.

3.2. Leaf $\delta^{13}C_p$ and WUE under Different Habitat Conditions

There were significant differences in the leaf $\delta^{13}C_p$ and WUE for the five tree species under the different habitat conditions (Figure 2). From the hillside to the peak, a positive trend was observed in the leaf $\delta^{13}C_p$ values of the five tree species. Consequently, the WUE values exhibited a corresponding increasing trend. Regardless of species, the relatively more negative leaf $\delta^{13}C_p$ values are mainly found at the hillside, but the relatively positive leaf $\delta^{13}C_p$ values are generated at the peak (Figure 2a). Thus, there was lower WUE at the hillside but higher WUE at the peak (Figure 2b).

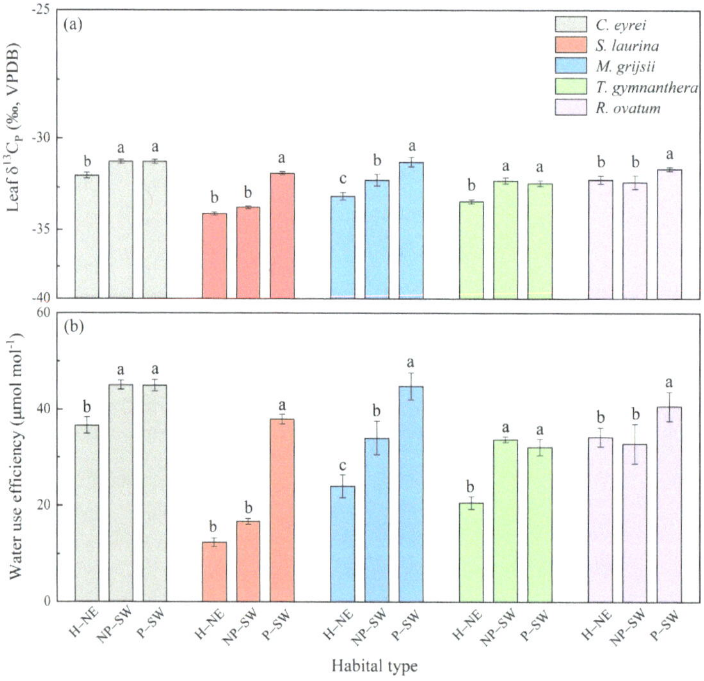

Figure 2. Leaf $\delta^{13}C_p$ (**a**) and WUE (**b**) of five tree species under different habitat conditions in a subtropical broadleaf evergreen forest on the western slope of Wuyi Mountain in southern China. Notes: H–NE: hillside on the northeast slope; NP–SW: near the top of the peak on the southwest figureslope; P–SW: the peak on the southwest slope. The different lowercase letters (a, b, c) mean that there is a significant difference among the treatments (H–NE, NP–SW, and P–SW) at $p < 0.05$.

3.3. Leaf Nutrients and Their Stoichiometric Ratios under Different Habitat Conditions

There were also significant differences in the leaf N and P contents and stoichiometric ratios under the different habitat conditions, but no significant differences in the leaf C contents (Figures 3 and 4). From the hillside to the peak, the leaf N and P contents of the five tree species showed an increasing trend (Figure 3), but the leaf C:N ratio, C:P ratio, and N:P ratio exhibited a decreasing trend (Figure 4). Specifically, the leaf N content ranged from 6.95 mg g^{-1} to 16.29 mg g^{-1}, the leaf P content ranged from 0.69 mg g^{-1} to 1.51 mg g^{-1}, the leaf C:N ratio ranged from 26.07 to 72.37, the leaf C:P ratio ranged from 290.56 to 744.28, and the leaf N:P ratio ranged from 8.77 to 14.98. Overall, the leaf N and P contents at the peak were higher than those at the hillside (Figure 3b,c), while the leaf C:N, C:P, and N:P ratios at the peak were lower than those at the hillside (Figure 4).

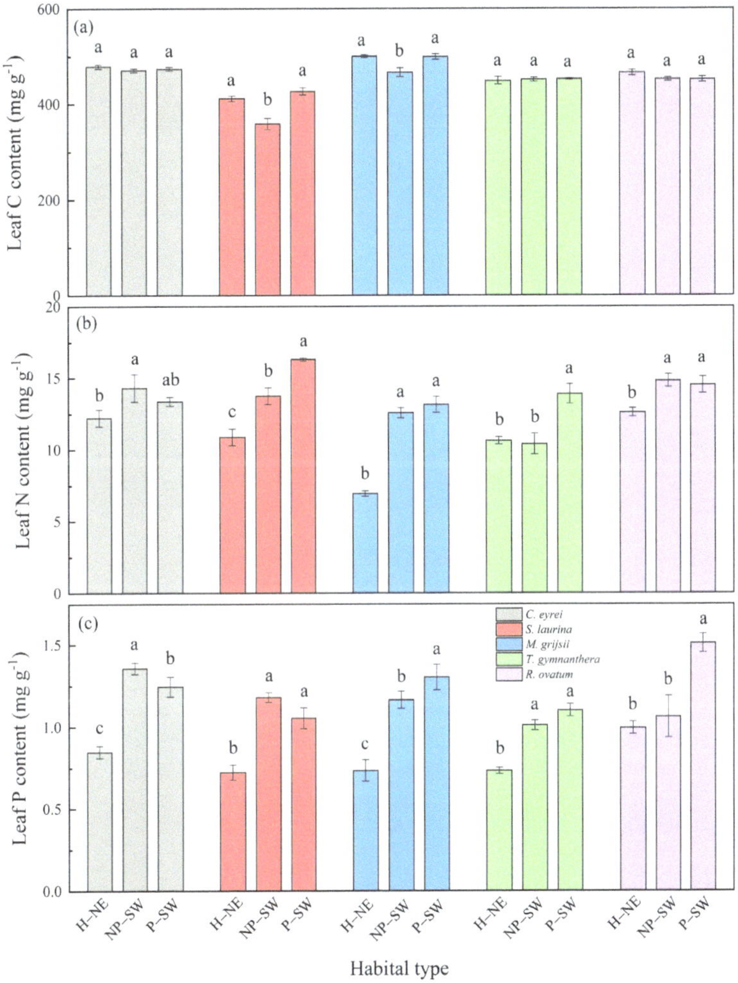

Figure 3. Leaf C content (**a**), N content (**b**), and P content (**c**) of five tree species under different habitat conditions in a subtropical broad-leaf evergreen forest on the western slope of Wuyi Mountain in southern China. Notes: H–NE: hillside on the northeast slope; NP–SW: near the top of the peak on the southwest slope; P–SW: the peak on the southwest slope. The different lowercase letters (a, b, c) mean that there is a significant difference among the treatments (H–NE, NP–SW, and P–SW) at $p < 0.05$.

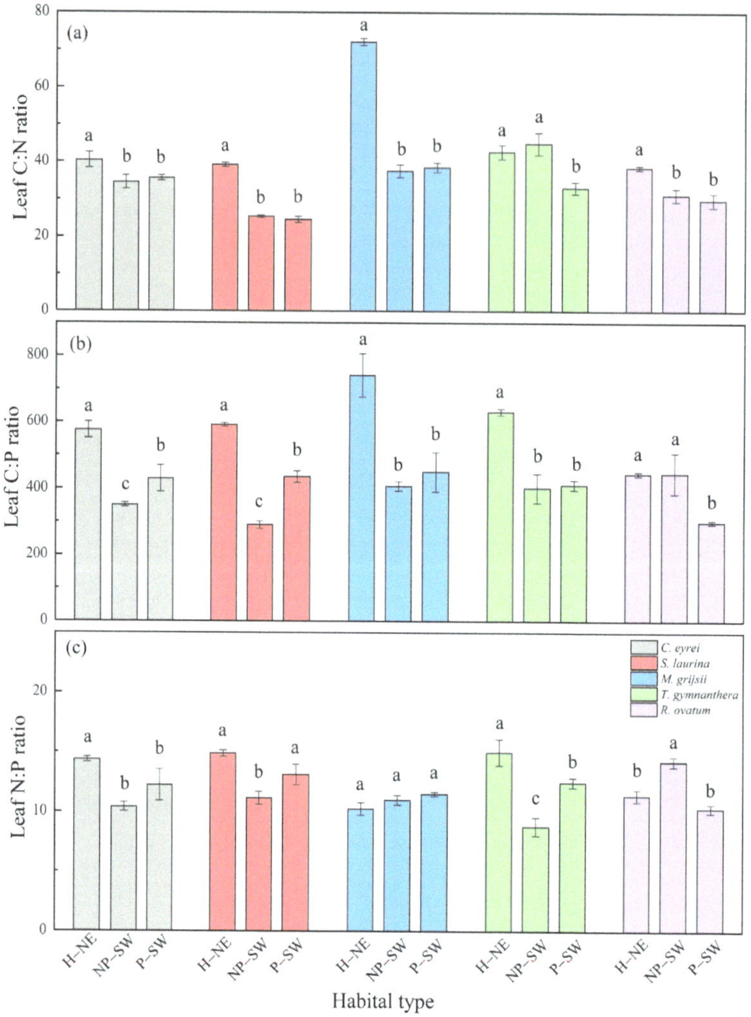

Figure 4. Leaf C:N ratio (**a**), C:P ratio (**b**), and N:P ratio (**c**) of five tree species under different habitat conditions in a subtropical broadleaf evergreen forest on the western slope of Wuyi Mountain in southern China. Notes: H–NE: hillside on the northeast slope; NP–SW: near the top of the peak on the southwest slope; P–SW: the peak on the southwest slope. The different lowercase letters (a, b, c) mean that there is a significant difference among the treatments (H–NE, NP–SW, and P–SW) at $p < 0.05$.

3.4. Relationships between Leaf Nutrients and WUE

The leaf nutrients (i.e., N, P, and N:P ratio) both had a significant relationship with WUE for three of the tree species, namely *C. eyrei*, *S. laurina*, and *T. gymnanthera*, while only either N or P had a significant relationship with WUE for the remaining two tree species, *M. grijsii* and *R. ovatum* (Figure 5). Specifically, the leaf N content was significantly positively correlated with WUE for *C. eyrei* ($r = 0.494$; Slope = 1.13; $p < 0.05$; Figure 5a), *S. laurina* ($r = 0.442$; Slope = 1.47; $p < 0.05$; Figure 5d), *M. grijsii* ($r = 0.691$; Slope = 2.33; $p < 0.01$; Figure 5g), and *T. gymnanthera* ($r = 0.728$; Slope = 2.66; $p < 0.01$; Figure 5j), but not for *R. ovatum* ($p > 0.05$; Figure 5m). The leaf P content was also significantly positively correlated with WUE for *C. eyrei* ($r = 0.759$; Slope = 13.82; $p < 0.01$; Figure 5b), *S. laurina* ($r = 0.899$;

Slope = 32.77; $p < 0.01$; Figure 5e), *T. gymnanthera* (r = 0.558; Slope = 18.89; $p < 0.05$; Figure 5k), and *R. ovatum* (r = 0.474; Slope = 7.98; $p < 0.05$; Figure 5n), but not for *M. grijsii* ($p > 0.05$; Figure 5h). However, the leaf N:P ratio was significantly negatively correlated with WUE for *C. eyrei* (r = −0.483; Slope = −0.83; $p < 0.05$; Figure 5c), *S. laurina* (r = −0.646; Slope = −1.85; $p < 0.01$; Figure 5f), and *T. gymnanthera* (r = −0.492; Slope = −4.09; $p < 0.05$; Figure 5l), but not for *M. grijsii* ($p > 0.05$; Figure 5i) and *R. ovatum* ($p > 0.05$; Figure 5o).

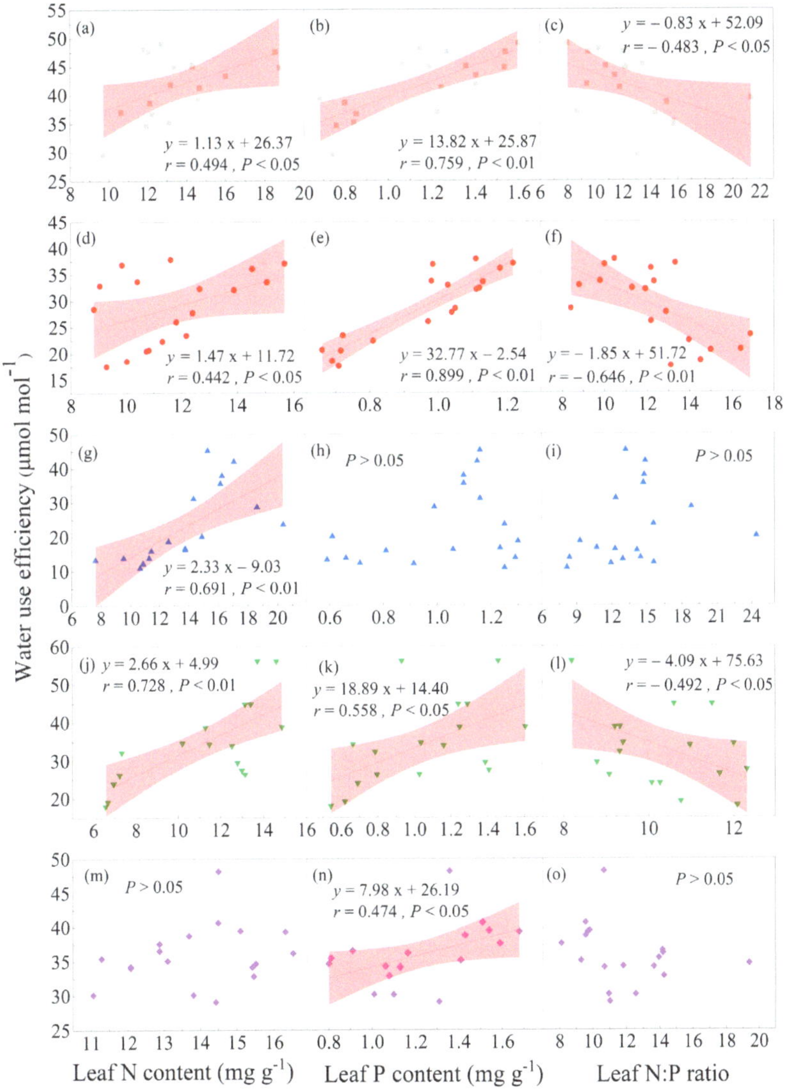

Figure 5. The relationship between WUE and the leaf N content, leaf P content, and leaf N:P of the five tree species in a subtropical broad-leaf evergreen forest on the western slope of Wuyi Mountain in southern China. Notes: the gray square represents *C. eyrei* (**a–c**), the red circle represents *S. laurina* (**d–f**), the blue triangle up represents *M. grijsii* (**g–i**), the green triangle down represents *T. gymnanthera* (**j–l**), and the pink diamond represents *R. ovatum* (**m–o**). The linear fits were used to test the coupling between WUE and leaf nutrients with the significance level set at 0.05. The shadow indicates 95% confidence intervals.

4. Discussion

4.1. Effect of Habitat Types on Leaf-Scale WUE

At the local scale, water availability and light conditions are considered the two most important environmental factors to influence the leaf-scale WUE, mainly by controlling the stomatal conductance or photosynthetic capacity [4,11,17]. In this subtropical forest, the $\delta^{13}C_p$ values of five tree species showed an increasing trend with decreasing soil moisture from the hillside to the peak (Table 1 and Figure 2), suggesting that soil water content has a more significant influence on leaf-scale WUE. Similarly, Huang et al. (2019) found that the leaf $\delta^{13}C_p$ values of trees gradually increased from the depression to the hilltop due to decreased soil moisture in a karst seasonal rainforest [17]. Zhang et al. (2017) also found that whether the instantaneous WUE derived from the gas exchange or the intrinsic WUE calculated from the carbon isotope model for the *Platycladus orientalis* seedling, both reached their maximum value at the lowest soil water content [16]. Thus, these two results are consistent with our study, implying that leaf-scale WUE is highly sensitive to soil water availability. The regulation mechanisms of soil water availability for the leaf-scale WUE may be the following: In habitats with limited soil moisture, plant defense cells may reduce intracellular solutes through metabolic processes as a response to water scarcity in the root systems. This can lead to an increase in water potential and subsequent water loss within cells [4]. In this case, part of the stomata will be closed due to dehydration, and this physiological strategy can not only reduce plant transpiration but also decrease the ratio of C_i/C_a, thereby resulting in a more positive value of $\delta^{13}C$ and higher WUE [22,23]. In contrast, stomatal conductance functions at its full capacity, resulting in a more negative value of $\delta^{13}C$ and lower WUE in a moist habitat [6].

In addition to water availability, light is another important environmental factor that can affect leaf-scale WUE [11,17]. This is mainly because photosynthetic processes such as chlorophyll distribution, leaf phototropism, and carboxylase activity are all closely related to light [32]. Considering that the insolation index at the peak may be higher than that of the hillside, more light will lead to an increase in CO_2 assimilated by photosynthesis, which may decrease stomatal conductance and result in a higher leaf $\delta^{13}C_p$ value and WUE for the high-altitude species than for the low-altitude species (Figure 2). Huang et al. (2019) also found that the leaf $\delta^{13}C_p$ value and WUE of *Sterculia monosperma* gradually increased from the depression to the peak with an increase in the insolation index [17]. Similarly, the $\delta^{13}C_p$ and WUE values were higher in leaves at the top of the canopy than in leaves at the lower canopy due to differences in light intensity in the forest canopy structure [11,32]. Therefore, this evidence suggested that habitat-induced differences in soil water availability and light intensity can significantly affect leaf-scale WUE values.

4.2. Effect of Habitat Types on Leaf Nutrients and Their Stoichiometric Ratios

Leaf C, N, and P contents could reflect the nutrient uptake status of the plants, which is mainly influenced by soil nutrients, temperature, precipitation, mycorrhizal, and succession time [12,19,20,33]. In this study, we found that the leaf N and P contents showed an increasing trend with an altitudinal gradient from the hillside to the peak for the five species, despite no significant difference in the leaf C content (Figure 3b,c). In general, as the most crucial element in the dry matter of the plant, C is primarily derived from CO_2 assimilation during photosynthesis; thus, it is not a limiting element for plant growth [7,20]. On the one hand, plants' N sources are influenced by the weathering of parent rocks and the decomposition of organic matter in the soil. On the other hand, plants can also obtain N sources from the atmosphere through biological and high-energy N-fixation [34]. In this region, the succession time of the subtropical broad-leaved evergreen forests dominated by *Castanopsis eyrei* at the peak was often longer than that at the hillside, which led to the soil organic matter and soil total N contents being greater at the peak than at the hillside (Table 1). Thus, we found that the leaf N content was increased with increasing altitude from the hillside to the peak (Figure 3b). However, a plant's P content is mainly derived from a parent rock's weathering; hence, the soil total P content is lower if the weathering time of the parent rock is longer [13]. In this case, the plant will suffer from a stronger P limitation, particularly in subtropical forests [35]. We found that the leaf P content showed an increasing trend from the hillside to the peak (Figure 3c). This is mainly because the chemical weathering of the parent rock decreases with increasing altitude on Wuyi Mountain [21]. This leads to increased soil P content in the high-altitude habitats, resulting in a higher leaf P content at the peak.

The leaf C:N and C:P ratios reflect the plant's ability to assimilate C as it absorbs N and P; hence, they are often used to indicate a plant's nutrient use efficiency [7,11,20]. In this subtropical forest, we found that the leaf C:N and C:P ratios decreased with an increase in altitudinal gradient, suggesting that the hillside has a higher nutrient utilization efficiency than the peak due to the relatively lower leaf N and P contents in the low-altitude habitat (Figure 4a,b). In addition, the leaf N:P ratio can be used to assess the nutrient limitation status of plants to some extent [13]. At the community level, Townsend et al. (2007) suggested that N limitation likely occurs at an N:P ratio < 14, with P limitation probable at an N:P ratio > 16, and maybe both N- and P limitations when 14 < N:P < 16 [36]. In this region, we found that the leaf N:P ratio of the five tree species was chiefly less than 14 at the peak (Figure 4c), suggesting that plant growth was generally limited by N in high-altitude habitats. Interestingly, the leaf N:P ratio of most of the tree species on the hillside ranged between 14 and 16 (Figure 4c), suggesting that the growth of plants in this habitat is likely co-limited both by N and P. However, the subtropical forest was often considered a P-limited ecosystem [35,36]. This is not quite consistent with our results, and the main reason for this may be related to the degree of soil development. As mentioned earlier, on Wuyi Mountain, there is a weaker weathering intensity of the parent rock on the high-altitude mountain than that of the low-altitude mountain [21]. Therefore, at the mountain peak, the soil has a higher P content, which results in plants experiencing N limitation (Table 1). On the contrary, there is a more serious loss in P in mineral soils on the hillside due to the stronger soil development degree; hence, plant growth in this habitat is often subjected to the co-limitation of N and P. These results further implied that the difference in soil development caused by habitats will significantly affect the leaf nutrients and their C N P stoichiometric ratios in the subtropical forest ecosystem.

4.3. The Coupling Relationships between WUE and Leaf Nutrients for the Five Tree Species

Theoretically, there is a close coupling relationship between intrinsic WUE and leaf traits in terrestrial ecosystems [14]. Leaf nutrients (e.g., N and P) influence intrinsic WUE mainly by modulating photosynthetic and transpiration rates [9,11,21,29]. N is an essential component of chlorophyll, proteins, and enzymes in plants, which can significantly impact plant photosynthesis [37]. It has been shown that an increased N content can promote chlorophyll and protein synthesis and increase leaf area, thereby increasing plant photosynthesis [14,27]. Moreover, increased N availability can also enhance the nutrient content of non-light organs, which will further increase cellular osmotic pressure and reduce water loss in leaves, thereby increasing WUE [38]. Similarly, the effects of P on leaf-scale WUE are mainly due to its influence on the Rubisco carboxylase, chlorophyll content, and corresponding photosynthetic rate [25]. Generally, P in mineral soils can be easily transported to the surface of the fine roots to increase the P concentration in the leaves under transpiration stress, thus causing a positive correlation between leaf P content and transpiration rate [27]. A recent study has proved that an increase in foliar P concentration could indirectly enhance the leaf-scale WUE in a subtropical forest [9]. The results of the above studies both support a positive correlation between WUE and both leaf N and P content, which is consistent with our research (Figure 5). In addition, the leaf N:P ratio was also closely correlated with intrinsic WUE [8,28]. It was found that the photosynthesis rate could reach its optimum value when the leaf N:P ratio was 12. Thus, the leaf N:P ratio was positively correlated with the photosynthesis rate at an N:P ratio < 12; however, the leaf N:P ratio was negatively correlated with the photosynthetic rate at an N:P ratio > 12 [39]. In our study, for most of the species, the leaf N:P ratio was greater than 12; hence, intrinsic

WUE was negatively correlated with the leaf N:P ratio (Figure 5). However, it is worth noting that the WUE of three of the tree species (i.e., *C. eyrei*, *S. laurina*, and *T. gymnanthera*) showed significant relationships both with the leaf N content, leaf P content, and leaf N:P ratio, while the WUE of the other two tree species (i.e., *M. grijsii* and *R. ovatum*) showed a significant positive correlation with only either the leaf N content or the leaf P content (Figure 5). We found that the increased rates of P for the three tree species were faster than those of N (i.e., the slope values of P were higher than those of N) with increased WUE from the hillside to the peak, meaning that P may be more important for modulating WUE for these tree species in the subtropic forest, which has been well-demonstrated in a recent study [9]. Thus, WUE was negatively correlated with the leaf N:P ratio for the three tree species (Figure 5). On the contrary, the increased rates of N for the other two tree species (i.e., *M. grijsii* and *R. ovatum*) were faster or the same as that of P, which would result in no significant correlations between WUE and the leaf N:P ratio for these two tree species. Thus, we suppose that this asynchrony among plants may be caused by differences in the rates of N or P uptake of species, which still requires further research in the future. Overall, our results suggested that N and P and their stoichiometry balance could co-modulate the water utilization for the five evergreen broad-leaved tree species in this region.

5. Conclusions

In this study, we investigated the leaf-scale WUE and leaf nutrients of five tree species (*C. eyrei*, *S. laurina*, *M. grijsii*, *T. gymnanthera*, and *R. ovatum*) in three different habitat types (i.e., hillside, near the top of the peak, and peak) on the western slope of the Wuyi Mountain in southern China. From the hillside to the peak, the leaf-scale WUE of the tree species increased gradually with the decrease in soil moisture and the increase in light. The leaf N and P contents showed an increasing trend due to an increase in soil nutrients, whereas the leaf C:N, C:P, and N:P ratios showed a decreasing trend. More importantly, we found that leaf-scale WUE was positively correlated with the leaf N and P contents but was negatively correlated with the leaf N:P ratio, especially for three of the five species (i.e., *C. eyrei*, *S. laurina*, and *T. gymnanthera*). These results indicated that changes in leaf-scale WUE and nutrients of evergreen broad-leaved tree species under different habitat types are mainly driven by differences in soil water availability, light, and soil development sequence, despite there being interspecific differences. It was also emphasized that leaf nutrients can be used as potential proxies to identify the variations in leaf-scale WUE for specific species. In summary, our findings are valuable for understanding the water-, carbon-, and nutrient-coupled relationships for the evergreen broad-leaved tree species, which are important in the face of global changes.

Author Contributions: K.-X.H.: investigation, writing—original draft; Z.-J.X., J.-C.W., H.W., H.-Q.Z., Z.-B.X., W.Z., J.-F.C. and L.-W.H.: investigation and data curation; J.-S.R., S.-S.X. and Y.Z.: methodology and reviewing; F.-X.S.: conceptualization, funding, writing—review and editing. All authors have read and agreed to the published version of the manuscript.

Funding: This study was supported by the National Natural Science Foundation of China (Nos. 42067049 and 32160357), the Double Thousand Plan of Jiangxi Province (jxsq2023102213), the Project for Forestry Science and Technology Extension Demonstration of Central Government-funded (JXTG(2023)15), the Jiangxi Provincial Natural Science Foundation (20212BAB203027), the Jiangxi Provincial Key Research and Development Program (20202BBGL73097), the Research Project of the Jiangxi Provincial Department of Forestry (CXZX(2022)38), and the Open Research Fund of Jiangxi Province Institute of Water Sciences (Grant Nos. 2022SKTR03 and 2022SKTR05).

Data Availability Statement: The data used to support the findings of this study are available from the corresponding author upon request.

Conflicts of Interest: The authors declare no conflict of interest.

References

1. Huang, M.T.; Piao, S.L.; Sun, Y.; Ciais, P.; Cheng, L.; Mao, J.F.; Poulter, B.; Shi, X.Y.; Zeng, Z.Z.; Wang, Y.P. Change in terrestrial ecosystem water-use efficiency over the last three decades. *Glob. Chang. Biol.* **2015**, *21*, 2366–2378. [CrossRef] [PubMed]
2. Driscoll, A.W.; Bitter, N.Q.; Sandquist, D.R.; Ehleringer, J.R. Multidecadal records of intrinsic water-use efficiency in the desert shrub *Encelia farinose* reveal strong responses to climate change. *Proc. Natl. Acad. Sci. USA* **2020**, *117*, 18161–18168. [CrossRef]
3. Wang, X.; Chen, G.; Wu, M.Q.; Li, X.Z.; Wu, Q.; Wang, P.; Zeng, H.; Yang, R.; Tang, X.L. Differences in the patterns and mechanisms of leaf and ecosystem-scale water use efficiencies on the Qinghai-Tibet Plateau. *Catena* **2023**, *222*, 106874. [CrossRef]
4. Ma, W.T.; Tcherkez, G.; Wang, X.M.; Schäufele, R.; Schnyder, H.; Yang, Y.; Gong, X.Y. Accounting for mesophyll conductance substantially improves ^{13}C-based estimates of intrinsic water-use efficiency. *New. Phytol.* **2021**, *229*, 1326–1338. [CrossRef]
5. Feng, X.H. Long-term c_i/c_a response of trees in western North America to atmospheric CO_2 concentration derived from carbon isotope chronologies. *Oecologia* **1998**, *117*, 19–25. [CrossRef]
6. Ale, R.; Zhang, L.; Li, X.; Raskoti, B.B.; Pugnaire, F.I.; Luo, T.X. Leaf $\delta^{13}C$ as an indicator of water availability along elevation gradients in the dry Himalayas. *Ecol. Indic.* **2018**, *94*, 266–273. [CrossRef]
7. ÅGREN, G.I. The C: N: P stoichiometry of autotrophs–theory and observations. *Ecol. Lett.* **2004**, *7*, 185–191. [CrossRef]
8. Yan, Z.B.; Kim, N.; Han, W.X.; Guo, Y.L.; Han, T.S.; Du, E.Z.; Fang, J.Y. Effects of nitrogen and phosphorus supply on growth rate, leaf stoichiometry, and nutrient resorption of *Arabidopsis thaliana*. *Plant Soil.* **2015**, *388*, 147–155. [CrossRef]
9. Hang, Z.Q.; Ran, S.S.; Fu, Y.R.; Wan, X.H.; Song, X.; Chen, Y.X.; Yu, Z.P. Functionally dissimilar neighbours increase tree water use efficiency through enhancement of leaf phosphorus concentration. *J. Ecol.* **2022**, *110*, 2179–2189. [CrossRef]
10. Hessen, D.O.; Ågrer, G.I.; Anerson, T.R.; Elser, J.J.; Ruiter, P.C. Carbon sequestration in ecosystems: The role of stoichiometry. *Ecology* **2004**, *85*, 1179–1192. [CrossRef]
11. Tian, J.Y.; Yuan, F.H.; Guan, D.X.; Wu, J.B.; Wang, A.Z. Water use efficiency and leaf nutrient characteristics of five major tree species in broadleaved Korean pine forest in Changbai Mountains, China. *Chin. J. Appl. Ecol.* **2022**, *33*, 304–310. (In Chinese with English Abstract)
12. Zheng, Z.M.; Lu, J.; Su, Y.Q.; Yang, Q.S.; Lin, Y.H.; Liu, H.M.; Yang, J.; Huang, H.; Wang, X.H. Differential effects of N and P additions on foliar stoichiometry between species and community levels in a subtropical forest in eastern China. *Ecol. Indic.* **2020**, *117*, 106537. [CrossRef]
13. Vitousek, P.M.; Porder, S.; Houlton, B.Z.; Chadwick, O.A. Terrestrial phosphorus limitation: Mechanisms, implications, and nitrogen-phosphorus interactions. *Ecol. Appl.* **2010**, *20*, 5–15. [CrossRef]
14. Rumman, R.; Atkin, O.K.; Bloomfield, K.J.; Eamus, D. Variation in bulk-leaf ^{13}C discrimination, leaf traits and water-use efficiency-trait relationships along a continental-scale climate gradient in Australia. *Glob. Chang. Biol.* **2018**, *24*, 1186–1200. [CrossRef] [PubMed]
15. Liu, J.; Su, Y.G.; Li, Y.; Huang, G. Shrub colonization regulates $\delta^{13}C_p$ enrichment between soil and vegetation in deserts by affecting edaphic variables. *Catena* **2021**, *203*, 105365. [CrossRef]
16. Zhang, Y.E.; Yu, X.X.; Chen, L.H.; Jia, G.D.; Zhao, N.; Li, H.Z.; Chang, X.M. Foliar water use efficiency of *Platycladus orientalis* sapling under different soil water contents. *Chin. J. Appl. Ecol.* **2017**, *28*, 2149–2154. (In Chinese with English Abstract)
17. Huang, F.Z.; Li, D.X.; Wang, B.; Xiang, W.S.; Guo, Y.L.; Wen, S.J.; Chen, T.; Li, X.K. Foliar stable carbon isotope composition and water use efficiency of plant in the Karst seasonal rain forest. *Chin. J. Appl. Ecol.* **2019**, *30*, 1833–1839. (In Chinese with English Abstract)
18. Gong, X.Y.; Ma, W.T.; Yu, Y.Z.; Fang, K.Y.; Yang, Y.S.; Tcherkez, G.; Adams, M.A. Overestimated gains in water-use efficiency by global forests. *Glob. Chang. Biol.* **2022**, *28*, 4923–4934. [CrossRef]
19. Lu, J.N.; Zhao, X.Y.; Wang, S.K.; Feng, S.; Ning, Z.Y.; Wang, R.X.; Chen, X.P.; Zhao, H.S.; Chen, M. Untangling the influence of abiotic and biotic factors on leaf C, N, and P stoichiometry along a desert-grassland transition zone in northern China. *Sci. Total. Environ.* **2023**, *884*, 163902. [CrossRef]
20. Li, L.; Liu, L.; Yu, Z.; Peñuelas, J.; Sardans, J.; Chen, Q.F.; Xu, J.B.; Zhou, G.Y. Carbon, nitrogen and phosphorus stoichiometry in natural and plantation forests in China. *Forests* **2022**, *13*, 755. [CrossRef]
21. Kong, L.L.; Lin, J.; Huang, Z.Q.; Yu, Z.P.; Xu, Z.K.; Liang, Y.F. Variations of water use efficiency and its relationship with leaf nutrients of different altitudes of Wuyi Mountains, China. *Chin. J. Appl. Ecol.* **2017**, *28*, 2102–2110. (In Chinese with English Abstract)
22. Farquhar, G.D.; O'Leary, M.H.; Berry, J.A. On the relationship between carbon isotope discrimination and the intercellular carbon dioxide concentration in leaves. *Funct. Plant. Biol.* **1982**, *9*, 121–137. [CrossRef]
23. Farquhar, G.D.; Ehleringer, J.R.; Hubick, K.T. Carbon isotope discrimination and photosynthesis. *Annu. Rev. Plant. Biol.* **1989**, *40*, 503–537. [CrossRef]
24. Adiredjo, A.L.; Navaud, O.; Muños, S.; Langlade, N.B.; Lamaze, T.; Grieu, P. Genetic control of water use efficiency and leaf carbon isotope discrimination in sunflower (*Helianthus annuus* L.) subjected to two drought scenarios. *PLoS ONE* **2014**, *9*, e101218. [CrossRef]
25. Talbi Zribi, O.; Abdelly, C.; Debez, A. Interactive effects of salinity and phosphorus availability on growth, water relations, nutritional status and photosynthetic activity of barley (*Hordeum vulgare* L.). *Plant Biology* **2011**, *13*, 872–880. [CrossRef]
26. Jacob, J.; Lawlor, D.W. Stomatal and mesophyll limitations of photosynthesis in phosphate deficient sunflower, maize and wheat plants. *J. Exp. Bot.* **1991**, *42*, 1003–1011. [CrossRef]

27. Chen, Y.H.; Han, W.X.; Tang, L.Y.; Tang, Z.Y.; Fang, J.Y. Leaf nitrogen and phosphorus concentrations of woody plants differ in responses to climate, soil and plant growth form. *Ecography* **2013**, *36*, 178–184. [CrossRef]
28. Huang, Z.Q.; Liu, B.; Davis, M.; Sardans, J.; Peñuelas, J.; Billings, S. Long-term nitrogen deposition linked to reduced water use efficiency in forests with low phosphorus availability. *New. Phytol.* **2016**, *210*, 431–442. [CrossRef]
29. Zeng, H.H.; Wu, J.E.; Liu, W.J. Water use efficiency and leaf nutrient contents of plants in jungle rubber agroforestry system. *Subtrop. Plant Sci.* **2019**, *48*, 125–133. (In Chinese with English Abstract)
30. Wu, Z.Y. *Vegetation of China*; Science Press: Beijing, China, 1980. (In Chinese)
31. Chen, H.M.; Shi, F.X.; Xu, J.W.; Liu, X.P.; Mao, R. Tree mycorrhizal type controls over soil water-extractable organic matter quantity and biodegradation in a subtropical forest of southern China. *Forest. Ecol. Manag.* **2023**, *535*, 120900. [CrossRef]
32. Hanba, Y.T.; Mori, S.; Lei, T.T.; Koike, T.; Wada, E. Variations in leaf $\delta^{13}C$ along a vertical profile of irradiance in a temperate Japanese forest. *Oecologia* **1997**, *110*, 253–261. [CrossRef]
33. Reich, P.B.; Oleksyn, J. Global patterns of plant leaf N and P in relation to temperature and latitude. *Proc. Natl. Acad. Sci. USA* **2004**, *101*, 11001–11006. [CrossRef] [PubMed]
34. Bloom, A.J. The increasing importance of distinguishing among plant nitrogen sources. *Curr. Opin. Plant Biol.* **2015**, *25*, 10–16. [CrossRef]
35. Cui, E.Q.; Lu, R.L.; Xu, X.N.; Sun, H.F.; Qiao, Y.; Ping, J.Y.; Qiu, S.Y.; Lin, Y.H.; Bao, J.H.; Yong, Y.T.; et al. Soil phosphorus drives plant trait variations in a mature subtropical forest. *Glob. Chang. Biol.* **2022**, *28*, 3310–3320. [CrossRef] [PubMed]
36. Townsend, A.R.; Cleveland, C.C.; Asner, G.P.; Bustamante, M.M. Controls over foliar N:P ratios in tropical rain forests. *Ecology* **2007**, *88*, 107–118. [CrossRef] [PubMed]
37. Yoo, S.D.; Greer, D.H.; Laing, W.A.; McManus, M.T. Changes in photosynthetic efficiency and carotenoid composition in leaves of white clover at different developmental stages. *Plant Physiol. Bioch.* **2003**, *41*, 887–893. [CrossRef]
38. Shangguan, Z.P.; Shao, M.A.; Dyckmans, J. Nitrogen nutrition and water stress effects on leaf photosynthetic gas exchange and water use efficiency in winter wheat. *Environ. Exp. Bot.* **2000**, *44*, 141–149. [CrossRef]
39. Garrish, V.; Cernusak, L.A.; Winter, K.; Turner, B.L. Nitrogen to phosphorus ratio of plant biomass versus soil solution in a tropical pioneer tree, *Ficus insipida*. *J. Exp. Bot.* **2010**, *61*, 3735–3748. [CrossRef]

forests

MDPI

Article

Characterization of *Pseudomonas* sp. En3, an Endophytic Bacterium from Poplar Leaf Endosphere with Plant Growth-Promoting Properties

Beiyan Deng, Ling Wu, Hongju Xiao and Qiang Cheng *

State Key Laboratory of Tree Genetics and Breeding, Key Laboratory of Forestry Genetics & Biotechnology of Ministry of Education, Co-Innovation Center for Sustainable Forestry in Southern China, Nanjing Forestry University, Nanjing 210037, China; dengbeiyan@njfu.edu.cn (B.D.); wuling@njfu.edu.cn (L.W.); xiaohongju@njfu.edu.cn (H.X.)

* Correspondence: chengqiang@njfu.edu.cn

Abstract: Growth-promoting endophytic bacteria possess substantial potential for sustainable agriculture. Here, we isolated an endophytic bacterium, *Pseudomonas* sp. En3, from the leaf endosphere of *Populus tomentosa* and demonstrated its significant growth-promoting effects on both poplar and tomato seedlings. The phosphorus solubilization and nitrogen fixation abilities of strain En3 were confirmed via growth experiments on NBRIP and Ashby media, respectively. Salkowski staining and HPLC-MS/MS confirmed that En3 generated indole-3-acetic acid (IAA). The infiltration of En3 into leaf tissues of multiple plants did not induce discernible disease symptoms, and a successful replication of En3 was observed in both poplar and tobacco leaves. Combining Illumina and Nanopore sequencing data, we elucidated that En3 possesses a circular chromosome of 5.35 Mb, exhibiting an average G + C content of 60.45%. The multi-locus sequence analysis (MLSA) and genome average nucleotide identity (ANI) supported that En3 is a novel species of *Pseudomonas* and constitutes a distinct phylogenetic branch with *P. rhizosphaerae* and *P. coleopterorum*. En3 genome annotation analysis revealed the presence of genes associated with nitrogen fixation, phosphate solubilization, sulfur metabolism, siderophore biosynthesis, synthesis of IAA, and ethylene and salicylic acid modulation. The findings suggest that *Pseudomonas* sp. En3 exhibits significant potential as a biofertilizer for crop and tree cultivation.

Keywords: poplar; endophytes; plant growth promotion; *Pseudomonas*; colonization; genome

1. Introduction

Endophytic bacteria can colonize plant tissues without eliciting any disease symptoms in host plants [1]. Many endophytic bacteria can promote plant growth by assisting the acquisition of essential nutrients, synthesizing phytohormones, or enhancing host resistance against pathogenic microorganisms. The mechanism of plant growth promotion via endophytic bacteria resembles that of rhizosphere bacteria residing in soil; however, their presence within plant tissues allows for more direct beneficial effects on the closely associated plant cells [2]. Therefore, the identification and characterization of endophytic plant growth-promoting (PGP) bacteria holds significant potential for application in sustainable agriculture and silviculture [3].

Pseudomonas is a diverse and complex bacterial genus that occurs in a broad range of ecological niches, including plant tissues [4]. Some endophytic *Pseudomonas* species are well known as PGP agents. For instance, the rice endophytic *P. stutzeri* A15 employs a nitrogenase complex to fix nitrogen [5], and three *Pseudomonas* isolates derived from *Miscanthus giganteus* can produce gluconic acid to solubilize inorganic phosphate compounds [6]. *P. mosselii* and *P. putida* W619 from the root tissues of wheat and poplar were confirmed to biosynthesize indole-3-acetic acid (IAA) via the tryptophan pathway [7,8].

P. fluorescens YsS6 and *P. migulae* 8R6 can reduce ethylene content in mini carnation fresh cut flowers via 1-aminocyclopropane-1-carboxylate (ACC) deaminase [9]. *P. fluorescens* BRZ63, *P. simiae* PICF7, and *P. putida* BP25, endophytes of oilseed rape, olive, and pepper, respectively, have been found to inhibit the growth of various pathogenic fungi or bacteria [10–12]. The endophytic bacterium *P. brassicacearum* Zy-2-1 from *Sphaerophysa salsula*, *P. fluorescens* Sasm05 from *Sedum alfredii* Hance, and *P. azotoformans* ASS1 from *Alyssum serpyllifolium* can protect plants from abiotic stress and expedite the phytoremediation process of metal-contaminated soils [13–15]. The colonization of host tissues by these endophytic *Pseudomonas* species continuously confers benefits to their hosts and presents significant potential for application as alternatives to chemical fertilizers and pesticides.

Poplar trees (*Populus* spp.) are widely cultivated worldwide for producing fiberboard, paper pulp, and construction lumber [16]. Several poplar endophytic bacteria derived from root and stem endosphere have been characterized for their growth-promoting abilities. Two strains from different *Burkholderia* species, *B. vietnamiensis* WPB and *B. pyrrocinia* JK-SH007, both isolated from poplar stem, can promote growth of bluegrass and poplar seedlings, respectively [17–19]. *Enterobacter* sp. strain 638 and *P. putida* W619 are primarily colonized in poplar roots and have been shown to promote biomass accumulation of poplar cuttings [8,20]. To date, there are no reports characterizing PGP bacteria from the poplar leaf endosphere.

In this study, *Pseudomonas* sp. En3 was isolated from the leaf endosphere of poplar, which can grow in the leaves of multiple plants without causing disease symptoms and promote the growth of poplar and tomato plants. Strain En3 exhibits auxin production, phosphate solubilization, and nitrogen fixation capabilities. Multi-locus sequence analysis (MLSA) and genome-wide comparison revealed it as a novel *Pseudomonas* species. Genome annotation analysis indicates its high potential for promoting plant growth.

2. Materials and Methods

2.1. Isolation of Endophytic Bacteria from Poplar Leaves

In July 2022, healthy leaves of *Populus tomentosa* within the campus of Nanjing Forestry University in Nanjing, Jiangsu Province, China (118.82° E, 32.08° N) were collected. The leaf discs were immersed in a 5% NaClO solution for 5 min and subsequently rinsed five times with sterile distilled water. The surface-disinfected leaf discs were placed on LB agar plates at 28 °C for 48 h. Bacteria observed at the periphery of the leaf discs were streaked, and single colonies were preserved. To validate the efficacy of the sterilization procedure, the final rinse water (100 μL) was spread onto LB agar plates and incubated under the same conditions.

2.2. Identification of Endophytic Bacteria

Genomic DNA of endophytic bacteria was extracted using the TIANamp Bacteria DNA Kit (Tiangen, Beijing, China). The 16S rRNA gene was amplified using primer pairs 27F (5′-AGAGTTTGATCMTGGCTCAG-3′) and 1492R (5′-GGTTACCTTGTTACGACTT-3′) [21]. PCR products were purified from the gel and ligated into the pMD19-T vector (Takara, Dalian, China) for Sanger sequencing. The obtained 16S rDNA sequences were subjected to a BLASTn search against the NCBI rRNA database and ezBiocloud (https://www.ezbiocloud.net/, accessed on 26 October 2023).

2.3. Plant Growth-Promoting Assays

The poplar cultivar 'Shanxin' (*Populus davidiana* × *P. bolleana*) was micropropagated on half-strength Murashige and Skoog (MS) medium. Two-week-old rooted seedlings were transplanted into pots containing nutrient soil and subsequently cultivated in a growth chamber under controlled conditions of a 16 h light/8 h dark photoperiod at 23 °C. One week post-transplantation, the seedlings were used in plant growth-promoting assays. Tomato plants (*Solanum lycopersicum*) were sowed under identical conditions in the same growth chamber; one-week-old tomato seedlings were used for plant growth-promoting assays. Bacterial cells were collected from LB plates and resuspended in sterile distilled water (OD600 = 0.5). The bacterial suspension was administered twice weekly (30 mL per pot), while control plants received an equal volume of sterile distilled water. The plant height (from apical meristem to the soil substrate) and fresh weight (whole plant after washing off soil substrate and blotting out water) of the plants were measured and recorded.

2.4. Plant Growth-Promoting Properties of Strain En3

The phosphate solubilizing activities of strain En3 were assessed by spotting 10 μL of cultures on the top of NBRIP plates (10 g D-glucose, 5 g $Ca_3(PO_4)_2$, 5 g $MgCl_2{\cdot}6H_2O$, 0.25 g $MgSO_4{\cdot}7H_2O$, 0.2 g KCl, 0.1 g $(NH_4)_2SO_4$, 15 g agar in 1 L distilled water) [22]. The presence or absence of a transparent halo around bacterial colonies was recorded.

To assess nitrogen-fixing ability, a 10 μL aliquot of the culture of En3 was inoculated onto Ashby nitrogen-free agar medium (5 g glucose, 5 g mannitol, 0.1 g $CaCl_2{\cdot}2H_2O$, 0.1 g $MgSO_4{\cdot}7H_2O$, 5 mg $Na_2MoO_4{\cdot}2H_2O$, 0.9 gK_2HPO_4, 0.1 g KH_2PO_4, 0.01 g $FeSO_4{\cdot}7H_2O$, 5 g $CaCO_3$, 15 g agar in 1 L distilled water) [23].

The Salkowski method was employed to evaluate the capacity of En3 for indole-3-acetic acid (IAA) production. Briefly, the bacteria were cultured in LB liquid medium supplemented with 0.5 mg/mL L-tryptophan until reaching OD600 of 1.5. The supernatants were then mixed with Salkowski reagent (1.2% $FeCl_3$ in 37% H_2SO_4) for 30 min, and the resulting color reaction was observed [24]. Positive controls included a solution containing 50 mg/L IAA and *P. syringae* pv. *tomato* DC3000 supernatant. Furthermore, culture supernatants from three independent experiments were subjected to high-performance liquid chromatography and tandem mass spectrometry (HPLC-MS/MS) analysis at RaySource Biotechnology (Nanjing, China) to determine the concentrations of IAA and IAA synthesis intermediates including indole-3-acetamide (IAM), indole-3-pyruvate (IPyA), and indole-3-acetonitrile (IAN).

2.5. Analysis of the Colonization of Strain En3 in Multiple Plants

Strain En3 colonization experiments were conducted using three-week-old tobacco (*Nicotiana benthamiana*), two-week-old tomato, and two-week-old oilseed rape (*Brassica napus*) plants grown in nutrient soil in growth chamber. Additionally, four-week-old poplar 'Shanxin' seedlings grown on half-strength MS medium were also used. Bacterial cells were suspended in 10 mM $MgCl_2$ and diluted to a concentration of 10^5 prior to infiltration into the leaves of tobacco, tomato, and oilseed rape using a syringe. For poplar 'Shanxin' seedlings, the bacterial suspension with 0.015% Silwet L-77 was infiltrated under continuous vacuum for 3 min at a pressure of 500 mmHg. Afterward, the seedlings were dried with sterile filter paper and planted in half-strength MS medium supplemented with 100 mg/L Cefotaxime.

One spontaneous rifampicin-resistant mutant from strain En3 was used for a growth curve assay to avoid contamination by environmental bacteria. Leaf discs were collected at 0, 2, or 4 days post-infiltration (dpi), and homogenized in 10 mM $MgCl_2$. The cfu/cm^2 for tobacco and cfu/mg for poplar were determined by plating serial dilutions of leaf extracts on LB plates containing 100 mg/L rifampicin. Three independent experiments were performed [25].

2.6. Genome Sequencing, Assembling, and Annotation

The genomic DNA of strain En3 was extracted using the TIANamp Bacteria DNA Kit (Tiangen, Beijing, China). The quality and concentration of DNA were verified on 1.5% agarose gels and using a Qubit 3.0 Fluorometer (Invitrogen, Carlsbad, CA, USA). Long-read and short-read sequencing DNA libraries were prepared with an SQK-LSK109 ligation kit (Oxford Nanopore Technologies (ONT), Oxford, UK) and Nextera DNA Flex Library Prep Kit (Illumina, San Diego, CA, USA), respectively, according to the manufacturer's protocols. The DNA libraries were then sequenced by the Oxford Nanopore PromethION sequencer (ONT) with 48 h runs and the Illumina NovaSeq 6000 platform (Illumina) with paired-end 150 bp read lengths by Benagen company (Wuhan, China). The En3 genome was assembled using Unicycler software (Version: 0.5.0) [26] with a combination of ONT and Illumina reads. Then, two rounds of error correction were performed on the assembly result based on the Illumina reads using Pilon (v.1.23) [27].

The coding sequences, tRNAs, and rRNAs were predicted by Prodigal, Aragorn, and RNAmmer in Prokka software (Version: 1.14.6) [28]. The predicted protein functions were further annotated using the National Center for Biotechnology Information (NCBI) non-redundant proteins (NR) and Kyoto Encyclopedia of Genes and Genomes (KEGG) databases.

2.7. Phylogenetic and Average Nucleotide Identity (ANI) Analyses

The DNA sequences of 16S rRNA, *gyrB*, *rpoD*, and *rpoB* genes from the representative type strains of *P. fluoresens* group, *P. syringae* group, *P. angulliseptica* group, *P. alcaligenes* group, *P. oleovorans* group, and *P. putida* group [29], as well as strain En3, were concatenated and aligned. A phylogenetic tree of the multilocus alignment was constructed with the neighbor-joining method using MEGA 7.0 [30]. The inferred phylogenies were tested by 1000 bootstrap replicates.

The average nucleotide identity (ANI) was calculated using the OrthoANIu algorithm [31]. The automated multi-locus species tree (autoMLST) [32] was also used for analyzing phylogeny and whole genome similarity.

3. Results

3.1. Isolation of Endophytic Bacteria from Poplar Leaves and Analysis of Plant Growth-Promoting Activity

We isolated four strains of endophytic bacteria (En1–En4) with different colony morphology from healthy *P. tomentosa* leaves through rigorous surface disinfection procedures. The 16S rRNA genes of these strains were sequenced and compared against the NCBI rRNA database and ezBiocloud, revealing top hits *Sphingomonas sanguinis* strain NBRC 13937^T (99.86% identity), *Sphingobium yanoikuyae* strain NBRC 15102^T (99.72% identity), *Pseudomonas azerbaijanorientalis* strain SWRI123^T (99.10% identity), and *Microbacterium testaceum* strain NBRC 12675^T (99.10% identity). Consequently, we designated them as *Sphingomonas* sp. En1 (GenBank accession no. OR511661), *Sphingobium* sp. En2 (OR511660), *Pseudomonas* sp. En3 (OR511655), and *Microbacterium* sp. En4 (OR511666) (Table 1).

Table 1. Sequence analysis of 16S rRNA of endophytic bacteria isolated from Poplar leaves.

Endophyte	Top Hit (Accession No.)	Sequence Length (bp)	% Identity	Genbank Accession No.
En1	*Sphingomonas sanguinis* NBRC 13937^T (NR_113637)	1409	99.86	OR511661
En2	*Sphingobium yanoikuyae* NBRC 15102^T (NR_113730)	1409	99.72	OR511660
En3	*Pseudomonas azerbaijanorientalis* SWRI123^T (CP077078)	1448	99.10	OR511655
En4	*Microbacterium testaceum* NBRC 12675^T (NZ_BJMLO1000022)	1444	99.10	OR511666

To assess the plant growth-promoting activity of endophytic bacteria, we continuously irrigated transplanted poplar 'Shanxin' seedlings with a suspension of these four strains. After one month of treatment, none of the four strains induced any disease symptoms in poplar seedlings; only *Pseudomonas* sp. En3 exhibited a significant ability to enhance the growth of poplar trees ($p < 0.01$) (Figure 1A). Compared to the control group (irrigated with H_2O), En3-irrigated poplar trees displayed a substantial increase in plant height by 54%

and fresh weight by 87% on average (Figure 1B,C). Furthermore, tomato seedlings treated with En3 for two weeks demonstrated remarkable growth-promoting effects ($p < 0.01$) in both plant height (an average increase of 67%) and fresh weight (an average increase of 171.1%) compared to the control group (Figure 1D–F). The fact that *Pseudomonas* sp. En3 promotes growth in both distantly related plant species suggests the possibility of general plant-promoting features.

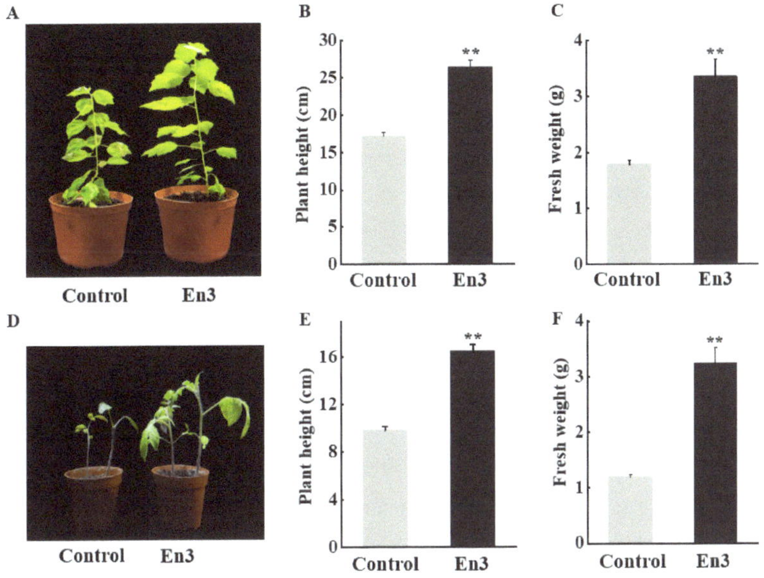

Figure 1. Plant growth-promoting activity of *Pseudomonas* sp. En3. (**A**) Morphology, (**B**) plant height, and (**C**) fresh weight of poplar 'Shanxin' were recorded after irrigation with an En3 suspension or water for one month. Data shown are mean ± SD (n = 12; ** $p < 0.01$ compared with control, Student's t-test). (**D**) Morphology, (**E**) plant height, and (**F**) fresh weight of tomato were recorded after irrigation with En3 suspension or water for two weeks. Values are means ± SD (n = 20; ** $p < 0.01$ compared with control, Student's t-test).

3.2. Analysis of PGP Features of Pseudomonas sp. En3

Phosphorus solubilization, nitrogen fixation, and IAA production are common features of PGP bacteria. The NBRIP medium, which incorporates insoluble $Ca_3(PO_4)_2$ as the sole phosphorus source, was used to evaluate the phosphorus solubilization capacity of strain En3. Following incubation at 28 °C for 5 days, colonies of *Pseudomonas* sp. En3 displayed a clear transparent halo around them, indicative of phosphorus solubilization (Figure 2A). In addition, En3 was able to grow well on nitrogen-free Ashby medium, suggesting its ability to fix atmospheric nitrogen (Figure 2B).

The Salkowski reagent was employed for the preliminary analysis of the capacity of strain En3 to synthesize indole derivatives. *P. syringae* pv. *tomato* DC3000, a model bacterial pathogen known for its ability to produce IAA, was selected as the positive control. The strains En3 and DC3000 were cultured to an equivalent concentration with an L-tryptophan supplement. Remarkably, both supernatants exhibited a discernible red color upon treatment with Salkowski's reagent; however, the supernatant of En3 displayed a more intense shade of red. Conversely, no obvious red coloration was observed in the supernatants of both En3 and DC3000 without L-tryptophan supplementation (Figure 2C). These findings suggest that En3 can produce indole derivatives in a tryptophan-dependent manner and to a greater extent than DC3000. To further resolve the indole derivatives produced by En3 and DC3000, we employed HPLC-MS/MS analysis to examine the supernatant of cultures with L-tryptophan supplement (Figures S1 and S2). Tables 2 and 3 demonstrate the presence of IAA in the culture supernatant of En3 and DC3000, with significantly higher levels observed in En3 (three biological replicates, $p < 0.01$), reaching a fold increase of 2.85. Notably, IAA biosynthesis intermediates (IAM, IPyA, and IAN) were also found at significantly higher levels in En3 compared to DC3000 (9.05-fold, 7.5-fold, and 8.81-fold, respectively).

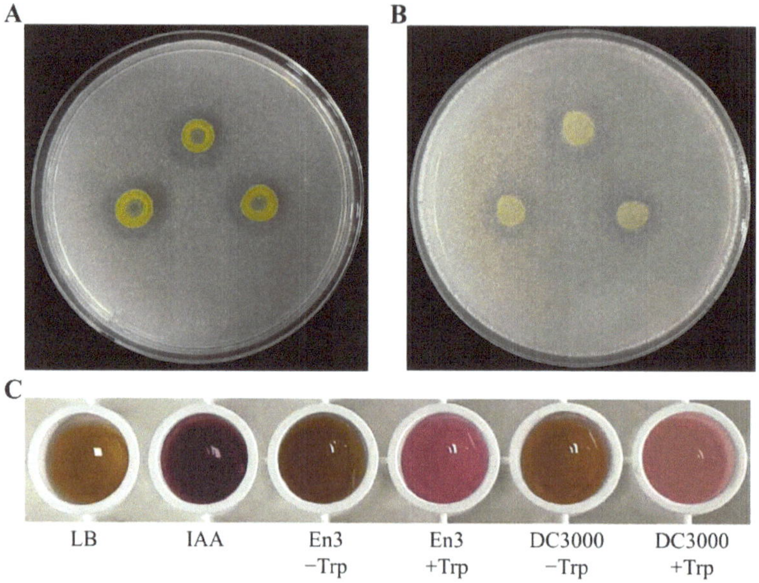

Figure 2. Characterization of *Pseudomonas* sp. En3 PGP features. (**A**) Phosphorous solution zones were observed when En3 was cultivated on NBRIP medium. (**B**) En3 grew on nitrogen-free Ashby medium. The photographs (**A**,**B**) were taken on the fifth day after inoculation. (**C**) Indole derivatives were detected using Salkowski reagent. LB, LB liquid medium; IAA, 50 mg/L IAA; En3 − Trp, culture supernatant of En3 without L-tryptophan; En3 + Trp, culture supernatant of En3 supplemented with 0.5 mg/mL L-tryptophan; DC3000−Trp, culture supernatant of DC3000 without L-tryptophan; DC3000 + Trp, culture supernatant of DC3000 supplemented with 0.5 mg/mL L-tryptophan.

Table 2. HPLC-MS/MS analysis of indole derivatives produced by En3 in LB medium supplemented with L-tryptophan.

Indole Derivatives	IAA	IAM	IPyA	IAN
Mass-to-charge ratio (M/Z)	176.2 ≥ 129.8	175.0 ≥ 130.0	202.0 ≥ 114.9	157.1 ≥ 130.0
Acqusion Standard	5.49	4.74	6.10	6.35
time(min) Sample	5.50	4.74	6.10	6.35
Concentration (ng/mL) [a]	7.947 ± 0.342	1137.607 ± 21.849	1487.677 ± 37.020	14.100 ± 0.420

[a] Values are average ± SD (n = 3).

Table 3. HPLC-MS/MS analysis of indole derivatives produced by DC3000 in LB medium supplemented with L-tryptophan.

Indole Derivatives	IAA	IAM	IPyA	IAN
Mass-to-charge ratio (M/Z)	176.2 ≥ 129.8	175.0 ≥ 130.0	202.0 ≥ 114.9	157.1 ≥ 130.0
Acqusion Standard	5.49	4.74	6.10	6.35
time(min) Sample	5.51	4.74	6.10	6.35
Concentration (ng/mL) [a]	2.787 ± 0.214	125.743 ± 6.579	198.283 ± 12.689	1.600 ± 0.249

[a] Values are average ± SD (n = 3).

3.3. Analyzing Colonization Activity of Pseudomonas sp. En3

The seedlings of poplar 'Shanxin' and tomato showed no disease symptoms when continuously exposed to the En3 suspension. Hence, we aimed to further investigate whether direct infiltration of En3 into leaf tissues could induce disease symptoms and whether En3 could reproduce in plants. The infiltration of an En3 suspension of 10^5 cfu/mL was performed via vacuum infiltration into micropropagated poplar seedlings, as well as via direct injection into leaves of tobacco, oilseed rape, and tomato. Plants were observed for 5 days. The infiltrated areas on the leaves of different plants retained their green coloration, while no significant changes were observed in the overall morphology of the infiltrated poplar plants (Figure 3A–D). Subsequently, a spontaneous mutant ($En3^{Rif}$) of En3 resistant to rifampicin was used for the analysis of bacterial populations grown in tobacco and poplar leaves. In poplar 'Shanxin', the $En3^{Rif}$ counts of 2 dpi (3.49 log cfu/mg) and 4 dpi (4.25 log cfu/mg) samples increased by 11- and 68-fold, respectively, compared to the $En3^{Rif}$ count of the 0 dpi (2.41 log cfu/mg) sample (Figure 3E). Similarly, the $En3^{Rif}$ counts in tobacco increased by 5-fold at 2 dpi (2.89 cfu/cm^2) and 12-fold at 4 dpi (3.23 cfu/cm^2) compared to 0 dpi (2.10 log cfu/cm^2) (Figure 3F). These findings suggest that the presence of En3 within the leaf tissue of multiple plant species does not induce disease symptoms and exhibits a capability for proliferation. The model pathogen *P. syringae* pv. *tomato* DC3000 is known to induce a hypersensitive response (HR) in *N. benthamiana*. Thus, we separately infiltrated *N. benthamiana* leaves with DC3000 and En3 at concentrations of 10^7 cfu/mL and 10^5 cfu/mL, respectively. It was observed that DC3000 triggered an HR, while En3 did not result in any symptoms (Figure S3), suggesting that the inoculation conditions used in this study effectively supported bacterial functionality.

Figure 3. Colonization activity of *Pseudomonas* sp. En3. (**A–D**) Phenotypes of poplar 'Shanxin' (**A**), tobacco (**B**), oilseed rape (**C**), and tomato (**D**) leaves infiltrated with 10^5 cfu/mL En3. Photographs were taken on 5 dpi. (**E**,**F**) Qualification of the $En3^{Rif}$ population in poplar 'Shanxin' (**E**) and tobacco (**F**). Different letters on the bars indicate significant differences ($p < 0.05$) (one-way ANOVA with three independent experiments).

3.4. General Features of the Pseudomonas sp. En3 Genome

Based on the trimmed Illumina (4.1 Gb) and ONT data (1.1 Gb), one circular chromosome of strain En3 of 5.35 Mb with 60.45% average G + C content was assembled. The sequencing depths were 207.14× (ONT) and 720.01× (Illumina). No plasmid was found. A total of 4807 protein-encoding sequences (CDSs) were predicted with an average length of 1003 bp, and this accounted for 90.1% of the total genome. In addition, the En3 genome had 67 tRNAs, 16 rRNA genes (5S, 16S, and 23S), 9 genomic islands, and 3 CRISPRs. The numbers of CDSs annotated in NR and KEGG databases were 4729 and 1746, respectively. A total of 501 predicted proteins had signal peptides and no transmembrane structures and were predicted to be secreted proteins (Table 4). The assembled and annotated sequences of En3 were deposited in GenBank (accession number CP124218).

Table 4. General features of the En3 genome.

Feature	Value
Genome size (bp)	5,350,973
Contig	1
GC content (%)	60.45
tRNA	67
rRNA (5S, 16S, 23S)	16 (6,5,5)
Protein-coding genes (CDS)	4807
Average CDS length (nt)	1003
CDS assigned to NR	4729
CDS assigned to KEGG	1746
Secreted protein	501

3.5. Phylogenetic Analysis

A BLASTN search against the NCBI rRNA database using the 16S rDNA sequence (1347 bp) of strain En3 revealed that the top 10 hits were strains from the *P. fluorescens* group (6) or the *P. syringae* group (4), with approximate nucleotide differences (15–17 bp, identity > 99%). Phylogenetic analysis based on the 16S rRNA gene of representative type strains from various *Pseudomonas* groups indicated that strain En3 did not closely cluster with any known *Pseudomonas* species or group (Figure S4). To determine the taxonomic position of En3 accurately, an MLSA based on concatenated 16S rRNA gene and housekeeping genes *gyrB*, *rpoB*, and *rpoD* was performed. The strain En3 clustered with *P. rhizosphaerae* and *P. coleopterorum* with a high bootstrap support (81%) in the MLSA. Multiple taxonomic studies of the genus *Pseudomonas* have demonstrated that *P. rhizosphaerae* and *P. coleopterorum* constitute a distinct branch, separate from any defined group within the genus [4,29].

The average nucleotide identity (ANI) is widely used to determinate similarities between the genomes and an ANI value of 95% is considered as the cutoff for the delineation of bacterial species [33]. ANI analysis between the genome of *Pseudomonas* sp. En3 and that of *Pseudomonas* representative type strains showed that strain En3 exhibits a high ANI value with *P. rhizosphaerae* and *P. coleopterorum* (78.53% and 78.60%), while it has a lower value with bacteria from other clades (<78.28%). This further supports the inferred phylogenetic relationship based on multi-locus sequences. The ANI value between the genomes of *P. rhizosphaerae* and *P. coleopterorum* was 90.82%, much higher than that between En3 and *P. rhizosphaerae* and *P. coleopterorum*, indicating that strain En3 may represent a novel orphan species. In addition, an AutoMLST analysis based on 82 concatenated sequences of housekeeping genes from closely related genomes also supports the closest phylogenetic relationship between strain En3 and *P. coleopterorum* (Figure 4). A culture of strain En3 was deposited in the China Center for Type Culture Collection (AB 2023153).

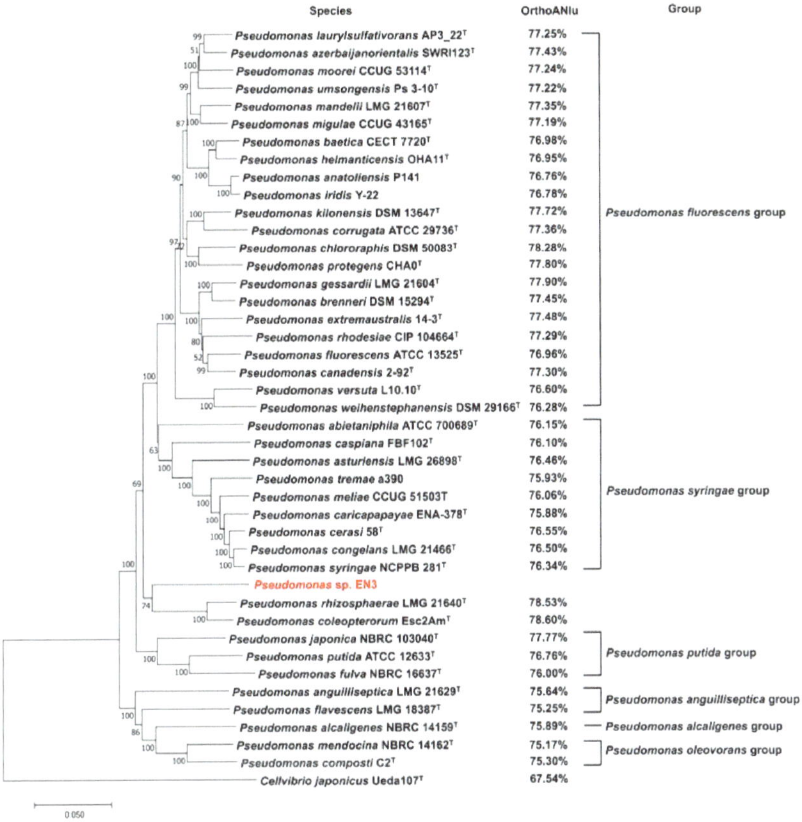

Figure 4. Neighbor-joining phylogenetic tree reconstruction based on the concatenated sequences of the four housekeeping universal genes (16S rRNA, *gyrB*, *rpoB*, and *rpoD*). Species, representative type strains from six *Pseudomonas* groups. OrthoANIu value, average nucleotide identity (ANI) based on a comparison of the En3 genome to that of Pseudomonas strains in the tree. Group, groups of *Pseudomonas*. Numbers at each node indicate the bootstrap percentage (n = 1000). The superscript letter "T" indicates the type strain.

3.6. Plant Growth-Promoting and Colonization Features of the Pseudomonas sp. En3 Genome

The gene annotation of En3 revealed that this bacterium possesses many genes and operons associated with plant growth-promoting activities. The En3 strains possess *iscU*, *iscS*, and *NRT* genes associated with nitrogen fixation; *gltB*, *gltD*, *nirB*, and *nirD* genes involved in ammonia assimilation; and *norR* and *hmp* genes related to nitrite stress for efficient nitrogen metabolism. For phosphorus metabolism, En3 has the PhoB/PhoR two-component system that senses the phosphorus concentration and the *phoU* and *pstABCS* operons involved in phosphate transport. We also observed the presence of the *ssuABCDE* operon, which is responsible for sulfonate transport, as well as the *cysAWUP* operon involved in sulfate transport, and *cysI* and *cysJ* genes associated with sulfite metabolism in the En3 genome. The En3 strain harbors genes responsible for the synthesis and transport of siderophores, including *entD*, *bfd*, *bfr*, *tonB*, *exbB*, *exbD*, and *fhuE*. As expected, the En3 strain harbors *trpABCDEG* operons and *iaaM* and *amiE* genes implicated in IAA biosynthesis. In addition, En3 possesses two ACC deaminases and one salicylate hydroxylase, which may be involved in the regulation of the plant hormones ethylene and salicylic acid (Table 5).

Table 5. Genes associated with PGP in the *Pseudomonas* sp. En3 genome.

PGP Activity	Gene	Gene Annotation	Locus
Nitrogen Metabolism	*iscU*	nitrogen fixation protein NifU and related proteins	ctg_01045
	iscS	cysteine desulfurase	ctg_01044
	NRT	MFS transporter, nitrate/nitrite transporter	ctg_02133
	gltB	glutamate synthase (NADPH) large chain	ctg_00450
	gltD	glutamate synthase (NADPH) small chain	ctg_00451
	nirB	nitrite reductase (NADH) large subunit	ctg_01700
	nirD	nitrite reductase (NADH) small subunit	ctg_01701
	norR	anaerobic nitric oxide reductase transcription regulator	ctg_01006
	hmp	nitric oxide dioxygenase	ctg_01007
Phosphate Metabolism	*phoB*	OmpR family, phosphate regulon response regulator	ctg_00055
	phoR	OmpR family, phosphate regulon sensor kinase	ctg_00056
	phoU	phosphate transport system protein	ctg_00061
	pstB	phosphate transport system ATP-binding protein	ctg_00062
	pstA	phosphate transport system permease protein	ctg_00063
	pstC	phosphate transport system permease protein	ctg_00064
	pstS	phosphate transport system substrate-binding protein	ctg_00065
Sulfur metabolism	*ssuB*	sulfonate transport system ATP-binding protein	ctg_00353
	ssuC	sulfonate transport system permease protein	ctg_00352
	ssuD	alkanesulfonate monooxygenase	ctg_00351
	ssuA	sulfonate transport system substrate-binding protein	ctg_00350
	ssuE	FMN reductase	ctg_00349
	cysA	sulfate transport system ATP-binding protein	ctg_00327
	cysW	sulfate transport system permease protein	ctg_00326
	cysU	sulfate transport system permease protein	ctg_00325
	cysP	sulfate transport system substrate-binding protein	ctg_00324
	cysI	sulfite reductase hemoprotein beta-component	ctg_02867
	cysJ	sulfite reductase flavoprotein alpha-component	ctg_01699
Siderophore	*tonB*	TonB-dependent transporters	ctg_00044, ctg_00494
	exbB	biopolymer transport protein	ctg_03795
	exbD	biopolymer transport protein	ctg_03794
	fhuE	outer membrane receptor for ferric coprogen	ctg_03457, ctg_02420, ctg_02095, ctg_00748
	entD	enterobactin synthetase component D	ctg_03886
	bfd	bacterioferritin-associated ferredoxin	ctg_01441
	bfr	bacterioferritin	ctg_01439, ctg_04538, ctg_00624
Plant Hormones	*trpA*	tryptophan synthase alpha chain	ctg_00189
	trpB	tryptophan synthase beta chain	ctg_00190
	trpE	anthranilate synthase component I	ctg_04614
	trpG	anthranilate synthase component II	ctg_04613
	trpD	anthranilate phosphoribosyltransferase	ctg_04612
	trpC	indole-3-glycerol phosphate synthase	ctg_04611
	iaaM	tryptophan 2-monooxygenase	ctg_04661
	amiE	amidase	ctg_02687, ctg_02400, ctg_03665
	NahG	salicylate hydroxylase	ctg_02543
	dcyD	1-aminocyclopropane-1-carboxylate deaminase	ctg_03709, ctg_00163

As En3 can colonize multiple plant leaves, we conducted a genomic analysis of its features involved in host–plant interaction. The genome of En3 contains chemotaxis system genes that recognize environmental cues, including two *cheABDVWY* operons, one *wspABCDEFR* operon, and 32 mcp protein genes. We identified a complete set of genes (*hcp*, *TssMLKJHGFECB*, *vgrG*, and *PAAR*) encoding the type VI secretion system in the En3 genome. The En3 strain also harbors gene clusters *hrpLJPVTGFDBZ* and *hrcVNRSTUJ*, which are responsible for the type III secretion system; however, it lacks *hrpA* and *hrpP* encoding the needle unit and needle-length regulator, respectively. Using the Effectidor prediction tool, seven type III secretion system effectors were predicted; three with annotations—avrE, CigR, and hopJ (Table S1).

4. Discussion

Pseudomonas represents one of the most diverse and ubiquitous bacterial genera among Gram-negative bacteria [34]. It encompasses a vast array of species, with 316 validly named species deposited in the List of Prokaryotic Names with Standing in Nomenclature (LPSN, https://lpsn.dsmz.de/genus/pseudomonas, accessed on 23 August 2023). Additionally, the number of *Pseudomonas* species in the LPSN database has consistently grown, as evidenced by the inclusion of 76 newly reported species in 2022. The 16S rRNA gene sequence is commonly used for the taxonomic classification of prokaryotes, although its discriminatory capacity often proves limited at the species level [35]. Multi-locus sequence analysis (MLSA), based on multiple housekeeping genes that are universal within the genus and have high sequence divergence, is, therefore, employed to elucidate the phylogenetic relationships among species within a given genus [36]. The commonly employed housekeeping genes for MLSA of the *Pseudomonas* genus include 16S rRNA, and the *rpoB*, *rpoD*, and *gryB* genes. Previous investigations have substantiated that the *Pseudomonas* phylogenetic groups defined via MLSA using these four housekeeping genes are concordant with outcomes derived from analyses involving a larger set of 100,120 housekeeping genes and whole-genome sequences [4].

In this study, we used the concatenated sequences of the four housekeeping genes from *Pseudomonas* sp. En3 and its closely related type strains to construct a phylogenetic tree. The type strains belonging to the *P. fluorescens*, *P. syringae*, *P. putida*, *P. anguilliseptica*, *P. oleovorans*, and *P. alcaligenes* groups were each clustered into distinct clades with robust bootstrap support. However, *Pseudomonas* sp. En3 was assigned within a distinct branch that includes *P. rhizosphaerae* and *P. coleopterorum*, which are not affiliated with any established *Pseudomonas* group. Furthermore, the results of OrthoANIu and AutoMLST based on a genome-wide comparison also supported that En3 had the highest similarity with *P. rhizosphaerae* and *P. coleopterorum*. The OrthoANIu value of En3 was found to be less than 78.60% when compared with all closely related *Pseudomonas* species, which is far below the cutoff of 95% ANI value used for species delineation [37]. Therefore, strain En3 can be classified as a novel *Pseudomonas* strain within the distinct branches of *P. rhizosphaerae* and *P. coleopterorum*.

The recent several taxonomic studies on *Pseudomonas* species provide support for the assignment of *P. rhizosphaerae* and *P. coleopterorum* as a distinct branch closely related to the groups of *P. putida*, *P. fluorescens*, *P. lutea*, and *P. syringae* [4,29,38]. The type strain LMG 21640^T of *P. rhizosphaerae*, isolated from the rhizospheric soil of grass growing in Spain, has been confirmed to possess phosphate solubilizing activity [39]. The *P. coleopterorum* $Esc2Am^T$, which was isolated from the bark beetle (*Hylesinus fraxini*), has been found to possess cellulase-producing activity [40]. In this study, we identified a new member of this distinct branch from the leaf endosphere of poplar and subsequently demonstrated its growth-promoting activity for young seedlings of both poplar and tomato.

Phosphorus is an indispensable nutrient for the growth and productivity of plants, playing a crucial role in various vital plant functions such as photosynthesis, respiration, and energy transfer [41]. Most phosphorus, however, exists in soil as inorganic phosphate with a very low soluble concentration, rendering it unavailable to plants. The species of the *Pseudomonas* genus have been demonstrated to possess the ability to solubilize insoluble phosphates. Among them, *P. frederiksbergensis* JW-SD2 efficiently solubilizes phosphates and significantly promotes the growth of poplar trees [42]. In this study, we demonstrated that *Pseudomonas* sp. En3 could efficiently dissolve tricalcium phosphate in the NBRIP medium, forming a clear dissolution zone. The analysis of the genome annotation further revealed that the En3 genome harbors a PhoB/PhoR two-component system, which functions as a phosphorus concentration sensor, and an integral pstABCS transporter responsible for facilitating phosphorus uptake. The speculation can be extended to suggest that En3 possesses the ability to assimilate diverse sources of phosphorus found in various environments.

Nitrogen is an essential element for plants, as it constitutes the fundamental components of nucleic acids and proteins. Despite its abundance in the atmosphere as nitrogen gas (N_2), it cannot be directly assimilated by plants [43]. Previous studies have confirmed that certain strains belonging to the *Pseudomonas* genus, such as *P. koreensis* CY4, *P. entomophila* CN11, *P. protegens* Pf-5 X940, and *P. stutzeri* A1501, possess nitrogen-fixing capabilities, which can enhance plant nutrient uptake and growth [44–46]. The growth of *Pseudomonas* sp. En3 on nitrogen-free Ashby medium suggests its capacity for atmospheric nitrogen fixation. Furthermore, genome annotation analysis revealed that En3 encodes iscU and iscS, which are thought to be involved in the assembly of [Fe-S] clusters of nitrogenase and are essential for nitrogen fixation [47]. The En3 genome also encodes multiple proteins involved in nitrogen metabolism, such as nirBD, which facilitates the conversion of nitrite to ammonia, and gltBD and hmp, which play a role in responding to nitrogen starvation [48]. The results obtained from genomic annotation and growth experiments provide support for the ability of the poplar endophytic bacterium En3 to solubilize phosphorus and fix nitrogen, indicating its potential as a biofertilizer for enhancing nutrient assimilation in forest and crop systems.

In this study, we discovered that strain En3 can enhance plant growth and colonize leaf tissues of various plants without inducing symptoms. The IAA produced by En3 may contribute to these two abilities. The major natural auxin of plants, IAA, is a crucial regulator of cell division, expansion, and differentiation, exerting significant influence on plant growth and development. Several PGP bacteria promote plant growth by producing IAA, for example, impairing the IAA synthesis of *P. putida* GR12-2 can diminish its capacity to enhance plant root growth [49]. On the other hand, IAA plays a pivotal role in mediating plant–microbe interactions. IAA can attenuate the salicylic acid-dependent immune response of plants and facilitate the colonization of certain biotrophic/hemi-biotrophic pathogens [50]. For instance, the disruption of the indole-3-acetaldehyde dehydrogenase genes resulted in impaired synthesis of IAA in *P. syringae* DC3000, leading to the reduced growth of DC3000 in *Arabidopsis thaliana* and the upregulation of the defense gene *PR1* [51].

The HPLC-MS/MS analysis in this study confirmed that strain En3 exhibited a significantly higher IAA production (2.85-fold) compared to DC3000 cultured under identical conditions, indicating that the concentration of IAA produced by En3 was sufficient to exert an inhibitory effect on the plant immune response. We also found that the levels of IAA biosynthesis intermediates (IAM, IPyA, and IAN) were significantly higher (>7.5-fold) in En3 compared to DC3000. In fact, the exogenous application of IAA intermediates can also elicit morphological alterations in plants; however, it remains unclear whether these alterations are attributable to the intermediates themselves or to IAA converted from these intermediates. For example, IAM treatment can induce *Arabidopsis* phenotypes resembling those of auxin overproduction mutants [52]. Therefore, the high levels of IAA and IAA intermediates in strain En3 may contribute to its plant growth-promoting activity.

Additionally, En3 harbors the genes encoding salicylate hydroxylase (NahG) and 1-aminocyclopropane-1-carboxylate (ACC) deaminases, which directly degrade salicylic acid and hydrolyze ethylene's precursor ACC, respectively. These genes may contribute to the asymptomatic colonization of En3 by regulating plant defense hormones. The En3 harbors components of the type III secretion system in its cytoplasm, inner membrane, periplasm, and outer membrane; however, it lacks the essential needle structure. This suggests that En3 may not possess the ability to inject effectors into plant cells but could potentially secrete effectors into the apoplast of plants to facilitate colonization. The type VI secretion system of bacteria can deliver effectors into prokaryotic or eukaryotic cells, potentially facilitating manipulation of the host plant or competition with other microorganisms [53]. The investigation of whether the type VI secretion system of En3 possesses such functions warrants further exploration.

5. Conclusions

We isolated an endophytic bacterium, *Pseudomonas* sp. En3, from the leaf endosphere of poplar. Molecular phylogenetic and whole genome comparison analyses confirmed that strain En3 is a novel species in the genus *Pseudomonas*, distinct from any previously defined *Pseudomonas* groups. The strain En3 could significantly promote the growth of poplar and tomato and propagate in a variety of plant tissues without causing disease symptoms. The strain En3 exhibited significant growth-promoting activity on both poplar and tomato seedlings and demonstrated the ability to colonize the leaf tissue of multiple plants without inducing disease symptoms. We conducted experimental validation to confirm the capabilities of En3 in phosphorus solubilization, nitrogen fixation, and IAA production. The analysis of the genome annotation of En3 unveiled a series of crucial genes and pathways associated with promoting plant growth. These findings provide support for the potential use of the En3 strain as a biofertilizer for various crops and trees.

Supplementary Materials: The following supporting information can be downloaded at: https://www.mdpi.com/article/10.3390/f14112203/s1, Figure S1: Mass spectra of IAA and IAM produced by En3 and DC3000; Figure S2: Mass spectrum of IAA synthesis intermediates produced by En3 and DC3000; Figure S3: *N. benthamiana* leaves infiltrated with different concentrations of *P. syringae* pv. *tomato* DC3000 and *Pseudomonas* sp. En3; Figure S4: Neighbor-joining phylogenetic tree reconstruction based on 16S rRNA genes; Table S1: Genes associated with colonization activity in *Pseudomonas* sp. En3 genome.

Author Contributions: Conceptualization, B.D., L.W., H.X. and Q.C.; methodology, B.D.; software, B.D.; validation, B.D., L.W. and H.X.; formal analysis, B.D.; investigation, B.D.; resources, Q.C.; data curation, B.D.; writing—original draft preparation, B.D.; writing—review and editing, Q.C.; visualization, Q.C.; supervision, Q.C.; project administration, Q.C.; funding acquisition, Q.C. All authors have read and agreed to the published version of the manuscript.

Funding: This work was supported by the National Natural Science Foundation of China (Grant No. 31870658).

Data Availability Statement: The assembled sequences of En3 were deposited in GenBank (accession number CP124218) and the Genome Warehouse in the National Genomics Data Center (NGDC) (accession number GWHEQCH00000000). A culture of strain En3 was deposited in the China Center for Type Culture Collection (AB 2023153).

Conflicts of Interest: The authors declare no conflict of interest.

References

1. Bacon, C.W.; Hinton, D.M. Bacterial endophytes: The endophytic niche, its occupants, and its utility. In *Plant-Associated Bacteria*; Gnanamanickam, S.S., Ed.; Springer: Dordrecht, The Netherlands, 2006; pp. 155–194.
2. Morales-Cedeño, L.R.; del Carmen Orozco-Mosqueda, M.; Loeza-Lara, P.D.; Parra-Cota, F.I.; de los Santos-Villalobos, S.; Santoyo, G. Plant growth-promoting bacterial endophytes as biocontrol agents of pre- and post-harvest diseases: Fundamentals, methods of application and future perspectives. *Microbiol. Res.* **2021**, *242*, 126612. [CrossRef] [PubMed]
3. Santoyo, G.; Moreno-Hagelsieb, G.; del Carmen Orozco-Mosqueda, M.; Glick, B.R. Plant growth-promoting bacterial endophytes. *Microbiol. Res.* **2016**, *183*, 92–99. [CrossRef] [PubMed]
4. Lalucat, J.; Mulet, M.; Gomila, M.; García-Valdés, E. Genomics in Bacterial Taxonomy: Impact on the Genus *Pseudomonas*. *Genes* **2020**, *11*, 139. [CrossRef] [PubMed]
5. Yan, Y.L.; Yang, J.; Dou, Y.T.; Chen, M.; Ping, S.Z.; Peng, J.P.; Lu, W.; Zhang, W.; Yao, Z.Y.; Li, H.Q.; et al. Nitrogen fixation island and rhizosphere competence traits in the genome of root-associated *Pseudomonas stutzeri* A1501. *Proc. Natl. Acad. Sci. USA* **2008**, *105*, 7564–7569. [CrossRef]
6. Oteino, N.; Lally, R.D.; Kiwanuka, S.; Lloyd, A.; Ryan, D.; Germaine, K.J.; Dowling, D.N. Plant growth promotion induced by phosphate solubilizing endophytic *Pseudomonas* isolates. *Front. Microbiol.* **2015**, *6*, 745. [CrossRef]

7. Emami, S.; Alikhani, H.A.; Pourbabaei, A.A.; Etesami, H.; Sarmadian, F.; Motessharezadeh, B. Assessment of the Potential of Indole-3-Acetic Acid Producing Bacteria to manage Chemical Fertilizers Application. *Int. J. Environ. Res. Public Health* **2019**, *13*, 603–611. [CrossRef]
8. Weyens, N.; Boulet, J.; Adriaensen, D.; Timmermans, J.P.; Prinsen, E.; Van Oevelen, S.; D'Haen, J.; Smeets, K.; van der Lelie, D.; Taghavi, S.; et al. Contrasting colonization and plant growth promoting capacity between wild type and a *gfp*-derative of the endophyte *Pseudomonas putida* W619 in hybrid poplar. *Plant Soil.* **2012**, *356*, 217–230. [CrossRef]
9. Ali, S.; Charles, T.C.; Glick, B.R. Delay of flower senescence by bacterial endophytes expressing 1-aminocyclopropane-1-carboxylate deaminase. *J. Appl. Microbiol.* **2012**, *113*, 1139–1144. [CrossRef]
10. Chlebek, D.; Pinski, A.; Zur, J.; Michalska, J.; Hupert-Kocurek, K. Genome Mining and Evaluation of the Biocontrol Potential of *Pseudomonas fluorescens* BRZ63, a New Endophyte of Oilseed Rape (*Brassica napus* L.) against Fungal Pathogens. *Int. J. Mol. Sci* **2020**, *21*, 8740. [CrossRef]
11. Montes-Osuna, N.; Cernava, T.; Gomez-Lama Cabanas, C.; Berg, G.; Mercado-Blanco, J. Identification of Volatile Organic Compounds Emitted by Two Beneficial Endophytic *Pseudomonas* Strains from Olive Roots. *Plants* **2022**, *11*, 318. [CrossRef]
12. Agisha, V.N.; Kumar, A.; Eapen, S.J.; Sheoran, N.; Suseelabhai, R. Broad-spectrum antimicrobial activity of volatile organic compounds from endophytic *Pseudomonas putida* BP25 against diverse plant pathogens. *Biocontrol. Sci. Technol.* **2019**, *29*, 1069–1089. [CrossRef]
13. Kong, Z.Y.; Deng, Z.S.; Glick, B.R.; Wei, G.H.; Chou, M.X. A nodule endophytic plant growth-promoting *Pseudomonas* and its effects on growth, nodulation and metal uptake in *Medicago lupulina* under copper stress. *Ann. Microbiol.* **2017**, *67*, 49–58. [CrossRef]
14. Chen, B.; Luo, S.; Wu, Y.J.; Ye, J.Y.; Wang, Q.; Xu, X.M.; Pan, F.S.; Khan, K.Y.; Feng, Y.; Yang, X. The effects of the endophytic bacterium *Pseudomonas fluorescens* Sasm05 and IAA on the plant growth and cadmium uptake of *Sedum alfredii* Hance. *Front. Microbiol.* **2017**, *8*, 2538. [CrossRef]
15. Ma, Y.; Rajkumar, M.; Moreno, A.; Zhang, C.; Freitas, H. Serpentine endophytic bacterium *Pseudomonas azotoformans* ASS1 accelerates phytoremediation of soil metals under drought stress. *Chemosphere* **2017**, *185*, 75–85. [CrossRef] [PubMed]
16. Bradshaw, H.; Ceulemans, R.; Davis, J.; Stettler, R. Emerging model systems in plant biology: Poplar (Populus) as a model forest tree. *J. Plant Growth Regul.* **2000**, *19*, 306–313. [CrossRef]
17. Xin, G.; Zhang, G.Y.; Kang, J.W.; Staley, J.T.; Doty, S.L. A diazotrophic, indole-3-acetic acid-producing endophyte from wild cottonwood. *Biol. Fertil. Soils* **2009**, *45*, 669–674. [CrossRef]
18. Ren, J.H.; Ye, J.R.; Liu, H.; Xu, X.L.; Wu, X.Q. Isolation and characterization of a new *Burkholderia pyrrocinia* strain JK-SH007 as a potential biocontrol agent. *World J. Microbiol. Biotechnol.* **2011**, *27*, 2203–2215. [CrossRef]
19. Liu, W.H.; Chen, F.F.; Wang, C.E.; Fu, H.H.; Fang, X.Q.; Ye, J.R.; Shi, J.Y. Indole-3-Acetic Acid in *Burkholderia pyrrocinia* JK-SH007: Enzymatic Identification of the Indole-3-Acetamide Synthesis Pathway. *Front. Microbiol.* **2019**, *10*, 2559. [CrossRef]
20. Taghavi, S.; Garafola, C.; Monchy, S.; Newman, L.; Hoffman, A.; Weyens, N.; Barac, T.; Vangronsveld, J.; van der Lelie, D. Genome survey and characterization of endophytic bacteria exhibiting a beneficial effect on growth and development of poplar trees. *Appl. Environ. Microbiol.* **2009**, *75*, 748–757. [CrossRef]
21. Prodhan, M.Y.; Rahman, M.B.; Rahman, A.; Akbor, M.A.; Ghosh, S.; Nahar, M.N.E.; Simo; Shamsuzzoha, M.; Cho, K.M.; Haque, M.A. Characterization of Growth-Promoting Activities of Consortia of Chlorpyrifos Mineralizing Endophytic Bacteria Naturally Harboring in Rice Plants—A Potential Bio-Stimulant to Develop a Safe and Sustainable Agriculture. *Microorganisms* **2023**, *11*, 1821. [CrossRef]
22. Nautiyal, C.S. An efficient microbiological growth medium for screening phosphate solubilizing microorganisms. *FEMS Microbiol. Lett.* **1999**, *170*, 265–270. [CrossRef] [PubMed]
23. Kizilkaya, R. Yield response and nitrogen concentrations of spring wheat (*Triticum aestivum*) inoculated with *Azotobacter Chroococcum* strains. *Ecol. Eng.* **2008**, *33*, 150–156. [CrossRef]
24. Bric, J.M.; Bostock, R.M.; Silverstone, S.E. Rapid in situ assay for indoleacetic acid production by bacteria immobilized on a nitrocellulose membrane. *Appl. Environ. Microbiol.* **1991**, *57*, 535–538. [CrossRef] [PubMed]
25. Sun, K.; Liu, J.; Gao, Y.Z.; Sheng, Y.H.; Kang, F.X.; Waigi, M.G. Inoculating plants with the endophytic bacterium *Pseudomonas* sp. Ph6-gfp to reduce phenanthrene contamination. *Environ. Sci. Pollut. Res.* **2015**, *22*, 19529–19537. [CrossRef]
26. Wick, R.R.; Judd, L.M.; Gorrie, C.L.; Holt, K.E. Unicycler: Resolving bacterial genome assemblies from short and long sequencing reads. *PLoS Comput. Biol.* **2017**, *13*, e1005595. [CrossRef]
27. Walker, B.J.; Abeel, T.; Shea, T.; Priest, M.; Abouelliel, A.; Sakthikumar, S.; Cuomo, C.A.; Zeng, Q.D.; Wortman, J.; Earl, A.M.; et al. Pilon: An integrated tool for comprehensive microbial variant detection and genome assembly improvement. *PLoS ONE* **2014**, *9*, e112963. [CrossRef]
28. Seemann, T. Prokka: Rapid prokaryotic genome annotation. *Bioinformatics* **2014**, *30*, 2068–2069. [CrossRef]
29. Lalucat, J.; Gomila, M.; Mulet, M.; Zaruma, A.; Garcia-Valdes, E. Past, present and future of the boundaries of the *Pseudomonas* genus: Proposal of *Stutzerimonas* gen. Nov. *Syst. Appl. Microbiol.* **2022**, *45*, 126289. [CrossRef]
30. Kumar, S.; Stecher, G.; Tamura, K. MEGA7: Molecular Evolutionary Genetics Analysis Version 7.0 for Bigger Datasets. *Mol. Biol. Evol.* **2016**, *33*, 1870–1874. [CrossRef]
31. Yoon, S.H.; Ha, S.M.; Lim, J.M.; Kwon, S.J.; Chun, J. A large-scale evaluation of algorithms to calculate average nucleotide identity. *Antonie Van Leeuwenhoek* **2017**, *110*, 1281–1286. [CrossRef]

32. Alanjary, M.; Steinke, K.; Ziemert, N. AutoMLST: An automated web server for generating multi-locus species trees highlighting natural product potential. *Nucleic. Acids. Res.* **2019**, *47*, W276–W282. [CrossRef] [PubMed]
33. Richter, M.; Rosselló-Móra, R. Shifting the genomic gold standard for the prokaryotic species definition. *Proc. Natl. Acad. Sci. USA* **2009**, *106*, 19126–19131. [CrossRef] [PubMed]
34. Peix, A.; Ramírez-Bahena, M.H.; Velázquez, E. Historical evolution and current status of the taxonomy of genus *Pseudomonas*. *Infect. Genet. Evol.* **2009**, *9*, 1132–1147. [CrossRef] [PubMed]
35. Mende, D.R.; Sunagawa, S.C.; Zeller, G.; Bork, P. Accurate and universal delineation of prokaryotic species. *Nat. Methods* **2013**, *10*, 881–884. [CrossRef]
36. Glaeser, S.P.; Kämpfer, P. Multilocus sequence analysis (MLSA) in prokaryotic taxonomy. *Syst. Appl. Microbiol.* **2015**, *38*, 237–245. [CrossRef]
37. Goris, J.; Konstantinidis, K.T.; Klappenbach, J.A.; Coenye, T.; Vandamme, P.; Tiedje, J.M. DNA-DNA hybridization values and their relationship to whole-genome sequence similarities. *Int. J. Syst. Evol. Microbiol.* **2007**, *57*, 81–91. [CrossRef]
38. Girard, L.; Lood, C.; Höfte, M.; Vandamme, P.; Rokni-Zadeh, H.; van Noort, V.; Lavigne, R.; De Mot, R. The Ever-Expanding Pseudomonas Genus: Description of 43 New Species and Partition of the *Pseudomonas* putida Group. *Microorganisms* **2021**, *9*, 1766. [CrossRef]
39. Peix, A.; Rivas, R.; Mateos, P.F.; Martínez-Molina, E.; Rodríguez-Barrueco, C.; Velázquez, E. *Pseudomonas rhizosphaerae* sp nov., a novel species that actively solubilizes phosphate in vitro. *Int. J. Syst. Evol. Microbiol.* **2003**, *53*, 2067–2072. [CrossRef]
40. Menéndez, E.; Ramírez-Bahena, M.H.; Fabryová, A.; Igual, J.M.; Benada, O.; Mateos, P.F.; Peix, A.; Kolařík, M.; García-Fraile, P. *Pseudomonas coleopterorum* sp nov., a cellulase-producing bacterium isolated from the bark beetle *Hylesinus fraxini*. *Int. J. Syst. Evol. Microbiol.* **2015**, *65*, 2852–2858. [CrossRef]
41. Malhotra, H.; Vandana; Sharma, S.; Pandey, R. Phosphorus nutrition: Plant growth in response to deficiency and excess. In *Plant Nutrients and Abiotic Stress Tolerance*; Hasanuzzaman, M., Fujita, M., Oku, H., Nahar, K., Hawrylak-Nowak, B., Eds.; Springer Nature Ltd.: Singapore, 2018; pp. 171–190.
42. Zeng, Q.W.; Wu, X.Q.; Wen, X.Y. Identification and characterization of the rhizosphere phosphate-solubilizing bacterium *Pseudomonas frederiksbergensis* JW-SD2, and its plant growth-promoting effects on poplar seedlings. *Ann. Microbiol.* **2016**, *66*, 1343–1354. [CrossRef]
43. Batista, M.B.; Dixon, R. Manipulating nitrogen regulation in diazotrophic bacteria for agronomic benefit. *Biochem. Soc. Trans.* **2019**, *47*, 603–614. [CrossRef] [PubMed]
44. Li, H.B.; Singh, R.K.; Singh, P.; Song, Q.Q.; Xing, Y.X.; Yang, L.T.; Li, Y.R. Genetic diversity of nitrogen-fixing and plant growth promoting *Pseudomonas* species isolated from sugarcane rhizosphere. *Front. Microbiol.* **2017**, *8*, 1268. [CrossRef] [PubMed]
45. Fox, A.R.; Soto, G.; Valverde, C.; Russo, D.; Lagares, A., Jr.; Zorreguieta, Á.; Alleva, K.; Pascuan, C.; Frare, R.; Ayub, N.D. Major cereal crops benefit from biological nitrogen fixation when inoculated with the nitrogen—Fixing bacterium *Pseudomonas protegens* Pf—5 X940. *Environ. Microbiol.* **2016**, *18*, 3522–3534. [CrossRef] [PubMed]
46. Ke, X.B.; Feng, S.; Wang, J.; Lu, W.; Zhang, W.; Chen, M.; Lin, M. Effect of inoculation with nitrogen-fixing bacterium *Pseudomonas stutzeri* A1501 on maize plant growth and the microbiome indigenous to the rhizosphere. *Syst. Appl. Microbiol.* **2019**, *42*, 248–260. [CrossRef]
47. Johnson, D.C.; Dos Santos, P.C.; Dean, D.R. NifU and NifS are required for the maturation of nitrogenase and cannot replace the function of *isc*-gene products in *Azotobacter vinelandii*. *Biochem. Soc. Trans.* **2005**, *33*, 90–93. [CrossRef]
48. Reitzer, L. Nitrogen assimilation and global regulation in *Escherichia coli*. *Annu. Rev. Microbiol.* **2003**, *57*, 155–176. [CrossRef]
49. Patten, C.L.; Glick, B.R. Role of *Pseudomonas putida* indoleacetic acid in development of the host plant root system. *Appl. Environ. Microbiol.* **2002**, *68*, 3795–3801. [CrossRef]
50. Gupta, R.; Anand, G.; Bar, M. Developmental Phytohormones: Key Players in Host-Microbe Interactions. *J. Plant Growth Regul.* **2023**, *42*, 1–22. [CrossRef]
51. McClerklin, S.A.; Lee, S.G.; Harper, C.P.; Nwumeh, R.; Jez, J.M.; Kunkel, B.N. Indole-3-acetaldehyde dehydrogenase-dependent auxin synthesis contributes to virulence of *Pseudomonas syringae* strain DC3000. *PLoS Pathog.* **2018**, *14*, e1006811. [CrossRef]
52. Gao, Y.B.; Dai, X.H.; Aoi, Y.K.; Takebayashi, Y.; Yang, L.P.; Guo, X.R.; Zeng, Q.W.; Yu, H.C.Z.; Kasahara, H.; Zhao, Y.D. Two homologous *INDOLE-3-ACETAMIDE* (*IAM*) *HYDROLASE* genes are required for the auxin effects of IAM in *Arabidopsis*. *J. Genet. Genom.* **2020**, *47*, 157–165. [CrossRef]
53. Bernal, P.; Llamas, M.A.; Filloux, A. Type VI secretion systems in plant—Associated bacteria. *Environ. Microbiol.* **2018**, *20*, 1–15. [CrossRef] [PubMed]

forests

MDPI

Article

Increased Vegetation Productivity of Altitudinal Vegetation Belts in the Chinese Tianshan Mountains despite Warming and Drying since the Early 21st Century

Yong Zhang, Chengbang An *, Lai Jiang, Liyuan Zheng, Bo Tan, Chao Lu, Wensheng Zhang and Yanzhen Zhang

Key Laboratory of Western China's Environmental Systems (Ministry of Education), College of Earth and Environmental Sciences, Lanzhou University, Lanzhou 730000, China; zynmg@outlook.com (Y.Z.)
* Correspondence: cban@lzu.edu.cn

Abstract: Gaining a deep understanding of how climate change affects the carbon cycle in dryland vegetation is of utmost importance, as it plays a pivotal role in shaping the overall carbon cycle in global ecosystems. It is currently not clear how plant communities at varying elevations in arid mountainous regions will respond to climate change in terms of their productivity. The aim of this study was to investigate the effect of climate change on vegetation productivity in different altitudinal vegetation belts of the Tianshan Mountains between 2000 and 2021, utilizing satellite-derived vegetation productivity and climate data. The findings suggest a notable increase in vegetation productivity across diverse altitudinal vegetation belts. The productivity of vegetation in the coniferous forest and alpine meadow belts displayed a notably higher interannual trend compared to other vegetation belts. Notably, an increase in vegetation productivity was accompanied by warming and drying. The productivity of altitudinal vegetation belts, however, appears to be resilient to current climate trends and was not significantly impacted by the severity of atmospheric drought. The trend of increased vegetation productivity was primarily driven by CO_2 fertilization. Our results highlight that the extent of climate change may need to reach a threshold to noticeably affect variations in vegetation productivity in arid mountainous.

Keywords: atmospheric drought; CO_2 fertilization; mountain ecosystem; vegetation carbon flux; global warming hiatus

1. Introduction

The investigation of carbon sinks in terrestrial ecosystems holds significant scientific value in the efforts to mitigate global anthropogenic warming. The primary objective of the United Nations Paris Agreement is to limit the increase in global average temperature to below 2.0 °C above preindustrial levels and pursue efforts to limit the increase to 1.5 °C in order to tackle the adverse impacts of climate change [1,2]. Because of the importance of fossil fuel energy use in economic development in some countries, reducing fossil fuel energy use plays a limited role in achieving this core objective [3]. Vegetation and soil carbon pools absorb carbon dioxide from the atmosphere; the amount absorbed annually is equivalent to 20–30% of the carbon dioxide emitted by humans [4]. This indicates the significant contribution of terrestrial ecosystems to the global carbon cycle [5]. Terrestrial ecosystems include vegetation and soil carbon pools [6]. All vegetation communities can fix approximately 120 Pg C from the atmosphere annually, and this carbon exchange is highly sensitive to climate change [7,8]. Gross primary productivity (GPP) and net primary productivity (NPP) are commonly employed metrics for assessing the magnitude and variability of vegetation productivity [9,10]. GPP represents the overall quantity of carbon that vegetation captures from the atmosphere through carbon assimilation, while NPP represents GPP minus the portion consumed by plant respiration, which forms plant biomass [11–13]. Quantitative analysis of changes in GPP and NPP in different plant communities can improve the understanding of the feedback of terrestrial ecosystem carbon cycles to current climate change.

There is evident spatial heterogeneity in both the distribution and variability of vegetation productivity. Biotic and abiotic factors have an impact on the spatial distribution of both NPP and GPP [14]. Forests and grasslands have remarkably higher carbon pools than shrubs and wetlands [15]. Globally, GPP and NPP decrease with an increase in soil nutrients [16]. Temperature, water availability, and solar radiation determine the meridional distributions of GPP and NPP [17], and those at low elevations are dependent on the carbon use efficiency; however, GPP and NPP at high elevations are mainly limited by temperature and humidity [18]. In addition to the spatial distribution, changes in vegetation productivity are regulated by several environmental factors. Meta-analyses have shown that warming had a positive effect on GPP in global grasslands and that carbon fluxes increased more in boreal region grasslands than in temperate and subarid grasslands [7,19]. At the regional scale, changes in vegetation productivity of different ecosystems on the Tibetan Plateau are mainly influenced by carbon dioxide (CO_2) concentration, temperature, and moisture; however, the main drivers of such changes are controversial [20–23]. Notably, warming and drying reduce the availability of soil moisture in arid regions, weakening vegetation productivity. Therefore, warming may negatively affect the GPP and NPP in arid regions [24–26].

The trend in vegetation productivity demonstrates inconsistent patterns both on a global and regional scale. During the period from 1982 to 2015, there was an increase in GPP across tropical rainforests, with the exception of the Congo rainforest [17]. Among them, the Amazon rainforest displayed a higher magnitude of increase. Additionally, GPP demonstrated an upward trend in North America and eastern China. Conversely, it decreased in the western mountain ranges of the United States, the Australian deserts, and western Central Asia. The variances in global GPP trends can be attributed to changes in temperature and humidity, which serve as the primary influencing factors. There has also been attention devoted to studying changes in montane vegetation productivity. From 1982 to 2019, the NPP of mountain vegetation in Yunnan Province, China, exhibited an overall increasing trend [27]. Among them, mountain forests demonstrated higher rates of increase. However, during the period from 2000 to 2012, the NPP in the Qilian Mountains, situated in the arid zone of China, experienced a decreasing trend primarily due to atmospheric drought [28]. On the Tibetan Plateau, the NPP of desert grassland and alpine grassland displayed an upward trend between 2001 and 2017, while deciduous broad-leaved forests experienced a decrease [29]. Additionally, severe drought in the years 2000 to 2010 led to reduced NPP in fir forests in the mountains of southern China [30].

Given their status as critical global biodiversity hotspots, mountain vegetation in arid zones is highly responsive to climate change [31,32]. Similarly, arid regions are sensitive and vulnerable to global change, and warming leads to remarkable changes in vegetation productivity [33]. Studying and identifying the changes and factors influencing vegetation productivity in arid mountainous areas can enhance our understanding of the crucial role that vegetation plays in the global carbon cycle within arid regions. The Tianshan Mountains, situated within the arid region of Central Asia, not only provide important ecosystem services for the surrounding plains but also serve as important carbon pools [34]. Relatively few studies have been conducted on vegetation productivity in the Tianshan Mountains, and several research findings indicate that changes in vegetation productivity in this area are closely related to air temperature, precipitation, and CO_2 fertilization [35,36]. Significantly, since the 1980s, the Tianshan Mountains have exhibited trends of warming and increased precipitation, and the wetting trend has an obvious elevation dependence [37]. Additionally, relatively complete altitudinal vegetation belts have developed in the Tianshan Mountains [38]. Nevertheless, there is currently a lack of comprehensive research on how climate change affects the productivity of various vegetation belts in the Tianshan Mountains.

To fill this knowledge gap, we used remote sensing data of ecosystem productivity (GPP and NPP) to investigate changes in productivity at different vegetation belts in the Tianshan Mountains over the 21st century and attempted to analyze the potential drivers of this trend. Based on these analyses, we aimed to determine (1) whether there is a consistent response to climate change among the productivity of different vegetation belts at different elevations and (2) whether warming and increased precipitation or warming and decreased precipitation are the primary factors responsible for the interannual variation in primary productivity among different vegetation belts at various altitudes in the Tianshan Mountains. In summary, our study offers novel insights into how mountain vegetation in arid regions responds to climate change and contributes to a deeper understanding of the role of arid mountain regions in the global vegetation carbon cycle.

2. Materials and Methods

2.1. Study Area

The Tianshan Mountains are the largest mountain range in the arid region of Central Asia, stretching across China, Kazakhstan, Kyrgyzstan, and Uzbekistan in an east-west direction (Figure 1a). Xinjiang, China, is home to sixty percent of the Tianshan Mountains. The Tianshan Mountains exhibit a temperate continental climate, with an average annual precipitation range of 27.4 to 673.2 mm (1990–2020) and average annual temperatures varying from −15.4 to 14.1 °C. The higher relative elevation ranges (−154 to 7439 m) have led to the formation of a relatively complete altitudinal vegetation belt in the Tianshan Mountains [39]. Ref. [40] provided clarification on the average elevation range and composition of altitudinal vegetation belts within the Tianshan Mountains (Figure 1b). The climatic characteristics of vegetation belts at high elevations (>1500 m) are characterized by warm and humid conditions, and those at low elevations (<1500 m) are characterized by warm and dry conditions (Figure 1b). Land cover data from the Tianshan Mountains indicate that the natural vegetation types mainly include forests (tree cover), grasslands, and sparse vegetation (Figure 1c). Grasslands and sparse vegetation have the largest distribution, and both have an area proportion of more than 40% (Figure 1c). Although only comprising 3.4% of the total area, forests serve an important role as carbon sinks.

2.2. Climate Data

Our previous research has shown a strong agreement between climate observatory data and reanalysis data (i.e., ERA5-Land and TerraClimate) from the Tianshan Mountains throughout the entire time series [41]. ERA5-Land is a comprehensive global climate reanalysis dataset designed for global and regional climate change studies. It offers a wide range of climate variables and boasts a spatial resolution of 0.1° × 0.1° (approximately 9 km) and a temporal resolution ranging from months to hours [42]. In this study, we used monthly temperatures from the dataset (https://cds.climate.copernicus.eu/cdsapp, accessed on 8 May 2023) to calculate the average growing season (i.e., May to September) and obtain growing season temperatures (MAT) each year in the Tianshan Mountains for the years 2000–2021. TerraClimate is a hydrological, climatic dataset that provides high-resolution coverage worldwide. The resolution it provides is 4 km spatially and 1 month temporally [43]. VPD, commonly used as a measure of atmospheric drought, is defined as the disparity between saturation vapor pressure and actual vapor pressure [44]. We utilized monthly precipitation and vapor pressure deficit (VPD) (https://www.climatologylab.org/terraclimate.html, accessed on 14 May 2023) data to calculate the sum of growing season precipitation (PRE) and the average of growing season VPD in the Tianshan Mountains throughout the study period.

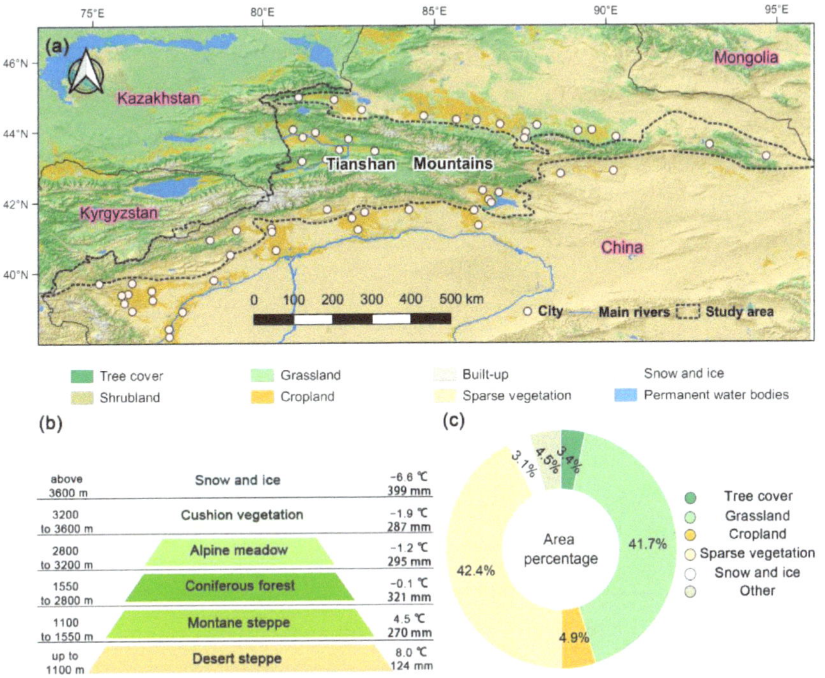

Figure 1. Composition and distribution of major vegetation types in the Tianshan Mountains. (**a**) Spatial distribution of land cover. (**b**) Composition and climatic characteristics (average annual precipitation and temperature) of altitudinal vegetation belts. (**c**) Percentage of area of major vegetation types. Land cover data were obtained from ESA WorldCover open access (https://worldcover2021.esa.int/, accessed on 3 May 2023).

2.3. *Atmospheric CO_2 Concentration*

Since the global distribution of atmospheric CO_2 is heterogeneous in time and space (e.g., seasonal and latitudinal differences) [45,46], it is essential to employ atmospheric CO_2 measurements from regional atmospheric background stations. This allows for a comprehensive examination of the effects of CO_2 concentration variations on regional vegetation carbon fluxes. The Mt. Waliguan (WLG) station is located in Qinghai Province, China, which belongs to the hinterland of the Eurasian continent, with geographic coordinates of 36.29° N, 100.89° E, and an altitude of 3810 m above sea level, and it is one of the observational background stations of the World Meteorological Organization's Global Atmospheric Observations project. The WLG station is far from areas of intensive human activity and has a predominantly semi-arid steppe vegetation type [47]. The Tianshan Mountains, situated in the hinterland of the Eurasian continent, are also relatively less impacted by human activities. Therefore, it is reliable to use the CO_2 concentration (ppm) measured by the WLG station to analyze the effects on carbon fluxes in the Tianshan Mountains. Here, we used monthly scale CO_2 concentrations to perform growing season averaging to obtain growing season CO_2 concentration data over the study epoch (https://community.wmo.int/en/activity-areas/gaw, accessed on 20 May 2023).

2.4. Land Cover Data

A global annual land cover map, which utilizes imagery from Sentinel-1 and Sentinel-2 satellites, was created by the ESA WorldCover team (https://worldcover2021.esa.int/, accessed on 1 June 2023). This land cover map has a spatial resolution of 10 m, contains 11 categories, and has an overall classification accuracy of 77% [48]. The higher classification accuracy ensures the reliability of extracting the range of altitudinal vegetation belts. In this study, we reclassified the ESA WorldCover data (version 2021) based on the average elevation range of each altitudinal vegetation belt (Table 1). Based on this process, the geographical boundaries of each altitudinal vegetation belt were extracted. The detailed spatial distribution pattern and reclassification information are presented in Figure 2 and Table 1.

Table 1. Reclassification information of major vegetation types based on ESA WorldCover data.

WorldCover Categories	Reclassification Categories	Elevation Ranges (m)	Abbreviations
Main Vegetation Types	**Altitudinal Vegetation Belts**		
Sparse vegetation	Cushion vegetation	3200–3600	CV
	Desert steppe	<1100	DS
Grassland	Alpine meadow	2800–3200	AM
	Montane steppe	1100–1550	MS
Tree cover	Coniferous forest	1550–2800	CF

Note: The specific elevational range of each altitudinal vegetation belt was obtained from the results of [40]. This range represents the optimal elevation for the distribution of each vegetation type.

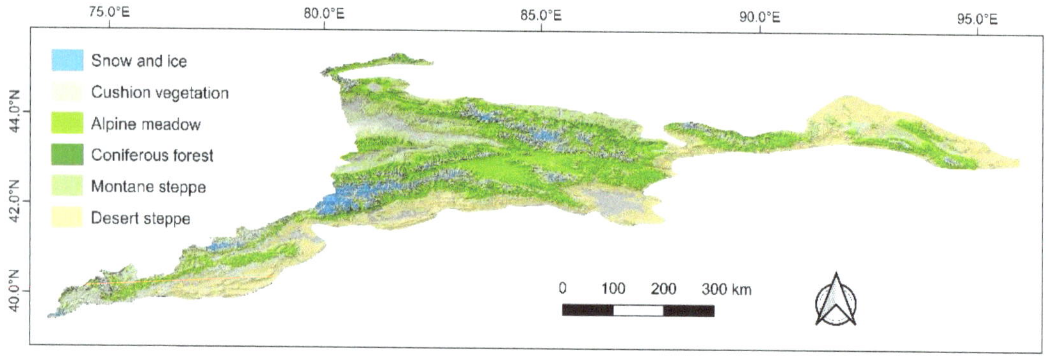

Figure 2. Spatial patterns of vegetation belts at different altitudes.

2.5. Elevation Data (DEM)

The DEMs derived from the Space Shuttle Radar Topography Mission (SRTM) have been extensively utilized as supplementary data in studies related to climatology, ecology, and geomorphology [49]. SRTM DEMs have a spatial resolution of 90 m, an average elevation error of 16 m, and cover 80% of the global area. We used SRTM DEMs (version 4) to acquire the spatial extent of different altitudinal vegetation belts (https://srtm.csi.cgiar.org/, accessed on 5 June 2023).

2.6. Primary Productivity

Vegetation productivity data (GPP and NPP) were obtained free of charge from the MODIS satellite image-derived MOD17A2H-based product. MOD17A2H version 06 offers GPP and NPP at 500 m spatial resolution and intervals of 8 days [50]. We preprocessed the NPP and GPP raster files by reprojecting them using the Google Earth Engine (the new coordinate system is World Geodetic System 1984). MOD17A2H also includes a Psn_QC band that provides quality information for GPP and NPP in the designated area.

By utilizing this band, we can effectively filter and acquire high-quality GPP and NPP data where the Psn_QC value is equal to 0. We summed all the GPP and NPP raster data separately for each growing season from 2000 to 2021. Finally, the growing season GPP (g C m^{-2} yr^{-1}) and NPP (g C m^{-2} yr^{-1}) for 22 years were obtained using the study area masking process.

The accuracy of MODIS-derived GPP from different biomes has been validated using 800 eddy covariance flux tower GPPs (FLUXNET GPP) worldwide. The validation results show good agreement between MODIS-derived GPP and FLUXNET GPP, with spatial and temporal variations [51]. In other words, MODIS GPP and NPP are reliable data to investigate temporal variations in vegetation productivity in diverse altitudinal vegetation belts in the Tianshan Mountains.

2.7. Statistical Analyses

We used the Shapiro-Wilk test to perform normality tests for all climate variables and vegetation productivity data. We found that all the variables were normally distributed. Simple linear regression involves modeling the relationship between two variables, namely the independent and dependent variables [52]. It is utilized to approximate the dependent variable through the use of the best-fit line. Simple linear regression was employed to assess interannual trends (slope (k)) in climate (i.e., MAT, PRE, and VPD) and vegetation productivity (GPP and NPP). The linear regression equation can be represented by the following formula:

$$y = b + kx \tag{1}$$

where y represents the dependent variable, which refers to the climatic variables and vegetation productivity chosen for this study. x represents the time period from 2000 to 2021. k and b represent the slope and intercept of the fitted curve, respectively. A positive slope denotes an upward trend in the time series data, while a negative slope indicates a downward trend. The absolute value of the slope indicates the magnitude of the trend per unit of time.

The Mann–Kendall test was utilized to assess the statistical significance of inter-annual trends in both the climate and vegetation productivity data. In this study, the Python third-party package "pymannkendall" was employed to calculate the p-value for the slope and perform the significance test. The pymannkendall is a Python package used for conducting the Mann–Kendall trend test, a widely adopted method for detecting trends in time series data and assessing the statistical significance of those trends. Using this Python package, we analyzed the spatial distribution of interannual temperature and precipitation trends in the Tianshan Mountains (see Figure 3). Additionally, we calculated the regional average climate trends across various altitudinal vegetation belts (see Figure 4).

Pearson's correlation coefficients were utilized to ascertain the correlation between climatic variables and vegetation productivity. We utilized OriginPro software (version 2023b, OriginLab Corp., Northampton, MA, USA) for the calculation of correlation coefficients and conducted a two-tailed significance test to assess the statistical significance of the correlation coefficients.

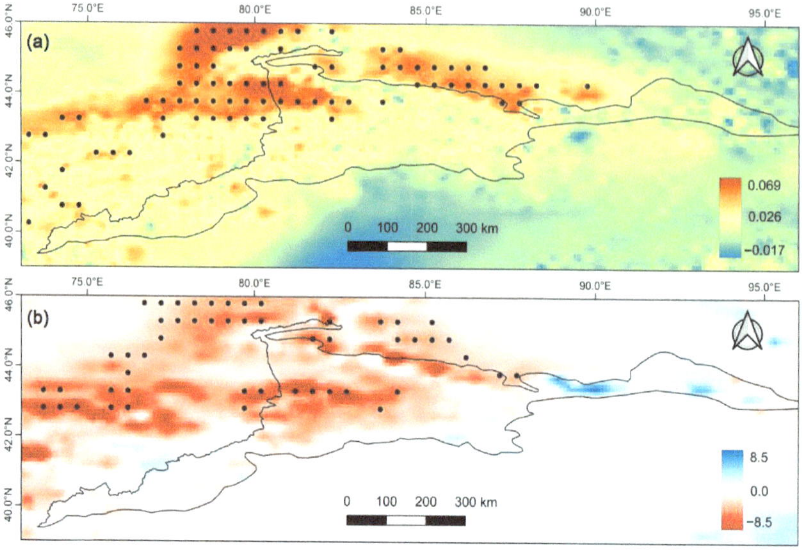

Figure 3. Climate tendency in the Tianshan Mountains from 2000 to 2021. (**a**) Interannual tendencies of MAT (°C yr^{-1}) and (**b**) PRE (mm yr^{-1}) in the Tianshan Mountains. Areas marked with black dots indicate statistically significant trends at the 95% confidence level.

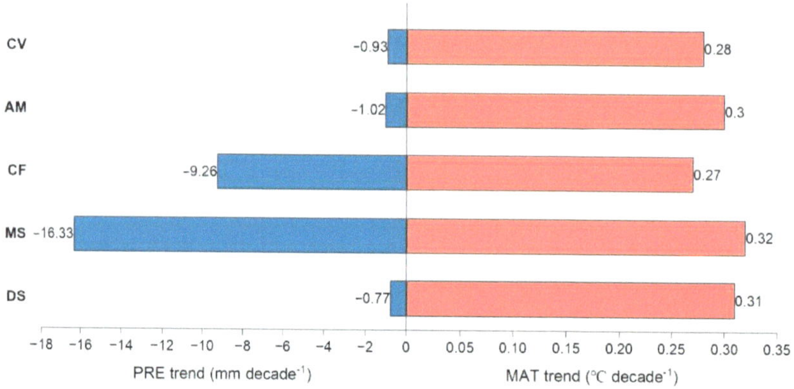

Figure 4. Average MAT (°C decade^{-1}) and PRE (mm decade^{-1}) trends across different altitudinal vegetation belts. The meaning of the abbreviation for each vegetation belt is shown in Table 1. The significance test was not passed by the MAT and PRE trends for all altitudinal vegetation belts ($p > 0.05$).

3. Results

3.1. Tendency of the Climate in the Tianshan Mountains

Tianshan Mountains showed warming and drying. Here, we analyzed the MAT and PRE trends in different altitudinal vegetation belts from 2000 to 2021. During the study period, the Tianshan Mountains experienced slight warming (Figure 3a). Only the northwestern part of the Tianshan Mountains experiences significant warming, accounting for a mere 9.4% of the entire area. The MAT trend ranged from −0.2 to 0.7 °C decade^{-1}, and the average MAT trend was 0.2 °C decade^{-1}. Further, PRE decreased slightly (Figure 3b). Only certain regions in the western section of the Tianshan Mountains exhibit noteworthy aridity. The precipitation trend ranged from −42.2 to 49.6 mm decade^{-1}, with an average trend of -0.2 mm decade^{-1}. The region experiencing a notable decrease in PRE covers just 6.5% of the total area.

The MAT and PRE trends in different altitudinal vegetation belts were consistent. Although divergent altitudinal vegetation belts showed slight warming ($p > 0.05$), the warming trends varied: they were higher in the lower elevation vegetation belts (i.e., DS and MS) than in the higher elevation belts (i.e., CF and CV) (Figure 4). Despite the minimal disparity in warming magnitude among all vegetation belts, the lower-elevation vegetation belts exhibit a slightly higher warming trend compared to their higher-elevation counterparts. Moreover, precipitation has slightly decreased across all altitudinal vegetation belts: MS and CF had the largest drying trend (-16.33 and -9.26 mm decade^{-1}), and CV and DS had the smallest (below 0.7 mm decade^{-1}).

Over the course of the study period, there was no significant trend in the VPD among various vegetation belts at different altitudes. Based on Figure 5, slightly increasing trends in VPD were observed across all altitudinal vegetation belts. Among them, DS and MS exhibited a higher rate of increase compared to other vegetation belts. However, AM did not show obvious inter-annual variation in VPD. It is important to note that the VPD trends in all vegetation belts did not meet the criteria for statistical significance.

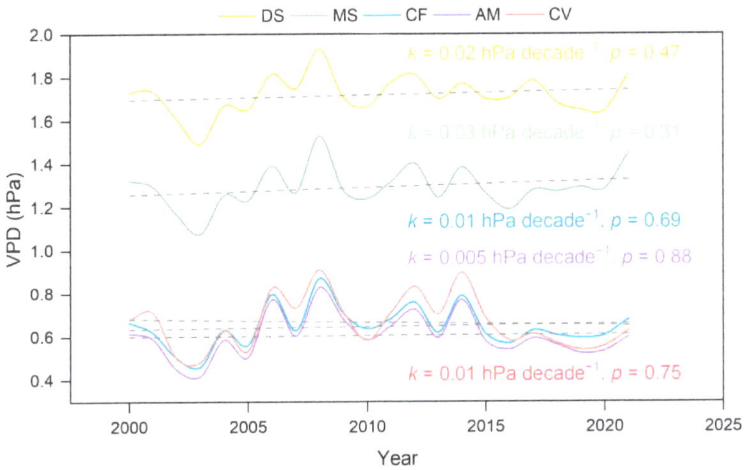

Figure 5. Mean VPD of different altitudinal vegetation belts during the growing season. Gray dashed lines are linear trend lines.

3.2. GPP Tendency in the Altitudinal Vegetation Belts Observed by Satellite

The GPP trend in the Tianshan Mountains showed remarkable spatial heterogeneity. The GPP trend ranged from -46.0 to 46.8 g C m^{-2} yr^{-1} and had an average trend of 1.7 g C m^{-2} yr^{-1} ($p < 0.05$) (Figure 6a). Notably, 15.1% of the regions experienced a decrease in GPP, and 84.9% experienced an increase in GPP (Figure 6a). Of these, the GPP trend is predominantly concentrated between 0 and 5 g C m^{-2} yr^{-1}, accounting for 75.5% of the total area. In the western part of the Tianshan Mountains (i.e., Ili River valley), where the GPP mainly decreased, all other areas showed an increase. To summarize, there was an overall increasing trend in the GPP of the Tianshan Mountains.

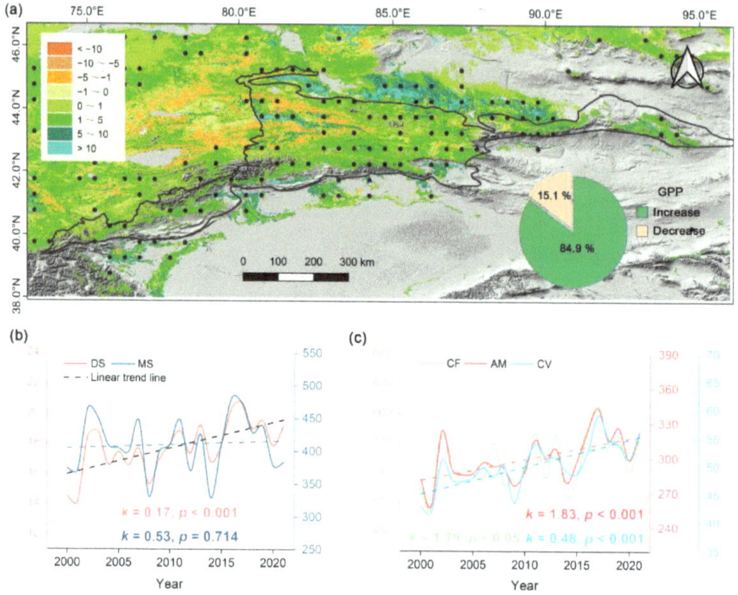

Figure 6. Temporal trend in GPP during the growing season from 2000 to 2021. (**a**) Spatial patterns of GPP trends (unit is g C m^{-2} yr^{-1}) during the growing season in the Tianshan Mountains. The area surrounded by the solid black line is the study area; the gray region is the non-vegetated area. The black dots represent statistically significant trends at the 0.05 level. (**b**) Interannual variability in regional mean GPP (y-axis unit is g C m^{-2} yr^{-1}) in lower elevation altitudinal vegetation belts. (**c**) Interannual variability in regional mean GPP (y-axis unit is g C m^{-2} yr^{-1}) in higher elevation altitudinal vegetation belts.

GPP increased in different altitudinal vegetation belts from 2000 to 2021. Specifically, in the lower elevation altitudinal vegetation belts, DS GPP exhibited a significant increase ($p < 0.001$) with a regional average trend of 0.17 g C m^{-2} yr^{-1} (Figure 6b). Nevertheless, although MS GPP showed a higher rate of increase, this trend was not significant ($p > 0.05$). GPP demonstrated a significant increase ($p < 0.05$) across all altitudinal vegetation belts at higher elevations (Figure 6c). CF and AM showed higher rates of increase, with an approximate regional average tendency of 1.8 g C m^{-2} yr^{-1}. The regional average tendency of CV was 0.5 g C m^{-2} yr^{-1}. Overall, the tendency of increasing GPP was higher in the higher elevation vegetation belts than in the lower elevation vegetation belts.

3.3. NPP Tendency in the Altitudinal Vegetation Belts Observed by Satellite

The spatial pattern of NPP trends exhibited similarity to the trends observed in GPP. The trend of NPP in the Tianshan Mountains ranged from −32.9 to 40.5 g C m^{-2} yr^{-1}, with a regional average trend of 1.4 g C m^{-2} yr^{-1} ($p < 0.05$) (Figure 7a). Similarly, lower NPP trends were observed in the western region of the study area. In contrast, the other regions showed an increase in NPP. Altogether, NPP increased in over 86.7% of the study area and decreased in over 13.3% of the study area. In which, the NPP trend is primarily concentrated between 0 and 5 g C m^{-2} yr^{-1}, representing 80.9% of the total area. The average NPP trend during the study epoch indicates that vegetation in the Tianshan Mountains acts as an important carbon sink.

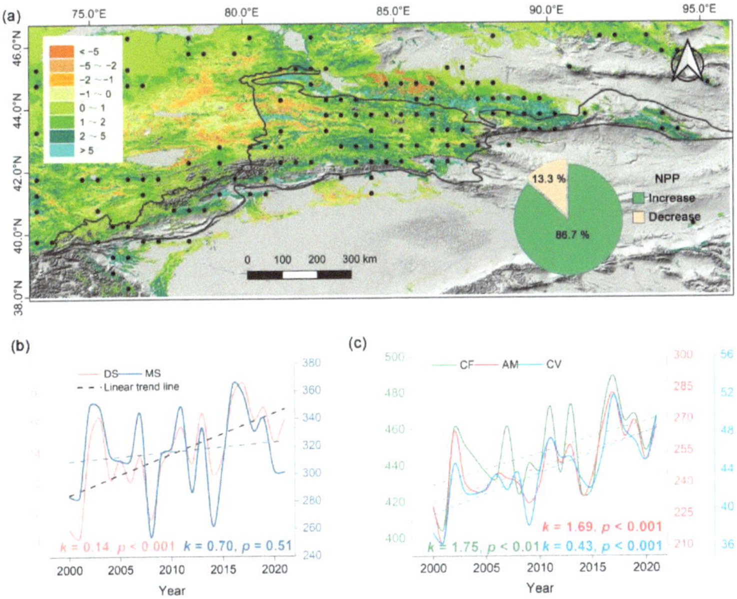

Figure 7. Spatial and temporal variations in growing season NPP from 2000 to 2021. (**a**) Spatial patterns of NPP trends (unit is g C m^{-2} yr^{-1}) during the study epoch in the Tianshan Mountains. The interpretation of the black dots in the figure remains consistent with that in Figure 6. (**b**) Interannual variability in regional mean GPP (y-axis unit is g C m^{-2} yr^{-1}) in lower elevation altitudinal vegetation belts. (**c**) Interannual variability in regional mean GPP (y-axis unit is g C m^{-2} yr^{-1}) in higher elevation altitudinal vegetation belts.

Except for that of MS, NPP in the other altitudinal vegetation belts increased considerably from 2000 to 2021. The annual growing season trends of NPP in DS and MS were 0.1 and 0.7 g C m^{-2} yr^{-1}, respectively; however, only the NPP trend in DS passed the 0.001 significance level (Figure 7b). CF and AM also had higher rates of increase in NPP, with regional average trends of 1.8 and 1.7 g C m^{-2} yr^{-1}, respectively (Figure 7c). Although the rate of increase in NPP in the CV group was less than that in the CF and AM groups, the trend was significant ($p < 0.001$). In general, altitudinal vegetation belts in the Tianshan Mountains act as carbon sinks, and the middle vegetation belts show higher carbon sink capacities than the lowest and highest belts due to denser vegetation cover.

3.4. The Impact of Current Climate Patterns

The effect of MAT on GPP variation in different vegetation belts was weak. The variations in GPP across different vegetation belts exhibited a positive correlation with temperature (Figure 8a). The correlation coefficients (0.42 and 0.39) of CF, AM, and MAT were higher relative to the other altitudinal vegetation belts (i.e., DS, MS, and CV), and the magnitude of the correlation coefficients between other altitudinal vegetation belts and MAT were similar (approximately 0.30). However, the GPP variation and MAT were not significantly correlated ($p > 0.05$). Similarly, GPP variations in all altitudinal vegetation belts were positively correlated with PRE (Figure 8a), but the extent to which precipitation affected GPP varied across altitudinal vegetation belts. Specifically, the interannual variability in GPP for the DS and MS was more sensitive to PRE variability ($r = 0.55, 0.8$; $p < 0.01, 0.001$, respectively). The impact of the PRE on interannual GPP variability in other altitudinal vegetation belts was not significant ($p > 0.05$). In general, the interannual variation in GPP in altitudinal vegetation belts exhibited positive effects from temperature and precipitation.

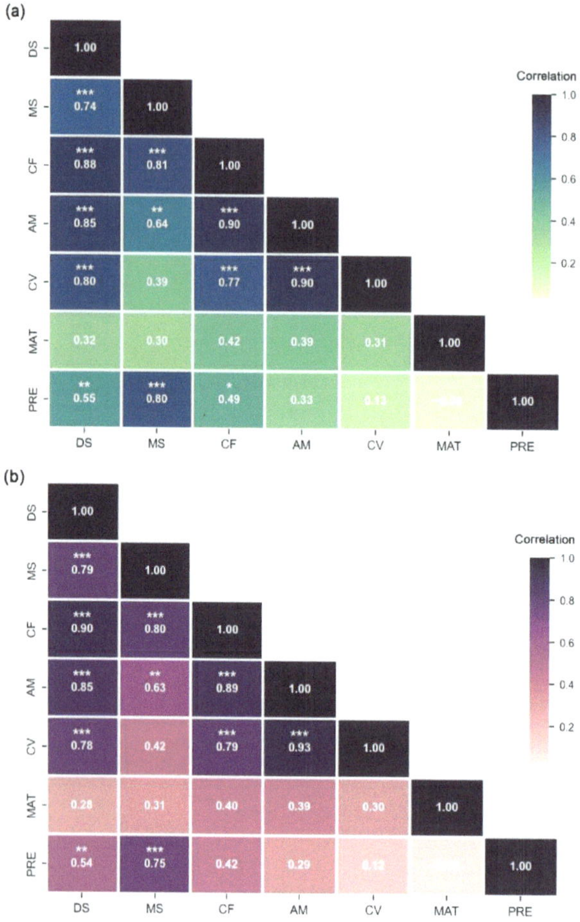

Figure 8. Correlation matrix plot of climatic variables and growing season vegetation productivity. (**a**) Correlation coefficients of GPP and climate variables for different altitudinal vegetation belts. (**b**) As in panel (**a**), but for NPP. Asterisks indicate that the correlation passed the significance test: $p < 0.05$ *, $p < 0.01$ **, $p < 0.001$ ***.

Interannual variations in NPP in the different altitudinal vegetation belts were also not sensitive to MAT. The correlation coefficients between NPP and MAT in the altitudinal vegetation belt ranged from 0.28 to 0.40; however, the correlation was not significant ($p > 0.05$) (Figure 8b). PRE also had a positive effect on NPP variability in all altitudinal vegetation belts. Of these, there was a significant positive correlation between the changes in NPP in the DS and MS and PRE ($r = 0.54, 0.75$; $p < 0.01, 0.001$), whereas those in other altitudinal vegetation belts were weakly positively correlated with the PRE. The findings indicate that the impact of PRE was felt more strongly in the vegetation carbon sinks in lower-elevation altitudinal belts from 2000 to 2021.

3.5. The Correlation between Changes in Vegetation Productivity and CO_2 Concentrations

All vegetation belts, except for MS, displayed significant correlations between changes in vegetation productivity and CO_2 concentrations. For both GPP and NPP, there was a significant positive correlation (with correlation coefficients exceeding 0.6, $p < 0.01$) between interannual changes in DS and CO_2 concentrations. Additionally, for vegetation belts at higher elevations, interannual changes in GPP and NPP were also significantly and positively correlated with CO_2 concentrations ($p < 0.01$), with CV displaying higher correlation coefficients ($r = 0.75$, $p < 0.001$). Interestingly, there was no significant correlation observed between changes in vegetation productivity at MS and CO_2 concentrations.

4. Discussion

4.1. Universality and Uniqueness of Vegetation Productivity Variations in the Tianshan Mountains

To comprehend carbon exchange between vegetated ecosystems and the atmosphere, it is crucial to monitor variations in vegetation productivity (GPP and NPP) in diverse plant communities and geographic scales [8]. The average global GPP increased remarkably from 2000 to 2019, dominated by warming and CO_2 fertilization [53]. However, vegetation productivity increased considerably in only 6.2% of the grassland areas and 7.3% of the evergreen broadleaf forest areas worldwide [9]. Few studies have analyzed changes in vegetation productivity by considering mountains as separate geographical units. From 2000 to 2016, warming dominated the increase in NPP in the high-elevation zone of the Hengduan Mountain region (subtropical monsoon climate); however, drying resulted in a decrease in NPP in the river valley region [54]. This indicates an elevational difference in the direction of vegetation productivity changes in the mountains. The Qilian Mountains (temperate continental climate) have shown decreasing trends in NPP and GPP since 2000, and the cooling trend led to a decrease in NPP after 2010 [55]. Collectively, these studies indicate that temperature is a crucial factor contributing to the alterations observed in vegetation productivity on both global and regional levels.

Our findings indicate a notable rise in GPP and NPP within the Tianshan Mountains, which aligns with global patterns. In addition, our results were in line with global and regional studies demonstrating that climate factors had positive impacts on GPP and NPP across various altitudinal vegetation belts. While our findings were consistent with the literature, we identified certain unique responses of various altitudinal vegetation belts in the Tianshan Mountains to climate change. We concluded that temperature and precipitation did not dominate the increase in GPP and NPP during the study period, as they did not change significantly (see Section 3.1); however, vegetation productivity increased significantly, suggesting that the contributions of temperature and precipitation were weak. Furthermore, although the temperature was positively correlated with vegetation productivity, the correlations were not significant ($p > 0.05$), which supports our conclusion. Although the GPP and NPP of DS and MS were significantly positively correlated with precipitation, the insignificant change in precipitation only affected the interannual fluctuation. Thus, we posit that GPP and NPP in different altitudinal vegetation belts are insensitive to current climate trends.

4.2. Why Vegetation Productivity Changes Are Insensitive to Temperature and Precipitation

The sensitivity of variations in vegetation productivity to warming diminished during the warming hiatus period. Temperature affects plant photosynthesis by influencing the rate of enzyme-catalyzed reactions [56]. Warming is expected to enhance the photosynthetic rate of vegetation and, thus, the net ecosystem carbon exchange [57]. In addition, the warming generally extends the duration of the growing season through advancements in plant leaf coloring or prolonged leaf drop, thereby enhancing vegetation productivity [20,58]. In general, warming has a positive impact on vegetation productivity. Notably, the rate of global warming did not increase remarkably from 1998 to 2012 (i.e., warming hiatus) because of the decrease in latent heat released from the surface to the atmosphere [59]. Nonetheless, the global atmospheric CO_2 concentration increased continuously during this period [60]. Despite the increasing trend (1982–2012) in global gross primary productivity (GPP), satellite observations showed a decrease of 87.2% in the rate of warming hiatus compared to the warming period (1982–1998). Furthermore, the correlation between GPP and temperature showed a decrease in sensitivity, shifting from a significant correlation during the warming period to an insignificant correlation during the warming hiatus [61]. Based on these findings, it can be strongly inferred that the positive impact of temperature on vegetation productivity diminished notably during the period of warming hiatus. Our results showed that although the rate of warming in different altitudinal vegetation belts remained high (0.27–0.32 °C decade^{-1}), the warming trend was not significant ($p > 0.05$). Similar to the relationship between global GPP and temperature, the non-significant increase in temperature since 2000 has led to a limited contribution of warming to vegetation productivity.

Decreased precipitation in the Tianshan Mountains did not negatively affect vegetation productivity. Leaf stomata are important channels of photosynthesis, respiration, and transpiration in plants [62]. During severe drought, plants reduce stomatal conductance to mitigate water loss [63]. The partial closure of plant leaf stomata attenuates the evaporative cooling effect, leading to plant death by heat stress, and reduces the photosynthetic rate and, consequently, plant carbon flux [64]. In general, severe droughts negatively affect vegetation productivity. In arid regions, VPD and soil moisture are considered the limiting factors affecting vegetation growth [65]. Global and regional studies have shown that increased VPD offsets or attenuates the positive effects of warming and CO_2 fertilization on vegetation productivity [65,66]. Our results revealed warming and precipitation reduction trends in the altitudinal vegetation belts, which may enhance atmospheric drought and negatively affect vegetation productivity. However, the VPD trends in the different altitudinal vegetation belts showed a slightly increasing trend ($p > 0.05$) from 2000 to 2021 (Figure 5), suggesting that the decrease in precipitation did not obviously affect the GPP and NPP trends. Further, the non-significant changes in growing season precipitation (Figure 4) suggest a limited role of precipitation in the GPP and NPP trends.

4.3. Factors Dominating the Tendency of Growing Vegetation Productivity in Altitudinal Vegetation Belts

Elevated CO_2 concentration can lead to increased productivity in vegetation. Solar radiation mainly regulates vegetation productivity changes in humid areas (higher cloud cover and water vapor content) [67]. Hence, the influence of solar radiation on the interannual trends of vegetation productivity in the altitudinal vegetation belts is not a topic that will be discussed. The consumption of fossil energy sources increases the concentration of atmospheric CO_2, which contributes to global warming while stimulating photosynthesis in plants (CO_2 fertilization) [68]. In general, CO_2 fertilization plays a significant role in augmenting the capacity of vegetation to absorb CO_2. Existing literature has demonstrated that CO_2 fertilization stimulates vegetative growth and facilitates the accumulation of biomass [68]. Therefore, we further investigated the sensitivity of vegetation productivity in altitudinal vegetation belts to changes in CO_2 during the growing season. GPP and NPP variations were significantly and positively correlated with CO_2 concentrations in all altitudinal vegetation belts, except for MS (Figure 9a,b). In which the GPP and NPP of CV were more sensitive than those of other areas to changes in CO_2 concentration. MS was not sensitive to changes in CO_2 concentration because there were no significant changes in GPP or NPP (Figures 6b and 7b). In summary, we conclude that CO_2 fertilization played a more important role than precipitation and temperature in determining the interannual trends of vegetation productivity across altitudinal vegetation belts in the study period.

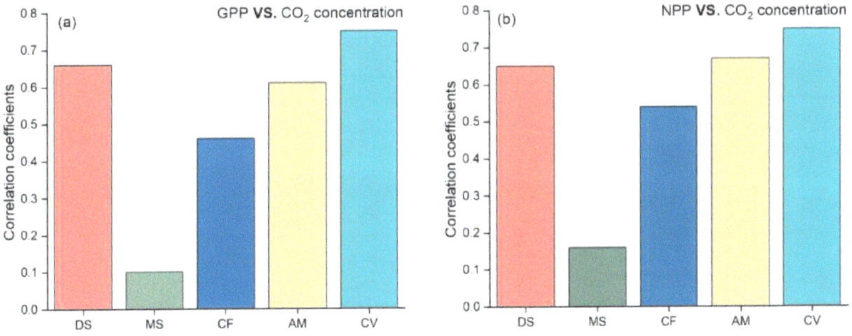

Figure 9. Correlation between vegetation productivity and CO_2 concentration (ppm) in this time series. (**a**) Correlation coefficients between GPP in altitudinal vegetation belts and CO_2 concentration for 2000–2021. (**b**) As in panel (**a**), but for NPP.

4.4. Limitations and Significance

The climate reanalysis dataset is a gridded dataset that refers to satellite and observational data and is created through data assimilation methods [69]. Nevertheless, 79.2% of the meteorological stations in the Tianshan Mountains are located below 2000 m above sea level [37], which indicates that the reanalysis data perform poorly in high mountainous areas. Therefore, there may be some uncertainty when using reanalysis data to investigate climate trends in higher-elevation altitudinal vegetation belts. In future research, we aim to enhance the reliability of climate trend analysis in the alpine vegetation belts by using ensemble averaging of multiple reanalysis datasets. Human activity has also driven changes in vegetation productivity [17], and the primary human activity in the Tianshan Mountains range is grazing. Therefore, additional research is required to assess the impact of grazing activities on vegetation productivity in grassland belts.

The soil organic carbon pool constitutes the most extensive carbon reservoir within terrestrial ecosystems; therefore, small variations in its storage may cause obvious fluctuations in atmospheric CO_2 concentrations [70]. Plant carbon input (NPP) is the main source of soil organic carbon [71], and an increase in plant carbon input can enhance its content and stock [72]. Our results showed that NPP increased significantly in all vegetation belts except MS, enhancing soil carbon sink capacity. Moreover, the vegetation belts situated at higher elevations tended to exhibit higher NPP, thereby enhancing their soil organic carbon content. In conclusion, increased NPP improved plant and soil carbon sequestration in the Tianshan Mountains.

5. Conclusions

Based on satellite-derived vegetation productivity (GPP and NPP) and climate variables, we examined changes in vegetation productivity and responses to current (2000–2021) climate trends in the Tianshan Mountains. The outcomes of our study demonstrate that the altitudinal vegetation belts in the Tianshan Mountains exhibited a slight warming and drying trend over the studied period of 2000–2021. The warming trends in different altitudinal vegetation belts ranged from 0.27 to 0.32 °C decade^{-1}, and precipitation trends ranged from -16.33 to -0.77 mm decade^{-1}. Vegetation productivity increased significantly in all altitudinal vegetation belts except MS, and the rate of increase was greatest (> 1.6 g C m^{-2} yr^{-1}) in CF and AM. The altitudinal vegetation belts exhibited a positive correlation between climate factors and both GPP and NPP. However, climatic factors did not dominate the trend of increasing vegetative productivity. Notably, warming and drying did not lead to a significant increase in the VPD, further suggesting that the role of climatic factors in vegetation productivity was weak. We conclude that the sustained rise in global CO_2 concentrations has dominated the significant increase in vegetation productivity in the altitudinal vegetation belts.

Author Contributions: Y.Z. (Yong Zhang): writing—original draft, validation, investigation and data curation; C.A.: funding acquisition and writing—original draft; L.J.: methodology, formal analysis, data curation and software; L.Z.: methodology and formal analysis; W.Z.: data curation, software and validation; B.T.: visualization, software and supervision; C.L.: data curation and software; Y.Z. (Yanzhen Zhang): formal analysis, data curation and validation. All authors have read and agreed to the published version of the manuscript.

Funding: National Natural Science Foundation of China [Grant number 42071102, 42220104001].

Data Availability Statement: No data was used for the research described in the article.

Acknowledgments: The authors would like to thank the ESA WorldCover team for providing land cover data at 10 m resolution (https://worldcover2021.esa.int/, accessed on 1 June 2023).

Conflicts of Interest: The authors declare that they have no known competing financial interests or personal relationships that could have appeared to influence the work reported in this paper.

References

1. Xu, L.; Yu, G.; He, N.; Wang, Q.; Gao, Y.; Wen, D.; Li, S.; Niu, S.; Ge, J. Carbon Storage in China's Terrestrial Ecosystems: A Synthesis. *Sci. Rep.* **2018**, *8*, 2806. [CrossRef] [PubMed]
2. Seneviratne, S.I.; Rogelj, J.; Séférian, R.; Wartenburger, R.; Allen, M.R.; Cain, M.; Millar, R.J.; Ebi, K.L.; Ellis, N.; Hoegh-Guldberg, O.; et al. The Many Possible Climates from the Paris Agreement's Aim of 1.5 °C Warming. *Nature* **2018**, *558*, 41–49. [CrossRef] [PubMed]
3. Duan, H.; Zhou, S.; Jiang, K.; Bertram, C.; Harmsen, M.; Kriegler, E.; van Vuuren, D.P.; Wang, S.; Fujimori, S.; Tavoni, M.; et al. Assessing China's Efforts to Pursue the 1.5 °C Warming Limit. *Science 1979* **2021**, *372*, 378–385. [CrossRef] [PubMed]
4. Tang, X.; Zhao, X.; Bai, Y.; Tang, Z.; Wang, W.; Zhao, Y.; Wan, H.; Xie, Z.; Shi, X.; Wu, B.; et al. Carbon Pools in China's Terrestrial Ecosystems: New Estimates Based on an Intensive Field Survey. *Proc. Natl. Acad. Sci. USA* **2018**, *115*, 4021–4026. [CrossRef] [PubMed]
5. Al-Yaari, A.; Wigneron, J.; Ciais, P.; Reichstein, M.; Ballantyne, A.; Ogée, J.; Ducharne, A.; Swenson, J.J.; Frappart, F.; Fan, L.; et al. Asymmetric Responses of Ecosystem Productivity to Rainfall Anomalies Vary Inversely with Mean Annual Rainfall over the Conterminous United States. *Glob. Chang. Biol.* **2020**, *26*, 6959–6973. [CrossRef]
6. Wang, L.; Gao, J.; Shen, W.; Shi, Y.; Zhang, H. Carbon Storage in Vegetation and Soil in Chinese Ecosystems Estimated by Carbon Transfer Rate Method. *Ecosphere* **2021**, *12*, e03341. [CrossRef]
7. Wang, N.; Quesada, B.; Xia, L.; Butterbach-Bahl, K.; Goodale, C.L.; Kiese, R. Effects of Climate Warming on Carbon Fluxes in Grasslands—A Global Meta-analysis. *Glob. Chang. Biol.* **2019**, *25*, 1839–1851. [CrossRef]
8. Sha, Z.; Bai, Y.; Li, R.; Lan, H.; Zhang, X.; Li, J.; Liu, X.; Chang, S.; Xie, Y. The Global Carbon Sink Potential of Terrestrial Vegetation Can Be Increased Substantially by Optimal Land Management. *Commun. Earth Environ.* **2022**, *3*, 8. [CrossRef]
9. Ding, Z.; Peng, J.; Qiu, S.; Zhao, Y. Nearly Half of Global Vegetated Area Experienced Inconsistent Vegetation Growth in Terms of Greenness, Cover, and Productivity. *Earths Future* **2020**, *8*, e2020EF001618. [CrossRef]
10. Robinson, N.P.; Allred, B.W.; Smith, W.K.; Jones, M.O.; Moreno, A.; Erickson, T.A.; Naugle, D.E.; Running, S.W. Terrestrial Primary Production for the Conterminous United States Derived from Landsat 30 m and MODIS 250 m. *Remote Sens. Ecol. Conserv.* **2018**, *4*, 264–280. [CrossRef]
11. Hu, Q.; Li, T.; Deng, X.; Wu, T.; Zhai, P.; Huang, D.; Fan, X.; Zhu, Y.; Lin, Y.; Xiao, X.; et al. Intercomparison of Global Terrestrial Carbon Fluxes Estimated by MODIS and Earth System Models. *Sci. Total Environ.* **2022**, *810*, 152231. [CrossRef] [PubMed]
12. He, Y.; Piao, S.; Li, X.; Chen, A.; Qin, D. Global Patterns of Vegetation Carbon Use Efficiency and Their Climate Drivers Deduced from MODIS Satellite Data and Process-Based Models. *Agric. For. Meteorol.* **2018**, *256–257*, 150–158. [CrossRef]
13. Tian, C.; Yue, X.; Zhou, H.; Lei, Y.; Ma, Y.; Cao, Y. Projections of Changes in Ecosystem Productivity under 1.5 °C and 2 °C Global Warming. *Glob. Planet. Chang.* **2021**, *205*, 103588. [CrossRef]
14. Khalifa, M.; Elagib, N.A.; Ribbe, L.; Schneider, K. Spatio-Temporal Variations in Climate, Primary Productivity and Efficiency of Water and Carbon Use of the Land Cover Types in Sudan and Ethiopia. *Sci. Total Environ.* **2018**, *624*, 790–806. [CrossRef]
15. He, Q.; Zhou, G.; Lü, X.; Zhou, M. Climatic Suitability and Spatial Distribution for Summer Maize Cultivation in China at 1.5 and 2.0 °C Global Warming. *Sci. Bull.* **2019**, *64*, 690–697. [CrossRef]
16. Zhang, Y.; Huang, K.; Zhang, T.; Zhu, J.; Di, Y. Soil Nutrient Availability Regulated Global Carbon Use Efficiency. *Glob. Planet. Chang.* **2019**, *173*, 47–52. [CrossRef]
17. Sun, Z.; Wang, X.; Yamamoto, H.; Tani, H.; Zhong, G.; Yin, S.; Guo, E. Spatial Pattern of GPP Variations in Terrestrial Ecosystems and Its Drivers: Climatic Factors, CO2 Concentration and Land-Cover Change, 1982–2015. *Ecol Inf.* **2018**, *46*, 156–165. [CrossRef]
18. Wei, D.; Qi, Y.; Ma, Y.; Wang, X.; Ma, W.; Gao, T.; Huang, L.; Zhao, H.; Zhang, J.; Wang, X. Plant Uptake of CO2 Outpaces Losses from Permafrost and Plant Respiration on the Tibetan Plateau. *Proc. Natl. Acad. Sci. USA* **2021**, *118*, e2015283118. [CrossRef]
19. Shi, L.; Lin, Z.; Tang, S.; Peng, C.; Yao, Z.; Xiao, Q.; Zhou, H.; Liu, K.; Shao, X. Interactive Effects of Warming and Managements on Carbon Fluxes in Grasslands: A Global Meta-Analysis. *Agric. Ecosyst. Environ.* **2022**, *340*, 108178. [CrossRef]

20. Yuan, F.; Liu, J.; Zuo, Y.; Guo, Z.; Wang, N.; Song, C.; Wang, Z.; Sun, L.; Guo, Y.; Song, Y.; et al. Rising Vegetation Activity Dominates Growing Water Use Efficiency in the Asian Permafrost Region from 1900 to 2100. *Sci. Total Environ.* **2020**, *736*, 139587. [CrossRef]
21. Wang, Y.; Hu, J.; Yang, Y.; Li, R.; Peng, C.; Zheng, H. Climate Change Will Reduce the Carbon Use Efficiency of Terrestrial Ecosystems on the Qinghai-Tibet Plateau: An Analysis Based on Multiple Models. *Forests* **2020**, *12*, 12. [CrossRef]
22. Lin, S.; Wang, G.; Feng, J.; Dan, L.; Sun, X.; Hu, Z.; Chen, X.; Xiao, X. A Carbon Flux Assessment Driven by Environmental Factors Over the Tibetan Plateau and Various Permafrost Regions. *J. Geophys. Res. Biogeosci* **2019**, *124*, 1132–1147. [CrossRef]
23. Kang, X.; Li, Y.; Wang, J.; Yan, L.; Zhang, X.; Wu, H.; Yan, Z.; Zhang, K.; Hao, Y. Precipitation and Temperature Regulate the Carbon Allocation Process in Alpine Wetlands: Quantitative Simulation. *J. Soils Sedim.* **2020**, *20*, 3300–3315. [CrossRef]
24. Song, L.; Li, Y.; Ren, Y.; Wu, X.; Guo, B.; Tang, X.; Shi, W.; Ma, M.; Han, X.; Zhao, L. Divergent Vegetation Responses to Extreme Spring and Summer Droughts in Southwestern China. *Agric. For. Meteorol.* **2019**, *279*, 107703. [CrossRef]
25. Chen, Y.; Feng, X.; Tian, H.; Wu, X.; Gao, Z.; Feng, Y.; Piao, S.; Lv, N.; Pan, N.; Fu, B. Accelerated Increase in Vegetation Carbon Sequestration in China after 2010: A Turning Point Resulting from Climate and Human Interaction. *Glob. Chang. Biol.* **2021**, *27*, 5848–5864. [CrossRef]
26. Liu, X.; Ma, Q.; Yu, H.; Li, Y.; Li, L.; Qi, M.; Wu, W.; Zhang, F.; Wang, Y.; Zhou, G.; et al. Climate Warming-Induced Drought Constrains Vegetation Productivity by Weakening the Temporal Stability of the Plant Community in an Arid Grassland Ecosystem. *Agric. For. Meteorol.* **2021**, *307*, 108526. [CrossRef]
27. He, Y.; Yan, W.; Cai, Y.; Deng, F.; Qu, X.; Cui, X. How Does the Net Primary Productivity Respond to the Extreme Climate under Elevation Constraints in Mountainous Areas of Yunnan, China? *Ecol. Indic.* **2022**, *138*, 108817. [CrossRef]
28. Yan, M.; Tian, X.; Li, Z.; Chen, E.; Li, C.; Fan, W. A Long-Term Simulation of Forest Carbon Fluxes over the Qilian Mountains. *Int. J. Appl. Earth Obs. Geoinf.* **2016**, *52*, 515–526. [CrossRef]
29. Zhang, Y.; Hu, Q.; Zou, F. Spatio-Temporal Changes of Vegetation Net Primary Productivity and Its Driving Factors on the Qinghai-Tibetan Plateau from 2001 to 2017. *Remote Sens.* **2021**, *13*, 1566. [CrossRef]
30. Wang, L.; Zhang, Y.; Berninger, F.; Duan, B. Net Primary Production of Chinese Fir Plantation Ecosystems and Its Relationship to Climate. *Biogeosciences* **2014**, *11*, 5595–5606. [CrossRef]
31. Rumpf, S.B.; Gravey, M.; Brönnimann, O.; Luoto, M.; Cianfrani, C.; Mariethoz, G.; Guisan, A. From White to Green: Snow Cover Loss and Increased Vegetation Productivity in the European Alps. *Science 1979* **2022**, *376*, 1119–1122. [CrossRef] [PubMed]
32. Xie, X.; Tian, J.; Wu, C.; Li, A.; Jin, H.; Bian, J.; Zhang, Z.; Nan, X.; Jin, Y. Long-Term Topographic Effect on Remotely Sensed Vegetation Index-Based Gross Primary Productivity (GPP) Estimation at the Watershed Scale. *Int. J. Appl. Earth Obs. Geoinf.* **2022**, *108*, 102755. [CrossRef]
33. Lian, X.; Piao, S.; Chen, A.; Huntingford, C.; Fu, B.; Li, L.Z.X.; Huang, J.; Sheffield, J.; Berg, A.M.; Keenan, T.F.; et al. Multifaceted Characteristics of Dryland Aridity Changes in a Warming World. *Nat. Rev. Earth Environ.* **2021**, *2*, 232–250. [CrossRef]
34. Lu, Y.; Xu, X.; Zhao, J.; Han, F. Spatiotemporal Evolution of Mountainous Ecosystem Services in an Arid Region and Its Influencing Factors: A Case Study of the Tianshan Mountains in Xinjiang. *Land* **2022**, *11*, 2164. [CrossRef]
35. Zhu, S.; Li, C.; Shao, H.; Ju, W.; Lv, N. The Response of Carbon Stocks of Drylands in Central Asia to Changes of CO2 and Climate during Past 35 years. *Sci. Total Environ.* **2019**, *687*, 330–340. [CrossRef]
36. Liu, L.; Guan, J.; Han, W.; Ju, X.; Mu, C.; Zheng, J. Quantitative Assessment of the Relative Contributions of Climate and Human Factors to Net Primary Productivity in the Ili River Basin of China and Kazakhstan. *Chin. Geogr. Sci.* **2022**, *32*, 1069–1082. [CrossRef]
37. Zhang, Y.; An, C.; Liu, L.; Zhang, Y.; Lu, C.; Zhang, W. High Mountains Becoming Wetter While Deserts Getting Drier in Xinjiang, China since the 1980s. *Land* **2021**, *10*, 1131. [CrossRef]
38. Zhang, Y.; Liu, L.; Liu, Y.; Zhang, M.; An, C. Response of Altitudinal Vegetation Belts of the Tianshan Mountains in Northwestern China to Climate Change during 1989–2015. *Sci. Rep.* **2021**, *11*, 4870. [CrossRef]
39. Zhang, Y.; An, C.-B.; Liu, L.-Y.; Zhang, Y.-Z.; Lu, C.; Zhang, W.-S. High-Elevation Landforms Are Experiencing More Remarkable Wetting Trends in Arid Central Asia. *Adv. Clim. Chang. Res.* **2022**, *13*, 489–495. [CrossRef]
40. Zhang, B.; Mo, S.; Wu, H.; Xiao, F. Digital Spectra and Analysis of Altitudinal Belts in Tianshan Mountains, China. *J. Mt. Sci.* **2004**, *1*, 18–28. [CrossRef]
41. Zhang, Y.; An, C.; Zheng, L.; Liu, L.; Zhang, W.; Lu, C.; Zhang, Y. Assessment of Lake Area in Response to Climate Change at Varying Elevations: A Case Study of Mt. Tianshan, Central Asia. *Sci. Total Environ.* **2023**, *869*, 161665. [CrossRef] [PubMed]
42. Roffe, S.J.; van der Walt, A.J. Representation and Evaluation of Southern Africa's Seasonal Mean and Extreme Temperatures in the ERA5-Based Reanalysis Products. *Atmos. Res.* **2023**, *284*, 106591. [CrossRef]
43. Abatzoglou, J.T.; Dobrowski, S.Z.; Parks, S.A.; Hegewisch, K.C. TerraClimate, a High-Resolution Global Dataset of Monthly Climate and Climatic Water Balance from 1958–2015. *Sci. Data* **2018**, *5*, 170191. [CrossRef] [PubMed]
44. Noguera, I.; Vicente-Serrano, S.M.; Peña-Angulo, D.; Domínguez-Castro, F.; Juez, C.; Tomás-Burguera, M.; Lorenzo-Lacruz, J.; Azorin-Molina, C.; Halifa-Marín, A.; Fernández-Duque, B.; et al. Assessment of Vapor Pressure Deficit Variability and Trends in Spain and Possible Connections with Soil Moisture. *Atmos. Res.* **2023**, *285*, 106666. [CrossRef]
45. Stephens, B.B.; Gurney, K.R.; Tans, P.P.; Sweeney, C.; Peters, W.; Bruhwiler, L.; Ciais, P.; Ramonet, M.; Bousquet, P.; Nakazawa, T.; et al. Weak Northern and Strong Tropical Land Carbon Uptake from Vertical Profiles of Atmospheric CO_2. *Science 1979* **2007**, *316*, 1732–1735. [CrossRef]

46. Sweeney, C.; Karion, A.; Wolter, S.; Newberger, T.; Guenther, D.; Higgs, J.A.; Andrews, A.E.; Lang, P.M.; Neff, D.; Dlugokencky, E.; et al. Seasonal Climatology of CO_2 across North America from Aircraft Measurements in the NOAA/ESRL Global Greenhouse Gas Reference Network. *J. Geophys. Res. Atmos.* **2015**, *120*, 5155–5190. [CrossRef]
47. Cheng, S.; An, X.; Zhou, L.; Tans, P.P.; Jacobson, A. Atmospheric CO2 at Waliguan Station in China: Transport Climatology, Temporal Patterns and Source-Sink Region Representativeness. *Atmos. Environ.* **2017**, *159*, 107–116. [CrossRef]
48. Wang, N.; Zhang, X.; Yao, S.; Wu, J.; Xia, H. How Good Are Global Layers for Mapping Rural Settlements? Evidence from China. *Land* **2022**, *11*, 1308. [CrossRef]
49. Zhao, X.; Su, Y.; Hu, T.; Chen, L.; Gao, S.; Wang, R.; Jin, S.; Guo, Q. A Global Corrected SRTM DEM Product for Vegetated Areas. *Remote Sens. Lett.* **2018**, *9*, 393–402. [CrossRef]
50. Wang, L.; Zhu, H.; Lin, A.; Zou, L.; Qin, W.; Du, Q. Evaluation of the Latest MODIS GPP Products across Multiple Biomes Using Global Eddy Covariance Flux Data. *Remote Sens.* **2017**, *9*, 418. [CrossRef]
51. Yao, J.; Liu, H.; Huang, J.; Gao, Z.; Wang, G.; Li, D.; Yu, H.; Chen, X. Accelerated Dryland Expansion Regulates Future Variability in Dryland Gross Primary Production. *Nat. Commun.* **2020**, *11*, 1665. [CrossRef] [PubMed]
52. Rukh, S.; Schad, T.; Strer, M.; Natkhin, M.; Krüger, I.; Raspe, S.; Eickenscheidt, N.; Hentschel, R.; Hölscher, A.; Reiter, P.; et al. Interpolated Daily Temperature and Precipitation Data for Level II ICP Forests Plots in Germany. *Ann. For. Sci.* **2022**, *79*, 47. [CrossRef]
53. Song, Y.; Jiao, W.; Wang, J.; Wang, L. Increased Global Vegetation Productivity Despite Rising Atmospheric Dryness Over the Last Two Decades. *Earths Future* **2022**, *10*, e2021EF002634. [CrossRef]
54. Wang, Y.; Dai, E.; Wu, C. Spatiotemporal Heterogeneity of Net Primary Productivity and Response to Climate Change in the Mountain Regions of Southwest China. *Ecol. Indic.* **2021**, *132*, 108273. [CrossRef]
55. Xu, H.; Zhao, C.; Wang, X. Spatiotemporal Differentiation of the Terrestrial Gross Primary Production Response to Climate Constraints in a Dryland Mountain Ecosystem of Northwestern China. *Agric. For. Meteorol.* **2019**, *276–277*, 107628. [CrossRef]
56. Moore, C.E.; Meacham-Hensold, K.; Lemonnier, P.; Slattery, R.A.; Benjamin, C.; Bernacchi, C.J.; Lawson, T.; Cavanagh, A.P. The Effect of Increasing Temperature on Crop Photosynthesis: From Enzymes to Ecosystems. *J. Exp. Bot.* **2021**, *72*, 2822–2844. [CrossRef]
57. Liberati, D.; Guidolotti, G.; Dato, G.; De Angelis, P. Enhancement of Ecosystem Carbon Uptake in a Dry Shrubland under Moderate Warming: The Role of Nitrogen-driven Changes in Plant Morphology. *Glob. Chang. Biol.* **2021**, *27*, 5629–5642. [CrossRef]
58. Liu, H.; Lu, C.; Wang, S.; Ren, F.; Wang, H. Climate Warming Extends Growing Season but Not Reproductive Phase of Terrestrial Plants. *Glob. Ecol. Biogeogr.* **2021**, *30*, 950–960. [CrossRef]
59. Liu, B.; Zhou, T. Atmospheric Footprint of the Recent Warming Slowdown. *Sci. Rep.* **2017**, *7*, 40947. [CrossRef]
60. Kosaka, Y.; Xie, S.-P. Recent Global-Warming Hiatus Tied to Equatorial Pacific Surface Cooling. *Nature* **2013**, *501*, 403–407. [CrossRef]
61. Ballantyne, A.; Smith, W.; Anderegg, W.; Kauppi, P.; Sarmiento, J.; Tans, P.; Shevliakova, E.; Pan, Y.; Poulter, B.; Anav, A.; et al. Accelerating Net Terrestrial Carbon Uptake during the Warming Hiatus Due to Reduced Respiration. *Nat. Clim. Chang.* **2017**, *7*, 148–152. [CrossRef]
62. Kusumi, K.; Hirotsuka, S.; Kumamaru, T.; Iba, K. Increased Leaf Photosynthesis Caused by Elevated Stomatal Conductance in a Rice Mutant Deficient in SLAC1, a Guard Cell Anion Channel Protein. *J. Exp. Bot.* **2012**, *63*, 5635–5644. [CrossRef] [PubMed]
63. Li, Y.; Li, H.; Li, Y.; Zhang, S. Improving Water-Use Efficiency by Decreasing Stomatal Conductance and Transpiration Rate to Maintain Higher Ear Photosynthetic Rate in Drought-Resistant Wheat. *Crop J.* **2017**, *5*, 231–239. [CrossRef]
64. Marchin, R.M.; Backes, D.; Ossola, A.; Leishman, M.R.; Tjoelker, M.G.; Ellsworth, D.S. Extreme Heat Increases Stomatal Conductance and Drought-induced Mortality Risk in Vulnerable Plant Species. *Glob. Chang. Biol.* **2022**, *28*, 1133–1146. [CrossRef] [PubMed]
65. Madani, N.; Parazoo, N.C.; Kimball, J.S.; Ballantyne, A.P.; Reichle, R.H.; Maneta, M.; Saatchi, S.; Palmer, P.I.; Liu, Z.; Tagesson, T. Recent Amplified Global Gross Primary Productivity Due to Temperature Increase Is Offset by Reduced Productivity Due to Water Constraints. *AGU Adv.* **2020**, *1*, e2020AV000180. [CrossRef]
66. Yuan, W.; Zheng, Y.; Piao, S.; Ciais, P.; Lombardozzi, D.; Wang, Y.; Ryu, Y.; Chen, G.; Dong, W.; Hu, Z.; et al. Increased Atmospheric Vapor Pressure Deficit Reduces Global Vegetation Growth. *Sci. Adv.* **2019**, *5*, aax1396. [CrossRef]
67. Chen, S.; Zhang, Y.; Wu, Q.; Liu, S.; Song, C.; Xiao, J.; Band, L.E.; Vose, J.M. Vegetation Structural Change and CO2 Fertilization More than Offset Gross Primary Production Decline Caused by Reduced Solar Radiation in China. *Agric. For. Meteorol.* **2021**, *296*, 108207. [CrossRef]
68. Wang, S.; Zhang, Y.; Ju, W.; Chen, J.M.; Ciais, P.; Cescatti, A.; Sardans, J.; Janssens, I.A.; Wu, M.; Berry, J.A.; et al. Recent Global Decline of CO2 Fertilization Effects on Vegetation Photosynthesis. *Science 1979* **2020**, *370*, 1295–1300. [CrossRef]
69. Huerta, A.; Bonnesoeur, V.; Cuadros-Adriazola, J.; Gutierrez, L.; Ochoa-Tocachi, B.F.; Román-Dañobeytia, F.; Lavado-Casimiro, W. PISCOeo_pm, a Reference Evapotranspiration Gridded Database Based on FAO Penman-Monteith in Peru. *Sci. Data* **2022**, *9*, 328. [CrossRef]
70. Kou, D.; Ma, W.; Ding, J.; Zhang, B.; Fang, K.; Hu, H.; Yu, J.; Wang, T.; Qin, S.; Zhao, X.; et al. Dryland Soils in Northern China Sequester Carbon during the Early 2000s Warming Hiatus Period. *Funct. Ecol.* **2018**, *32*, 1620–1630. [CrossRef]

71. Huang, J.; Liu, W.; Yang, S.; Yang, L.; Peng, Z.; Deng, M.; Xu, S.; Zhang, B.; Ahirwal, J.; Liu, L. Plant Carbon Inputs through Shoot, Root, and Mycorrhizal Pathways Affect Soil Organic Carbon Turnover Differently. *Soil. Biol. Biochem.* **2021**, *160*, 108322. [CrossRef]
72. Dintwe, K.; Okin, G.S. Soil Organic Carbon in Savannas Decreases with Anthropogenic Climate Change. *Geoderma* **2018**, *309*, 7–16. [CrossRef]

forests

MDPI

Article

The Effect of Curcin Protein and Jatropha Plantation on Soil Fungi

Zhiping Lai, Bingbing Zhang, Xianfei Niu, Rui Ma, Ting Wang, Cheng Cheng, Yingying Ren, Xueying Wang, Na Hu, Nan Jiang and Ying Xu *

Key Laboratory of Bio-Resources and Eco-Environment of Ministry of Education, College of Life Sciences, Sichuan University, Chengdu 610065, China; laizhiping@stu.scu.edu.cn (Z.L.); niuxianfei@stu.scu.edu.cn (X.N.); mr0825@126.com (R.M.)

* Correspondence: xuying@scu.edu.cn

Abstract: *Jatropha curcas* is widely planted as a highly drought-resistant biodiesel feedstock. Curcin protein is one of the Jatropha ribosomal inactivation proteins with broad-spectrum antifungal activity that may enter the soil ecosystem as a result of large-scale Jatropha cultivation and affect fungi and various enzymatic activities in the soil. In this research, the influence of curcin protein and Jatropha planting on soil fungi was investigated, and the levels of curcin in various tissues and organs of Jatropha were measured with an enzyme-linked immunosorbent assay. It was found that the content of curcin in seed kernels reaches 2 mg/g, which is much higher than that in other tissues. After the seeds have fallen into the soil, the level of curcin in the soil rises rapidly, reaching 59.22 μg/g soil and 67.49 μg/g soil in different soil samples, respectively. It then falls by more than 99% within six days. High-throughput sequencing technology was used to study the soils treated with different concentrations of curcin, and the results of the soil fungal alpha diversity index analysis showed that the fungal communities did not change significantly, but the abundance of each fungal community changed significantly. The degree of influence of different concentrations of curcin treatment on the abundance of the soil dominant fungal community were investigated for concentrations of 0.5 μg/g, 50 μg/g and 5 μg/g, and showed that concentrations of 0.5 μg/g and 50 μg/g are more likely to change fungal community structure in soil, and with the increasing extension of the treatment time, they may be detrimental to the conservation of soil ecosystems. Internal transcribed spacer (ITS) sequencing of soil fungi from Jatropha planted and unplanted areas in four regions with different climate types showed that Jatropha planting significantly altered the soil fungal communities in each region. There was a negative impact on soil fungal communities in tropical maritime monsoon and subtropical dry and hot monsoon climates, while a positive impact was observed in subtropical monsoon and tropical highland monsoon climates due to Jatropha cultivation. In conclusion, Jatropha plantations and curcin protein have an impact on soil fungi and thereby affect the ecological system of the soil.

Keywords: *Jatropha curcas*; ribosome-inactivating proteins; curcin; soil enzyme activities; soil fungi

1. Introduction

Soil is the most biologically diverse and the most active energy-exchanging and material-cycling layer of life in the surface system of the Earth, with the capacity to sustain plants and animals, maintain and improve water and air quality and support healthy human life. Soil fungi convert much of the decaying organic matter into plant-available nutrients through humification and mineralization. Fungi make up the major part of the microbial biomass and are most versatile in decomposing organic residues. They transform a larger portion of decaying plant residue into available nutrients. For instance, mycorrhizal fungi accelerate leaf litter decomposition in a northern hardwood forest [1]. They also increased plant uptake of nutrients such as nitrogen, phosphorus and potassium by influencing plant's root morphology and physiological characteristics, such as mycorrhizal fungi [2,3]. However, there are also many fungi that are causal agents of plant diseases.

Fungi pose a significant threat to both plant and animal biodiversity [4]. The richness of fungal decomposers was consistently and positively associated with ecosystem stability worldwide, while the opposite pattern was found for the richness of fungal plant pathogens, particularly in grasslands [5]. The relative stability of the structure of the soil fungal community therefore largely determines the balance of the soil ecosystem.

Plant growth had an important influence on changes in the populations of inter-rooted microorganisms, and any change in the composition or quantity of root secretions had the possibility to affect the soil microbiota [6–8]. For example, root exudates from slow- and fast-growing *Arabidopsis thaliana* plants have distinct effects on both microbial community structure and soil nutrients available to plants [9]. Even after harvest, the decomposition of plant litter continued to release new gene products into the environment [10], potentially changing the inter-root environment for long periods of time and thus having many effects on the soil environment. For example, Araujo et al. have showed a significant reduction in soil microbial biomass following the conversion of native savannah to eucalyptus forest [11]. The transgenic cotton line (25C-1) was significantly enriched in two beneficial bacteria, *Arthrobacter* and *Sphingomonas*, after continuous cultivation, which had a positive impact on the soil ecology [12]. The study by Kumar et al. investigated that changing land use (natural woodland, rotational cropland, and cropland) had significant effects on soil microbial biomass, with natural woodland having more biomass, cropland having medium biomass, and rotational cropland having less biomass [13]. This suggests that land use change can have a considerable impact on the soil microbial community.

Ribosome inactivating proteins (RIPs) are a group of active proteins widely found in plants and fungi that can disrupt ribosome structure, leading to ribosome inactivation and inhibition of protein biosynthesis [14–16]. Many studies have shown that RIPs have a variety of biological activities, such as antiviral, insecticidal, immunological and antifungal activities [17–20].

Jatropha curcas, a small, semi-fleshy tree or large shrub of the family *Euphorbia*, has been widely planted in tropical and subtropical regions for its high drought and barren tolerance as a biodiesel feedstock [21]. By 2012, China had approximately 200,000 hectares of Jatropha resources (including natural forests) [22]. There is a type I ribosome-inactivating protein curcin in Jatropha [23], which not only strongly inhibits protein synthesis in cell-free systems, but also has an inhibitory effect on plant pathogenic fungi at certain concentrations [24], such as *Saccharomyces cerevisiae*, *Rhizoctonia solani*, *Pyricularia oryzae*, *Gibberella zeae* and *Sclerotinia sclerotiorum* [24].

The Jatropha ribosomal inactivation protein curcin may enter the soil ecosystem through plant debris, seeds, pollen and root secretions because of large-scale cultivation of Jatropha and affect non-target organisms, fungi and various enzymatic activities in the soil, thus affecting soil ecological diversity and stability. The study aimed to examine the diverse impacts of Jatropha plantations and curcin on soil ecosystems.

2. Materials and Methods

2.1. Soil for Curcin Protein In Vitro Treatment Experiments

The experimental soil was collected from an oilseed rape field in Jinjiang District, Chengdu, Sichuan Province (longitude: 30°34′44″ E, latitude: 104°9′25″ N, altitude: 503 m), with a sample area of 23 m^2 and a maximum temperature of 16 °C and a minimum temperature of 12 °C on a cloudy day. The local area belongs to the subtropical monsoon climate zone, with hot and rainy summers and warm and humid winters. The average annual temperature is 16.6 °C, sunshine 1156.7 h and rainfall 966.9 mm. A total of 2 kg soil samples were taken from the surface layer at a depth of 1–15 cm via isometric sampling. Surface vegetation was removed, fresh soil was sieved to 2 mm and the soil was mixed thoroughly.

2.2. Sample Collection of Jatropha Cultivation Areas

As Jatropha is a tropical plant, the soils from Jatropha plantations and non-plantation areas were collected in four different tropical or subtropical climate regions, Jinhexiang, Sichuan Province (longitude: 101°42′ E; latitude: 27°06′ N, elevation: 2500 m), Xichang, Sichuan Province (longitude: 101°57′ E, latitude: 27°42′ N, elevation: 2000 m), Yuanmou, Yunnan Province (longitude: 101°49′ E; latitude: 25°50′ N, elevation: 1312 m) and Haikou, Hainan Province (longitude: 109°75′ E; latitude: 19°88′ N, elevation: 126 m). Among them, Jinhexiang belongs to the subtropical monsoon climate; Xichang belongs to the tropical highland monsoon climate; Yuanmou belongs to the subtropical dry and hot monsoon climate and Haikou belongs to the tropical maritime monsoon climate. Approximately 1 kg of fresh soil was collected at a depth of 1–20 cm using the five-point sampling method, mixed and passed through a 2 mm sieve after removing dead branches and leaves on the surface, and soil samples were taken in triplicate for a total of 24 samples.

2.3. Curcin Treatment Soil Experiment

Based on our previous studies [25] and research on the antifungal activity of curcin in vitro, three concentrations of 0.5 μg/g, 5 μg/g and 50 μg/g were prepared. 20 g of soil was weighed into 100 mL sterilized triangular flasks, 1 mL each of 10 μg/mL, 100 μg/mL and 1000 μg/mL protein solution was pipetted into the triangular flasks with the same volume of ddH20 as the blank control, a total of four groups of three replicates each, stirred well with a sterilized glass rod and the triangular flasks were sealed with a triangular flask sealing film with air holes. The sealed triangular flasks were incubated in the dark at 25 °C in an artificial climate incubator with humidity adjusted to 60% and samples were collected on the 1st, 3rd, 7th and 12th days after incubation.

2.4. DNA Extraction and Sequencing

A total of 48 samples were taken from soils treated with different concentrations of curcin protein, and a total of 18 samples were collected from soils in different climate regions with Jatropha plantation and non-plantation areas. The methods used to collect the samples are identical to those outlined in Section 2.2. The DNA of samples were extracted using the MP Biomedical Fast DNA soil sample extraction kit (Mobio Laboratories, Carlsbad, CA, USA) according to its instructions, and the samples were tested and then sent to BGI Genomics for ITS sequencing of the fungi. Raw data were spliced using FLASH (version 1.2.11) software, low quality data were filtered using Trimmomatic (version 0.33) software and chimeras were removed using UCHIME (version 8.1) software. Sequences were clustered using USEARCH (version 10.0). The OTU sequences were compared with the Unite Release 7.2 database using RDP classifier (version 2.2) software and species annotation was performed to cluster the samples.

2.5. Determination Method of Enzyme Activity

Soil sucrase enzyme activity was determined with a colorimetric method using 3,5-dinitrosalicylic acid [26]. Soil acid phosphatase activity was determined with a colorimetric method using disodium phenyl phosphate [27]. Soil urease activity was determined with the sodium phenol-sodium hypochlorite colorimetric method [28]. Soil catalase activity was determined with the potassium permanganate colorimetric method [27].

2.6. Determination of the Degradation Rate of the Jatropha Ribosome Inactivation Protein Curcin in Soil

Due to the high content of curcin in the seed kernel of Jatropha and its value for use in transgenic plants, the dynamics of its accumulation and degradation in the soil is an important indicator of its bioenvironmental risk. In order to simulate the degradation process of Jatropha seed kernels in the soil in a natural environment, soils from Jinhexiang, Sichuan Province (where Jatropha actually grows) and Lvyangcun, Chengdu, Sichuan

Province (where Jatropha does not grow, but with fertile soil and rich in microorganisms) were selected for comparison.

(1) Soil from Jinhexiang and Lvyangcun was taken and passed through a 20 mesh sieve.
(2) A total of 0.1 g of Jatropha seeds ground into powder with liquid nitrogen was added to 0.5 g of soil; 0.4 mL of distilled water was added, mixed with a pellet pestle (RS-Sigma) for 5 s, centrifuged at 3000 rpm for 30 s at room temperature, and the supernatant was aspirated to obtain a homogeneous mixture of soil, seeds and water. Twenty-one replicates were performed for the soil at each site (set of seven time points, three replicates per set).
(3) EP tubes were placed in a thermostat, maintained at 30 °C and 80% relative humidity and incubated in the dark. Samples were taken on days 0, 6, 12, 18, 24, 30 and 36 of degradation and three replicates were taken at each time point.
(4) Soil curcin protein was extracted and ELISA assays were performed.

2.7. Data Analysis and Processing

Diversity indices were mainly calculated using the R package Vegan (https://cran.r-project.org/package=vegan, accessed on 15 June 2023), and data processing, statistical analysis and graph plotting were performed in SPSS26, R (v4.2.0) and GraphPad Prism 8.0.1.

3. Results

3.1. Degradation of Curcin Protein in Soil and the Effect on Enzyme Activities

3.1.1. Possible Pathways of Curcin into the Soil and Its Degradation Dynamics in the Soil

The endosperm of Jatropha seeds is relatively rich in curcin protein [29], but other parts, such as the leaves, may also contain curcin. In order to investigate the possible pathways for curcin to enter the soil, five Jatropha trees were randomly selected from the Jatropha planting area and 100 g each of seeds, green leaves and roots were collected during the fruiting period, and 100 g each of dried and fallen Jatropha fruits and yellow leaves were collected from the plantation. The results showed that the content of curcin in seed kernels was up to 2 mg/g tissue, 13 μg/g in green leaves, 15 μg/g in yellow leaves, 2 μg/g in the fruit peel and 3 μg/g in roots, and the content of curcin in seed kernels was much higher than that in leaves, roots and pericarp (Figure 1A). It is therefore suggested that curcin enters the soil mainly through the seed kernel, in addition to being secreted by the root system, and may affect the balance of soil ecosystems.

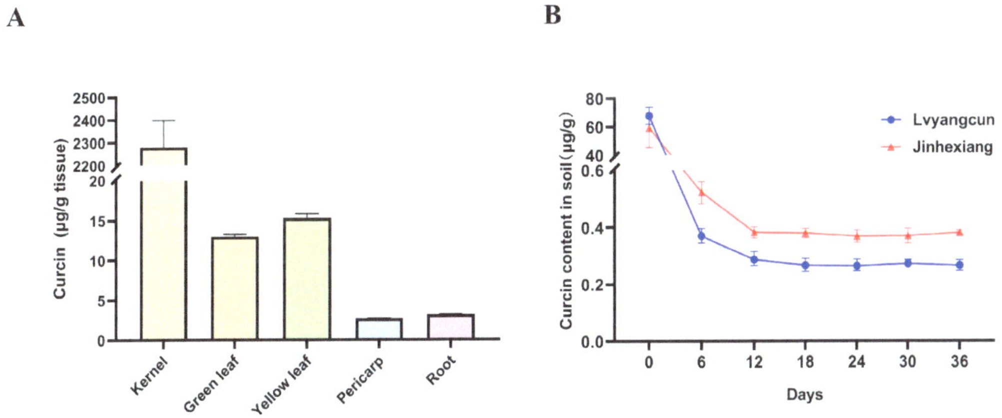

Figure 1. Possible sources of curcin in soils and its degradation dynamics. (**A**) Contents of curcin in different tissues of Jatropha and (**B**) Degradation trends of curcin in soil from Jinhexiang and Lvyangcun.

3.1.2. Degradation Dynamics of Curcin Protein in the Soil

In order to determine the amount of curcin in the soil around Jatropha in its natural growing state, the inter-root soil and the growing soil were collected at three different locations, three replicates of each location, ground to a fine powder, extracted curcin and assayed with monoclonal antibody indirect enzyme-linked immunosorbent assay (ELISA). Curcin was found to be below the minimum detection limit of the method in both inter-root soil and growing soil. The reason for this is thought to be that when seeds, leaves, fruits and other tissues enter the soil, curcin is rapidly degraded by biological factors (e.g., bacteria, fungi, etc.) or adsorbed by the soil.

Protein can combine with soil particles, increasing retention time. Some research has shown that clay minerals, humic acid and organic mineral aggregates in soil can adsorb Bt toxin and retain it, and the bound toxin still retains insecticidal activity, which has varying degrees of influence on soil organisms and soil enzymes [30–32]. Different masses of curcin were added to the soil, and the carrying capacity of the soil for curcin was determined with sodium dodecyl sulfate-polyacrylamide gel electrophoresis (SDS-PAGE) and Western Blot. When the addition of curcin was less than 40 mg curcin/g soil, there was no curcin band in the supernatant, which indicated that curcin was completely adsorbed by the soil; when the addition of curcin reached 40 mg curcin/g soil, the curcin band appeared in the supernatant, i.e., the adsorption of curcin by the soil reached saturation, and the non-adsorbed part was free in the supernatant, which indicated that the carrying capacity of the soil for curcin is about 40 mg/g soil. Therefore, there is a high likelihood of curcin retention in areas where *Jatropha curcas* is cultivated and in the inter-root soils where curcin was not detected.

In order to simulate the degradation process of seed kernels in the soil in a natural environment, the initial results of curcin were 59.22 μg/g soil and 67.49 μg/g soil after adding a mixture of Jatropha seed kernels to soils sampled from Jinhexiang and Lvyangcun, respectively, and after entering the soil, the content of curcin would decrease by more than 99% to less than 0.6 μg/g soil within six days and stabilize after 12 days, with curcin levels remaining at 0.4 μg/g soil in the Jinhexiang soil and around 0.3 μg/g soil in the Lvyangcun soil at a much faster degradation rate (Figure 1B). Therefore, the litter of the Jatropha plant, such as its seeds, can release curcin protein into the soil when it enters the soil and remain in the soil for a period of time.

3.1.3. Effect of Curcin on Soil Enzyme Activities

Soil enzymes were involved in almost all biochemical reactions in the soil, and their activity was closely related to soil properties, soil type and environmental conditions, characterizing the vigor of material metabolism in the soil. At the same time, soil enzymes influenced soil microbial populations and community structure, and were widely used as important indicators for assessing soil quality and soil biological activity [33,34].

To investigate the effect of curcin protein on soil enzyme activities after soil incorporation, the effect of curcin on soil sucrase, acid phosphatase, urease and catalase activities, respectively, was measured. In general, from the experimental results, curcin had a greater effect on soil sucrase and urease activities, showing a trend of promotion, then inhibition and finally promotion on sucrase and total inhibition on urease, while the effects on acid phosphatase and catalase were not obvious (Figure 2).

We hypothesize that curcin proteins are able to enter the soil through the Jatropha litter and remain there for a certain period of time, potentially resulting in a modified soil environment through the impact on soil fungi and enzyme activity.

3.2. Effect of Curcin Protein on Soil Fungi

3.2.1. Soil Fungi ITS Sequencing Quality Assessment Results

The soils treated with different concentrations of curcin protein and DNA in soils from different regions of Jatropha plantation and non-plantation areas were extracted for ITS sequencing of the fungus, and the OTU coverage ranged from 99.89% to 100%, and the values indicated the probability of species detection. The OTU coverage of 24 samples from soils of Jatropha planting and non-planting areas in different regions was greater than 99.5%, and the rest of the indicators also met the criteria. The dilution curves of the two groups of samples tended to be flat, indicating that the sample sequences were adequate. In conclusion, the quality of the sequencing data is satisfactory and can be used for the next data analysis.

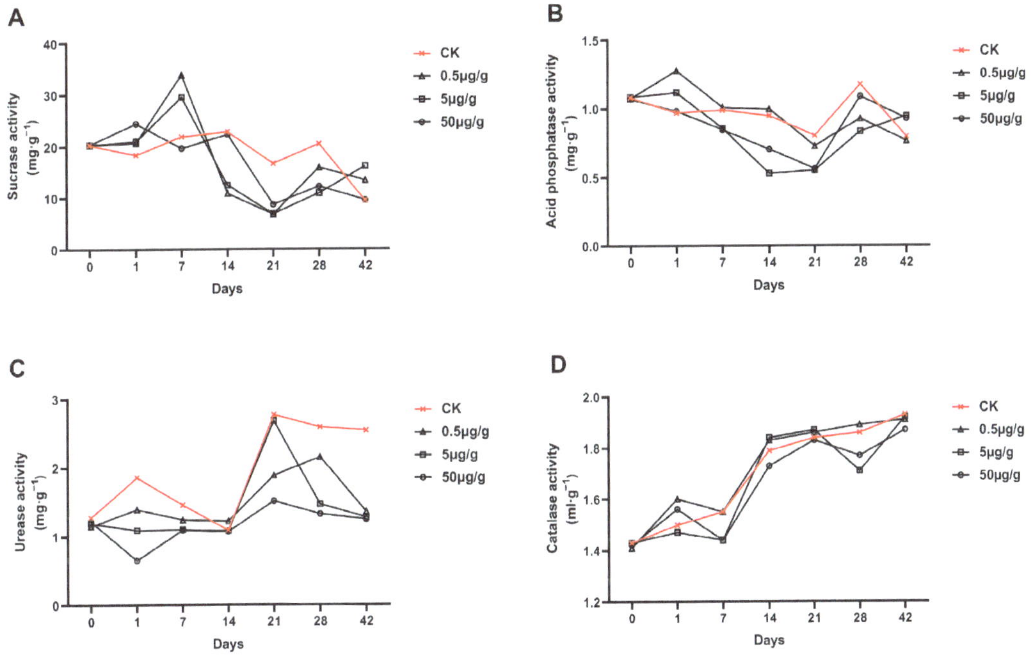

Figure 2. Effect of different concentrations of curcin on activities of enzymes in soil. (**A**) sucrase, (**B**) acid phosphatase, (**C**) urease and (**D**) catalase. CK, untreated soil sample that was used as the control.

3.2.2. Effect of Different Concentrations of Curcin Treatment on Soil Fungal Diversity

To explore the effect of curcin protein on soil fungi, soils were treated with different concentrations of curcin in vitro and fungal diversity indices were examined. The soil samples with 0.5 μg/g of curcin showed a significant increase in fungal abundance in the soil and no significant change in diversity at the early stage of treatment, and a significant decrease in fungal abundance and a significant increase in diversity at the late stage of treatment. A concentration of 5 μg/g of curcin in soil samples resulted in a significant increase in soil fungal abundance at the early stage and no significant change in the late stage. A concentration of 50 μg/g of curcin in soil samples showed an overall trend of increase in the OUT number of soil fungi after treatment; the Chao1 index increased in the early stage of treatment and showed no significant change in the late stage; the Shannon index showed an overall decreasing trend (Table 1). In general, different concentrations of curcin protein treatment caused different changes in the abundance and diversity of soil fungi, with 50 μg/g soil curcin having a more significant effect on soil fungi in the early stage, 0.5 μg/g soil curcin having a significant effect in the later stage, and 5 μg/g soil curcin having a relatively small effect on soil fungi. Concentrations of 0.5 μg/g and 50 μg/g are more likely to change fungal community structure in the soil, and with an increasing extension of the treatment time, they may be detrimental to the conservation of soil ecosystems.

Table 1. Alpha diversity index of soil fungi treated with curcin at different concentrations.

Day	Treatment	OTU	Chao1	ACE	Shannon	Simpson
1	CK	183.5 ± 0.5 f	221.1 ± 15.4 d	284.8 ± 2.6 c	4.21 ± 0.01 bc	0.031 ± 0.000 b
	C0.5	277.0 ± 3.07 bc	303.7 ± 11.3 b	298.0 ± 13.7 b	4.28 ± 0.05 bc	0.030 ± 0.004 b
	C5	300.0 ± 0.1 b	326.2 ± 4.2 b	321.1 ± 6.4 b	4.47 ± 0.01 ab	0.028 ± 0.00 b
	C50	275.5 ± 3.5 bc	302.7 ± 0.3 b	289.8 ± 4.0 bc	3.57 ± 0.01 d	0.08 ± 0.01 a
3	CK	229.5 ± 3.5 de	268.5 ± 2.5 c	284.4 ± 10.0 c	4.32 ± 0.06 b	0.021 ± 0.003 b
	C0.5	211.5 ± 0.5 e	313.8 ± 2.3 b	479.7 ± 9.7 a	4.44 ± 0.02 ab	0.021 ± 0.001 b
	C5	263.0 ± 4.0 cd	261.4 ± 12.6 c	265.4 ± 14.5 c	4.47 ± 0.01 ab	0.023 ± 0.00 b
	C50	235.5 ± 4.5 de	302.7 ± 0.3 b	275.6 ± 8.3 c	4.06 ± 0.08 c	0.06 ± 0.02 ab
7	CK	236.0 ± 12 de	248.1 ± 15.1 cd	257.6 ± 18.8 c	3.96 ± 0.08 c	0.034 ± 0.001 b
	C0.5	274.0 ± 10.0 c	268.5 ± 1.5 c	292.2 ± 13.2 bc	4.37 ± 0.08 ab	0.024 ± 0.001 b
	C5	293.5 ± 20.5 bc	297.6 ± 23.6 bc	254.8 ± 38.7 c	4.56 ± 0.08 a	0.024 ± 0.00 b
	C50	240.5 ± 4.5 d	244.0 ± 6.0 cd	241.4 ± 4.1 c	4.07 ± 0.00 c	0.039 ± 0.01 ab
12	CK	463.0 ± 1.0 a	489.8 ± 1.9 a	488.8 ± 1.9 a	3.56 ± 0.005 de	0.08 ± 0.002 a
	C0.5	236.5 ± 3.5 de	224.3 ± 13.8 d	244.1 ± 7.9 c	4.08 ± 0.005 c	0.03 ± 0.001 b
	C5	475.5 ± 18.5 a	508.0 ± 20.1 a	501.8 ± 22.8 a	3.60 ± 0.25 d	0.077 ± 0.05 a
	C50	469.0 ± 3.0 a	508.0 ± 5.1 a	501.3 ± 0.9 a	3.33 ± 0.01 e	0.08 ± 0.00 a

Note: Any difference between groups with the same marker letter is not significant, any difference with different marker letters is significant; C0.5: 0.5 μg/g curcin treatment, C5: 5 μg/g curcin treatment, C50: 50 μg/g curcin treatment and CK: control treatment.

3.2.3. Effect of Different Concentrations of Curcin Treatment on the Composition of Soil Fungal Communities

The sequencing results showed that Ascomycota was the dominant fungal phylum in the soils collected for this study, accounting for about 70% of the abundance. *Penicillium* and *Kazachstania* were the most abundant dominant fungal genera, *Mortierella* and *Purpureocillium* were the dominant fungal genera, *Aspergillus*, *Thermoascus*, *Metacordyceps*, *Rhizopus* and *Solicoccozyma* were the next dominant fungal genera (Figure 3).

After treating the soil with 0.5 μg/g of curcin, the overall abundance of the dominant fungal genus decreased in the first day treatment group and increased in the remaining three treatment groups. Soil dominant fungal genera showed similar trends in the first and third day treatment groups: the abundance of *Kazachstania* increased and the abundance of *Penicillium* and *Mortierella* decreased significantly; the abundance of soil dominant fungal genera showed the opposite trend in the seventh day treatment group; on the 12th day, a new dominant fungal genus *Saccharomyces* appeared and the abundance of *Kazachstania* increased, while the abundance of the remaining dominant fungal genera decreased (Figure 3).

After treatment of the soil with 5 μg/g of curcin, the total abundance of the dominant fungal phylum decreased only in the treatment group on the seventh day, but the change in the dominant fungal phylum was not significant in the treatment group in the remaining time periods; the overall abundance of the dominant fungal genus decreased in the treatment groups in different time periods, and the abundance of the genus *Kazachstania* increased and the genus *Penicillium* decreased in the treatment groups on the 1st and 12th day; there was no significant change in the abundance of the dominant fungal genus in the third day treatment group and in the seventh day treatment group, the abundance of the genus *Kazachstania* decreased and the abundance of the genera *Penicillium* and *Mortierella* increased (Figure 3).

After treatment of the soil with 50 μg/g of curcin, the overall abundance of both the dominant fungal phylum and the dominant fungal genus in the soil showed an increasing trend. The abundance of Ascomycota in the dominant fungal phylum increased significantly, but the abundance of Basidiomycota decreased significantly; the abundance of *Penicillium* in the dominant fungal genus showed a significant increase (Figure 3).

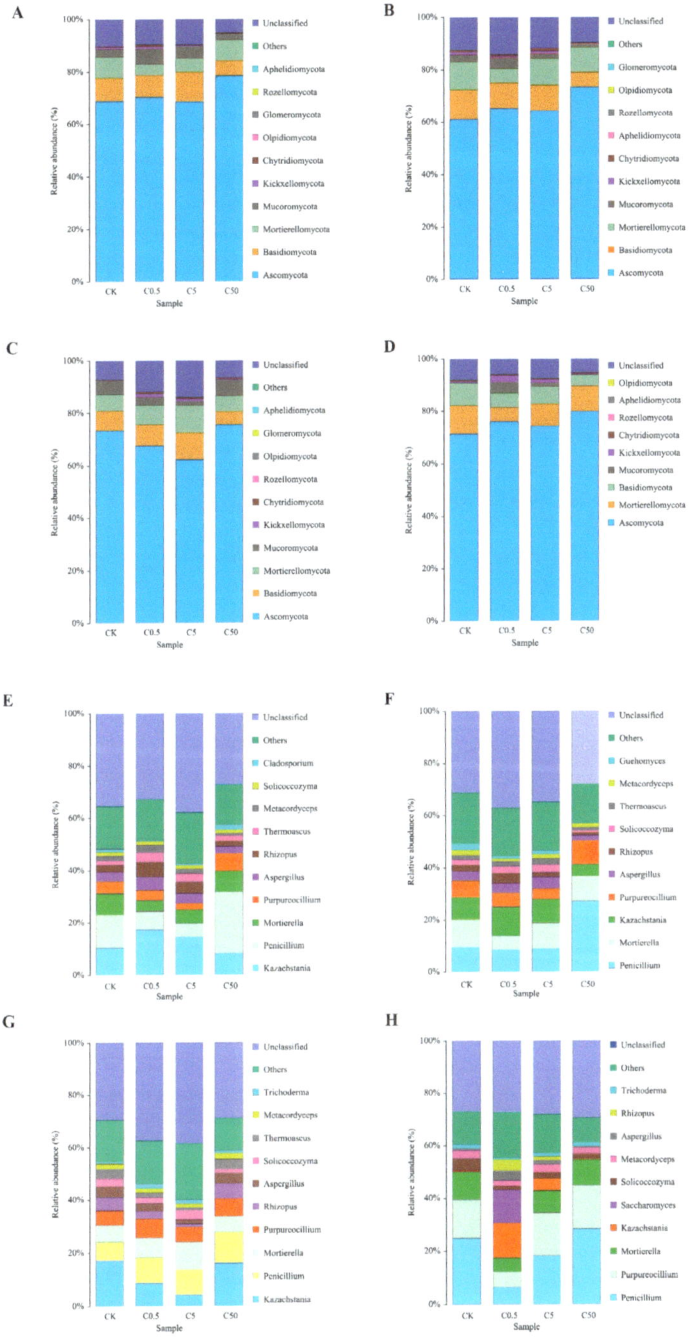

Figure 3. Composition of fungal communities treated with different concentrations of curcin. (**A–D**): 1 day, 3 days, 7 days and 12 days (phylum level); (**E–H**): 1 day, 3 days, 7 days and 12 days (genus level). C0.5: 0.5 μg/g curcin treatment, C5: 5 μg/g curcin treatment, C50: 50 μg/g curcin treatment and CK: control treatment.

3.2.4. Indicator Groups of Soil Fungal Communities Treated with Different Concentrations of Curcin

The different samples at each concentration were divided into four groups according to treatment duration and the fungal communities from phylum to genus in the samples were identified with LEfSe (Line Discriminant Analysis (LDA) Effect Size) analysis and the indicator groups detected in the CK and different concentrations of curcin treatment groups were presented with branching plots (Figure 4A,C,E,G); at the same time, histograms (Figure 4B,D,F,H) showed the communities that play an important role in fungal diversity (LDA score > 3). After one day of treatment, only the 0.5 μg/g soil curcin and 5 μg/g soil curcin treatment groups had significantly different fungal communities, whereas after 7 and 12 days, significantly changed fungal communities, i.e., indicator groups, were present in all four treatment groups at all concentrations.

Fungal indicator groups were more abundant in the 12th day's samples than in the other three groups with significant changes in the 0.5 μg/g soil curcin treatment group for the genera *Kazachstania, Aspergillus, Thermoascus* and *Saccharomyces* as indicator groups; in the 5 μg/g soil curcin treatment group, the abundance of fungi varied less, with only two indicator groups: *Fusarium* and *Papiliotrema*; in the 50 μg/g soil curcin treatment group, the indicator groups were *Penicillium, Gymnopilus* and *Beauveria*. Among them, *Kazachstania, Aspergillus, Saccharomyces* and *Penicillium* were the dominant fungal genera. It was found that the effect of 50 μg/g soil curcin on the dominant soil fungal genera increased with increasing treatment duration. After 12 days of treatment, the 5 μg/g soil curcin treatment had less effect on the soil fungal community than the 0.5 μg/g soil and 50 μg/g soil curcin treatments.

The effects of different concentrations of curcin treatments on soil fungal communities were all different, with the 0.5 μg/g and 5 μg/g soil curcin treatment groups having significantly different fungal communities, while the 50 μg/g soil curcin treatment group had the most significant change in fungal communities. It was also found that the effect of 50 μg/g soil curcin on dominant soil fungi increased with increasing treatment time. These results can contribute to a better understanding of the effects of curcin on soil microbial diversity and provide theoretical support for biological control in agricultural production.

3.3. Analysis of Soil Fungal Communities in Jatropha Planting and Non-Planting Areas in Different Climatic Regions

3.3.1. Comparison of Soil Fungal Diversity in Jatropha Planting and Non-Planting Areas in Different Climatic Regions

The ITS sequences of fungi from 24 soil samples collected from different regions were sequenced, and the results showed that the OTU coverage of soil samples from all regions was greater than 99.5%. The abundance and diversity of soil fungi in the soil samples from the planting areas of the tropical maritime monsoon climate (Haikou) and the subtropical dry and hot monsoon climate (Yuanmou) were lower than those from the non-planting areas, but the differences were not obvious. Unlike other localities where the abundance and diversity of fungi in soil samples from planting area and non-planting areas were consistent, the abundance of fungi in soil samples from the tropical highland monsoon climate (Xichang) showed that the non-planting area was greater than the planting area, and the diversity of fungi showed that the planting area was greater than the non-planting area. Noteworthy, the abundance and diversity of soil fungi in the soil samples from the subtropical monsoon climate plantation area (Jinhexiang) were significantly higher than those from the non-plantation area (Figure 5). Jatropha cultivation affects soil fungal communities, with varying trends observed in areas with different climate types.

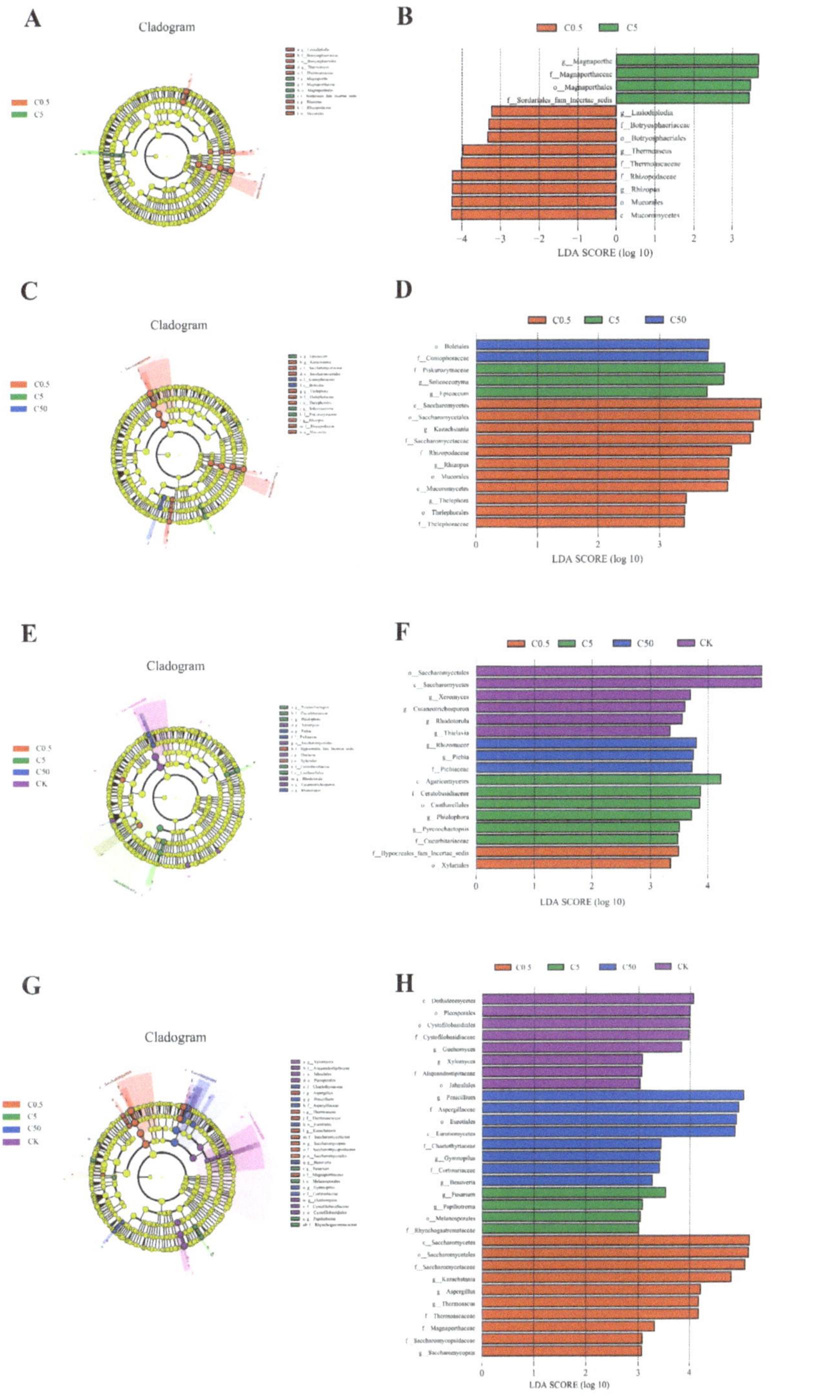

Figure 4. Phylogenetic distribution of soil fungal communities and linear discriminant analysis. Phylogenetic distribution of soil fungal communities at 1 day (**A**), 3 days (**C**), 7 days (**E**) and 12 days (**G**)

for four different treatments. Linear discriminant analysis was used to derive the associated fungal communities scoring ≥ 3 in the four different treatment groups on 1 day (**B**), 3 days (**D**), 7 days (**F**) and 12 days (**H**). Nodes of different colors represent each fungal community that had a significant effect on inter-taxon variation, yellow nodes had a non-significant effect, and node diameters represent the abundance of the community. c, o, f and g indicate class, order, family and genus, respectively; C0.5: 0.5 μg/g soil curcin treatment, C5: 5 μg/g soil curcin treatment, C50: 50 μg/g soil curcin treatment and CK: control treatment.

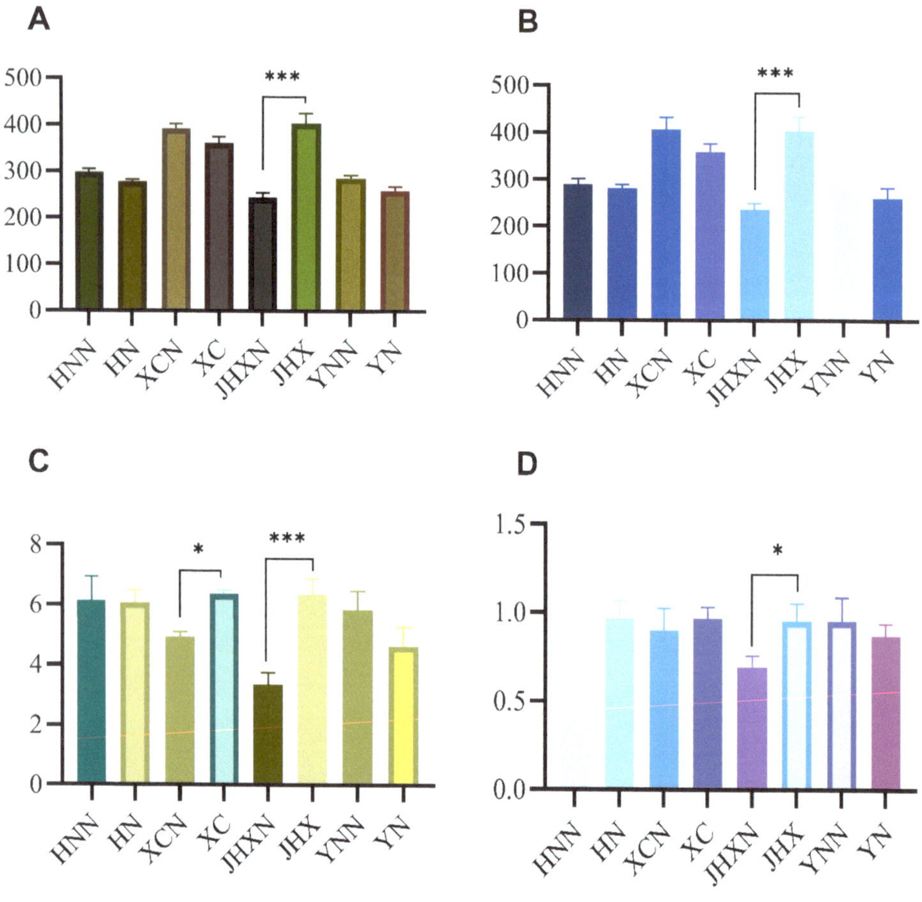

Figure 5. Alpha diversity of soil fungal in different areas. (**A**) Chao1, (**B**) ACE, (**C**) Shannon and (**D**) Simpson. HN: Haikou, Hainan planting area; HNN: Haikou, Hainan non-planting area; XC: Xichang planting area; XCN: Xichang non-planting area; JHX: Jinhexiang planting area; JHXN: Jinhexiang non-planting area; YN: Yuanmou, Yunnan planting area and YNN: Yuanmou, Yunnan non-planting area. Note: ***: $p < 0.001$. *: $p < 0.05$.

3.3.2. Soil Fungal Community Composition in Jatropha Planting and Non-Planting Areas in Different Climatic Regions

Soil fungal sequencing results from a total of 24 samples from different locations showed that at the phylum level, the abundance of Ascomycota (51.76%–94.20%) was absolutely dominant in different soil samples, and it was more abundant in non-plantation areas than in plantation areas in the tropical maritime monsoon climate (Hainan), tropical highland monsoon climate (Xichang) and subtropical monsoon climate (Jinhexiang), while in the subtropical dry and hot monsoon climate (Yunnan) soil samples it showed higher abundance in planting areas than that in non-planting areas. The abundance of Ascomycota, Basidiomycota and Chytridiomycota found among in the tropical maritime monsoon climate (Hainan) planting area (59.21%, 14.27% and 0.16%) versus the non-planting area (81.60%, 3.25% and 1.75%), in the tropical highland monsoon climate (Xichang) planting area (51.76%, 6.67% and 0.74%) versus the non-planting area (72.19%, 23.27% and 0.25%), in the subtropical monsoon climate (Jinhexiang) planting area (80.52%, 7.19% and 0.56%) versus non-planting area (94.20%, 2.30% and 0.10%), and in the subtropical dry and hot monsoon climate (Yunnan) planting area (75.98%, 9.34% and 0.25%) versus non-planting area (60.33%, 26.56% and 4.59%) varied considerably (Figure 6A). At the genus level, the dominant fungal genera with higher abundance in the different soil samples differed and the composition of the fungal communities varied considerably. The fungal genera with the highest abundance in the planting and non-planting areas differed in soil samples from all areas except Jinhexiang (Figure 6B). The large differences in the dominant fungal phylum and dominant fungal genus between planting and non-planting areas suggested that Jatropha planting had a large impact on the dominant fungal community in the soil.

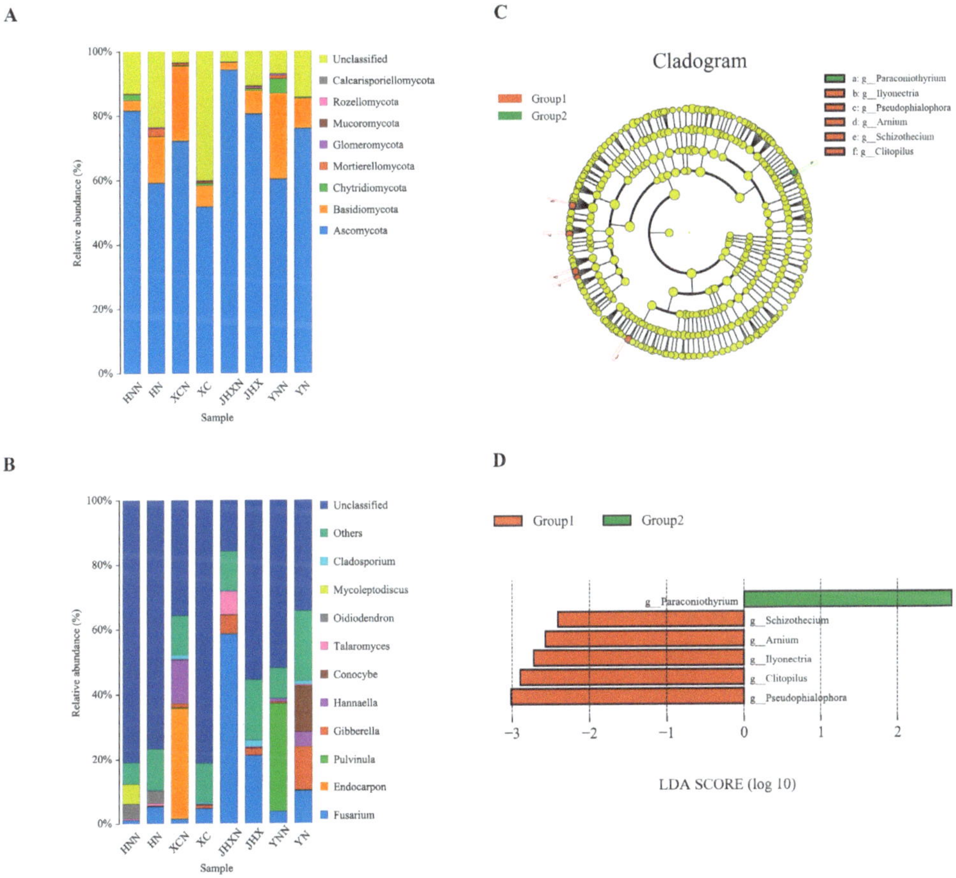

Figure 6. Composition, phylogenetic distribution and linear discriminant analysis of soil fungal communities in Jatropha planting and non-planting areas. Composition of soil fungal communities at different locations on (**A**) phylum level and (**B**) genus level. (**C**) Phylogenetic distribution of soil fungal communities in the two different treatments. (**D**) Associated fungal communities scoring ≥2 in the two treatment groups were derived using linear discriminant analysis.

Analysis of the significance of differences between groups revealed species that were only significantly different at the genus level when the LDA value = 2, and were mainly found in samples from all planting areas, including *Schizothecium*, *Arnium*, *Ilyonectria*, *Clitopilus* and *Pseudophialophora*, while only one species, *Paraconiothyruim*, was significantly different between the different non-planting area samples (Figure 6C,D). More species changed significantly in abundance in planting areas than that in non-planting areas, suggesting that Jatropha planting had a great impact on soil fungal composition.

4. Discussion

4.1. Retention and Degradation of Curcin Protein in Soil

Ribosomal inactivating proteins act as antifungal proteins; their broad-spectrum antifungal properties have great potential for agricultural pest and disease control [35]. Curcin is valuable for applying in transgenic plants because of its fungal resistance, and curcin has been demonstrated to be biologically active in both cell-free and whole animal systems [29]. There are no relevant studies on the retention and degradation dynamics of curcin proteins in soil. Saxena [36] and Stotzky [32,37] found that Bt toxins released into the soil by transgenic Bt plants and soil Bt bacteria are rapidly adsorbed by soil particles and humic acid to form bound Bt toxins that are difficult for microorganisms to degrade and thus remain biologically active for a longer period of time, and compete with other substances (i.e., heavy metal) in the environment for adsorption sites on mineral surfaces [38,39]. We hypothesized that curcin proteins and BT toxins may share similar degradation mechanisms.

Experimental results have shown that after Jatropha seeds are incorporated into the soil, the level of curcin is reduced by more than 99% within a short period of time. Curcin levels in soil are not always at extremely low levels and it is possible that after seeds, leaves, fruits and other tissues are introduced into the soil, due to biological factors (e.g., microorganisms) or some physical factors (e.g., ultraviolet light, temperature, humidity, etc.), the plant tissues decompose rapidly, the cells break down and curcin enters the soil, causing a rapid increase in the local soil curcin levels. Perhaps the adsorption of soil particles makes it difficult to extract curcin, leading to a rapid reduction in its content in a short period of time, while the degradation of curcin by soil microorganisms is the main reason for its continued reduction over time. Studies on the degradation of BT proteins in soil by Helassa et al. showed that a rapid decrease in the amount of extractable toxin was observed during the first 14 days, with a decrease of $86 \pm 7.5\%$ of the initial value, followed by a slower decrease [40]. The degradation trends of curcin protein and BT toxin in soil are similar, but the degradation rate of curcin is faster and reaches a steady state earlier.

After six days of degradation, the curcin levels stabilized, probably due to the gradual loss of moisture in the soil and the slowing of microbial activity. Although the initial addition was the same, differences were observed in the amount of curcin detected in the Jinhexiang and Lvyangcun soils. It is speculated that the reason for this may be firstly, the different extraction efficiency of the method for curcin in different soils, and secondly, it may be that curcin degrades at different rates in different soils and that soil properties (e.g., pH, mineral content composition, microbial community, etc.) affect the rate of protein degradation [41]. Jatropha seeds are mostly harvested at maturity as a feedstock for biodiesel production, but residues from other parts of the plant have a greater chance of entering the soil. Although tissues such as leaves, roots and fruit bark do not contain as much curcin as seed kernels', their long-term accumulation in the soil cannot be ignored. Residues from these parts of the plant may still be an important factor in the accumulation of curcin in soil and their effects on soil ecology deserve further investigation.

Curcin is capable of persisting in soil at low concentrations over extended periods of time, and the long-term efficacy of pest control using curcin may be influenced by several factors, consisting of the extent of its degradation, the nature of the degradation products and their distribution in soil. In cases where the degradation products retain antifungal and antipest properties, curcin may still confer resistance to fungi and pests to some degree. Further research is required to affirm this.

4.2. Assessment of the Effect of Exogenous Curcin Protein on Soil Enzyme Activity and Fungal Communities

Soil enzymes are involved in almost all soil biochemical reactions and are now widely used as important indicators of soil quality and soil biological activity [33,34]. Sucrase is an important hydrolytic enzyme in soils, affecting the decomposition and conversion of soil organic carbon; acid phosphatase is the main enzyme in the phosphorus cycle and is an indicator of soil organic phosphorus mineralization and biological activity, converting organic phosphorus to inorganic phosphorus as an effective nutrient; urease promotes the hydrolysis of nitrogenous organic matter and is closely related to the formation and effectiveness of nitrogen in soils [42,43]. Soil catalase activity, which correlates with soil respiration intensity and soil microbial activity, is effective in preventing hydrogen peroxide toxicity and is an important indicator of soil microcosm environment [44]. The results of this study showed that the Jatropha ribosomal inactivation protein curcin had a greater effect on soil sucrase and urease activities under laboratory conditions, and it is speculated that Jatropha cultivation may also alter soil enzyme activities through curcin protein.

Among the dominant fungal communities in curcin-treated soils, *Penicillium* and *Aspergillus* were mostly soil saprophytic fungi with high reproductive capacity and high environmental adaptability. The same saprophytic fungi were also known as *Mortierella* and *Rhizopus*. *Rhizopus* was an important producer of lipase [45,46], and a common pathogenic genus, causing a variety of plant diseases such as soft rot [47]. *Purpureocillium* was an important biocontrol fungus and was very effective against root-knot nematodes that lead to serious damage to cash crops [48]. There is a certain antagonism between the different flora in the soil, and the beneficial flora and pathogenic flora restrain each other to bring the soil ecology into balance, increasing the defenses and resistance of the soil ecosystem, reducing the occurrence of pests and diseases and preventing soil pollution, which is conducive to the healthy functioning of the soil ecosystem.

Zhaolei et al. [49] employed T-RFLP fingerprinting to examine the impact of varied Cry1Ac toxin concentrations on the diversity of soil fungal communities. The findings indicate that soil fungal community structure remained unaltered over the incubation period of 100 days. These outcomes bear partial similarity to those of the current study. In this study, 0.5 μg/g soil curcin resulted in a significant reduction in soil fungal OUT after 12 days of treatment compared to control samples, but a significant increase in the Shannon index, indicating that application of exogenous curcin protein reduced the abundance of fungal communities but increased the number of fungal species at the taxonomic level, thereby increasing the diversity of fungal communities. It is noteworthy that the protein treatment did not significantly alter the dominant species composition of the soil fungal community, but rather the abundance of some of the dominant species, and the effect was more pronounced in the 0.5 μg/g soil curcin treatment group.

4.3. Analysis of Soil Fungal Communities in Jatropha Planting and Non-Planting Areas in Different Regions

Studies have shown that 20%–50% of the photosynthetic assimilation products of plants are transferred to the below-ground fraction, and most of them are released to the root zone in the form of organic and inorganic secretions, which affect the composition of the soil microbial community [50]. This conclusion was confirmed by Corey et al. [51] who added a plant root secretion to the test group for incubation and obtained a soil fungal community that was very similar to that of the soil in which the plant was grown, suggesting that inter-root secretions are a mechanism for regulating the composition of the soil fungal community. Because curcin protein can alter the composition of soil fungal communities and can enter the soil in a variety of ways, the differences in soil fungal communities between Jatropha planting areas and non-planting areas in different climatic regions were investigated. The results showed that Jatropha cultivation in four different climatic regions had a major effect on the dominant fungal community in the soil. This indicates that Jatropha cultivation has an impact on the composition of soil fungi.

Kumar's study [52] showed that the diversity of soil fungi is generally correlated with the physicochemical properties of the soil, and the differences in the physicochemical properties and soil texture of the soil samples from the three sites may be one of the reasons for the differences in the effects of Jatropha cultivation on the fungal communities of each site. Additionally, the differences in the effects of Jatropha on soil fungal communities were also related to the age and species of Jatropha. It has been shown that plants release different amounts of material into the environment through root secretions, pollen and plant residues during different growth periods [53], and the growth status of Jatropha may vary from region to region, with different effects on soil fungal communities. This study did not sample the soil physicochemical properties of the non-planting areas of the site. Therefore, the study cannot reflect the effect of Jatropha planting on soil physicochemical properties, nor can it be ruled out that different physicochemical properties cause differences in soil fungal communities, which will be incorporated into the experimental design of the following study.

4.4. Shortcomings and Prospects

This study used fungal ITS sequencing to analyze changes in the structure of soil fungal communities. Compared to previous studies that used PLFA to analyze the effects of Jatropha on soil microorganisms [54], the target population is more refined (only soil fungi were studied) and the research methods are more advanced (PLFA can only analyze quantitatively dominant microbial communities, whereas fungal ITS sequencing can detect low abundance microorganisms and is more accurate). While most studies of fungal community composition have a long treatment period, this study focuses on changes in fungal community composition in the short term after treatment.

This study is only a preliminary exploration of the dynamics of curcin degradation in soil, and further research is needed to clarify the detailed process of curcin degradation. For example, the sampling time points between 0 and 6 days should be added to explore the process of curcin degradation from 100% to less than 1%, and the sampling time after 36 days should be extended to explore the maximum time that curcin can remain in soil within the detection limit. In addition, the degradation dynamics of curcin can be studied under specific microbial populations, different UV irradiation conditions and different temperatures and humidity.

The detection methods of high convenience and high sensitivity are urgently in need. For samples measurements, updated extraction methods (e.g., bioassay) and experimental analysis methods are needed to eliminate the bias caused by the determination of curcin protein concentration. In reality, our research on curcin protein and Jatropha cultivation is short-term, which may result in underestimating or overestimating the impact on non-target organisms. Hence, extended observations are imperative for comprehending the impact on soil. Overall, the use of curcin protein as a biopesticide may require additional considerations in terms of its effects on soil microbial communities and nutrient cycling. Further research is needed to fully understand the long-term impacts of curcin protein on soil ecosystems and to evaluate its use as a sustainable agricultural tool.

5. Conclusions

Curcin proteins can enter the soil through various means and persist for an extended period, resulting in certain impacts on soil enzyme activities and fungal communities. Furthermore, Jatropha farming elicits divergent effects on soil fungi under varied climatic conditions.

The results showed that the curcin content in Jatropha seed kernels is up to 2 mg/g tissue, which is much higher than in other tissues. The soil carrying capacity for curcin is approximately 40 mg/g. After Jatropha seeds were incorporated into the soil, the level of curcin in the soil increased rapidly and decreased by more than 99% within six days, after which it stabilized, while the rate of curcin degradation in the soil varied from soil to soil. Curcin had a more significant effect on the fungal community at a later stage after entering the soil. Concentrations of 0.5 μg/g and 50 μg/g cause more structural changes in soil fungal communities than 5 μg/g curcin, which may disrupt the balance between soil microorganisms and be detrimental to the conservation of soil ecosystems.

The effects of Jatropha cultivation on the abundance and diversity of soil fungi were decreasing in the subtropical dry and hot monsoon climate and the tropical maritime monsoon climate and increasing in the subtropical monsoon climate, and effects on soil fungi in the tropical highland monsoon climate were decreasing in abundance and increasing in diversity; the dominant fungal phylum in each region remained unchanged, all being Ascomycota, but the dominant fungal genus with the highest abundance changed, except in Jinhexiang. Various soil properties vary in different climate types, resulting in contrasting impacts of Jatropha cultivation on soil fungi. There was a detrimental impact on soil fungal communities in tropical maritime monsoon and subtropical dry and hot monsoon climates. It is recommended that land managers in these regions prioritize the risk level of soil organisms in Jatropha cultivation areas. Conversely, there was a beneficent effect in subtropical monsoon and tropical highland monsoon climates, thus offering potential for soil utilization in Jatropha plantation areas within these regions.

Further research is required to investigate the impact of Jatropha cultivation and curcin proteins on soil microorganisms. To enhance utilization and preservation, fundamental research is necessary to examine the underlying mechanisms, such as investigating the enduring consequences of curcin on soil microorganisms.

Author Contributions: Conceptualization, Z.L. and Y.X.; Data curation, Z.L., B.Z. and N.J.; Funding acquisition, Y.X.; Investigation, Z.L., B.Z., X.N., C.C. and R.M.; Methodology, Z.L., B.Z., T.W., N.J. and N.J.; Resources, X.N., R.M., T.W., C.C., N.J. and N.H.; Visualization, Z.L., Y.R. and X.W.; Writing—original draft, Z.L. and B.Z.; Writing—review & editing, X.N., C.C., Y.R. and Y.X. All authors have read and agreed to the published version of the manuscript.

Funding: This research was funded by the National Nature Science Foundation, China (No. 31870315).

Data Availability Statement: The data in this study are available on request from the corresponding author.

Conflicts of Interest: The authors declare no conflict of interest.

References

1. Lang, A.K.; Jevon, F.V.; Vietorisz, C.R.; Ayres, M.P.; Matthes, J.H. Fine roots and mycorrhizal fungi accelerate leaf litter decomposition in a northern hardwood forest regardless of dominant tree mycorrhizal associations. *New Phytol.* **2021**, *230*, 316–326. [CrossRef] [PubMed]
2. Jiang, Y.; Wang, W.; Xie, Q.; Liu, N.; Liu, L.; Wang, D.; Zhang, X.; Yang, C.; Chen, X.; Tang, D.; et al. Plants transfer lipids to sustain colonization by mutualistic mycorrhizal and parasitic fungi. *Science* **2017**, *356*, 1172–1175. [CrossRef] [PubMed]
3. Bonfante, P.; Genre, A. Mechanisms underlying beneficial plant-fungus interactions in mycorrhizal symbiosis. *Nat. Commun.* **2010**, *1*, 48. [CrossRef] [PubMed]
4. Fisher, M.C.; Henk, D.A.; Briggs, C.J.; Brownstein, J.S.; Madoff, L.C.; McCraw, S.L.; Gurr, S.J. Emerging fungal threats to animal, plant and ecosystem health. *Nature* **2012**, *484*, 186–194. [CrossRef]
5. Liu, S.; García-Palacios, P.; Tedersoo, L.; Guirado, E.; van der Heijden, M.G.A.; Wagg, C.; Chen, D.; Wang, Q.; Wang, J.; Singh, B.K.; et al. Phylotype diversity within soil fungal functional groups drives ecosystem stability. *Nat. Ecol. Evol.* **2022**, *6*, 900–909. [CrossRef]
6. Lei, F.; Fu, J.; Zhou, R.; Wang, D.; Zhang, A.; Ma, W.; Zhang, L. Chemotactic response of Ginseng bacterial soft-rot to Ginseng root exudates. *Saudi J. Biol. Sci.* **2017**, *24*, 1620–1625. [CrossRef]
7. Hannula, S.E.; de Boer, W.; van Veen, J.A. Do genetic modifications in crops affect soil fungi? a review. *Biol. Fertil. Soils* **2014**, *50*, 433–446. [CrossRef]
8. Sasse, J.; Martinoia, E.; Northen, T. Feed Your Friends: Do Plant Exudates Shape the Root Microbiome? *Trends Plant Sci.* **2018**, *23*, 25–41. [CrossRef]
9. Zhao, M.; Zhao, J.; Yuan, J.; Hale, L.; Wen, T.; Huang, Q.; Vivanco, J.M.; Zhou, J.; Kowalchuk, G.A.; Shen, Q. Root exudates drive soil-microbe-nutrient feedbacks in response to plant growth. *Plant Cell Environ.* **2021**, *44*, 613–628. [CrossRef]

10. Gebhard, F.; Smalla, K. Monitoring field releases of genetically modied sugar beets for persistence of transgenic plant dna and horizontal gene transfer. *FEMS Microbiol. Ecol.* **2019**, *28*, 261–272. [CrossRef]
11. Araujo, A.S.F.; Silva, E.F.L.; Nunes, L.; Carneiro, R.F.V. The Effect of Converting Tropical Native Savanna to *Eucalyptus Grandis* Forest on Soil Microbial Biomass. *Land Degrad. Dev.* **2010**, *21*, 540–545. [CrossRef]
12. Tian, W.H.; Yi, X.L.; Liu, S.S.; Zhou, C.; Wang, A.Y. Effect of transgenic cotton continuous cropping on soil bacterial community. *Ann. Microbiol.* **2020**, *70*, 61. [CrossRef]
13. Ashutosh Kumar, S.; Xiao-Jin, J.; Bin, Y.; Junen, W.; Apurva, R.; Chunfeng, C.; Jitendra, A.; Pingyuan, W.; Wenjie, L.; Nandita, S. Biological indicators affected by land use change, soil resource availability and seasonality in dry tropics. *Ecol. Indic.* **2020**, *115*, 106369. [CrossRef]
14. Weise, C.; Schrot, A.; Wuerger, L.T.D.; Adolf, J.; Gilabert-Oriol, R.; Sama, S.; Melzig, M.F.; Weng, A. An unusual type I ribosome-inactivating protein from *Agrostemma githago* L. *Sci. Rep.* **2020**, *10*, 15377. [CrossRef] [PubMed]
15. Amirzadeh, N.; Moghadam, A.; Niazi, A.; Afsharifar, A. Recombinant anti-HIV MAP30, a ribosome inactivating protein: Against plant virus and bacteriophage. *Sci. Rep.* **2023**, *13*, 2091. [CrossRef]
16. Liu, J.; Wen, D.; Song, X.; Su, P.; Lou, J.; Yao, D.; Zhang, C. Evolution and natural selection of ribosome-inactivating proteins in bacteria, fungi, and plants. *Int. J. Biol. Macromol.* **2023**, *248*, 125929. [CrossRef]
17. Citores, L.; Ferreras, J.M. Biological Activities of Ribosome-Inactivating Proteins. *Toxins* **2023**, *15*, 35. [CrossRef]
18. Bolognesi, A.; Bortolotti, M.; Maiello, S.; Battelli, M.G.; Polito, L. Ribosome-Inactivating Proteins from Plants: A Historical Overview. *Molecules* **2016**, *21*, 1627. [CrossRef]
19. Stirpe, F. Ribosome-inactivating proteins. *Toxicon Off. J. Int. Soc. Toxinol.* **2004**, *44*, 371–383. [CrossRef]
20. Citores, L.; Iglesias, R.; Gay, C.; Ferreras, J.M. Antifungal activity of the ribosome-inactivating protein BE27 from sugar beet (*Beta vulgaris* L.) against the green mould *Penicillium digitatum*. *Mol. Plant Pathol.* **2016**, *17*, 261–271. [CrossRef]
21. Montes, J.M.; Melchinger, A.E. Domestication and Breeding of *Jatropha curcas* L. *Trends Plant Sci.* **2016**, *21*, 1045–1057. [CrossRef] [PubMed]
22. Long, L. Review and prospects of Jatropha biodiesel industry in China. *Renew. Sustain. Energy Rev.* **2012**, *16*, 2178–2190.
23. Qin, S.; Wang, X.; Han, P.; Lai, Z.; Ren, Y.; Ma, R.; Cheng, C.; Wang, T.; Xu, Y. LRP1-Mediated Endocytosis May Be the Main Reason for the Difference in Cytotoxicity of Curcin and Curcin C on U2OS Osteosarcoma Cells. *Toxins* **2022**, *14*, 771. [CrossRef]
24. Qin, X.; Shao, C.; Hou, P.; Gao, J.; Lei, N.; Jiang, L.; Ye, S.; Gou, C.; Luo, S.; Zheng, X.; et al. Different functions and expression profiles of curcin and curcin-L in *Jatropha curcas* L. *Z. Naturforschung C J. Biosci.* **2010**, *65*, 355–362. [CrossRef]
25. Hu, N.; Yang, Q.; Li, C.-Y.; Zhang, M.; Xu, Y.; Chen, F. Extraction of Total DNAs from Soil Samples of *Jatropha curcas* L. Available Regions and Validation by Multiplex-PCR Method. *J. Sichuan Univ. Nat. Sci. Ed.* **2016**, *53*, 683–688.
26. Miller, G.L. Use of Dinitrosalicylic Acid Reagent for Determination of Reducing Sugar. *Anal. Chem.* **1959**, *31*, 420–428. [CrossRef]
27. Alef, K.E.; Nannipieri, P.E. *Methods in Applied Soil Microbiology and Biochemistry*; Elsevier: Amsterdam, The Netherlands, 1995; pp. 569–576.
28. Zhang, Y.; Zhou, G.; Wu, N. A Review of Studies on Soil Enzymology. *J. Trop. Subtrop. Bot.* **2004**, *12*, 83–90.
29. He, W.; King, A.J.; Khan, M.A.; Cuevas, J.A.; Ramiaramanana, D.; Graham, I.A. Analysis of seed phorbol-ester and curcin content together with genetic diversity in multiple provenances of *Jatropha curcas* L. from Madagascar and Mexico. *Plant Physiol. Biochem.* **2011**, *49*, 1183–1190. [CrossRef]
30. Crecchio, C.; Stotzky, G. Insecticidal activity and biodegradation of the toxin from *Bacillus thuringiensis* subsp. kurstaki bound to humic acids from soil. *Soil Biol. Biochem.* **1998**, *30*, 463–470. [CrossRef]
31. Crecchio, C.; Stotzky, G. Biodegradation and insecticidal activity of the toxin from *Bacillus thuringiensis* subsp kurstaki bound on complexes of montmorillonite-humic acids-Al hydroxypolymers. *Soil Biol. Biochem.* **2001**, *33*, 573–581. [CrossRef]
32. Stotzky, G. Persistence and biological activity in soil of insecticidal proteins from *Bacillus thuringiensis* and of bacterial DNA bound on clays and humic acids. *J. Environ. Qual.* **2000**, *29*, 691–705. [CrossRef]
33. Liu, G.; Zhang, X.; Wang, X.; Shao, H.; Yang, J.; Wang, X. Soil enzymes as indicators of saline soil fertility under various soil amendments. *Agric. Ecosyst. Environ.* **2017**, *237*, 274–279. [CrossRef]
34. Dick, R.P.; Burns, R.G. A Brief History of Soil Enzymology Research. In *Methods of Soil Enzymology*; Soil Science Society of America: Fitchburg, WI, USA, 2011.
35. Zhu, F.; Zhou, Y.K.; Ji, Z.L.; Chen, X.R. The Plant Ribosome-Inactivating Proteins Play Important Roles in Defense against Pathogens and Insect Pest Attacks. *Front. Plant Sci.* **2018**, *9*, 146. [CrossRef] [PubMed]
36. Saxena, D.; Stotzky, G.; Flores, S. Insecticidal toxin in root exudates from Bt corn. *Nature* **1999**, *402*, 480. [CrossRef] [PubMed]
37. Stotzky, G. Persistence and biological activity in soil of the insecticidal proteins from *Bacillus thuringiensis*, especially from transgenic plants. *Plant Soil* **2005**, *266*, 77–89. [CrossRef]
38. Hung, T.P.; Truong, L.V.; Binh, N.D.; Frutos, R.; Quiquampoix, H.; Staunton, S. Persistence of detectable insecticidal proteins from *Bacillus thuringiensis* (Cry) and toxicity after adsorption on contrasting soils. *Environ. Pollut.* **2016**, *208*, 318–325. [CrossRef]
39. Zhou, X.; Li, H.; Liu, D.; Hao, J.; Liu, H.; Lu, X. Effects of toxin from *Bacillus thuringiensis* (Bt) on sorption of Pb (II) in red and black soils: Equilibrium and kinetics aspects. *J. Hazard. Mater.* **2018**, *360*, 172–181. [CrossRef]
40. Helassa, N.; M'Charek, A.; Quiquampoix, H.; Noinville, S.; Déjardin, P.; Frutos, R.; Staunton, S. Effects of physicochemical interactions and microbial activity on the persistence of Cry1Aa Bt (*Bacillus thuringiensis*) toxin in soil. *Soil Biol. Biochem.* **2011**, *43*, 1089–1097. [CrossRef]

41. Liu, J.; Liang, Y.-S.; Hu, T.; Zeng, H.; Gao, R.; Wang, L.; Xiao, Y.-H. Environmental fate of Bt proteins in soil: Transport, adsorption/desorption and degradation. *Ecotoxicol. Environ. Saf.* **2021**, *226*, 112805. [CrossRef]
42. Liang, Y.; Yang, Y.; Yang, C.; Shen, Q.; Zhou, J.; Yang, L. Soil enzymatic activity and growth of rice and barley as influenced by organic manure in an anthropogenic soil. *Geoderma* **2003**, *115*, 149–160. [CrossRef]
43. Zhang, H.S.; Zai, X.M.; Wu, X.H.; Qin, P.; Zhang, W.M. An ecological technology of coastal saline soil amelioration. *Ecol. Eng.* **2014**, *67*, 80–88. [CrossRef]
44. Woese, C.R. Bacterial evolution. *Microbiol. Rev.* **1987**, *51*, 221–271. [CrossRef] [PubMed]
45. Yu, X.W.; Xu, Y.; Xiao, R. Lipases from the genus Rhizopus: Characteristics, expression, protein engineering and application. *Prog. Lipid Res.* **2016**, *64*, 57–68. [CrossRef] [PubMed]
46. Helal, S.E.; Abdelhady, H.M.; Abou-Taleb, K.A.; Hassan, M.G.; Amer, M.M. Lipase from *Rhizopus Oryzae* R1: In-depth characterization, immobilization, and evaluation in biodiesel production. *J. Genet. Eng. Biotechnol.* **2021**, *19*, 1. [CrossRef]
47. Chen, X.; Guo, S.; Li, X.; Wang, H. New Endophytic Fungal Strain, i.e., *Rhizopus fungi* Dhs96 Useful for Preventing and Treating Soft Rot Diseases of Dendrobium, Seedling Nursery Beds or Seedling of Dendrobium Plant, and Planting Dendrobium Plant. CN102776127-A.; CN102776127-B, CN102776127-A 14 Nov 2012 C12N-001/14 201324 Pages: 7 Chinese CN102776127-B 10 Jul 2013 C12N-001/14 201371 Chinese. 2013. Available online: https://www.webofscience.com/wos/diidw/full-record/DIIDW:2013C38135 (accessed on 15 October 2023).
48. Liu, R.; Bao, Z.X.; Li, G.H.; Li, C.Q.; Wang, S.L.; Pan, X.R.; Zhang, K.Q.; Zhao, P.J. Identification of Nematicidal Metabolites from *Purpureocillium lavendulum*. *Microorganisms* **2022**, *1*, 1343. [CrossRef]
49. Li, Z.; Bu, N.; Chen, X.; Cui, J.; Xiao, M.; Song, Z.; Nie, M.; Fang, C. Soil incubation studies with Cry1Ac protein indicate no adverse effect of Bt crops on soil microbial communities. *Ecotoxicol. Environ. Saf.* **2018**, *152*, 33–41. [CrossRef]
50. Korenblum, E.; Dong, Y.; Szymanski, J.; Panda, S.; Jozwiak, A.; Massalha, H.; Meir, S.; Rogachev, I.; Aharoni, A. Rhizosphere microbiome mediates systemic root metabolite exudation by root-to-root signaling. *Proc. Natl. Acad. Sci. USA* **2020**, *117*, 3874–3883. [CrossRef]
51. Broeckling, C.D.; Broz, A.K.; Bergelson, J.; Manter, D.K.; Vivanco, J.M. Root exudates regulate soil fungal community composition and diversty. *Appl. Environ. Microbiol.* **2008**, *74*, 738–744. [CrossRef]
52. Kumar, U.; Saqib, H.S.A.; Islam, W.; Prashant, P.; Patel, N.; Chen, W.; Yang, F.; You, M.; He, W. Landscape Composition and Soil Physical-Chemical Properties Drive the Assemblages of Bacteria and Fungi in Conventional Vegetable Fields. *Microorganisms* **2022**, *10*, 1202. [CrossRef]
53. Li, Y.; Wang, C.; Ge, L.; Hu, C.; Wu, G.; Sun, Y.; Song, L.; Wu, X.; Pan, A.; Xu, Q.; et al. Environmental Behaviors of Bacillus thuringiensis (Bt) Insecticidal Proteins and Their Effects on Microbial Ecology. *Plants* **2022**, *11*, 1212. [CrossRef]
54. Chaudhary, D.R.; Lorenz, N.; Dick, L.K.; Dick, R.P. FAME Profiling and Activity of Microbial Communities During *Jatropha curcas* L. Residue Decomposition in Semiarid Soils. *Soil Sci.* **2011**, *176*, 625–633. [CrossRef]

forests

MDPI

Article

Thinning Promotes Soil Phosphorus Bioavailability in Short-Rotation and High-Density *Eucalyptus grandis* × *E. urophylla* Coppice Plantation in Guangxi, Southern China

Xiangsheng Xiao [1,2], Izhar Ali [1,2], Xu Du [1,2], Yuanyuan Xu [1,2], Shaoming Ye [1,2] and Mei Yang [1,2,*]

1 Guangxi Colleges and Universities Key Laboratory for Cultivation and Utilization of Subtropical Forest Plantation, Guangxi University, Nanning 530004, China; 2109302017@st.gxu.edu.cn (X.X.); izharali@gmail.com (I.A.); 1809303003@st.gxu.edu.cn (X.D.); yuanyuanxu_2016@163.com (Y.X.); yshaoming@163.com (S.Y.)

2 Key Laboratory of National Forestry and Grassland Administration on Cultivation of Fast-Growing Timber in Central South China, College of Forestry, Guangxi University, Nanning 530004, China

* Correspondence: fjyangmei@126.com

Abstract: Thinning can improve soil nutrient supply, but the effects of thinning on soil phosphorus (P) contents and bioavailable mechanisms in high-density and short-rotation *Eucalyptus* coppice forests are not well reported. Therefore, we conducted five intensities of thinning treatments, which were 83% (283 tree ha^{-1}, T1), 66% (566 tree ha^{-1}, T2), 50% (833 tree ha^{-1}, T3), 33% (1116 tree ha^{-1}, T4), and 0% (1665 tree ha^{-1}) in a 2nd 6-year-old *E. grandis* × *E. urophylla* coppice plantation with 8 years as a rotation, investigated soil nutrient contents, microbial biomass, and extracellular enzyme activities of 0–20 and 20–40 cm soil layers after two years of thinning, and analyzed the relationship between available phosphorus (AP) and other indicators. The results showed that soil total phosphorus (TP) contents in 2nd *Eucalyptus* coppice plantations were lower than in native forest ecosystems, but T1 significantly increased ($p < 0.05$) TP by 81.42% compared to CK of 0–20 cm, whereas T2 and T3 improved available phosphorus (AP) by 86.87%–212.86% compared to CK. However, soil organic carbon (SOC), dissolved organic carbon (DOC), total nitrogen (TN), and alkaline hydrolysable nitrogen (AN) were not significantly different ($p < 0.05$) among all treatments. According to the analysis, soil TP contents were significantly positively related ($p < 0.001$) to SOC; soil total nutrients and DOC contents had the highest standardized total effect on AP; meanwhile, the quotient of microbial biomass directly conducted soil AP contents. These results highlighted that thinning can be used to alleviate soil P shortages by promoting multinutrient and biological cycles in *Eucalyptus* coppice forests.

Keywords: *Eucalyptus* plantation; thinning; soil nutrients; phosphorus bioavailability; microbial biomass; extracellular enzyme activities

1. Introduction

In China, the Guangxi area is a major *Eucalyptus* plantation area, with approximately 4.5 million hectares accounting for 45% of the total *Eucalyptus* plantation area in this country [1]. To obtain the maximum profit, the high-density (over 1110 tree ha^{-1}) and short-rotation (5–10 years) management modes have become increasingly common in *Eucalyptus* plantation management in Guangxi. This plantation ecosystem suffers severe impacts from human activities and then leads to weaker ecological and social functions than native forest ecosystems [2], such as a decrease in the diversity of understory plant communities, degradation of aggregates, nutrient losses, etc. [3–6]. Meanwhile, some improved *Eucalyptus* species, which are usually used in plantations, require 2–4 times more nutrients than what is returned by litter in this mode [7–9]. Therefore, the contradiction between declining soil fertility and the rapid growth of *Eucalyptus* is fiercer after the intraspecific competition starts in this silvicultural system, suggesting some stand density control technologies are necessary in high-density *Eucalyptus* plantations.

Thinning decreases stand and canopy densities and can (1) relieve nutrients and space competition in high-density plantations, (2) promote plant regeneration, resumption, and litter input [10,11], and (3) ultimately benefit soil nutrient accumulation, cycling, and bioavailability [12–15]. In a previous study, the functions of thinning in sustainable management of *Eucalyptus* plantations were discussed, which included improved fertilizer utilization efficiency, increased stand size classes, and enhanced productivity after thinning to mitigate land degradation [16]. However, most thinning research has focused on coniferous timber plantations, such as Chinese fir (*Cunninghamia lanceolata*) and Masson pine (*Pinus massoniana*) in southern China [17,18]. As a result, it remains unclear how thinning affects the soil nutrient characteristics and biochemical cycles in high-density and short-rotation *Eucalyptus* plantations.

In low-latitude regions, soil bioavailable P is relatively scarce due to leach and inorganic P adsorbed by Fe and Al ions, which makes P often the limiting element for the structure and function of forest ecosystems in tropical and subtropical areas [19–21]. Although P fertilizer input is a common method for supplying soil P in short-rotation plantations, the bioavailable P fraction fixed in acid soil is the main reason for low plant availability after fertilizing [22,23]. This implies that soil TP may be sufficient, but very low bioavailability limits P utilization in plantations (including *Eucalyptus* plantations) in the acid soil area of southern China. Plants mainly utilize some low molecular organic P and inorganic P fractions, which are connected with microbial biomass turnover and originate from phosphatase hydrolysis, respectively [23–25]. A former report found microorganisms that live in low-TP conditions had a stronger ability to capture P from stable minerals [26]. These P fractions are immobilized in microbial biomass and absorbed by mycorrhizal-root associations during the growth-death cycle of microbes [27–29]. Meanwhile, microorganisms and plant fine root exude phosphatase enzymes for hydrolysis of stable organic P fractions; however, their bioactivity and efficiency cannot be maintained for a long time [23,24]. The synthesis and exudation of phosphate enzymes must be continued for microbes and plants to absorb inorganic P. These processes consume a lot of carbon (C), nitrogen (N), and other mineral nutrients [30] and are impacted by forest environmental changes, soil physico-chemical properties, and human activities [24,31], which suggests that some silviculture actions can influence soil P plant availability and available mechanisms [15,32,33]. Therefore, it is meaningful to study the effects of thinning on soil P bioavailability and cycling in high-density and short-rotation *Eucalyptus* plantations. Such research could potentially inhibit soil P deposition, increase the utilization efficiency of P fertilizers, and reduce pollution during production, shipping, and use of P fertilizers.

Eucalyptus grandis × *E. urophylla* is one of the most common species in short-rotation and high-density *Eucalyptus* plantations in the Guangxi Zhuang Autonomous Region of China. It is usually regenerated by the coppice method, which can lead to land degradation [34]. In this study, an *E. grandis* × *E. urophylla* plantation with an initial density of 1665 tree ha^{-1} and 8 years of rotation was used as an experimental stand. We applied different thinning intensity treatments to a 6-year-old of the 2nd generation that was regenerated by coppice after clear cutting conducted on the 1st generation. Two years late (the age of the stands was 8 years old; it is the end of the second rotation), we collected soil samples from the 0–20 and 20–40 cm layers and measured soil nutrient content, microbial biomass, and enzyme activity. Specifically, two research questions were addressed: (1) Does thinning increase soil nutrient content and bioavailability, particularly P? (2) Which indicators play a significant role in soil P plant availability turnover in *E. grandis* × *E. urophylla* coppice forests? To answer these questions, three hypotheses were developed: **Hypothesis 1**: *Thinning improves soil AP content in E. grandis × E. urophylla coppice forests.* **Hypothesis 2**: *Microbial biomass phosphorus (MBP) is the most important factor affecting AP content in E. grandis × E. urophylla coppice forests.* **Hypothesis 3**: *Microbial biomass turnover is more important than enzyme hydrolysis for soil P bioavailability in*

E. grandis × *E. urophylla coppice forests*. This study aimed to provide insights into the effects of thinning on soil nutrient dynamics and to identify key indicators of soil P bioavailable turnover in *E. grandis* × *E. urophylla* coppice forests.

2. Materials and Methods

2.1. Study Site

The study was conducted at Qipo Forest Farm (108°43′–108°44′ E, 23°37′–23°38′ N) in Nanning city, Guangxi Zhuang Autonomous Region, within the southern subtropical hill area of China (Figure 1). The altitude of the study site is approximately 300 m. The climate in this region is subtropical monsoon, with annual mean, maximum, and minimum temperatures of 21.4 °C, 39.0 °C, and −2.2 °C, respectively. The annual total precipitation is 1300 mm and concentrates in the summer, with approximately 300 days of accumulated temperature over 10 °C per year and an annual total accumulated temperature of about 7200 °C. The dominant soil type in the area is Latosol, with a pH range of about 4.0–5.0, developed from granite or milestone, and a depth of about 80–90 cm. The understory plant community is composed of several common species, including *Miscanthus floridulus*, *Microlepia hancei*, *Pteris semipinnata*, *Rubus cochinchinensis*, and *Maesa japonica*.

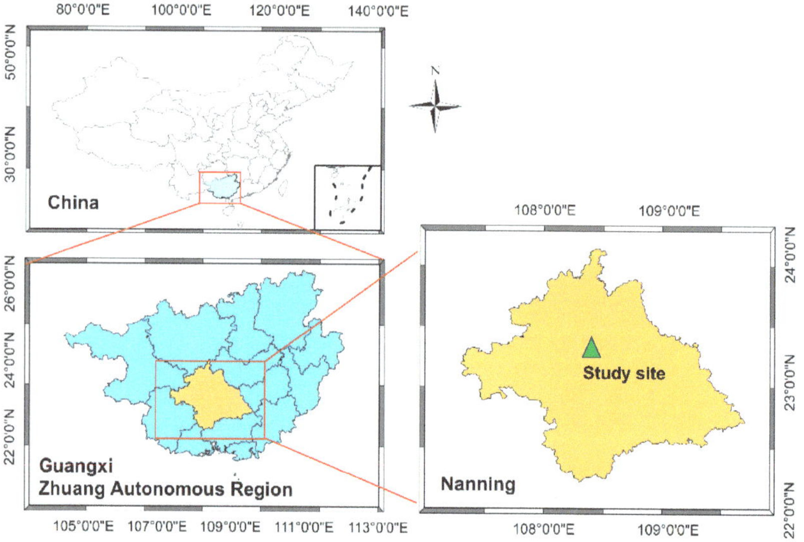

Figure 1. The location of the study site.

2.2. Experimental Design

The high density and short rotation *E. urophylla* × *E. grandis* DH32-28 experimental plantation located on a sloping land of sedimentary hill with an initial density of 1665 trees ha^{-1} and 8 years as a rotation. This forest, 1st generation, was planted by saplings in 2005, and all of the timber was harvested by clear cutting in April 2013. Some silvicultural practices were used to regenerate: (1) stumps were protected by spraying disinfector to promote coppice budding when harvesting was complete; (2) after two months, the 3–4 stronger coppices were retained, which budded from different sides on a stump with similar height and diameter; meanwhile, other coppices were removed; and (3) two months later again, we selected the best one, which would be cultivated to 2nd stand, from the retained coppices and removed others on every stump. In every August during 2013–2015, 0.5 kg of NPK compound fertilizer was applied to every tree at four spots around the trunk; the distance from the fertilizing spots to the trunk was about 30 cm. In March 2019, we established fifteen 400 m^2 (20 × 20 m) plots with similar growth conditions, slop aspect, slop gradient, and understory plant community at this 2nd *Eucalyptus* coppice (Table 1). The distance between each single plot was 30 m. Five thinning treatments with different intensities were assigned to fifteen plots through a completely randomized design. Each thinning-intensity treatment was replicated three times. The 5 thinning intensities were 83% (T1), 66% (T2), 50% (T3), 33% (T4), and 0% (CK), and the stand densities were 283, 566, 833, 1116, and 1665 tree ha^{-1} after thinning, respectively (Table 1). We selectively thinned weakened trees under the canopy, and no logging slash, including root, bark, branch, and leaf, was removed. After thinning, no further silviculture activity was conducted, and a 10 m buffer zone was established around every plot.

Table 1. Basic information of *E. grandis* × *E. urophylla* coppices forest on different thinnings.

Indexes	T1	T2	T3	T4	CK
Slop gradient (°)	25	24	23	27	25
Slop aspect	Northwest	Northwest	Northwest	Northwest	Northwest
Elevation (m)	302.1	299.8	304.5	303.9	304.6
Stand density (tree ha^{-1})	283	566	833	1116	1665
Canopy density	0.35	0.55	0.70	0.82	0.96
Average tree height (m)	26.23	23.87	23.58	25.37	24.67
Average diameter at breast height (cm)	24.17	19.53	19.60	19.43	16.77
Stock volume (m^3 ha)	113.86	148.41	199.56	259.27	365.60
Soil bulk density (g cm^{-3})	1.28	1.31	1.38	1.41	1.47
Soil total porosity (%)	45.56	43.43	41.69	40.42	37.14
Soil water content (%)	14.14	15.18	12.08	11.96	12.60
Soil pH	4.29	4.09	4.17	4.11	4.25

Note: T1—83% intensity thinning; T2—66% intensity thinning; T3—50% intensity thinning; T4—33% intensity thinning; CK—Control.

2.3. Sampling and Analysis

In September 2021, we measured the growth characteristics of *Eucalyptus* trees and canopy densities at every plot (Table 1). Simultaneously, we set nine sample dots following an "S" pattern in the middle of planting lines within each plot according to the method outlined by Dang et al. [11]. We carefully removed the litter from the soil surface and subsequently collected one soil sample from each designated point in both the 0–20 and 20–40 cm layers. The nine soil cores within a plot and layer were mixed together to create a composite sample. A total of 30 mixed soil composites (5 treatments × 3 plots of a treatment × 2 layers) were stored in 4 °C boxes with dry ice. In the laboratory, we fully mixed all the soil composites again, removed any residual debris and litter, passed them through a 2 mm griddle, and then divided them into two sub-amples. One sub-sample was processed by air drying, grinding, and sieving using a 0.25 mm griddle for measuring abiotic indicators. The other sub-sample was stored at 4 °C in a refrigerator for biotic indicator determination.

SOC and TN were measured using an elemental analyzer (Vario EL III; Elementar, Frankfurt, Germany) [35]. Additionally, AN was determined using the previous method described by Wang et al. [36]. The soil samples were digested and processed by 0.5 mol NaOH, steam distillation in succession, and then titration with boric acid. To measure TP, the soil sample was digested with $HClO_4$-H_2SO_4 mixed liquor, and the supernatant was absorbed and analyzed using the molybdenum blue colorimetric method at 880 mm (UV-3600i Plus, SHIMADZU, Hadano, Japan) [37,38]. Furthermore, microbial biomass carbon, nitrogen, and phosphorus (MBC, MBN, and MBP) were analyzed using the chloroform fumigation-extraction method [37]. The soil samples were incubated in darkness in an incubator (MJ-250-I, Jiecheng Experimental Instrument CO, LTD, Shanghai, China) for 48 h and then fumigated and extracted before being extracted with 0.5 mol L^{-1} K_2SO_4 for MBC and MBN and with Bray-1 (0.03 mol L^{-1} NH_4F-0.025 mol L^{-1} HCL) for MBP [39]. The oscillator shaking method (WSZ-100AR, Yiheng, Shanghai, China) was used during extraction. The C and N contents in the supernatant were measured using a TOC/TN analyzer (Multi N/C 3100, Analytic Jena, Jena, Germany), and the P content was analyzed using the molybdenum blue colorimetric method (UV-3600i Plus, SHIMADZU, Japan). DOC and AP were measured as the C and P contents of the leaching solutions before fumigation, respectively [35]. A universal conversion factor of 0.45, 0.54, and 0.4 was used to calculate MBC, MBN, and MBP, respectively [40]. The quotients of microbial biomass carbon, nitrogen, and phosphorus (qMBC, qMBN, and qMBP) are expressed as the ratios of MBC:SOC, MBN:TN, and MBP:TP, respectively. The activities of soil β-glucosidase (BG) and acid phosphatase (ACP) were determined using the microplate technique as previously described by Saiya-Cork et al. [41]. Whereas the previous method described by Fei et al. was used to measure urease (URE) activity [42], it was presented by the ammoniacal N content of 5 g of fresh soil dissolved in 10 mL of urea solution (10%, *w/w*) and 10 mL of citrate buffer (pH = 6.7), incubated at 38 °C under dark conditions for 24 h.

2.4. Statistical Analysis

SPSS 18.0 (IBM, Chicago, IL, USA) was used for statistical analysis. A one-way of variance (ANOVA) was conducted to test statistical significance among treatments, followed by Duncan's multiple range test ($p < 0.05$) for various comparisons. The data were presented as mean ± standard difference. Meanwhile, the Pearson correlation was analyzed by this software. The ggplot2 in R version 4.2.1 and the R Studio 2022.07.0 interface were used for regression analysis. Based on the results of Pearson correlation, partial least squares path modeling (PLS-PM) was conducted using the pls-pm package to explore the direct or indirect connections of soil nutrients (including SOC, DOC, TN, and TP), enzymes of C and N cycling (including BG and URE), quotients of microbial biomass (including qMBC and qMBP), ACP, MBP, and AP [35,43]. Finally, we used Origin Pro 2022 (Originlab, Northampton, MA, USA) and Microsoft PowerPoint 2016 (Microsoft, Redmond, WA, USA) for drawing figures.

3. Results

3.1. Response of Soil Nutrients Content to Thinning Intensity in *E. urophylla* × *E. grandis* Coppice Stands

The soil nutrient content of different thinning treatments from different soil layers is presented in Table 2. Soil total nutrient content (SOC, TN, TP, and AN) increased and then decreased with increasing thinning intensity at the 0–20 cm layer. However, no significant differences ($p < 0.05$) were observed for these nutrients between the different thinned *E. urophylla* × *E. grandis* coppice stands in two layers. Soil TP in the 0–20 cm layer was significantly higher ($p < 0.05$) in the T1 treatment (81.41%) compared to the CK treatment, as well as in the T4 and T3 treatments (92.26%) (Table 2), but there were no significant differences ($p < 0.05$) observed in the 20–40 cm layer among the five treatments. This suggests that the stronger thinning intensities benefited the soil P content in the surface layer. In the 0–20 cm soil layer, the trend of AP content was observed as follows: T2 > T3 > T1 > CK > T4. Both T2 and T3 exhibited significantly higher ($p < 0.05$) AP levels than CK by 130.56% and 99.36%, respectively (Table 2). Similarly, these two treatments, T2 and T3, displayed significantly higher ($p < 0.05$) AP contents than T4 by 132.47% and 101.30%, respectively (Table 2). In the 20–40 cm layer, the trend was T2 > T1 > T3 > T4 > CK, with significantly ($p < 0.05$) increased soil AP content in the T2 treatment compared to the T4 and CK treatments by 142.31% and 215.00%, respectively (Table 2).

Table 2. Changes in soil nutrient content and the *p*-value and *F*-value of different thinned *E. urophylla* × *E. grandis* coppices forests.

Indexes	Layers (cm)	T1	T2	T3	T4	CK	*F*	*p*
SOC ($g\ kg^{-1}$)	0–20	11.62 ± 3.62 ns	12.90 ± 2.51 ns	10.72 ± 1.18 ns	9.59 ± 1.30 ns	9.46 ± 0.42 ns	0.917	0.491
	20–40	8.55 ± 2.58 ns	7.06 ± 1.66 ns	6.56 ± 0.13 ns	6.16 ± 0.10 ns	7.37 ± 1.40 ns	0.866	0.516
DOC ($mg\ kg^{-1}$)	0–20	20.40 ± 5.48 ns	23.53 ± 5.17 ns	18.01 ± 5.05 ns	12.79 ± 3.37 ns	17.74 ± 1.38 ns	1.733	0.219
	20–40	8.89 ± 2.43 ns	7.75 ± 1.00 ns	10.11 ± 3.16 ns	5.54 ± 0.83 ns	7.87 ± 2.05 ns	1.307	0.332
TN ($g\ kg^{-1}$)	0–20	0.74 ± 0.19 ns	0.78 ± 0.15 ns	0.63 ± 0.04 ns	0.60 ± 0.09 ns	0.61 ± 0.04 ns	0.987	0.457
	20–40	0.62 ± 012 ns	0.49 ± 0.13 ns	0.47 ± 0.01 ns	0.48 ± 0.02 ns	0.55 ± 0.05 ns	1.170	0.381
AN ($mg\ kg^{-1}$)	0–20	38.76 ± 5.35 ns	37.73 ± 7.41 ns	34.65 ± 1.09 ns	33.88 ± 1.09 ns	34.39 ± 0.36 ns	0.564	0.694
	20–40	34.14 ± 7.23 ns	30.29 ± 6.42 ns	26.44 ± 0.96 ns	24.90 ± 1.45 ns	33.11 ± 1.66 ns	1.645	0.238
TP ($g\ kg^{-1}$)	0–20	0.38 ± 0.08 a	0.31 ± 0.09 ab	0.20 ± 0.02 b	0.20 ± 0.02 b	0.21 ± 0.05 b	3.573	0.047 *
	20–40	0.26 ± 0.07 ns	0.21 ± 0.06 ns	0.18 ± 0.03 ns	0.19 ± 0.00 ns	0.20 ± 0.02 ns	0.821	0.541
AP ($mg\ kg^{-1}$)	0–20	1.29 ± 0.40 bc	1.79 ± 0.38 a	1.55 ± 0.27 ab	0.77 ± 0.07 c	0.78 ± 0.19 c	5.044	0.017 *
	20–40	0.93 ± 0.27 ab	1.26 ± 0.42 a	0.76 ± 0.27 ab	0.52 ± 0.07 b	0.40 ± 0.09 b	3.498	0.049 *

Note: The lowercase letter indicates a significant difference ($p < 0.05$) between different thinning treatments in the same layer. T1—83% intensity thinning; T2—66% intensity thinning; T3—50 intensity thinning; T4—33% intensity thinning; CK—Control; SOC—Soil organic carbon; DOC—Dissolved organic carbon; TN—Total nitrogen; AN—Alkaline hydrolysable nitrogen; TP—Total phosphorus; AP—Available phosphorus; * and ns—significant difference at $p < 0.05$ and non-significant, respectively. (mean ± SD, $n = 3$).

3.2. Response of Soil Microbial Biomass to Thinning Intensity in E. urophylla × E. grandis Coppice Stands

The qMBC and MBC in soil showed the same trend (decreased and then increased with an increase in thinning intensity) among treatments except for the 20–40 cm layer of T4 (Figure 2a,b). Among soil layers, MBC was enhanced in 0–20 cm soil layers across thinning treatments (Figure 2a). Whereas qMBC showed the highest value in the 20–40 cm layer among all thinning treatments (Figure 2b). Among the treatments, T2 treatment had the lowest qMBC among all thinning treatments, with a significant difference ($p < 0.05$) of 47.34% and 44.81% compared to CK in the 0–20 cm and 20–40 cm layers, respectively (Figure 2b, Table 3). Moreover, T2 resulted in the lowest qMBN, which was significantly lower ($p < 0.05$) than T1, T3, T4, and CK by 39.13%, 36.04%, 40.09%, and 36.51%, respectively (Figure 2d, Table 3). Interestingly, soil MBP and qMBP in the 20–40 cm layer showed an increase followed by a decrease with stand density increasing. T2 was significantly higher ($p < 0.01$) than other treatments, showing an increase of 91.23%–189.45% in MBP and 67.55%–211.47% in qMBP (Figure 2e,f, Table 3). This suggests that T2 benefits soil P accumulation in microorganisms. However, thinning had no significant influence ($p < 0.05$) on MBN, MBP, and qMBP in the 0–20 cm layer (Figure 2c,e,f, Table 3), or on qMBN in the 20–40 cm layer (Figure 2d, Table 3).

Table 3. The *F*-value and *p*-value of one-way ANOVA of MBC, qMBC, MBN, qMBN, MBP, qMBP, BG, URE, and ACP in different thinning treatments and layers.

Indexes	Layers (cm)	*F*	*p*
MBC	0–20	8.528	0.003 **
	20–40	15.242	<0.001 **
qMBC	0–20	5.202	0.016 *
	20–40	5.782	0.011 *
MBN	0–20	1.239	0.355
	20–40	0.403	0.802
qMBN	0–20	4.163	0.031 *
	20–40	0.613	0.663

Table 3. *Cont.*

Indexes	Layers (cm)	*F*	*p*
MBP	0–20 20–40	2.155 8.875	0.148 0.003 **
qMBP	0–20 20–40	0.191 8.046	0.937 0.004 **
BG	0–20 20–40	3.604 3.892	0.046 * 0.037 *
URE	0–20 20–40	0.344 1.416	0.409 0.298
ACP	0–20 20–40	0.989 1.232	0.457 0.357

Note: * and ** stand significance differences at $p < 0.05$ and $p < 0.01$, respectively.

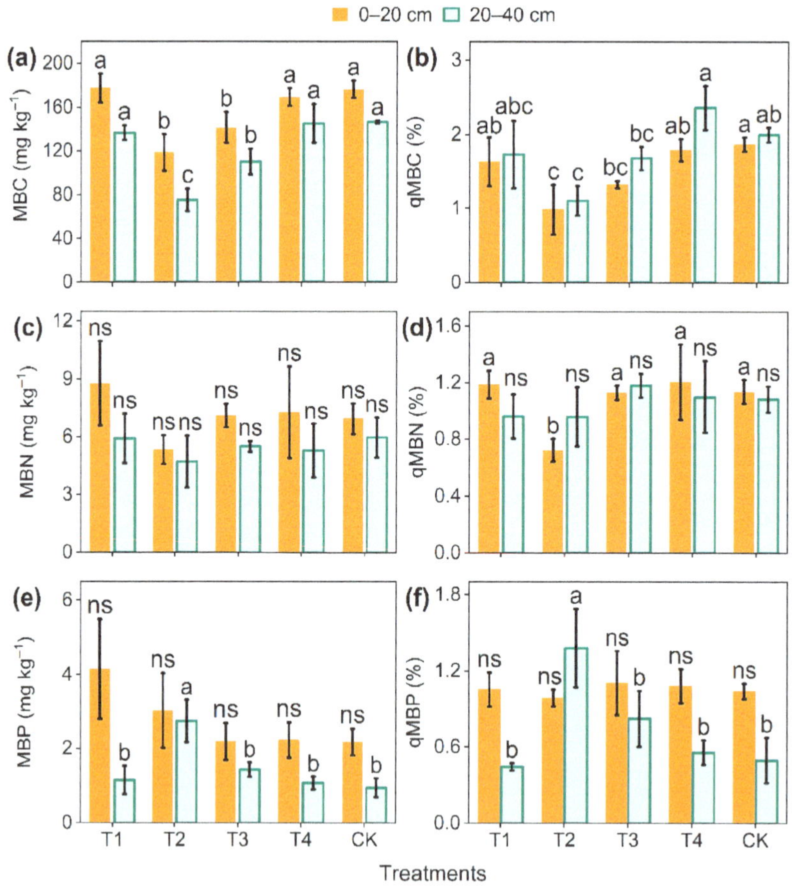

Figure 2. Changes in soil microbial biomass carbon (MBC) (**a**), nitrogen (MBN) (**c**), phosphorus (MBP) (**e**), the quotient of microbial biomass carbon (qMBC) (**b**), nitrogen (qMBN) (**d**), and phosphorus (qMBP) (**f**) of different thinned *E. grandis* × *E. urophylla* coppices forests (mean ± SD, $n = 3$). Note: ns—no significant difference. Different lower-case letters indicate significant differences among the treatments ($p < 0.05$).

3.3. *Response of Soil Enzyme Activities to Thinning Intensity in E. urophylla × E. grandis Coppice Stands*

Among soil enzyme activities, only BG activity was significantly affected ($p < 0.05$) by various thinning treatments (Figure 3a, Table 3), while the URE and ACP activities were not significantly ($p < 0.05$) influenced (Figure 3b, c, Table 3). BG activity was increased first and then decreased as thinning intensity increased, following the order T2 > T3 > T1 > CK > T4 and T2 > T3 > T4 > CK > T1 in 0–20 cm and 20–40 cm layers, respectively (Figure 3a). BG activity in T2 was higher than other treatments by 16.37%–40.35% in the 0–20 cm layer and by 10.10%–63.22% in the 20–40 cm layer (Figure 3a).

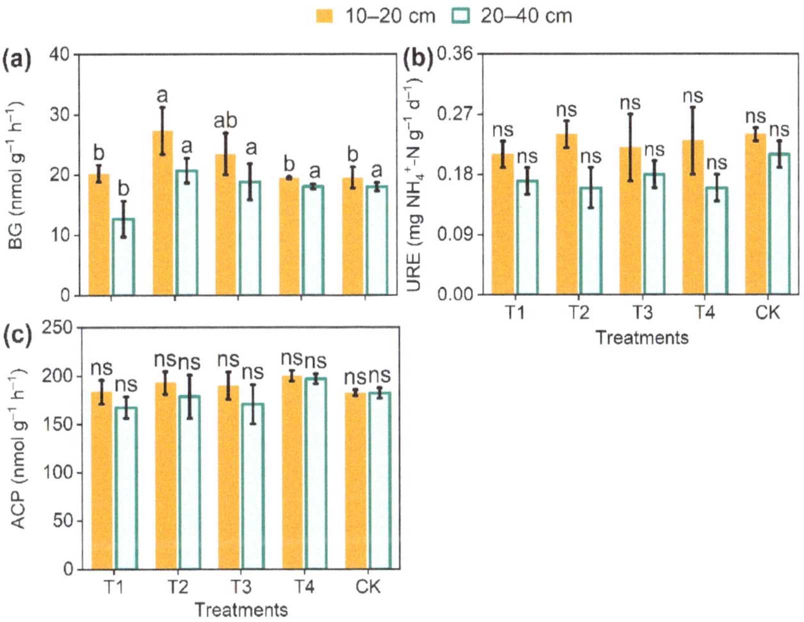

Figure 3. Effects of various thinning intensities on soil β-Glucosidase (BG) (**a**), urease (URE) (**b**), and acid phosphatase (ACP) (**c**) activities in *E. grandis* × *E. urophylla* coppices forests (mean ± SD, $n = 3$). Note: ns: no significant difference. Different lower-case letters indicate significant differences among the treatments ($p < 0.05$).

3.4. *The Influence of Soil Nutrients, Microbial Biomass, and Enzyme Activity on P Availability*

Regression and Pearson correlation analysis indicated that AP was significantly positively correlated with SOC ($R^2 = 0.476$), DOC ($R^2 = 0.457$), TN ($R^2 = 0.413$), AN ($R^2 = 0.404$), TP ($R^2 = 0.353$), BG ($R^2 = 0.425$), MBP ($R^2 = 0.48$) and qMBP ($R^2 = 0.203$) ($p < 0.05$) (Figures 4a–f and 5c,f, Table S1), while significantly negatively correlated with qMBC ($R^2 = 0.636$) and qMBN ($R^2 = 0.143$) ($p < 0.05$) (Figure 5d,e, Table S1). These results suggested that soil AP was affected by soil nutrients and microbial biomass indicators in *E. grandis* × *E. urophylla* coppice forests. Increasing nutrient supply and qMBP can promote soil P turnover from a stable to a bioavailable fraction.

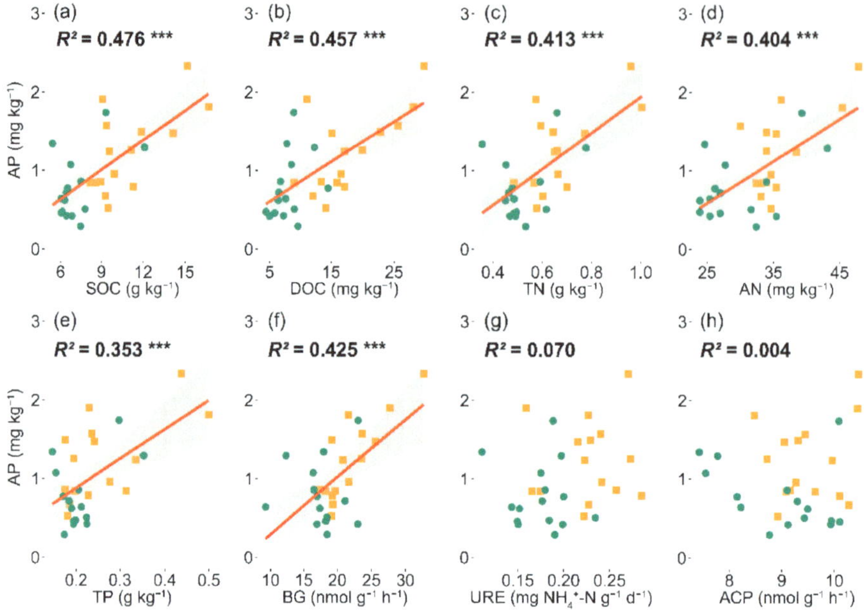

Figure 4. Regression analysis between AP and soil nutrients (**a–e**) and enzyme activities (**f–h**). Note: ***: $p < 0.001$. The confidence interval and fitting line were revealed at $p < 0.05$. Orange and depth green points were data in the 0–20 and 20–40 cm layers, respectively.

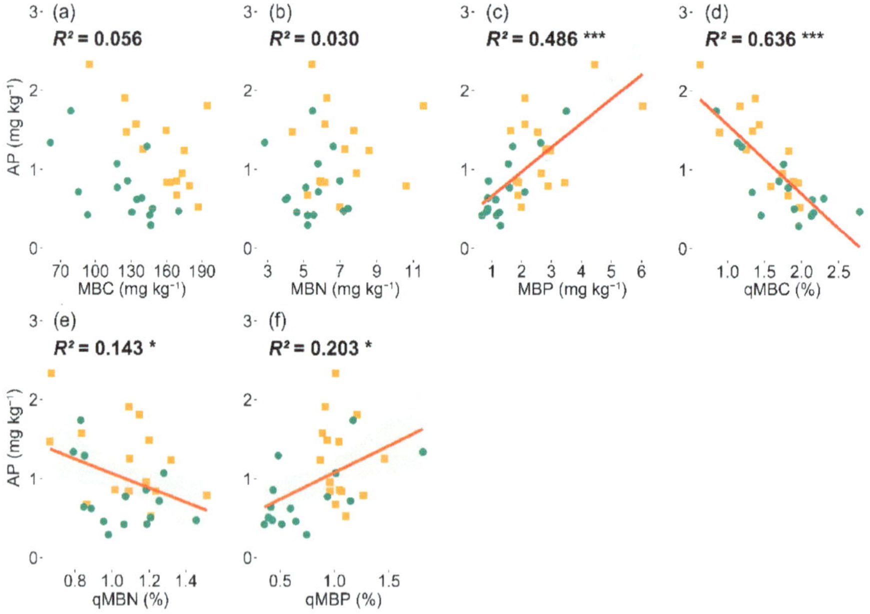

Figure 5. Regression analysis between AP and to microbial biomass (**a–c**) and quotient of microbial biomass (**d–f**). Note: *: $p < 0.05$, ***: $p < 0.001$. The confidence interval and fitting line were revealed at $p < 0.05$. Orange and depth green points were data in 0–20 and 20–40 cm layers, respectively.

The PLS-PM model revealed that soil nutrients had the highest standardized effect on AP, as shown in Figure 6b. The direct and indirect effects of soil nutrients through MBP, ACP, and the quotient of microbial biomass on AP were significant ($p < 0.05$) (Figure 6a). These results imply soil nutrient content was the most important factor affecting P bioavailability in *Eucalyptus* coppice forest. The quotient of microbial biomass was significantly affected by soil nutrients ($p < 0.05$), an enzyme of C and N cycling ($p < 0.01$), ACP ($p < 0.01$), and MBP ($p < 0.001$) directly (Figure 6a). However, the direct effects of enzymes C and N cycling, ACP, and MBP on AP were not significant ($p < 0.05$) (Figure 6a). Meanwhile, AP was significantly ($p < 0.01$) directly affected by the quotient of microbial biomass, suggesting that the effects of enzymes of C and N cycling, ACP, and MBP on AP were mainly through an indirect pathway (Figure 6). The two parameters of the quotient of microbial biomass, qMBC and qMBP, had different influences on AP, with qMBC having an adverse effect and qMBP having a positive impact (Figures 5d,f and 6a, Table S1). These results indicate that different indicators of microbial quotient had a varied influence on soil AP in an *E. grandis* × *E. urophylla* coppice forest.

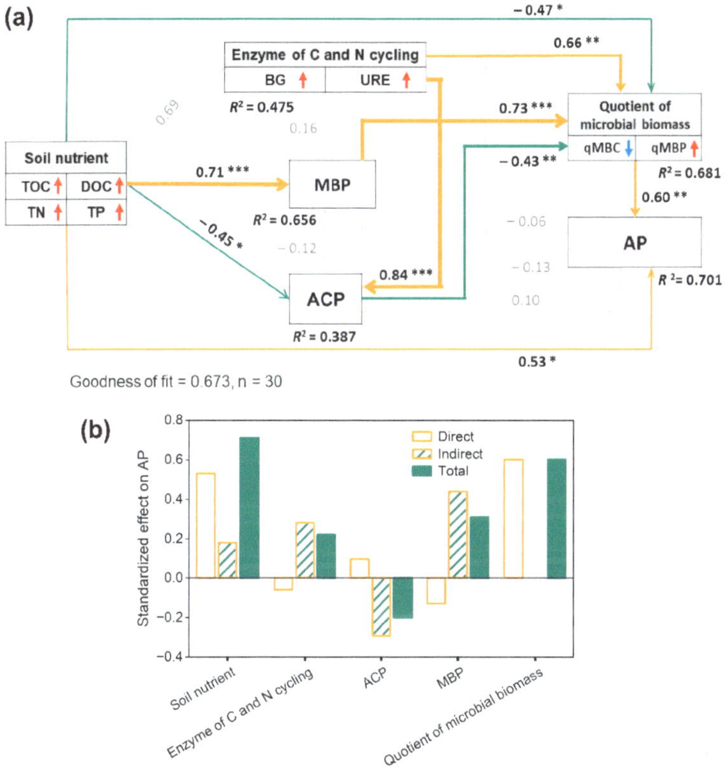

Figure 6. Partial least squares path model (PLS-PM) of soil nutrient, enzyme activity, microbial biomass, quotient of microbial biomass, and AP (**a**), and standardized direct, indirect, and total effect on AP of these indicators (**b**). Note: The numbers of the arrowed lines are path coefficients. Orange and deep green solid arrows indicate positive and negative significant ($p < 0.05$) paths, respectively. Gray dotted arrows represent a non-significant ($p < 0.05$) path. Red and blue arrows indicate positive and negative relationships, respectively, between measured variables and associated variables in measurement models of PLS-PM. *: $p < 0.05$, **: $p < 0.01$, ***: $p < 0.001$.

4. Discussion

4.1. Soil Nutrients of Different Intensity Thinned E. grandis × E. urophylla Coppice Plantations

Previous studies suggested that increasing soil SOC and TN after thinning requires an extended period [44,45]. In the present study, soil SOC, DOC, TN, and AN were not significantly influenced ($p < 0.05$) (Table 2) in different thinned *E. grandis* × *E. urophylla* coppice forests after two years. These results confirmed the regular fitting of the *Eucalyptus* plantation in Guangxi. Thinning of *Eucalyptus* plantations in short rotation can lead to a decrease in stand density, an increase in canopy openness, and destruction of the understory plant community and soil structure, which can accelerate the loss of soil C and nutrients, especially some labile C and N fractions, due to an increase in soil temperature, priming effect, and enzyme activity [27,46–48]. In the long term, thinning benefits plant community structure and promotes the soil nutrient cycle [11]. However, thinning provides *Eucalyptus* trees with more space for growth, enhancing their photosynthesis and nutrient uptake. Meanwhile, 6-year-old to 8-year-old *Eucalyptus* trees in the current study were in a rapid growth period; these accelerated soil nutrients utilized by trees and reduced litter input [49], which prolonged the process of increasing soil C and N contents in thinned *Eucalyptus* plantations. The SOC, DOC, TN, and AN trends were varied in two layers. These four indicators were higher in the T2 and T3 treatments than CK in the 0–20 cm layer but lower in the 20–40 cm layer (Table 2). The results comply with the rule of soil nutrients content in different layers that aggregate on the soil surface in subtropical plantations [50,51] and may be related to spatial variation of litter input and organism nutrients demand: (1) the understory vegetation root system may not develop entirely in these stands, and litter is mainly concentrated on the soil surface [52]; (2) meanwhile, *Eucalyptus* trees usually uptake nutrients from the depth of the soil layer, and microorganisms indicate higher N demand in soil below 20 cm in subtropical forests [50]. Therefore, the amount of soil nutrients consumed is higher in the 20–40 cm layer compared to the 0–20 cm layer.

Qiao et al. reported that the average TP content of native broad-leaved forest in the 0–20 cm and 20–40 cm layers was 0.38 ± 0.12 g kg^{-1} and 0.31 ± 0.11 g kg^{-1}, respectively, in Tiantong National Park, a subtropical region similar to our study area [53]. Our findings indicated that the soil TP contents of the CK treatment were 0.21 ± 0.05 g kg^{-1} and 0.20 ± 0.02 g kg^{-1} in these two layers (Table 2). This suggests that soil P accumulations in our 2nd generation *E. grandis* × *E. urophylla* coppice plantation were lower than in the native forest ecosystem. It is well known that there is more soil nutrient consumption and erosion in short-rotation and high-density *Eucalyptus* plantations than in native forest ecosystems [4,54,55]. Soil TP increased in both the T1 and T2 treatments after two years (Table 2), suggesting that thinning, especially high-intensity thinning, can promote soil P levels. Bedrock and mineral conditions are essential factors that affect soil TP content [56], indicating that TP is relatively stable in a specific ecosystem and that limited research exists. Therefore, litter input is an essential source for the soil P pool [20]. Our study found that soil TP is significantly positively ($p < 0.001$) related to SOC (Table S1), which suggests that increased litter input is the main reason for soil P accumulation in the 0–20 cm layer of T1 and T2 treatments. Soil AP content in the T2 and 0–20 cm layer of T3 was significantly higher than in the CK ($p < 0.05$) (Table 2), suggesting that moderate thinning intensities can improve P plant availability in *E. grandis* × *E. urophylla* coppice forest, thus confirming that thinning can improve soil AP content in *E. grandis* × *E. urophylla* coppice forest.

4.2. Soil Microbial Biomass and Enzyme Activity of Different Intensity Thinned E. grandis × E. urophylla Coppice Plantations

The results showed that soil MBN and MBP were significantly positively correlated ($p < 0.01$) with soil SOC, DOC, TN, AN, and TP (Table S1). However, MBC did not show a significant correlation ($p < 0.05$) with these soil nutrients (Table S1), suggesting that MBN and MBP were controlled by soil nutrient supply, while MBC was impacted by changes in the understory plant community and root traits after thinning [27,57]. The changes in plant structure after thinning were found to be a major reason affecting soil fungi [58,59], such as community, distribution, structure, and physiological activity. Fungi have a higher C proportion than bacteria in the soil and are the main fraction of soil MBC [11,60,61]; hence, the MBC of our stand soil changed after thinning. On a global scale, the average qMBN and qMBP in tropical and subtropical forest soils are 3.08% and 6.32%, respectively [62]. In our study, the soil average qMBN and qMBP of *E. grandis* × *E. urophylla* coppice plantations were 1.07% and 0.90%, respectively (Figure 2d,f), which are only a part of 34.74% and 14.24%, respectively, on the global scale. These results may be related to worse soil nutrient loss, physico-chemical properties, microbial multifunctionality, and increasing generations in *Eucalyptus* plantations [5,63,64]. This implied that the proportions of MBN and MBP in soil TN and TP were lower. Regression, Pearson correlation, and PLS-PM analysis suggested a significant positive relationship between qMBP and AP ($p < 0.05$) (Figures 5f and 6, Table S1). These results implied that the relatively inferior qMBP was a reason for low P bioavailability [23].

Soil BG activities rose initially and then decreased with thinning intensity in the current study (Figure 3a); this trend was opposite to the findings of Chen et al., who reported lower BG activity in low-intensity thinning treatment and higher activity in high-intensity treatment than CK in Chinese fir plantations [65]. These varied trends in BG activities suggest that tree species are one of the reasons affecting enzyme activity in thinned plantations. There are different forest structures, light environments, plant communities, and soil nutrient cycling characteristics in these plantations, which are constructed by various broad-leaved or coniferous tree species [66,67]. We found that BG activity had a very significant positive correlation ($p < 0.01$) with SOC and DOC (Table S1), suggesting that it was affected by substrate quantity and quality [68]. However, according to the results found by Zhou et al., N and P hydrolytic enzyme activity increased after thinning in the larch (*Larix olgensis*) plantation [48]. While our study found no significant difference ($p < 0.05$) in URE and ACP activity in *E. grandis* × *E. urophylla* coppice forest (Figure 3b,c), this result was reported in other tropical and subtropical plantations, such as teak (*Tectona grandis*) [69]. In nutrient-faulty acid soils, N and P absorbed by Fe and Al ions limit hydrolytic reactions of URE and ACP, and uptake of N and P from soil organic matter through mycorrhiza is a more efficient way for plants [19,22]. Therefore, URE and ACP activity in bulk soil may not be essential indicators for N and P uptake in *E. grandis* × *E. urophylla*, and matched studies between rhizosphere and bulk soils should be conducted in *Eucalyptus* plantations.

4.3. Function of Soil Total Nutrient and DOC on P Bioavailability in E. grandis × E. urophylla Coppice Forest

We observed that the standardized total effect on AP of soil nutrients (SOC, DOC, TN, and TP) was higher than enzyme activities, MBP, and the quotient of microbial biomass (Figure 6b). These results suggested soil total nutrients and DOC contents were the most important indicators for P bioavailability. Furthermore, C, N, and P are major elements required for microbial activity, especially heterotrophs. Adequate soil nutrient supply promotes continual turnover of soil nutrients and enzyme activity, thereby accelerating soil P plant availability [14,31]. Additionally, higher soil total nutrient levels promote plant growth, leading to improved litter input and long-term maintenance of forest soil fertility [70–72]. Moreover, DOC is the most effective energy resource for soil microbes and can affect soil microbial community and function [73]. Previously, several studies reported different effects of soil total nutrient and DOC on the soil P biochemical cycle; for example, Margalef et al. (2017) documented that TN was the principal factor that influenced the spatial gradient change of phosphatase activity on a global scale [31], and Zhang et al. found that long-term fertilizer input improved soil SOC and TN contents, resulting in increased MBP in the Loess Plateau [74]. Some fractions of DOC, such as low-molecular-weight organic acids, can occupy or change positions of soil particles that bind inorganic P, inhibiting the absorption of bioavailable P fractions and consequently increasing soil AP content [75].

4.4. Effects of Soil Microbial Biomass Turnover and Enzyme Activity on P Availability

Extracellular enzyme activities and microbial biomass turnover are two major pathways of soil P bioavailability [19,31]. Through regression, Pearson correlation, and PLS-PM analysis, we found that the relationship between MBP and AP was stronger than that between ACP and AP (Figures 4h, 5c and 6a, Table S1). Furthermore, the total standardized effects on AP of MBP and ACP were 0.31 and 0.20, respectively (Figure 6b), suggesting that MBP turnover had a stronger influence on AP than ACP in 2nd generation *Eucalyptus* coppice plantations [76]. Soil AP is utilized by microbes for their metabolism [77], but we found that the direct standardized effect on AP of MBP was not significant ($p < 0.05$) in our *Eucalyptus* plantation (Figure 6a). These results confirmed that microbial biomass turnover was more important than enzyme hydrolysis for soil AP in *E. grandis* × *E. urophylla* coppice forest. Additionally, the direct standardized effect of MBP on AP was negative and not significant ($p < 0.05$), but its indirect standardized effect on AP was positive, and the pathway MBP—the quotient of microbial biomass—AP was highly significant ($p < 0.01$) (Figure 6). This implies that regulating qMBP was the main way that MBP affected AP in the *Eucalyptus* coppice plantation. These results highlighted that accelerating P relative accumulation in microbial biomass effectively promotes soil P bioavailable turnover.

The effects of two quotients of microbial biomass indicators on AP were varied, with qMBC and qMBP having a significant negative correlation ($p < 0.01$) and a significant positive correlation ($p < 0.05$) with AP, respectively (Figures 5d,f and 6a, Table S1). These results suggest that the increases in MBC to SOC ratio and MBP to TP ratio had negative and positive effects on soil P bioavailability, respectively, in our *Eucalyptus* stands. On the one hand, C is a restricted element for soil microbes, according to former studies. Soil microbe activity and priming effect are maintained by an abundant C supply. Conversely, soil microbes are in inactive dormancy if soil C is insufficient, so soil nutrients cycle and turnover through microbial biomass are controlled by SOC contents [14,78]. The qMBC is an important indicator that reflects the ability of soil C supply; a lower qMBC means specific quantities of microbes can gain more C sources relatively from soil [79,80], and then some activities related to soil P bioavailability, which are mediated by microbial turnover, can be sustained [24,81]. Therefore, the relationship between qMBC and AP was significantly negative ($p < 0.001$) (Figure 5d, Table S1). We found that T2 treatment and the 0–20 cm layer of T3 treatment were significantly lower ($p < 0.05$) than T4 and CK treatments in qMBC (Figure 2b). These results imply that soil microbes can gain more energy resources from these treatments and layers, which benefits soil P bioavailability. Soil AP content was significantly enhanced ($p < 0.05$) in these two treatments (Table 2). On the other hand, soil microbes can uptake organic P and immobilize it in microbial biomass as nucleic acid, phosphatide, ATP, etc., and return it to the soil in small molecular organic P such as metabolic exude and residual body. These can be taken up by plant-mycorrhizal fungi systems [24,82,83]. Furthermore, this suggests that enhancing the MBP-to-TP ratio can promote soil P bioavailable turnover. This is the major reason for the significant positive relationship of qMBP with AP in *Eucalyptus* coppice plantations (Figure 5f, Table S1).

The regulation of the global soil P cycle and bioavailability are regulated by hydrolytic enzymes; however, the function of hydrolytic enzymes varies depending on the ecosystem [68,77,84]. For example, Zhu et al. found that standardized total effects on bioavailable-P of ACP were negative in subalpine coniferous (*Abies fabri*) forests in the eastern edge of the Tibetan Plateau [68]. This phenomenon also emerged in our study and the *E. grandis* × *E. urophylla* coppice forest (Figure 6). Lu et al. conducted a county-scale study in China to determine the role of biotic factors on soil bioavailable P fraction, and they concluded that other hydrolytic enzymes, in addition to ACP, that link with soil P mineral processes, e.g., phytase, were in an essential location in soil P bioavailable turnover, especially in some land use types that were disturbed by human activities [56]. Short rotation and frequent fertilizer input make *E. grandis* × *E. urophylla* plantations an ecosystem seriously influenced by industrialized forest management. Therefore, using ACP alone may not accurately reflect the function of hydrolytic enzymes related to soil

P mineralization. According to Lu et al., this was the main reason soil ACP showed no significant difference ($p < 0.05$) along five intensities of thinning treatments (Figure 3c) [56]. Furthermore, we found soil BG activity had a significant ($p < 0.01$) positive relationship with AP (Figure 4f, Table S1). In the PLS-PM model, the enzymes involved in C and N cycles (BG and URE) were key variations on AP through ACP and the quotient of microbial biomass (Figure 6). This suggests that increasing BG and URE activities can indirectly promote soil P bioavailability in *Eucalyptus* plantations. High BG and URE activities can (1) supply sufficient nutrients for soil microbes [14], (2) promote soil P accumulation in microbial biomass and enzyme synthesis [85], and (3) increase the microbial turnover rate [76]. These results suggested a tight relationship between C, N, and P cycles in the soil of *E. grandis* × *E. urophylla* coppice plantations, which has been confirmed in grassland and cropland [27,85–87].

5. Conclusions

In conclusion, our study found no significant differences ($p < 0.05$) in SOC, DOC, TN, and AN among five thinned-intensity treatments. However, we observed that the T1 stand had a significantly higher ($p < 0.05$) TP content in the 0–20 cm layer; meanwhile, the T2 and 20–40 cm layers of T3 had significantly higher ($p < 0.05$) AP contents. Additionally, our analysis implies that various soil indicators can be used to assess P bioavailability. Firstly, soil SOC, TN, TP, and DOC contents had the highest standardized effect on AP. Secondly, microbial turnover was the key process in the soil P cycle, and increasing qMBP benefited P bioavailability, while lower qMBP restricted soil AP content in *Eucalyptus* plantations. Lastly, soil P plant bioavailability is closely connected with soil C and N supply and cycle. These findings, which confirmed hypotheses 1 and 3, and rejected hypotheses 2, suggest that relatively heavy thinned intensity can accelerate soil P content and bioavailable turnover and highlight the primary functions of microbial biomass turnover and soil nutrient supply in the P plant-available mechanism in *Eucalyptus* coppice plantations. Therefore, some silviculture practices, such as prolonging rotation and introducing broad leaf or nitrogen-fixing tree species, should be used to inhibit soil nutrients loss and increase litter input for maintaining soil nutrient bioavailable turnover after thinning. More future research should focus on the influences of thinning soil nutrients and biochemical processes in the long term.

Supplementary Materials: The following supporting information can be downloaded at: https://www.mdpi.com/article/10.3390/f14102067/s1, Table S1. Pearson correlation matrix of soil nutrient, microbial biomass, and enzyme activity in *E. grandis* × *E. urophylla* sprout forests.

Author Contributions: Funding acquisition, S.Y. and M.Y.; Investigation, X.D. and Y.X.; Writing—original draft, X.X.; Writing—review and editing, I.A. All authors have read and agreed to the published version of the manuscript.

Funding: This work was supported by the National Natural Science Foundation of China (32371856) and the Key Projects of Guangxi Natural Science Foundation (2021GXNSFDA196003).

Data Availability Statement: Data will be made available upon request.

Acknowledgments: This work was supported by National Natural Science Foundation of China (32371856) and Key Projects of Guangxi Natural Science Foundation (2021GXNSFDA196003). We sincerely thank Qipo Forest Farm which gave help in plots selecting and soil sampling. And we sincerely thank Hongxiang Wang for beneficial advice in writing.

Conflicts of Interest: The authors have no relevant financial or non-financial interests to disclose.

References

1. Yang, G.R.; Wen, M.J.; Deng, Y.S.; Su, X.L.; Jiang, D.H.; Wang, G.; Chen, Y.W.; Chen, G.J.; Yu, S.F. Occurrence patterns of black water and its impact on fish in cutover areas of *Eucalyptus* plantations. *Sci. Total Environ.* **2019**, *693*, 133393. [CrossRef] [PubMed]
2. Lemessa, D.; Mewded, B.; Legesse, A.; Atinfau, H.; Alemu, S.; Maryo, S.; Tilahun, S. Do *Eucalyptus* plantation forests support biodiversity conservation. *For. Ecol. Manag.* **2022**, *523*, 120492. [CrossRef]

3. Wen, Y.G.; Ye, D.; Chen, F.; Liu, S.R.; Liang, H.W. The changes of understory plant diversity in continuous cropping system of *Eucalyptus* plantations, South China. *J. For. Res.* **2010**, *15*, 252–258. [CrossRef]
4. Wang, J.Y.; Deng, Y.S.; Li, D.Y.; Liu, Z.F.; Wen, L.L.; Huang, Z.G.; Jiang, D.H.; Lu, Y.P. Soil aggregate stability and its response to overland flow in successive *Eucalyptus* plantations in subtropical China. *Sci. Total Environ.* **2022**, *807*, 151000. [CrossRef]
5. Xu, Y.X.; Li, C.; Zhu, Y.L.; Wang, Z.C.; Zhu, W.K.; Wu, L.C.; Du, A.P. The shifts in soil microbial community and association network induced by successive planting of *Eucalyptus* plantations. *For. Ecol. Manag.* **2022**, *505*, 119877. [CrossRef]
6. Li, X.; Ye, D.; Liang, H.; Zhu, H.; Qin, L.; Zhu, Y.; Wen, Y. Effects of successive rotation regimes on carbon stocks in *Eucalyptus* plantations in subtropical China measured over a full rotation. *PLoS ONE* **2015**, *10*, e0132858. [CrossRef]
7. Laclau, J.P.; Deleporte, P.; Ranger, J.; Bouillet, J.P.; Kazotti, G. Nutrient dynamics throughout the rotation of *Eucalyptus* clonal stands in Congo. *Ann. Bot.* **2003**, *91*, 879–892. [CrossRef]
8. Foltran, E.C.; Rocha, J.H.T.; Bazani, J.H.; Gonçalvesa, J.L.M.; Rodriguesd, M.; Pavinatod, P.; Valdugae, G.R.; Errof, J.; Garcia-Mina, J.M. Phosphorus pool responses under different P inorganic fertilizers for a eucalyptus plantation in a loamy Oxisol. *For. Ecol. Manag.* **2019**, *435*, 170–179. [CrossRef]
9. McMahona, D.E.; Vergützb, L.V.; Valadares, S.V.; Silvab, I.R.D.; Jacksona, R.B. Soil nutrient stocks are maintained over multiple rotations in Brazilian *Eucalyptus* plantations. *For. Ecol. Manag.* **2019**, *448*, 364–375. [CrossRef]
10. Tsai, H.C.; Chiang, J.M.; McEwan, R.W.; Lin, T.C. Decadal effects of thinning on understory light environments and plant community structure in a subtropical forest. *Ecosphere* **2018**, *9*, e02464. [CrossRef]
11. Dang, P.; Gao, Y.; Li, J.L.; Yu, S.C.; Zhao, Z. Effects of thinning intensity on understory vegetation and soil microbial communities of a mature Chinese pine plantation in the Loess Plateau. *Sci. Total Environ.* **2018**, *630*, 171–180. [CrossRef] [PubMed]
12. Chen, X.; Page-Dumroese, D.; Lv, R.H.; Wang, W.W.; Li, G.L.; Liu, Y. Interaction of initial litter quality and thinning intensity on litter decomposition rate, nitrogen accumulation and release in a pine plantation. *Silva Fenn.* **2014**, *48*, 1211. [CrossRef]
13. Çömez, A.; Güner, Ş.T.; Tolunay, D. The effect of stand structure on litter decomposition in *Pinus sylvestris* L. stands in Turkey. *Anna. For. Sci.* **2021**, *78*, 19. [CrossRef]
14. Camenzind, T.; Hattenschwiler, S.; Treseder, K.K.; Lehmann, A.; Rillic, A. Nutrient limitation of soil microbial processes in tropical forests. *Ecol. Monogr.* **2018**, *88*, 4–21. [CrossRef]
15. Qiu, X.C.; Wang, H.B.; Peng, D.L.; Liu, X.; Yang, F.; Lia, Z.; Cheng, S. Thinning drives C:N:P stoichiometry and nutrient resorption in *Larix principis-rupprechtii* plantations in North China. *For. Ecol. Manag.* **2020**, *462*, 117984. [CrossRef]
16. Forrester, D.I. Growth responses to thinning, pruning and fertiliser application in *Eucalyptus* plantations: A review of their production ecology and interactions. *For. Ecol. Manag.* **2018**, *310*, 336–347. [CrossRef]
17. Zhou, L.; Cai, L.; He, Z.; Wang, R.W.; Wu, P.F.; Ma, X.Q. Thinning increases understory diversity and biomass, and improves soil properties without decreasing growth of Chinese fir in southern China. Environ. *Sci. Pollut. Res.* **2016**, *23*, 24135. [CrossRef] [PubMed]
18. Shen, Y.; Cheng, R.; Xiao, W.; Yang, S.; Guo, Y.; Wang, N.; Zeng, L.X.; Lei, L.; Wang, X.R. Labile organic carbon pools and enzyme activities of *Pinus massoniana* plantation soil as affected by understory vegetation removal and thinning. *Sci. Rep.* **2018**, *8*, 573. [CrossRef]
19. Lambers, H. Phosphorus acquisition and utilization in plants. *Annu. Rev. Plant Biol.* **2022**, *73*, 17–42. [CrossRef]
20. Chen, X.; Chen, H.Y.H.; Chang, S.X. Meta-analysis shows that plant mixtures increase soil phosphorus availability and plant productivity in diverse ecosystems. *Nat. Ecol. Evol.* **2022**, *6*, 1112–1121. [CrossRef]
21. Han, T.T.; Ren, H.; Hui, D.F.; Wang, J.; Lu, H.F.; Liu, Z.F. Light availability, soil phosphorus and different nitrogen forms negatively affect the functional diversity of subtropical forests. *Glob. Ecol. Conserv.* **2020**, *24*, e01334. [CrossRef]
22. Clarholm, M.; Skyllberg, U.; Rosling, A. Organic acid induced release of nutrients from metal-stabilized soil organic matter—The unbutton model. *Soil Biol. Biochem.* **2015**, *84*, 168–176. [CrossRef]
23. Menezes-Blackburn, D.; Giles, C.; Darch, T.; George, T.S.; Blackwell, M.; Stutter, M.; Shand, C.; Lumsdon, D.; Cooper, P.; Wendler, R.; et al. Opportunities for mobilizing recalcitrant phosphorus from agricultural soils: A review. *Plant Soil* **2018**, *427*, 5–16. [CrossRef] [PubMed]
24. Richardson, A.E.; Simpson, R.J. Soil microorganisms mediating phosphorus availability update on microbial phosphorus. *Plant Physiol.* **2011**, *156*, 989–996. [CrossRef]
25. Xu, D.; Dell, B.; Malajczuk, N.; Gong, M. Effects of P fertilisation and ectomycorrhizal fungal inoculation on early growth of eucalypt plantations in southern China. *Plant Soil* **2001**, *233*, 47–57. [CrossRef]
26. Kurth, J.K.; Albrecht, M.; Karsten, U.; Glaser, K.; Schloter, M.; Schulz, S. Correlation of the abundance of bacteria catalyzing phosphorus and nitrogen turnover in biological soil crusts of temperate forests of Germany. *Biol. Fertil. Soils* **2021**, *57*, 179–192. [CrossRef]
27. Chen, C.; Chen, H.Y.H.; Chen, X.; Huang, Z.Q. Meta-analysis shows positive effects of plant diversity on microbial biomass and respiration. *Nat. Commun.* **2019**, *10*, 1332. [CrossRef]
28. Reichart, T.; Rammig, A.; Fuchslueger, L.; Lugli, L.F.; Quesada, C.A.; Fleischer, K. Plant phosphorus-use and -acquisition strategies in Amazonia. *New Phytol.* **2022**, *234*, 1126–1143. [CrossRef]
29. Tedersoo, L.; Bahram, M.; Zobel, M. How mycorrhizal associations drive plant population and community biology. *Science* **2020**, *367*, eaba1223. [CrossRef]

30. Wang, J.P.; Chen, G.R.; Ji, S.H.; Zhong, Y.Q.; Zhao, Q.; He, Q.Q.; Wu, Y.H.; Bing, H.J. Close relationship between the gene abundance and activity of soil extracellular enzyme: Evidence from a vegetation restoration chronosequence. *Soil Biol. Biochem.* **2023**, *177*, 108929. [CrossRef]
31. Margalef, O.; Sardans, J.; Fernández-Martínez, M.; Molowny-Horas, R.; Janssens, I.A.; Ciais, P.; Goll, D.; Richter, A.; Obersteiner, M.; Asensio, D.; et al. Global patterns of phosphatase activity in natural soils. *Sci. Rep.* **2017**, *7*, 1337. [CrossRef]
32. Hu, B.; Yang, B.; Pang, X.Y.; Bao, W.K.; Tian, G.L. Responses of soil phosphorus fractions to gap size in a reforested spruce forest. *Geoderma* **2016**, *279*, 61–69. [CrossRef]
33. Xue, W.; Zhang, W.; Chen, Y. Heavy thinning temporally reduced soil carbon storage by intensifying soil microbial phosphorus limitation. *Plant Soil* **2023**, *484*, 33–48. [CrossRef]
34. Zhou, X.G.; Zhu, H.G.; Wen, Y.G.; Goodale, U.M.; Zhu, Y.L.; Yu, S.F.; Li, C.T.; Li, X.Q. Intensive management and declines in soil nutrients lead to serious exotic plant invasion in *Eucalyptus* plantations under successive short-rotation regimes. *Land Degrad. Dev.* **2022**, *31*, 297–310. [CrossRef]
35. Deng, X.; Cheng, F.; Li, M.; He, P.; Shen, L.; Liu, H.Y. Effect of different decay classes of *Eucalyptus* stump substrates on microbial resource limitation and carbon-use efficiency. *Plant Soil* **2022**, *478*, 651–669. [CrossRef]
36. Wang, Y.L.; Wu, P.N.; Qiao, Y.B.; Li, Y.M.; Liu, S.M.; Gao, C.K.; Liu, C.S.; Shao, J.; Yu, H.L.; Zhao, Z.H.; et al. The potential for soil C sequestration and N fixation under different planting patterns depends on the carbon and nitrogen content and stability of soil aggregates. *Sci. Total Environ.* **2023**, *897*, 165430. [CrossRef]
37. Carter, M.R.; Gregorich, E.G. *Soil Sampling and Methods of Analysis*, 2nd ed.; CRC Press: Boca Raton, FL, USA, 2008.
38. Watanade, F.S.; Olsen, S.R. Test of an ascorbic acid methods for determination phosphorus in water and $NaHCO_3$ extracts from soil. *Soil Sci. Soc. Am. J.* **1965**, *29*, 677–678. [CrossRef]
39. Wu, J.; He, Z.L.; Wei, W.X.; O'Donnell, A.G.; Syers, J.K. Quantifying microbial biomass phosphorus in acid soil. *Bio. Fert. Soil* **2000**, *32*, 500–507. [CrossRef]
40. Buttler, A.; Teuscher, R.; Deschamps, N.; Gavazov, K.; Bragazza, L.; Mariotte, P.; Schlaepfer, R.; Jassey, V.E.J.; Freund, L.; Cuartero, J.; et al. Impacts of snow-farming on alpine soil and vegetation: A case study from the Swiss Alps. *Sci. Total Environ.* **2023**, *903*, 166225. [CrossRef]
41. Saiya-Corka, K.R.; Sinsabaugha, R.L.; Zak, D.R. The effects of long term nitrogen deposition on extracellular enzyme activity in an *Acer saccharum* forest soil. *Soil Biol. Biochem.* **2002**, *34*, 1309–1315. [CrossRef]
42. Fei, Y.F.; Huang, S.Y.; Zhang, H.B.; Tong, Y.Z.; Wena, D.S.; Xia, X.Y.; Han Wang, H.; Luo, Y.M.; Barceló, D. Response of soil enzyme activities and bacterial communities to the accumulation of microplastics in an acid cropped soil. *Sci. Total Environ.* **2020**, *707*, 135634. [CrossRef]
43. Tian, S.Y.; Zhu, B.J.; Yin, R.; Wang, M.W.; Jiang, Y.J.; Zhang, C.Z.; Li, D.M.; Chen, X.Y.; Kardol, P.; Liu, M.J. Organic fertilization promotes crop productivity through changes in soil aggregation. *Soil Biol. Biochem.* **2022**, *165*, 108533. [CrossRef]
44. Zhou, T.; Wang, C.K.; Zhou, Z.H. Thinning promotes the nitrogen and phosphorous cycling in forest soils. *Agric. For. Meteorol.* **2021**, *311*, 108665. [CrossRef]
45. Zhang, X.Z.; Guan, D.X.; Li, W.B.; Sun, D.; Jin, C.G.; Yuan, F.H.; Wang, A.Z.; Wu, J.B. The effects of forest thinning on soil carbon stocks and dynamics: A meta-analysis. *For. Ecol. Manag.* **2018**, *429*, 36–43. [CrossRef]
46. Melilloa, J.M.; Butlera, S.; Johnson, J.; Tang, J. Soil warming, carbon–nitrogen interactions, and forest carbon budgets. *Proc. Natl. Acad. Sci. USA* **2011**, *108*, 9508–9512. [CrossRef]
47. Kuzyakov, Y.; Blagodatskaya, E. Microbial hotspots and hot moments in soil: Concept & review. *Soil Biol. Biochem.* **2015**, *83*, 184–199. [CrossRef]
48. Zhou, Z.H.; Wang, C.K.; Jin, Y.; Sun, Z.H. Impacts of thinning on soil carbon and nutrients and related extracellular enzymes in a larch plantation. *For. Ecol. Manag.* **2019**, *450*, 117523. [CrossRef]
49. Xu, Y.X.; Du, A.P.; Wang, Z.C.; Zhu, W.K.; Li, C.; Wu, L.C. Effects of different rotation periods of *Eucalyptus* plantations on soil physiochemical properties, enzyme activities, microbial biomass and microbial community structure and diversity. *For. Ecol. Manag.* **2020**, *456*, 117683. [CrossRef]
50. Liu, J.B.; Chen, J.; Chen, G.S.; Guo, J.F.; Li, Y.Q. Enzyme stoichiometry indicates the variation of microbial nutrient requirements at different soil depths in subtropical forests. *PLoS ONE* **2020**, *15*, e0220599. [CrossRef]
51. Yu, Y.; Yang, J.; Zeng, S.; Wu, D.M.; Jacobs, D.F.; Sloan, J.L. Soil pH, organic matter, and nutrient content change with the continuous cropping of *Cunninghamia lanceolata* plantations in South China. *J. Soils Sediments* **2017**, *17*, 2230–2238. [CrossRef]
52. Yang, H.J.; Pan, C.; Wu, Y.; Qing, S.Q.; Wang, Z.B.; Wang, D.H. Response of understory plant species richness and tree regeneration to thinning in *Pinus tabuliformis* plantations in northern China. *For. Ecosyst.* **2023**, *10*, 100105. [CrossRef]
53. Qiao, Y.; Wang, J.; Liu, H.M.; Huang, K.; Yang, Q.S.; Lu, R.L.; Yan, L.M.; Wang, X.H.; Xia, J.Y. Depth-dependent soil C-N-P stoichiometry in a mature subtropical broadleaf forest. *Geoderma* **2020**, *370*, 114357. [CrossRef]
54. Berhe, A.A.; Barnes, R.T.; Six, J.; Marín-Spiotta, E. Role of soil erosion in biogeochemical cycling of essential elements: Carbon, nitrogen, and phosphorus. *Annu. Rev. Earth Planet. Sci.* **2018**, *46*, 521–548. [CrossRef]
55. Alewell, C.; Ringeval, B.; Ballabio, C.; Robinson, D.A.; Panagos, P.; Borrelli, P.Q. Global phosphorus shortage will be aggravated by soil erosion. *Nat. Commun.* **2020**, *11*, 4546. [CrossRef]
56. Lu, J.L.; Jia, P.; Feng, S.W.; Wang, Y.T.; Zheng, J.; Ou, S.N.; Wu, Z.H.; Liao, B.; Shu, W.S.; Liang, J.L. Remarkable effects of microbial factors on soil phosphorus bioavailability: A country-scale study. *Glob. Chang. Biol.* **2022**, *28*, 4459–4471. [CrossRef] [PubMed]

57. Wu, L.J.; Wang, B.; Wu, Y.; Chen, D.M. Regional-scale evidence that determinants of soil microbial biomass and N mineralization depend on sampling depth and layer on the Mongolian Plateau. *Catena* **2022**, *213*, 106180. [CrossRef]
58. Zhang, J.; Quan, C.X.; Ma, L.L.; Chu, G.W.; Liu, Z.F.; Tang, X.L. Plant community and soil properties drive arbuscular mycorrhizal fungal diversity: A case study in tropical forests. *Soil Ecol. Lett.* **2021**, *3*, 52–62. [CrossRef]
59. Taki, H.; Inoue, T.; Tanaka, H. Responses of community structure, diversity, and abundance of understory plants and insect assemblages to thinning in plantations. *For. Ecol. Manag.* **2010**, *259*, 607–613. [CrossRef]
60. See, C.R.; Keller, A.B.; Hobbie, S.E.; Pett-Ridge, J. Hyphae move matter and microbes to mineral microsites: Integrating the hyphosphere into conceptual models of soil organic matter stabilization. *Glob. Chang. Biol.* **2022**, *28*, 2527–2540. [CrossRef] [PubMed]
61. Camargo, A.P.; Souza, R.S.C.; Jose, J.; Gerhardt, I.R.; Dante, R.A.; Mukherjee, S.; Huntemann, M.; Kyrpides, N.C.; Carazzolle, M.F.; Arruda, P. Plant microbiomes harbor potential to promote nutrient turnover in impoverished substrates of a Brazilian biodiversity hotspot. *ISME J.* **2022**, *17*, 354–370. [CrossRef]
62. Xu, X.F.; Thornton, P.E.; Post, W.M. A global analysis of soil microbial biomass carbon, nitrogen and phosphorus in terrestrial ecosystems. *Glob. Ecol. Biogeogr.* **2013**, *22*, 737–749. [CrossRef]
63. Chu, S.S.; Ouyang, J.H.; Liao, D.D.; Zhou, Y.D.; Liu, S.S.; Shen, D.C.; Wei, X.H.; Zeng, S.C. Effects of enriched planting of native tree species on surface water flow, sediment, and nutrient losses in a *Eucalyptus* plantation forest in southern China. *Sci. Total Environ.* **2019**, *675*, 224–234. [CrossRef] [PubMed]
64. Chen, F.L.; Zheng, H.; Zhang, K.; Ouyang, Z.Y.; Li, H.L.; Wu, B.; Shi, Q. Soil microbial community structure and function responses to successive planting of *Eucalyptus*. *J. Environ. Sci.* **2013**, *25*, 2102–2111. [CrossRef]
65. Chen, X.L.; Chen, H.Y.H.; Chen, X.; Wang, J.; Chen, B.; Wang, D.; Guan, Q.W. Soil labile organic carbon and carbon-cycle enzyme activities under different thinning intensities in Chinese fir plantations. *Appl. Soil Ecol.* **2016**, *107*, 162–169. [CrossRef]
66. Huang, Y.X.; Wu, Z.J.; Zong, Y.Y.; Li, W.Q.; Chen, F.S.; Wang, G.G.; Li, J.J.; Fang, X.M. Mixing with coniferous tree species alleviates rhizosphere soil phosphorus limitation of broad-leaved trees in subtropical plantations. *Soil Biol. Biochem.* **2022**, *175*, 108853. [CrossRef]
67. Lin, Y.; Xia, C.K.; Wu, G.Y.; Wang, F.C.; Wang, S.N.; Liu, Y.Q.; Chen, F.S. Replanting of broadleaved trees alters internal nutrient cycles of native and exotic pines in subtropical plantations of China. *For. Ecosyst.* **2022**, *9*, 100067. [CrossRef]
68. Adetunji, A.T.; Lewu, F.B.; Mulidzi, R. The biological activities of β-glucosidase, phosphatase and urease as soil quality indicators: A review. *J. Soil Sci. Plant Nutr.* **2017**, *17*, 794–807. [CrossRef]
69. Zhang, Q.Q.; Zhou, J.J.; Huang, G.H.; Yang, G.; Liu, G.F.; Liang, K.N. Effects of thinning intensity on soil quality and growth of teak plantation. *For. Res.* **2021**, *34*, 127–134. (In Chinese)
70. Prommer, J.; Wanek, W.; Hofhansl, F.; Trojan, D.; Offre, P.; Urich, T.; Schleper, C.; Sassmann, S.; Kitzle, B.; Soja, G.; et al. Biochar decelerates soil organic nitrogen cycling but stimulates soil nitrification in a temperate arable field trial. *PLoS ONE* **2014**, *9*, e86388. [CrossRef]
71. Witzgall, K.; Vidal, A.; Schubert, D.I.; Höschen, C.; Schweizer, S.A.; Buegger, F.; Pouteau, V.; Chenu, C.; Mueller, C.W. Particulate organic matter as a functional soil component for persistent soil organic carbon. *Nat. Commun.* **2021**, *12*, 4115. [CrossRef]
72. Chenua, C.; Angersb, D.A.; Barré, P.; Derrien, D.; Arrouays, D.; Balesdent, J. Increasing organic stocks in agricultural soils: Knowledge gaps and potential innovations. *Soil Tillage Res.* **2019**, *188*, 41–52. [CrossRef]
73. Yang, X.; Zhang, Y.; Liu, Q.; Guo, J.H.; Zhou, Q.C. Progress in the interaction of dissolved organic matter and microbes (1991–2020): A bibliometric review. *Environ. Sci. Pollut. R.* **2022**, *29*, 16817–16829. [CrossRef]
74. Zhang, S.L.; Li, Z.J.; Liu, J.M.; Li, Q.H.; Yang, X.Y. Long-term effects of straw and manure on crop micronutrient nutrition under a wheat-maize cropping system. *J. Plant Nutr.* **2015**, *38*, 742–753. [CrossRef]
75. Bolan, N.S.; Naidu, R.; Mahimairaja, S.; Baskaran, S. Influence of low-molecular-weight organic acids on the solubilization of phosphates. *Biol. Fertil. Soils* **1994**, *18*, 311–319. [CrossRef]
76. Achat, D.L.; Morel, C.; Bakker, M.R.; Augusto, L.; Pellerin, S.; Gallet-Budynek, A.; Gonzalez, M. Assessing turnover of microbial biomass phosphorus: Combination of an isotopic dilution method with a mass balance model. *Soil Biol. Biochem.* **2010**, *42*, 2231–2240. [CrossRef]
77. Jiang, C.; Zhu, B.; Zeng, H. Soil extracellular enzyme stoichiometry reflects the unique habitat of karst tiankeng and helps to alleviate the P-limitation of soil microbes. *Ecol. Indic.* **2022**, *144*, 109552. [CrossRef]
78. Wang, Z.Q.; Zhao, M.Y.; Yan, Z.B.; Yang, Y.H.; Niclas, K.J.; Huang, H.; Mipan, T.D.; He, X.J.; Hu, H.F.; Wright, J. Global patterns and predictors of soil microbial biomass carbon, nitrogen, and phosphorus in terrestrial ecosystems. *Catena* **2022**, *211*, 1060377. [CrossRef]
79. Insam, H.; Domsch, K.H. Relationship between soil organic carbon and microbial biomass on chronosequences of reclamation sites. *Microb. Ecol.* **1988**, *15*, 177–188. [CrossRef]
80. Kaschuk, G.; Alberton, Q.; Hungria, M. Three decades of soil microbial biomass studies in Brazilian ecosystems: Lessons learned about soil quality and indications for improving sustainability. *Soil Biol. Biochem.* **2010**, *42*, 1–13. [CrossRef]
81. Heuck, C.; Weig, A.; Spohn, M. Soil microbial biomass C:N:P stoichiometry and microbial use of organic phosphorus. *Soil Biol. Biochem.* **2015**, *85*, 119–129. [CrossRef]
82. Zhao, Y.; Li, Y.L.; Yang, F. Critical review on soil phosphorus migration and transformation under freezing-thawing cycles and typical regulatory measurements. *Sci. Total Environ.* **2021**, *751*, 141614. [CrossRef]

83. Zhang, D.S.; Kuzyakov, Y.; Zhu, H.T.; Alharbi, H.; Li, H.B.; Rengel, Z. Increased microbial biomass and turnover underpin efficient phosphorus acquisition by *Brassica chinensis*. *Soil Tillage Res.* **2022**, *223*, 105492. [CrossRef]
84. Zhu, H.; Bing, H.J.; Wu, Y.H.; Sun, H.Y.; Zhou, J. Low molecular weight organic acids regulate soil phosphorus availability in the soils of subalpine forests, eastern Tibetan Plateau. *Catena* **2021**, *203*, 105328. [CrossRef]
85. Hacker, M.; Ebeling, A.; Gessler, A.; Gleixner, G.; Macé, O.G.; Kroon, H.D.; Lange, M.; Mommer, L.; Eisenhauer, N.; Ravenek, J.; et al. Plant diversity shapes microbe-rhizosphere effects on P mobilisation from organic matter in soil. *Ecol. Lett.* **2015**, *18*, 1356–1365. [CrossRef] [PubMed]
86. Gu, Y.; Wang, P.; Kong, C.H. Urease, invertase, dehydrogenase and polyphenoloxidase activities in paddy soil influenced by allelopathic rice variety. *Eur. J. Soil Biol.* **2009**, *45*, 436–440. [CrossRef]
87. Vazqueza, E.; Benitoa, M.; Masaguera, A.; Espejo, R.; Díaz-Pinés, E.; Teutscherov, N. Long-term effects of no tillage and Ca-amendment on the activity of soil proteases and β-glucosidase in a Mediterranean agricultural field. *Eur. J. Soil Biol.* **2019**, *95*, 103135. [CrossRef]

forests

MDPI

Article

Diversity and Structure of Soil Microbial Communities in Chinese Fir Plantations and *Cunninghamia lanceolata–Phoebe bournei* Mixed Forests at Different Successional Stages

Weiyang Li [1,†], Huimin Sun [2,†], Minmin Cao [1], Liyan Wang [3], Xianghua Fang [4] and Jiang Jiang [1,*]

1 Co-Innovation Center for Sustainable Forestry in Southern China, Jiangsu Province Key Laboratory of Soil and Water Conservation and Ecological Restoration, Nanjing Forestry University, 159 Longpan Road, Nanjing 210037, China; liweiyang2023@outlook.com (W.L.); healer_cao@outlook.com (M.C.)
2 Center for Eco-Environmental Research, Nanjing Hydraulic Research Institute, Nanjing 210029, China; huimin_sun@outlook.com
3 Jiangxi Academy of Forestry and Nanchang Urban Ecosystem Research Station, Nanchang 330013, China; wangliyan052@163.com
4 Longquan Nature Conservation Centre, Hundred Mountains Group National Forest Park, Longquan 323700, China; selina_cui@outlook.com
* Correspondence: ecologyjiang@gmail.com
† These authors contributed equally to this work.

Abstract: *Cunninghamia lanceolata* is an important species in plantations and is widely planted in sub-tropical regions of China because of its fast-growing and productive characteristics. However, the monoculture planting is carried out in the pursuit of economic value. This planting mode has led to problems such as the exhaustion of soil fertility, decrease in vegetation diversity, and decrease in woodland productivity. In order to restore soil fertility and increase timber production, the introduction of broad-leaved tree species to plantations is an effective transformation model. Understanding how forest age changes and stand structure differences drive the composition and diversity of soil microbial communities is helpful in understanding the trend of soil–microbial changes in plantations and evaluating the effects of the introduction of broad-leaved tree species in soil–plant–microbial ecosystems in plantations. Therefore, the purpose of our study is to investigate the effects of forest age and pure forest conversion on *C. lanceolata–P. bournei*-mixed forest soil microbial community structure and diversity by detecting soil nutrients, enzyme activities, and soil microbial 16S and ITS rRNA gene sequencing. According to the findings, the diversity and abundance of bacterial communities in *C. lanceolata* plantations of different ages increased first and then decreased with the increase in forest age, and the max value was in the near-mature forest stage. The fungal abundance decreased gradually with stand age, with the lowest fungal diversity at the near-mature stand stage. During the whole growth process, the bacterial community was more limited by soil pH, nitrogen, and phosphorus. After introducing *P. bournei* into a Chinese fir plantation, the abundance and diversity of the bacterial community did not improve, and the abundance of the fungal community did not increase. However, soil nutrients, pH, and fungal community diversity were significantly improved. The results of these studies indicate that the introduction of broad-leaved tree species not only increased soil nutrient content, but also had a significant effect on the increase in the diversity of soil fungal communities, making the microbial communities of mixed forests more diverse.

Keywords: Chinese fir; forest age and stand structure; soil nutrient; soil microbial community

1. Introduction

Chinese fir (*Cunninghamia lanceolata*) is a fast-growing coniferous tree that has been widely cultivated for timber production in southern China [1]. The species makes an important contribution to environmental protection, the provision of substantial timber products, and ecological construction. Due to the rising demand for timber, Chinese fir plantations usually grow in a monoculture pattern accompanied by a continuous crop rotation [2]. These practices decrease soil fertility and ecosystem productivity because of the excessive afforestation density [3]. For example, Zhang Hong et al. [4] found that the soil phosphorus content gradually decreased with the increase in planting generations. An in-depth understanding of the reasons for declines in soil quality in C. lanceolata plantations is extremely important for the sustainable management of these plantations in China.

Governments and forest managers have been investigating forest management practices that can improve ecosystem functions and services [5]. The replacement of monocultures with mixed plantations is an effective afforestation method to mitigate soil erosion and enhance forest productivity [6–8]. Recently, several native broad-leaved trees were introduced into monoculture *C. lanceolata* plantations to improve soil status [8]. Phoebe bournei is an economically important, valuable, and rare broad-leaved native tree, and is required in mixed forests [1,9]. A mixed forest can increase the diversity of plants, which in turn positively affects microbial activity and influences soil microorganisms in a number of ways [10]. It affects soil microbes in several ways, for example, by accelerating litter decomposition and changing root exudates [11–13]. Not only are soil microorganisms an important driving factor for plants to obtain available nutrients for growth [14,15], they are also very sensitive to changes in soil properties [15,16].

The conversion of stand types may impact the structure of soil microbial communities by influencing soil environment [17,18]. In addition, soil microorganisms can influence forest productivity by participating in nutrient cycling [19] and soil organic matter turnover [20], while ensuring the reproduction of microbial communities. Therefore, it is necessary to understand the relationship and interaction between tree species characteristics, soil nutrient changes, and microbial community structure characteristics for the sustainable management of coniferous and broad-leaved mixed forests [21,22].

Furthermore, bacteria and fungi play different and important roles in biogeochemical cycles [23], because the structure and diversity of microbial communities may respond differently to soils in different ecosystems [24]. For example, fungi mainly decompose refractory cellulose and lignin [25,26]. Bacteria are mainly responsible for nutrient cycling [27]. However, poor knowledge exists about the differences in the responses of bacterial and fungal communities after the transformation of Chinese fir plantations into mixed forests and their relationship with soil characteristics.

Current studies in Chinese fir plantations mainly focus on soil properties, differences in soil properties [28], nutrient cycling rates [29], and microbial structure [30] between *C. lanceolata* plantation forests under monoculture and mixed plantation patterns. Wang et al. found that soil nutrients and microorganisms were enhanced after constructing a mixed forest [31]. Lei et al. found that soil microbial diversity was significantly higher in a near-natural forest versus in a *C. lanceolate* plantation [32]. Chu et al. reported that diversity and abundance of soil microbial communities in natural secondary forests was significantly higher than that in plantations [33].

Moreover, the significant effect of age on soil properties is widely recognized. However, there are few reports on the effects of stand-type changes in soil characteristics, as well as few studies that further research impacts on the diversity, composition, and abundance of microbial communities, determining the key driving factors (soil chemical factors and enzyme activities) affecting soil microbial communities in the single and mixed cultivation of Chinese fir (*C. lanceolata* × *P. bournei*). Moreover, there are relatively few studies on the relationship between microbial community structure and composition, as well as soil properties and enzymes. This greatly limits us in understanding the reaction of microbial communities to forest age and stand type. Therefore, an in-depth understanding of the effects of stand age and stand type on microbial communities will help us to make more rational plantation management methods.

Here, we aimed to fill the knowledge gaps in the effect of stand age and stand type on soil fertility and microbial diversity, composition, and structure in Chinese fir plantations.

We also aimed to advance our understanding of the influence of stand age on microbial structure and composition and explore key limiting factors that affect the soil microbial community structure. Therefore, our specific objectives were to determine: (1) how soil nutrients and enzyme activities vary with stand age and stand type; (2) how microbial diversity and composition changes with stand age and stand type; and (3) the key factors that affect the structure of soil microbial communities in plantation forests.

2. Material and Methods

2.1. Description of the Field Trial and Soil Sample Collection

The research site is located in Guanshan forest farm, Yongfeng County, Ji'an City, Jiangxi Province (115°17′–115°56′ E, 26°38′–27°32′ N). The climate is humid and is in the mid subtropical zone, with abundant rainfall and sufficient light. The average annual temperature is 18 °C and the average annual rainfall is 1627.3 mm. Red soil is a typical local soil type with the largest area. The main planting species are *C. lanceolata*, Slash pine, Moso bamboo, *P. bournei*, and Liquidambar formosana. Three main sample plots adjacent to Guanshan forest farm are selected: Dazhou Mountain Branch (115°29′7″ E, 27°14′48″ N), Dongmaokeng branch (115°33′10″ E, 27°10′43″ N), and Yuyuan branch (115°34′39″ E, 27°14′13″ N).

In December 2017, we sampled *C. lanceolata* plantations of over-ripe forest (OR), near-mature forest (NM), and young forest (YF), which had the same soil type, site conditions, and disturbance history. We also sampled a mixed forest of *C. lanceolata* and *P. bournei* with two forest ages, one near-mature *C. lanceolata* forest mixed with 5-year-old *P. bournei* (NM + PB05), and a near-mature *C. lanceolata* forest mixed with 10-year-old *P. bournei* (NM + PB10). Each sample plot was divided into three 20 m × 20 m quadrants, and we selected three trees randomly in each quadrant. The plant and animal residues were removed from the soil surface within 30 cm from the stem of *C. lanceolata*, and the soil profile was dug to a depth of 0–30 cm. The roots were taken near the main root system with fine roots, put into sterile self-sealing bags, and shaken vigorously for 1 min to separate the roots from the soil. We collected total 45 soil samples, which was used as rhizosphere soil, and were stored in a low-temperature refrigerator and transported to the laboratory for analysis.

2.2. Soil Properties and Soil Enzyme Activities Analysis

The samples used for microbial sequencing analysis were stored in a −70 °C ultra-low temperature refrigerator. The soil samples used for the remaining test indicators were placed in an oven, dried at a temperature of 105° to a constant weight, and then filtered using a 2 mm mesh sieve to remove impurities such as plant roots and stones.

All soil indicators are in accordance with the standard agreement. The glass electrode of PHSJ-3D acidity meter produced by Shanghai Rex Company was suspended in a 1 mol/L KCL solution (*w*: *v*, 1: 5) to determine the pH value of rhizosphere soil. Total carbon and total nitrogen were determined using an elemental analyzer (Vario EL III, Elementar, Nakano, Tokyo). SOC was measured using an elemental analyzer after removing carbonates from air-dried samples acidified with dilute hydrochloric acid (5%). Soil total phosphorus: acid soluble-molybdenum antimony anti-colorimetric determination. Soil available phosphorus: hydrochloric acid–ammonium fluoride extraction method. Soils' total potassium: Hydrofluoric acid–perchloric acid digestion method. Soils' available potassium: Ammonium acetate extraction method. Soils' available nitrogen: Petri dish diffusion method [34–38].

All enzyme activities were determined in accordance with the standard agreement [39]. Soil sucrase: 3,5-dinitrosalicylic acid colorimetric method was used to determine the quality of glucose produced by 1 g of soil cultured at 37 °C for 24 h to characterize the activity of invertase. Urease: Using a sodium salicylate-sodium dichloroisocyanurate colorimetric method, 1 g of soil was cultured at 37 °C for 2 h. Acid phosphatase: Measured by p-nitrophenyl disodium phosphate colorimetric method, 1 g soil was cultured at 37 °C for 1 h.

2.3. DNA Extraction, PCR Amplification, and Illumina MiSeq Sequencing of Soil Microorganisms

According to the manufacturer 's instructions, PowerSoil DNA extraction kit (Mo Bio Laboratories Inc., Carlsbad, CA, USA) was used for rhizosphere soil DNA extraction. The purity of the sample DNA was detected by a NanoPhotometer spectrophotometer, and the concentration was detected by Qubit2.0 fluorometer. The V3–V4 region of the 16S rDNA gene of rhizosphere soil bacteria were amplified with primers 341F (5′-CCTACGG NGGCWGCAG-3′) and 805R (5′-GACTACHVGGGT ATCTAATCC-3′), while the rhizosphere soil fungi were amplified with primers ITS1 (5′-ACC TGCGGARGGATCA-3′) and B58S3 (5′-GATCCRTTGYTRAAAGTT-3′). TaKaRa EXtaq enzyme was used during the amplification process to ensure the success and accuracy of the amplification, and each sample was subjected to three PCR reactions. After preliminary quantification using Qubit2.0, the DNA was diluted to 1 ng/uL with sterile water. All PCR reactions were using PhusionÂ® High-Fidelity PCR Master Mix (New England BioLabs, Beverly, MA, USA).

Firstly, the mixed PCR products with equal density ratio were purified by GeneJET gel extraction kit (Thermo Scientific, Waltham, MA, USA). Then, NEB Next Ultra DNA Library Prep Kit for Illumina (NEB, Ipswich, MA, USA) was used to generate sequencing libraries and index codes, and Qubit @ 2.0 fluorescence analyzer (Life Technologies, Carlsbad, CA, USA) and Agilent Bioanalyzer 2100 system were used for the quality assessment. Finally, the library was sequenced using the 250 pair-end protocol of the Illumina HiSeq2500 instrument.

In order to obtain a credible target sequence and facilitate the subsequent analysis, the sequence obtained by sequencing was first filtered and removed for low-quality base, Ns, and linker contamination sequences. Then, the PEAR method [40] was used to splice the corresponding sequence fragments of the paired-end sequencing, and the quality control process of the spliced sequence was removed according to QIIME 1.8.0 to remove the quality filtering of the original label and remove 'chi-mera'. Finally, the sequences with 97% similarity as the threshold were assigned to the OTUs, and the representative sequences of each OTU were classified using the RDP classifier or Closed Reference against the Greengenes 16S rRNA reference database [41–43]. The work was provided by Annoroad Gene Technology Co., Ltd. Beijing, China. Finally, we obtained 1,023,685 bacterial sequences and 477,975 fungal sequences from all rhizosphere soil samples.

2.4. Statistical Analysis

One-way analysis of variance in SPSS software (IBM SPSS Statistics 25.0; IBM Corp., Armonk, NY, USA) was used to test the differences in soil properties and enzyme activities, as well as diversity (Shannon, Simpson) and abundance (OTU, Chao1) among different communities. The Wayne diagram was drawn using InteractiVenn (http://www.interactivenn.net (accessed on 13 March 2023)) to show the common and unique OTUs of each community [44]. Linear discriminant analysis (LDA) and effect size (LEfSe; http://huttenhower.sph.harvard.edu/galaxy/root?tool_id=PICRUSt_normaliz (accessed on 13 March 2023)) was used to identify differences in bacterial/fungal taxa among communities [45]. The PCoA analysis in Canoco 5.0 software was used to show the differences in microbial population structure by the relative abundance of OTU. A fungal and bacterial Mantel test analysis using Spearman and Pearson functions of the Rstudio Vegan package revealed factors affecting the abundance of soil microbial communities in fir plantations and mixed forests.

3. Results

3.1. Soil Chemical Properties and Enzyme Activities

We found that SOC, TK, and AK contents gradually increased with the stand age of *C. lanceolata* plantations, with significant differences in TOC between the young forest and near-mature forest, non-significant differences in the SOC contents among the developmental stages, significant differences in TK contents among developmental stages, and significant differences in AK contents between the young forest and near-mature forest. pH gradually decreased with increasing stand age, with significant differences among the developmental stages (Table 1). TN, AN, TP, and AP contents showed a "V"-shaped trend with the increasing age of fir stands, with significant differences between NM and OR for TN and AN for all developmental stages, and TP and AP for NM and the rest of the stages. S-SC activity increased gradually with increasing stand age, but there were no significant differences among all developmental stages ($p < 0.05$, Table 1).

The soil TOC and TN contents differed significantly between NM and NM + PB10 in the Chinese fir forest and mixed stands of fir trees at the same developmental stage, but SOC content was not significantly different. Soil AN, and TP contents were significantly lower in the Chinese fir forest than in mixed forests. Soil AP content was significantly lower than that of NM + PB05 and higher than that of NM + PB10. Soil TK and AK content and pH were significantly lower in the Chinese fir forest than in mixed forests. The ACP activity was significantly lower than that of NM + PB05 and not significantly different from that of NM + PB10. S-SC was significantly lower in the Chinese fir forest than in mixed forests. SUE was significantly higher in the Chinese fir forest than in mixed forests ($p < 0.05$, Table 1).

3.2. *Characterization of Diversity of Soil Microbial Community*

The bacterial community richness and diversity (Table 2) showed a significant difference in differently aged *C. lanceolata* forests. OTU, ACE, and chao1 showed an overall upward trend in growth and then decline. Bacterial abundance mainly reached the highest in the near-mature forest stage, but both the Shannon and Simpson indices showed no difference (Table 2, $p < 0.05$).

In the *C. lanceolata* plantation (near-mature forest) and *C. lanceolata–P. bournei* mixed forest of the same stand age, we found that plantations were higher in bacterial diversity than mixed forests. The OTU, ACE, and chao1 indices showed a tendency to increase with the increasing stand age in the NM + PB05 forest and NM + PB10 forest, but there was no significant difference (Table 2, $p < 0.05$ for all pairs).

Shannon and Simpson indices showed a decreasing and then increasing trend with stand age, and a significant difference was found between the younger and near-mature stages. In mixed forests, the Shannon and Simpson indices did not vary significantly. The diversity indices of fir plantations at the same developmental stage were lower than those of mixed forests, and the differences in indices were significant (Table 3, $p < 0.05$ for all pairs).

For the fungal community, community richness was significantly different in differently aged *C. lanceolata* forests. Chao1 index values gradually decreased with the increasing stand age and reached the lowest value at the over-ripe stand stage. In the mixed forest, it had an upward trend of gradually increasing in abundance with the increasing forest age (Table 3, $p < 0.05$ for all pairs).

Table 1. Soil physicochemical characteristics and enzyme activities in the different ages of Chinese fir forest and mixed forests.

	TOC (g/kg)	SOC (mg/g)	TN (g/kg)	AN (mg/g)	TP (g/kg)	AP (mg/g)	TK (g/kg)	AK (mg/g)	pH	ACP ($mg \cdot g^{-1} \cdot h^{-1}$)	S-SC ($mg \cdot g^{-1} \cdot h^{-1}$)	SUE ($mg \cdot g^{-1} \cdot h^{-1}$)
YF	9.06 ± 0.79 a	0.94 ± 0.51 a	1.29 ± 0.10 bc	60.76 ± 8.08 bc	0.22 ± 0.56 b	0.667 ± 0.09 b	7.47 ± 0.50 a	22.69 ± 4.10 a	4.67 ± 0.20 c	0.46 ± 0.01 a	3.27 ± 0.52 a	28.38 ± 8.50 b
NM	14 ± 0.72 b	1.14 ± 0.87 ab	0.68 ± 0.03 a	33.32 ± 5.27 a	0.07 ± 0.00 a	0.37 ± 0.10 b	12.68 ± 0.25 b	33.41 ± 3.37 b	4.19 ± 0.71 b	0.52 ± 0.02 ab	4.72 ± 0.80 a	58.11 ± 9.60 d
OR	14.7 ± 0.45 bc	1.21 ± 0.05 ab	1.13 ± 0.15 b	56.04 ± 6.99 b	0.264 ± 0.34 b	0.56 ± 0.05 ab	15.5 ± 0.95 c	36.62 ± 6.70 b	3.85 ± 0.07 a	0.50 ± 0.01 a	5.10 ± 0.37 a	41.63 ± 0.09 c
NM + PB05	14.7 ± 1.4 bc	1.18 ± 0.29 ab	0.75 ± 0.15 a	72.38 ± 10.72 c	0.3719 ± 0.00 c	4.04 ± 0.13 d	14.9 ± 1.45 c	49.32 ± 0.51 c	4.62 ± 0.08 c	0.70 ± 0.11 c	11.81 ± 3.53 b	13.18 ± 1.65 a
NM + PB10	16.1 ± 1.3 c	1.34 ± 0.76 b	1.35 ± 0.05 c	51.36 ± 10.59 b	0.226 ± 0.23 b	2.73 ± 0.20 c	15.13 ± 0.86 c	34.23 ± 5.89 b	4.63 ± 0.13 c	0.56 ± 0.10 bc	14.14 ± 1.23 b	16.73 ± 1.79 a

Note: One-way ANOVA was used for analysis, $p < 0.05$. Different letters represent significant differences between groups. Data: mean ± SD ($n = 3$). TOC: total organic carbon; SOC: soil organic carbon; TN: total nitrogen; AN: available nitrogen; TP: total phosphorus; AP: available phosphorus; TK: total potassium; AK: available potassium; ACP: acid phosphatase; S-SC: sucrose; SUE: urease; YF: Young forest; NM: Near-mature forest; OR: Overripe forest.; NM + PB05: Near-mature *C. lanceolata* plantation mixed with 5 years old *P. bournei*; NM + PB10: Near-mature *C. lanceolata* plantation mixed with 10-year-old *P. bournei*.

Table 2. The bacterial diversity in different ages of Chinese fir plantations and mixed forests.

Treatment	OTU	ACE	Chao1	Shannon	Simpson
YF	1671.67 ± 89.07 b	1988.67 ± 60.43 c	2010.67 ± 71.93 c	5.81 ± 0.23 a	0.88 ± 0.40 a
NM	1944.67 ± 131.80 a	2319.67 ± 111.79 a	2342.00 ± 112.93 a	6.02 ± 0.12 a	0.64 ± 0.02 a
OR	1761.00 ± 69.46 ab	2071.00 ± 125.41 bc	2101.67 ± 155.00 bc	6.02 ± 0.17 a	0.62 ± 0.13 a
NM + PB05	1831.00 ± 100.54 ab	2172.33 ± 155.99 abc	2210.33 ± 144.67 abc	5.95 ± 0.17 a	0.67 ± 0.17 a
NM + PB10	1889.33 ± 165.24 ab	2279.67 ± 140.63 ab	2311.00 ± 93.21 ab	5.97 ± 0.15 a	0.67 ± 0.08 a

Note: YF: Young forest; NM: Near-mature forest; OR: Overripe forest.; NM + PB05: Near-mature *C. lanceolata* plantation mixed with 5 years old *P. bournei*; NM + PB10: Near-mature *C. lanceolata* plantation mixed with 10-year-old *P. bournei*. Different letters represent significant differences in Chinese fir plantations and mixed forests of different ages ($p < 0.05$).

Table 3. The fungal diversity in different ages of Chinese fir plantations and mixed forests.

Treatment	Chao1	Observed Species	PD Whole Tree	Shannon	Simpson
YF	779.93 ± 93.18 ab	676.00 ± 96.02 a	173.13 ± 20.75 a	6.40 ± 0.60 a	0.97 ± 0.02 a
NM	681.17 ± 80.70 ab	549.00 ± 131.53 a	157.88 ± 23.00 a	3.34 ± 1.95 b	0.57 ± 0.33 b
OR	584.43 ± 217.54 b	487.33 ± 202.21 a	145.97 ± 51.81 a	3.93 ± 2.32 ab	0.67 ± 0.33 ab
NM + PB05	768.49 ± 33.36 ab	697.67 ± 47.60 a	193.93 ± 8.98 a	6.23 ± 0.71 a	0.94 ± 0.05 ab
NM + PB10	826.32 ± 130.59 a	687.00 ± 150.12 a	189.86 ± 29.42 a	5.10 ± 1.31 ab	0.87 ± 0.13 ab

Note: YF: Young forest; NM: Near-mature forest; OR: Overripe forest.; NM + PB05: Near-mature *C. lanceolata* plantation mixed with 5 years old *P. bournei*; NM + PB10: Near-mature *C. lanceolata* plantation mixed with 10-year-old *P. bournei*. Different letters represent significant differences in Chinese fir plantations and mixed forests of different ages ($p < 0.05$).

3.3. Differences in the Taxonomic Level Characteristics of Soil Microbial Communities

All sequences were assigned to 48,980 OTUs across all samples and categorized to the bacterial domain, which includes 27 bacterial phyla, 70 classes, 143 orders, 243 families, and 375 genera. The most common phylum in OTUs was *Proteobacteria*, accounting for 32.87%, followed by *Acidobacteria* (OTUs: 28.24%), *Chloroflexi* (OTUs: 15.02%), *Actinobacteria* (OTUs: 9.55%), *Planctomycetes* (OTUs: 4.09%), *Verrucomicrobia* (OTUs: 3.03%), and *Firmicutes* (OTUs: 2.39%). A detailed abundance of phyla was shown in Figure 1.

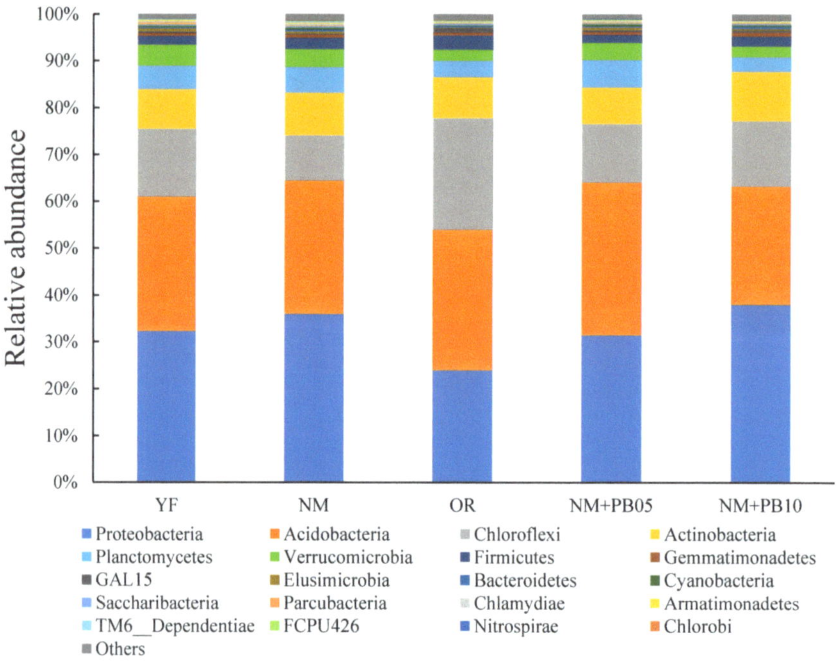

Figure 1. The relative abundance of bacterial phylum in young (YF), near-mature (NM), and over-ripe (OR) Chinese fir plantation; near-mature Chinese fir plantation mixed with 5-year-old *P. bournei* (NM + PB05); Chinese fir plantation mixed with 10-year-old *P. bournei* (NM + PB10).

Lefse analysis showed differences in the abundance of certain taxa in soil samples (Figure 2). When comparing microbial communities in pure stands of fir trees of different ages, we found that the order *Sp. Fcpu453* and the class *Betaproteobacteria* were the dominant species of the YF stands. In the communities of the OR stands, the family *Sp. HSB_ OF53_ F07* and the family *Sp. KF_ JG30_ B3* were the dominant species, which were significantly higher than other microbial communities. The bacterial community of the NM forest had no dominant species (Figure 2a). By comparing the microbial communities of fir plantations of different ages, we found that the order *Acidimicrobiales*, the family *Gemmatimonadaceae*, and the phylum *Gemmatimonadetes* were the dominant species in the NM + PB10 mixed forest communities. In the NM forest community, the class *Chlamydiae*, the phylum *Chlamydiae*, and the order *Chlamydiales* were the main dominant species affecting the soil bacterial community. In contrast, there was no high abundance of bacteria in the mixed forest community (Figure 2b).

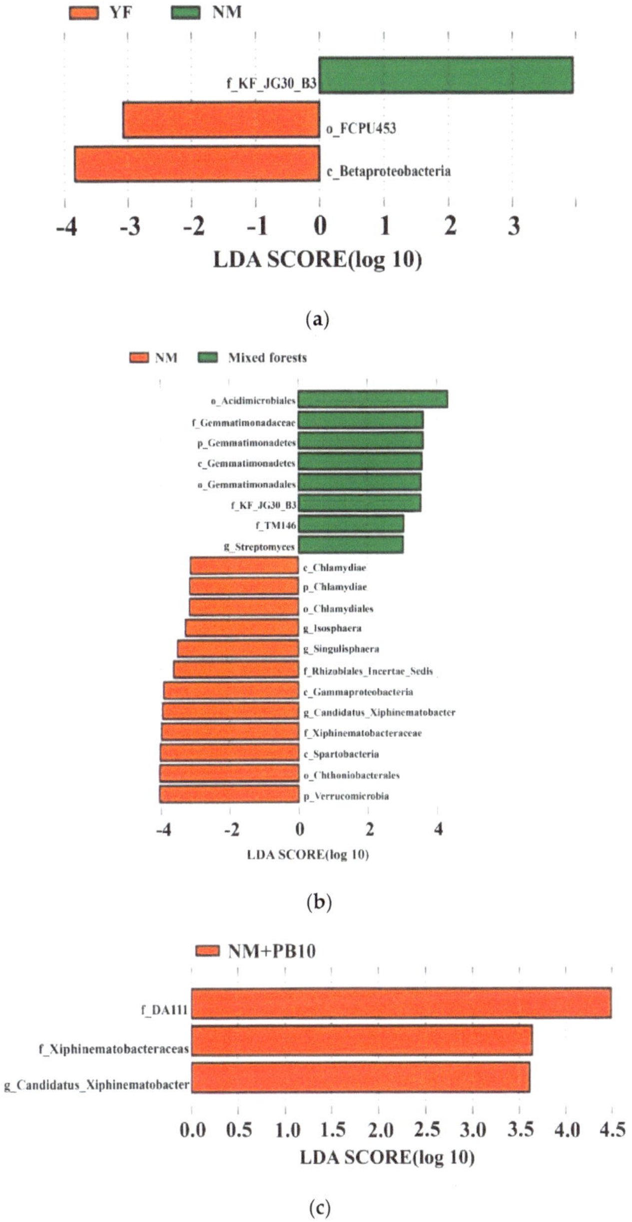

Figure 2. Lefse analysis of bacterial taxonomic levels in *C. lanceolata* plantations and Chinese fir-*P. bournei* mixed forests. (**a**) Lefse analysis of bacterial taxonomic levels in *C. lanceolata* plantations of different forest ages; (**b**) Lefse analysis of bacterial taxonomic levels in *C. lanceolata* plantations and Chinese fir-*P. bournei* mixed forests at the same developmental stage; and (**c**) Lefse analysis of bacterial taxonomic levels in Chinese fir–*P. bournei* mixed forests of different forest ages. Abbreviations: p: phylum, c: class, o: order, f: family, and g: genus. LDA > 3.5, $p < 0.05$.

Lefse analysis showed that the abundance of certain taxa differed in samples from mixed forests of different ages. In the NM + PB10 community, the family *DA111*, the family

Xiphinematobacteraceae, and the genus *Candidatus_Xiphinematobacter* were the dominant species. In contrast, there was no high abundance of bacteria in the NM + PB05 community (Figure 2c).

In the fungi community (Figure 3), all sequences were classified to the fungi domain and assigned to 31,865 OTUs across all samples, including 11 fungi phyla, 19 classes, 20 orders, 19 families, and 20 genera. The phylum with the most OTUs was *Ascomycota* (54.12%), followed by *Basidiomycota* (10.81%) and *Mortierellomycota* (2.34%).

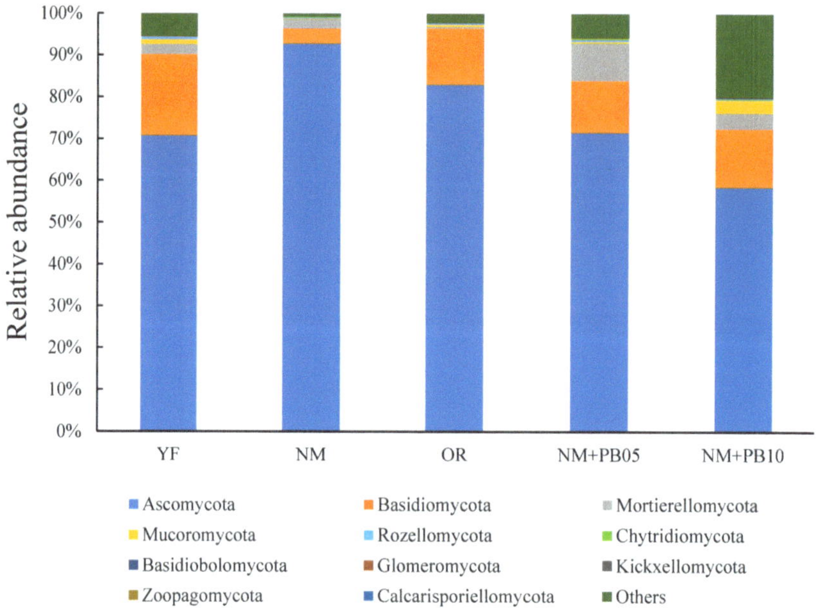

Figure 3. The relative abundance of fungi phylum in young (YF), near-mature (NM), and over-ripe (OR) Chinese fir plantations; near-mature Chinese fir plantation mixed with 5-year-old *P. bournei* (NM + PB05); Chinese fir plantation mixed with 10-year-old *P. bournei* (NM + PB10).

Lefse analysis showed differences in great numbers of dominant taxa in the soil samples. The abundance of fungal taxa in fir plantation forest samples differed between stand age and stand type. The genus *Ceratobasidium* and the genus *Leptodontidium* were the dominant species in the NM community. The order *Sordariales*, the order *Chaetothyriales*, and the order *Agaricales* were dominant in the YF community. There were no dominant species in the OR community (Figure 4a). The phylum *Basidiomycota*, the class *Cystobasidiomycetes*, the genus *Bannoa*, and the order *Xylariales* were the dominant species in the mixed forests community. In the NM community, the family *Clavulinaceae*, the family *Cordycipitaceae*, and the class *Archaeorhizomycetes* were the main species affecting the community (Figure 4b).

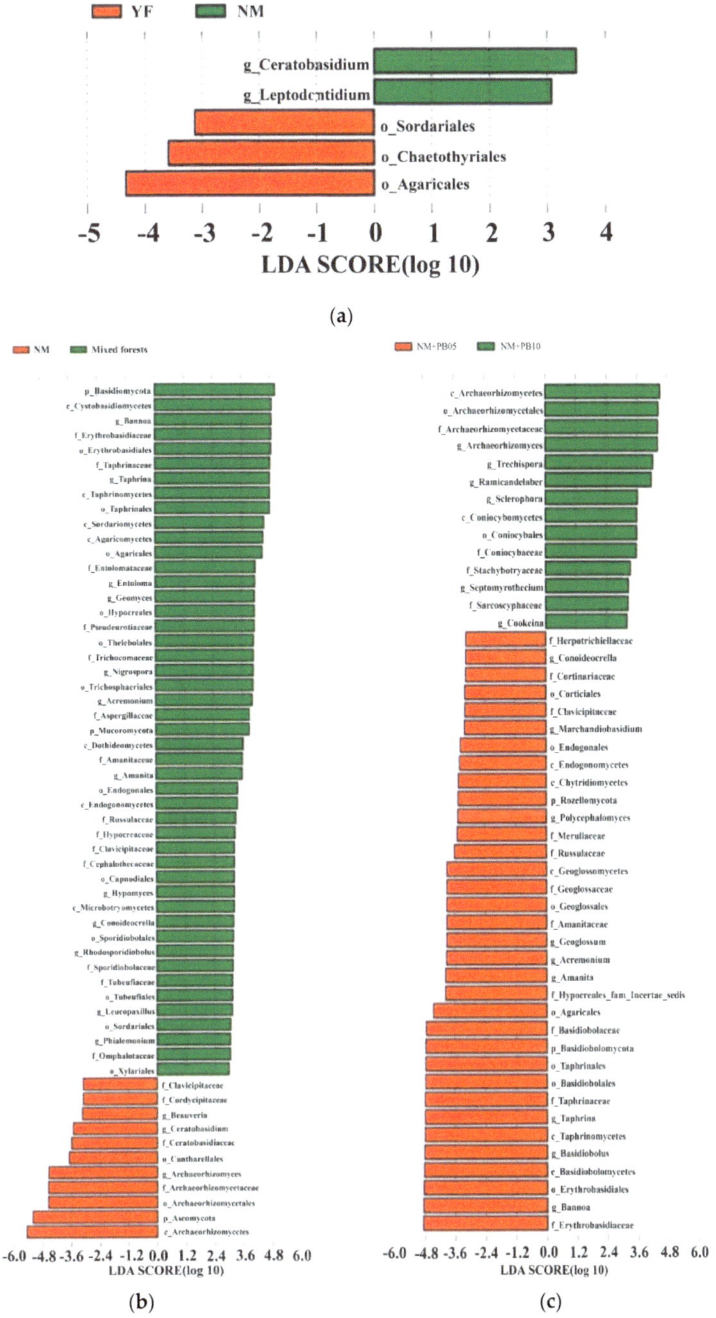

Figure 4. Lefse analysis of fungal taxonomic levels in pure and mixed cedar-heath forests. (**a**) Lefse analysis of fungal taxonomic levels in *C. lanceolata* plantations of different ages; (**b**) Lefse analysis of fungal taxonomic levels in plantations and mixed forests at the same developmental stage; (**c**) Lefse analysis of fungal taxonomic levels in mixed *C. lanceolate–P. bournei* forests of different ages. Abbreviations: p: phylum, c: class, o: order, f: family, and g: genus. LDA > 3.5, $p < 0.05$.

Lefse analysis showed that the abundance of certain taxa differed in samples from mixed forests of different forest ages. We found that the class *Archaeorhizomycetes*, the order *Archaeorhizomycetales*, and the family *Erythrobasidiaceae*, as well as other bacteria, were more abundant in the NM + PB05 mixed forest community. The family *Herpotrichiellaceae*, the genus *Conoideocrella*, the family *Corticiales*, and other bacteria were more abundant in the NM + PB10 mixed forest community (Figure 4c).

3.4. Structure of Microbial Community at OUT Level

In all treated samples, there were 12,503 OUT sequences in the bacterial community, and the bacterial community shared by all communities accounted for 11.4% of the OUT. The bacterial community of the Chinese fir plantation at different ages accounted for 47% of the total. The bacterial community of the mixed forest accounted for 53% of the OTUs (Figure 5a).

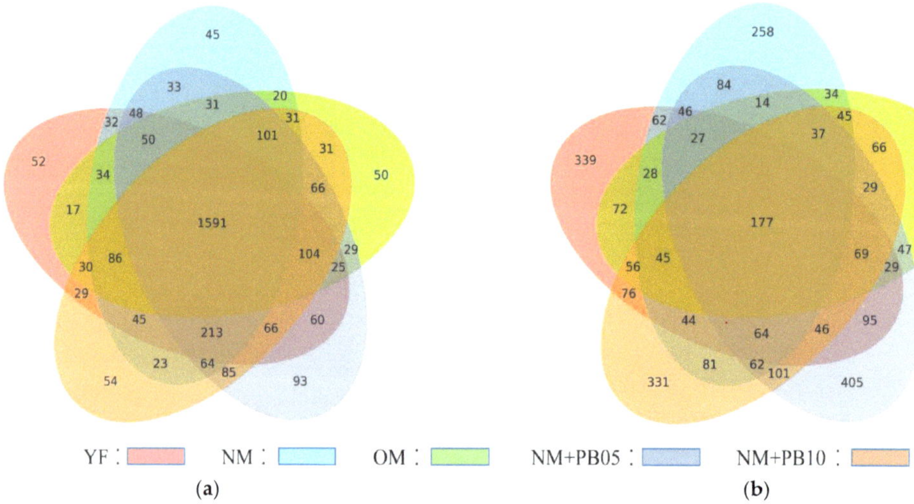

Figure 5. Venn plot shows the unique and shared OTUs among samples from Chinese fir plantation and mixed forest of different forest ages in the bacterial (**a**) and fungal (**b**) communities.

There were 6017 OTU sequences in all samples of fungal communities, of which all fungal communities had 3% of the same OTU. The fungal community of the *C. lanceolata* plantation at different ages accounted for 45% of the total communities, and the fungal community in the mixed forest accounted for 55% of the total communities (Figure 5b).

The first and second axes of the PCoA analysis based on the OTUs' data of differently aged Chinese fir plantations' soil bacterial communities explained 20.08% and 39.42% of the variance, respectively, suggesting that soil bacterial communities differed along the stand age gradient (Figure 6a). The PCoA analysis Axis1 explained 27.39% of the variance and the Axis2 explained 18.45% of the variance, illustrating that there were differences in the rhizosphere soil bacterial communities between Chinese fir plantations and Chinese fir–*P. bournei* mixed forests (Figure 6b).

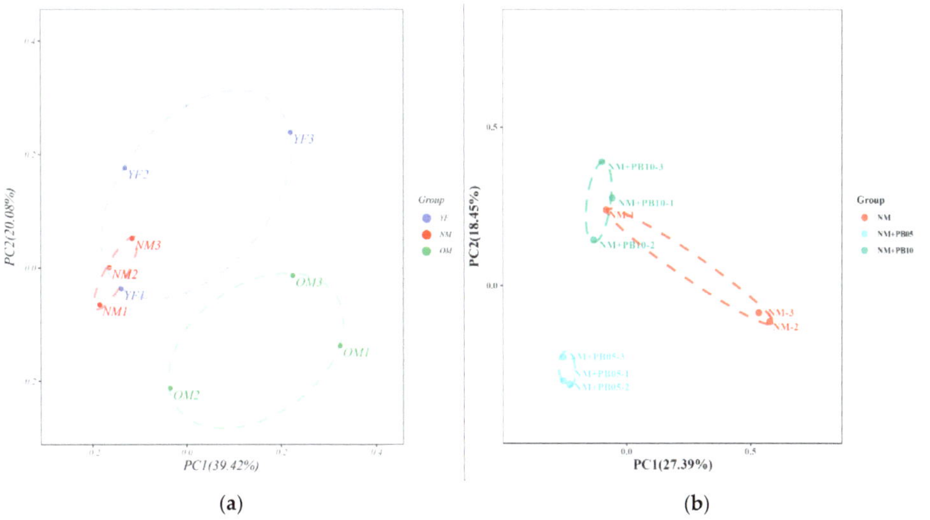

Figure 6. The PCoA results explained 59.47% of observed variation in rhizosphere soil bacteria communities in Chinese fir plantations of different ages (**a**) and 45.84% of observed variation in rhizosphere soil bacterial communities of Chinese fir–P. bournei mixed forests at the same developmental stage (**b**). The color of the points represents different soil bacterial communities (PerMANOVA test, $p < 0.05$).

Based on the OTUs data from differently aged Chinese fir plantations' rhizosphere soil fungal communities, the first and second PCoA axes explained 34.04% and 19.94%, respectively, of the variance. Thus, soil fungal communities differed along a gradient of forest age (Figure 7a). In a PCoA analysis of rhizosphere soil fungal communities in Chinese fir plantations and Chinese fir–*P. bournei* mixed forests at the same developmental stage, the first axis explained 32.36% of the variation and the second axis explained 23.09% of the variation, respectively (Figure 7b).

3.5. Identification of Key Drivers Affecting Microbial Community Structure

Soil microbial community is very sensitive to changes in soil factors, the age of Chinese fir plantation forests, and the construction of mixed forests. Soil properties continually change, and the microbial community's feedback to its factors differ, resulting in changes in the abundance of the microbial community (Figure 6). In the young forest stage, the soil bacterial community abundance was strongly influenced by soil pH (r = 0.99 **, Figure 8A). TK and pH had the greatest influence on the soil bacterial community abundance in the near-mature forest stage (r = 0.98 **, Figure 8B), and TP and ACP had the greatest influence in the over-ripe forest stage (r = 0.99 **, Figure 8C). AN dominated the soil bacterial community abundance in the mixed forest (r = 0.99 **, Figure 8D).

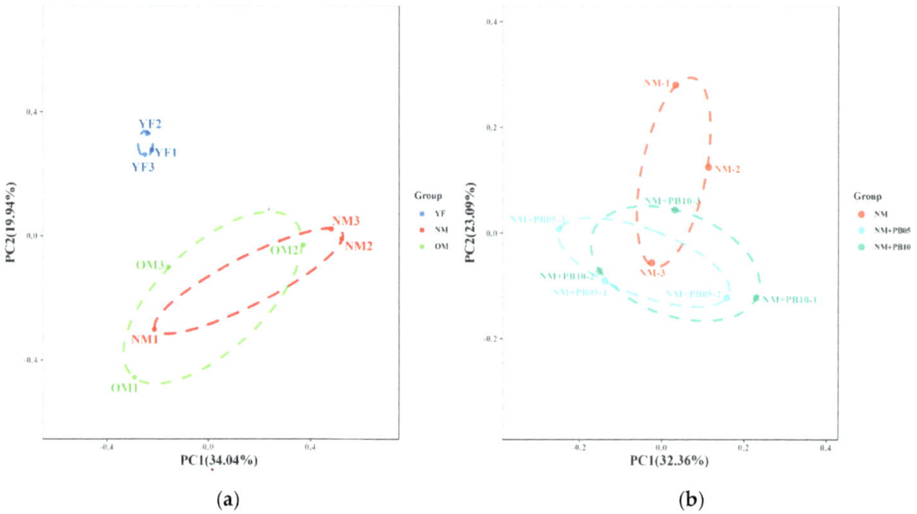

Figure 7. According to the results of PcoA, there is 53.98% of observed variation in rhizosphere soil fungal communities of differently aged Chinese fir plantations (a) and 55.45% of observed variation of Chinese fir–P. bournei mixed forests at the same developmental stage (b). The color of the dots represents the different soil bacterial communities.

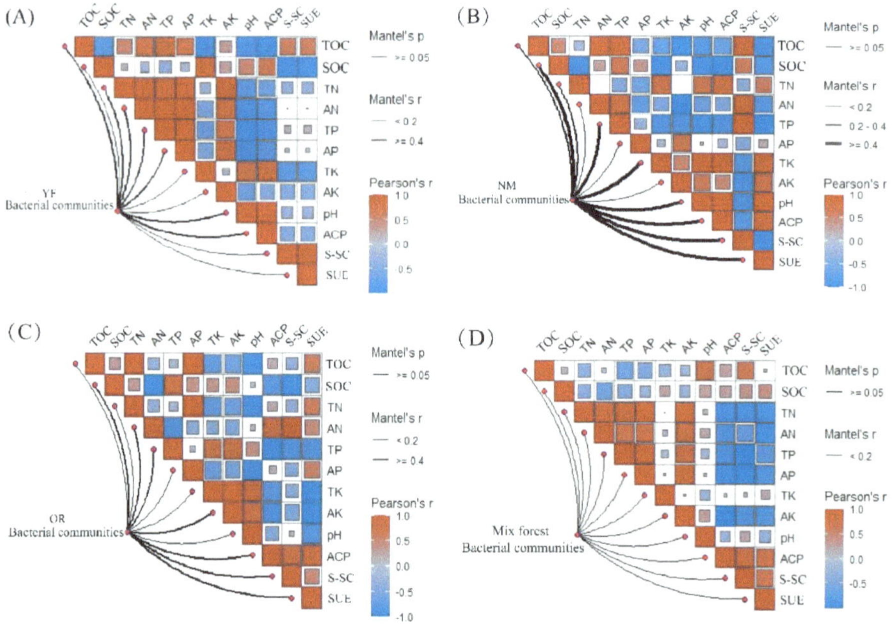

Figure 8. Identification of key driving factors affecting bacterial community structure in different forest ages and stand types. (A) Correlation of soil bacterial communities with soil nutrients and enzyme activities at the young forest; (B) Correlation of soil bacterial communities with soil nutrients and enzyme activities in near-mature forests; (C) Correlation of soil bacterial communities with soil

nutrients and enzyme activities in overripe forests; (**D**) Correlation of soil bacterial communities with soil nutrients and enzyme activities in Chinese fir–P. bournei mixed forests. Note: Significant correlation at Mantel's r ≥ 0.4. Correlation is significant at 0.2 ≤ Mantel's r ≤ 0.4. Data: means ± SD (*n* = 3). Color gradient indicates Spearman's correlation coefficient; the side width corresponds to the correlation coefficient for distance calculated by Mantel; color shade indicates permutations *p*-value for statistical significance. Same is below.

Soil factors affecting the soil fungal community abundance were different at each developmental stage of the plantation forests; thus, Mantel test analyses were used to identify soil factors affecting soil fungal community abundance at each developmental stage. We found that soil fungal community abundance was strongly influenced by TK at the young forest stage (r = 0.99 **, Figure 9A); and TN had the greatest influence on soil fungal community abundance at the near-mature forest stage (r = 0.99 **, Figure 9B), and SOC, AN, and S-SC at the over-ripe forest stage (r = 0.49 **, Figure 9C). The soil bacterial community abundance in the mixed forest was most affected by AN (r = 0.99 **, Figure 9D).

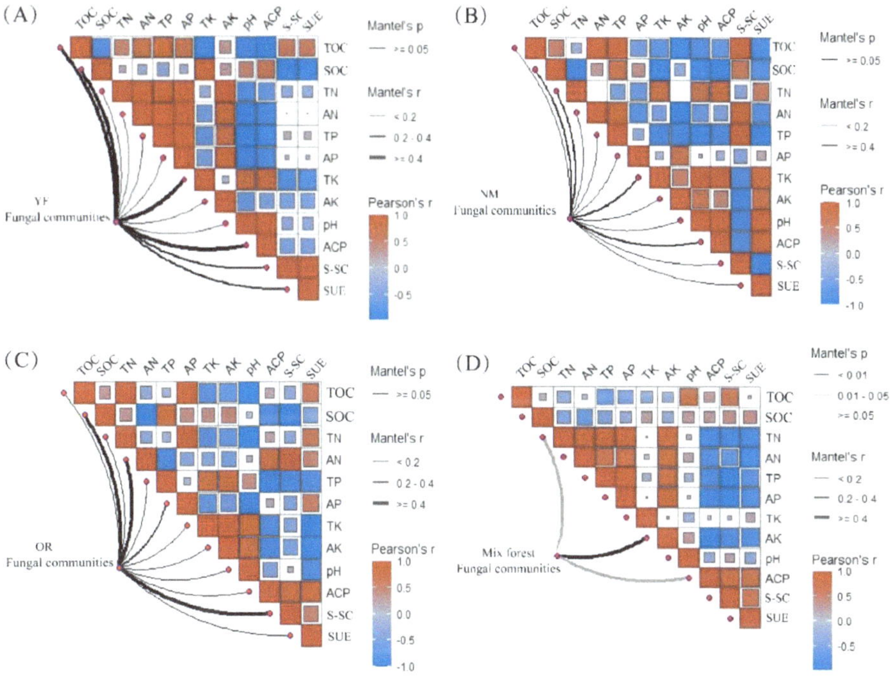

Figure 9. Identification of key driving factors affecting fungal community structure in different forest ages and stand types. (**A**) Correlation of soil bacterial communities with soil nutrients and enzyme activities at the young forest; (**B**) Correlation of soil bacterial communities with soil nutrients and enzyme activities in near-mature forests; (**C**) Correlation of soil bacterial communities with soil nutrients and enzyme activities in overripe forests; (**D**) Correlation of soil bacterial communities with soil nutrients and enzyme activities in Chinese fir–P. bournei mixed forests. Note: Significant correlation at Mantel's r ≥ 0.4. Correlation is significant at 0.2 ≤ Mantel's r ≤ 0.4. Data: means ± SD (*n* = 3). Color gradient indicates Spearman's correlation coefficient; the side width corresponds to the correlation coefficient for distance calculated by Mantel; color shade indicates permutations *p*-value for statistical significance.

4. Discussion

4.1. Analysis of the Effects of Forest Age and Stand Type on Rhizosphere Soil Chemical Properties and Enzyme Activity

We observed a continuous increase in TOC and SOC as the forests aged from 10 to 45 years old (Table 1). There is a similar trend to those reported by Deng et al. and Yu et al., who found that SOC concentration continued to increase with the increase in Chinese fir plantation age [3,46]. However, there are also many conclusions inconsistent with the results of this study. For example, Wei et al. found that SOC content decreased significantly in the near-mature forest stage and increased slightly in the over-mature forest stage [47]. Chen et al. found that SOC content continued to decline with the increasing forest age [48]. In this study, the TK and AK contents showed the same trend as SOC and varied significantly among different ages. TP, AP, TN, and AN all declined to the lowest trend in the near-mature forest stage and varied significantly. These observations are consistent with the research by Liu et al., who found that rhizosphere soil N and P contents changed in a V-shaped pattern with increasing forest age [49].

The initial low C content may be due to the fact that a large amount of wood was taken from the forest floor at the end of the last harvest period through slash burning, which caused a large amount of C to be burned off [50]. With the increasing forest age, the understory detritus continues to accumulate; however, the decomposition rate is slow and soil C gradually accumulates. Thus, the soil nutrients (N and P) consumed by trees before 25–30 years will exceed the returned soil nutrients [51,52]. This result can also be explained by the reduction in understory vegetation, which is related to the premature closure of the canopy during the growth of trees [48,53]. We observed a continuous decrease in understory vegetation from the young forest to near-mature forest, which supports the above explanation.

The K content increased with the increasing stand age. The number of plant roots and the amount of soil microorganisms increased, which promoted the conversion of insoluble mineral state potassium to water-soluble, exchangeable potassium. Moreover, the increase in AMF promoted the accumulation of K. In addition, Zhang Yuan et al. [54] found, in their study on gooseberry, that the demand for fast-acting K was higher during the juvenile period of plant growth, which may also be the cause of the younger stands having opposing trends in pH and K. In a study by Li et al. [55], it was found that certain phosphorus-solubilizing bacteria existed in the inter-root soil of cedar, and that in the long-term phosphorus-deficient environment of cedar, the content of phosphorus-solubilizing bacteria was higher. Moreover, phosphate-solubilizing bacteria can produce acidic substances and reduce soil pH.

The soil nutrient content and soil pH of the mixed forest were significantly higher than those of the planted forest. Zhongming et al. [56] and Joshi B [57] had similar results as ours; thus, it is recommended to construct a mixed forest while extending the rotation period.

4.2. Rhizosphere Soil Microbial Diversity and Abundance Influenced by Tree Age and Stand Type

In this study, after the introduction of *P. bournei* into a Chinese fir plantation, the α-diversity of the soil microbial community was greatly affected, which was manifested as the influence of the α-diversity of the soil bacterial community being greater than that of the soil fungal community (Tables 2 and 3). Moreover, the soil microbial community structure was obviously separated (Figures 4 and 5). Diakhate et al. found that bacterial α-diversity in millet and peanut intercropping was significantly affected by the planting pattern, which was consistent with the results of this study [58]. Similarly, Debenport et al. found that the diversity of bacterial communities increased significantly after the introduction of *P. reticulatum* or *G. senegalensis* into millet [59]. One of the reasons for this result is that bacterial communities are more sensitive than fungal communities in the early stages of stand-type change [60]. Studies have shown that in the early stages of forest development, environmental changes have a greater impact on soil bacterial communities, and soil bacterial communities change faster and are more sensitive than fungal communities [61,62]. Another reason is that the change in microbial species richness and diversity is not necessarily affected by the change in the structure of microbial communities, because changes in certain groups may be compensated by changes in other groups [63,64].

In *C. lanceolata* plantation forests of different stand ages, the quantity of OTUs of the soil bacterial community, first increased and then decreased, as stands aged (Tables 1 and 2). This may be due to the fact that controlled burning before afforestation caused the biomass that was not taken out of the forest floor to return to the soil, and after entering the fast-growing stage, the soil nutrients declined rapidly, and *C. lanceolata* continuously released organic acids through the root system as a way to enhance bacterial activity in order to obtain sufficient nutrients [15,65]. This is also corroborated by the fact that both acid phosphatase and urease activities reached their maximum at the near-mature forest stage. This is consistent with the monotonic increase in urease activity over 15–35 years of redwood plantations, as reported by Kang et al. [66].

The abundance of the fungal community continued to decline with stand age, with diversity being lowest at the near-mature forest stage (Tables 1 and 3), which may result because of two reasons. First, because of the severe soil acidification due to acidic substances produced by lower nutrient content *C. lanceolata* root secretions and microbial decomposition of difficult-to-decompose materials, the abundance of fungal species intolerant of acidic environments declined [67,68]. Second, as it grows, the forest canopy closes, light decreases, water and heat conditions within the forest become poorer, and the diversity of understory vegetation decreases. This is consistent with the results of Zhang et al. [69].

4.3. Effect of Stand Age and Stand Type on the Keystone Taxa

We identified site-specific OTUs in pure and *C. lanceolata* × *P. bournei* forests (Tables 1–3), which may serve as key taxa to explain the variation in microbial communities between plantations. At the level of bacterial phylum, *Chloroflexi* and *Firmicutes* were higher in pure and mixed stands of fir (Figure 1). It has been suggested that *Chloroflexi* and *Firmicutes* may play an important role in participating in carbon and nitrogen cycling, using recalcitrant carbon fixation [70,71], and have the ability to degrade and fix nitrogen from organic matter, lignin, and cellulose [72,73]. However, the abundance of *Chloroflexi*, *Firmicutes*, and other strains did not differ significantly in pure and mixed Chinese fir–*P. bournei* forests. This may be because during the dry season, the amount of apoplankton increased in the pure Chinese fir forest, and the dissolved oxygen content in the soil increased, which favored the growth of aerobic microorganisms and higher microbial biomass.

As for fungi, the introduction of broad-leaf trees not only increased the diversity of fungal communities but also significantly increased the relative abundance of *Basidiomycota* compared to pure stands of Chinese fir. Although previous studies confirmed that *Ascomycota* is more capable of hydrolyzing cellulose than *Basidiomycota* [74], we found no difference in *Ascomycota* abundance between pure and mixed stands of Chinese fir. Thus, *Basidiomycota* played a greater role in mixed stands than in pure stands (Figures 3 and 4b). The study by Y. Liang et al. found that *Basidiomycota* often formed mycorrhizae in symbiosis with plants to promote nutrient uptake and plant growth [75]. Therefore, the higher abundance of Basidiomycota in the mixed forest may be due to the higher number of symbiotic bacteria in the mixed forest.

4.4. Different Stand Age and Forestry Management Practices Might Influence Soil Microbial Communities via Soil Properties

Most of the studies considered soil factors and vegetation as the main factors affecting microbial community structure. Nielsen et al. [76] found that changes in soil properties had strong effects on bacterial and archaeal communities. Soil pH is the main factor affecting soil bacterial diversity [77,78]. The bacterial community diversity showed an upward then downward trend in different stand ages of Chinese fir plantations, and according to the Mantel test analysis, soil pH was significantly correlated with bacterial community diversity and had the greatest effect on bacterial community ($r = 0.99$ **). These findings may be due to the increased nutrient stress on Chinese fir trees as they aged, which releases root secretions and organic acids. These secretions have an attractive effect on some microorganisms, and these chemotactic bacteria can accumulate and multiply in large numbers in the inter-rhizosphere. For example, Li et al. found that when intercropping between grasses and legumes, the roots of legumes such as broad beans (*Vicia faba*) acidified the soil by releasing organic acids and proteins, which activated insoluble phosphorus and thus promoted phosphorus uptake by grasses such as maize (*Zea mays*) [79]. This process also resulted in no significant difference in bacterial community diversity between *C. lanceolata* plantation and *C. lanceolata–P. bournei* mixed forests at the near-mature stage. Liu et al. showed that the loss of soil bacterial diversity was mainly due to the decrease in soil pH [80]. Thus, bacterial diversity decreases in the overripe forest stage.

There are two main reasons affecting the diversity of soil fungal communities in fir plantations of different stand ages. One reason is soil nutrients. According to the Mantel test analysis, the fungal community diversity at different stand ages was initially the most influenced by potassium ($r = 0.99$ **) and gradually shifted to nitrogen ($r = 0.49$ **). Maisto et al. [81] showed that forest soil nutrients are mainly derived from apoplastic matter. The apoplastic material of Chinese fir plantations is mainly needles, which contain a large amount of lignin, cellulose, and tannin, which have relatively high C/N, leading to soil N deficiency and inhibition of fungal activity [82]. The other explanation is soil pH. Although pH did not have a direct effect on fungal community diversity according to the Mantel test analysis, this may be because the decrease in soil pH drives aluminum toxicity, which is detrimental to microbial growth [83].

Soil pH has a significant effect on soil microbial communities. In plantation forests, soil bacterial diversity tended to increase and then decrease with a gradual increase in soil acidity, a result that is consistent with the findings of F. Wu et al. [84]. Fungal diversity decreased significantly from the young stage to the near-mature stage and slightly increased after the overripe stage. The analysis was related to the gradual increase in soil acidity and the gradual decrease in nutrients in the plantation forest [85], which was consistent with the results of Dennis [86]. In mixed forests, soil acidity is maintained in a weakly acidic state, and the broad-leaf tree *P. bournei* may improve the quality of litter inputs, while microorganisms can participate in nutrient cycling through litter decomposition, leading to higher decomposition rates and more nutrients being released into the soil [87], which in turn affects forest ecology. This process is consistent with the results reported by Xia et al. on the introduction of Elaeocarpus decipiens Hemsl and Michelia macclurei broad-leaf trees into Chinese fir monoculture plantations, where the growth of Chinese fir was promoted and accompanied by changes in soil microorganisms [88].

5. Conclusions

From the perspective of microbiology, the results of this study are presented. The introduction of *P. bournei* into a Chinese fir plantation did not affect the number and diversity of bacterial communities in rhizosphere soil under a single planting pattern, but the diversity of fungal communities was twice that of the Chinese fir plantation. Rhizosphere soil nutrients, acid phosphatase, and sucrase were also significantly higher than in the Chinese fir plantation. Although the soil nutrients increased slightly at the over-ripening stage, there was still a considerable gap compared with the mixed forest. Therefore, in order to ensure the sustainable development of Chinese fir plantations, more attention should be paid to the middle and late stages of the growth of Chinese fir plantations rather than the early stages. In addition, the construction of mixed forest models should be used as the main forestry management method.

Author Contributions: Conceptualization, W.L. and J.J.; methodology, W.L., M.C. and J.J.; software, W.L. and H.S.; resources, L.W. and X.F.; writing—original draft preparation, W.L. and H.S.; writing—review and editing, W.L., H.S. and J.J.; visualization, W.L. and H.S.; supervision, J.J.; project administration, W.L. and J.J. All authors have read and agreed to the published version of the manuscript.

Funding: This study was supported by the National Natural Science Foundation of China (32071612) and the Baishanzu National Park Scientific Research Project (2022JBGS03, 2021ZDLY01).

Data Availability Statement: Data available on request due to restrictions surrounding, e.g., privacy or ethics. The data presented in this study are available on request from the corresponding author.

Conflicts of Interest: The authors declare no conflict of interest.

References

1. Wang, Z.C.; He, G.X.; Hou, Z.H.; Luo, Z.; Chen, S.X.; Lu, J.; Zhao, J. Soil C:N:P stoichiometry of typical coniferous (*Cunninghamia lanceolata*) and/or evergreen broadleaved (*Phoebe bournei*) plantations in south China. *For. Ecol. Manag.* **2021**, *486*, 118974. [CrossRef]
2. Bi, J.; Blanco, J.A.; Seely, B.; Kimmins, J.P.; Ding, Y.; Welham, C. Yield decline in Chinese-fir plantations: A simulation investigation with implications for model complexity. *Can. J. For. Res.* **2007**, *9*, 1615–1630. [CrossRef]
3. Yu, Y.C.; Yang, J.Y.; Zeng, S.C.; Wu, D.M.; Douglass, F.J.; Sloan, L. Soil pH, organic matter, and nutrient content change with the continuous cropping of *Cunninghamia lanceolata* plantations in South China. *J. Soils Sediments* **2016**, *9*, 2230–2238. [CrossRef]
4. Zhang, H.; Yu, J.D.; Li, H.Y.; Zhang, Y.L.; Pan, F.; Zhou, C.F.; Liu, A.Q. Characteristics of Soil Phosphorus in Cunninghamia lanceolata Plantations with Different Planting Rotations. *For. Res.* **2021**, *1*, 10–18.
5. Kaarakka, L.; Cornett, M.; Domke, G.; Ontl, T.; Dee, L.E. Improved forest management as a natural climate solution: A review. *Ecol. Solut. Evid.* **2021**, *2*, e12090. [CrossRef]
6. Ding, K.; Zhang, Y.; Wang, L.; Ge, S.; Zhang, Y.; Yang, Q.; Huang, H.H.; Tong, Z.K.; Zhang, J.H. Forest conversion from pure to mixed *Cunninghamia lanceolata* plantations enhances soil multifunctionality, stochastic processes, and stability of bacterial networks in subtropical southern China. *Plant Soil.* **2023**, *488*, 411–429. [CrossRef]
7. Wang, Q.K.; Wang, S.L.; Huang, Y. Comparisons of litterfall, litter decomposition and nutrient return in a monoculture *Cunninghamia lanceolata* and a mixed stand in southern China. *For. Ecol. Manag.* **2008**, *3–4*, 1210–1218. [CrossRef]
8. Zhao, J.Z.; Xie, D.M.; Wang, D.Y.; Deng, H.B. Current status and problems in certification of sustainable forest management in China. *Env. Manag.* **2011**, *6*, 1086–1094. [CrossRef]
9. Wang, W.F.; Wei, X.H.; Liao, W.M.; Blanco, J.A.; Liu, Y.Q.; Liu, S.R.; Liu, G.H.; Zhang, L.; Guo, X.M.; Guo, S.M. Evaluation of the effects of forest management strategies on carbon sequestration in evergreen broad-leaved (*Phoebe bournei*) plantation forests using FORECAST ecosystem model. *For. Ecol. Manag.* **2013**, *300*, 21–32. [CrossRef]
10. Khlifa, R.; Paquette, A.; Messier, C.; Reich, P.B.; Munson, A.D. Do temperate tree species diversity and identity influence soil microbial community function and composition? *Ecol. Evol.* **2017**, *19*, 7965–7974. [CrossRef]
11. Wardle, D.A.; Bardgett, R.D.; Klironomos, J.N.; Setälä, H.; van-der Putten, W.H.; Wall, D.H. Ecological Linkages Between Aboveground and Belowground Biota. *Science* **2004**, *5677*, 1629–1633. [CrossRef] [PubMed]
12. Shi, S.J.; Herman, D.J.; He, Z.J.; Pett-Ridge, J.; Wu, L.Y.; Zhou, J.Z.; Firestone, M.K. Plant roots alter microbial functional genes supporting root litter decomposition. *Soil Biol. Biochem.* **2018**, *127*, 90–99. [CrossRef]
13. de Vries, F.T.; Griffiths, R.I.; Knight, C.G.; Nicolitch, O.; Williams, A. Harnessing rhizosphere microbiomes for drought-resilient crop production. *Science* **2020**, *6488*, 270–274. [CrossRef] [PubMed]
14. Bastida, F.; Eldridge, D.J.; García, C.; Png, G.K.; Bardgett, R.D.; Delgado-Baquerizo, M. Soil microbial diversity–biomass relationships are driven by soil carbon content across global biomes. *ISME J.* **2021**, *15*, 2081–2091. [CrossRef]
15. Hartmann, M.; Frey, B.; Mayer, J.; Mäde, P.; Widmer, F. Distinct soil microbial diversity under long-term organic and conventional farming. *ISME J.* **2015**, *5*, 1177–1194. [CrossRef]
16. Tian, W.; Wang, L.; Li, Y.; Zhuang, K.M.; Li, G.; Zhang, J.B.; Xiao, X.J.; Xi, Y.G. Responses of microbial activity, abundance, and community in wheat soil after three years of heavy fertilization with manure-based compost and inorganic nitrogen. *Agric. Ecosyst. Environ.* **2015**, *213*, 219–227. [CrossRef]
17. Zhang, P.; Guan, P.; Hao, C.; Yang, J.; Xie, Z.; Wu, D. Changes in assembly processes of soil microbial communities in forest-to-cropland conversion in Changbai Mountains, northeastern China. *Sci. Total Environ.* **2022**, *818*, 151738. [CrossRef]
18. Meng, M.J.; Lin, J.; Guo, X.P.; Liu, X.; Wu, J.S.; Zhao, Y.P.; Zhang, J.C. Impacts of forest conversion on soil bacterial community composition and diversity in subtropical forests. *CATENA* **2019**, *175*, 167–173. [CrossRef]
19. Dai, Z.M.; Xiong, X.Q.; Zhu, H.; Xu, H.J.; Li, X.H.; Tang, C.; Xu, J.M. Association of biochar properties with changes in soil bacterial, fungal and fauna communities and nutrient cycling processes. *Biochar* **2021**, *3*, 239–254. [CrossRef]
20. Das, S.; Deb, S.; Sahoo, S.S.; Sahoo, U.K. Soil microbial biomass carbon stock and its relation with climatic and other environmental factors in forest ecosystems: A review. *Acta Ecol. Sin.* **2023**, *in press*. [CrossRef]
21. Xu, Z.Y.; Hu, Z.H.; Jiao, S.; Bell, S.M.; Xu, Q.; Ma, L.L.; Chen, J. Depth-dependent effects of tree species identity on soil microbial community characteristics and multifunctionality. *Sci. Total Environ.* **2023**, *878*, 162972. [CrossRef] [PubMed]
22. Bach, L.H.; Grytnes, J.A.; Halvorsen, R.; Ohlson, M. Tree influence on soil microbial community structure. *Soil Biol. Biochem.* **2010**, *11*, 1934–1943. [CrossRef]
23. Liu, Y.Y.; Dong, L.Z.; Zhang, H.J.; Deng, Y.Y.; Hu, B.; Wang, W. Distinct roles of bacteria and fungi in mediating soil extracellular enzymes under long-term nitrogen deposition in temperate plantations. *For. Ecol. Manag.* **2023**, *529*, 12068. [CrossRef]

24. Baldrian, P.; Kolarik, M.; Stursova, M.; Kopecky, J.; Valaskova, V.; Vetrovsky, T.; Žifčáková, L.; Šnajdr, J.; Rídl, J.; Vlček, Č.; et al. Active and total microbial communities in forest soil are largely different and highly stratified during decomposition. *ISME J.* **2012**, *2*, 248–258. [CrossRef]
25. Colombo, F.; Macdonald, C.A.; Jeffries, T.C.; Powell, J.R.; Singh, B.K. Impact of forest management practices on soil bacterial diversity and consequences for soil processes. *Soil Biol. Biochem.* **2016**, *94*, 200–210. [CrossRef]
26. Kjøller, A.; Struwe, S. Microfungi in ecosystems: Fungal occurrence and activity in litter and soil. *Oikos* **1982**, *39*, 391–422. [CrossRef]
27. Hayat, R.; Ali, S.; Amara, U.; Khalid, R.; Ahmed, I. Soil beneficial bacteria and their role in plant growth promotion: A review. *Ann. Microbiol.* **2010**, *4*, 579–598. [CrossRef]
28. Zhou, L.; Sun, Y.J.; Saeed, S.; Zhang, B.; Luo, M. The difference of soil properties between pure and mixed Chinese fir (*Cunninghamia lanceolata*) plantations depends on tree species. *Glob. Ecol. Conserv.* **2020**, *22*, e01009. [CrossRef]
29. Guo, J.H.; Feng, H.L.; Mcnie, P.; Liu, Q.Y.; Xu, X.; Pan, C.; Yan, K.; Feng, L.; Goitom, E.A.; Yu, Y.C. Species mixing improves soil properties and enzymatic activities in Chinese fir plantations: A meta-analysis. *CATENA* **2023**, *220*, 106723. [CrossRef]
30. Ding, K.; Zhang, Y.; Liu, H.; Yang, X.; Zhang, J.; Tong, Z. Soil bacterial community structure and functions but not assembly processes are affected by the conversion from monospecific *Cunninghamia lanceolata* plantations to mixed plantations. *Appl. Soil Ecol.* **2023**, *185*, 104775. [CrossRef]
31. Wang, Q.K.; Wang, S.L. Soil microbial properties and nutrients in pure and mixed Chinese fir plantations. *J. For. Res.* **2008**, *2*, 131–135. [CrossRef]
32. Lei, J.; Wu, H.; Li, X.; Guo, W.; Duan, A.; Zhang, J. Response of rhizosphere bacterial communities to Near-Natural forest management and tree species within Chinese fir plantations. *Microbiol. Spectr.* **2023**, *11*, e02328-22. [CrossRef] [PubMed]
33. Chu, H.Y.; Li, R.N.; Li, J.W.; Zhong, Z.Q.; Liu, X.F.; Li, Y.Q. Effects of forest conversion on soil microbial community structure. *Chin. J. Appl. Environ. Biol.* **2019**, *1*, 23–28.
34. Li, J.; Cooper, J.M.; Lin, Z.A.; Li, Y.T.; Yang, X.D.; Zhao, B.Q. Soil microbial community structure and function are significantly affected by long-term organic and mineral fertilization regimes in the North China Plain. *Appl. Soil Ecol.* **2015**, *96*, 75–87. [CrossRef]
35. Cui, Y.X.; Fang, L.C.; Deng, L.; Guo, X.B.; Han, F.; Ju, W.L.; Wang, X.; Chen, H.S.; Tan, W.F.; Zhang, X.C. Patterns of soil microbial nutrient limitations and their roles in the variation of soil organic carbon across a precipitation gradient in an arid and semi-arid region. *Sci. Total Environ.* **2019**, *658*, 1440–1451. [CrossRef] [PubMed]
36. Sparks, D.L.; Helmke, P.A.; Page, A.L.; Loeppert, R.H.; Soltanpour, P.N.; Tabatabai, M.A.; Johnston, C.T.; Sumner, M.E. *Methods of Soil Analysis: Part 3. Chemical Methods*, 1st ed.; The Soil Science Society of America, American Society of Agronomy: Madison, WI, USA, 1996; pp. 65–90.
37. Allen, S.E.; Grimshaw, H.M.; Parkinson, J.A.; Quarmby, C. *Chemical Analysis of Ecological Materials, 1st ed*; Institute Terrestrial Ecology, Merlewood Research Station: Cumbria, UK, 1977; pp. 494–495.
38. Wu, J.; Joergensen, R.G.; Pommerening, B.; Chaussod, R.; Brookes, P.C. Measurement of soil microbial biomass C by fumigation-extraction—An automated procedure. *Soil Biol. Biochem.* **1990**, *8*, 1167–1169. [CrossRef]
39. Zhou, L.K.; Zhang, Z.M. Determination method of soil enzyme activity. *China J. Soil Sci.* **1980**, *5*, 37–38.
40. Zhang, J.J.; Kobert, K.; Flouri, T.; Stamatakis, A. PEAR: A fast and accurate Illumina Paired-End reAd mergeR. *Bioinformatics* **2014**, *5*, 614–620. [CrossRef]
41. Ahn, J.; Sinha, R.; Pei, Z.H.; Dominianni, C.; Wu, J.; Shi, J.X.; Goedert, J.J.; Hayes, R.B.; Yang, L.Y. Human gut microbiome and risk for colorectalcancer. *J. Natl. Cancer Inst.* **2013**, *24*, 1907–1911. [CrossRef] [PubMed]
42. Caporaso, J.G.; Kuczynski, J.; Stombaugh, J.; Bittinger, K.; Bushman, F.D.; Costello, E.K.; Fierer, N.; Peña, A.G.; Goodrich, J.K.; Gordon, J.I.; et al. QIIME allows analysis of high-throughput community sequencing data. *Nat. Methods* **2010**, *5*, 335–336. [CrossRef]
43. Vasileiadis, S.; Puglisi, E.; Arena, M.; Cappa, F.; Cocconcelli, P.S.; Trevisan, M. Soil Bacterial Diversity Screening Using Single 16S rRNA Gene V Regions Coupled with Multi-Million Read Generating Sequencing Technologies. *PLoS ONE* **2012**, *8*, e42671.
44. Heberle, H.; Meirelles, G.V.; da Silva, F.R.; Minghim, R. InteractiVenn: A web-based tool for the analysis of sets through Venn diagrams. *BMC Bioinform.* **2015**, *1*, 169. [CrossRef]
45. Segata, N.; Izard, J.; Waldron, L.; Gevers, D.; Miropolsky, L.; Garrett, W.S.; Huttenhower, C. Metagenomic biomarker discovery and explanation. *Genome Biol.* **2011**, *6*, R60. [CrossRef]
46. Deng, L.; Wang, K.B.; Chen, M.L.; Shangguan, Z.P.; Sweeney, S. Soil organic carbon stroage capacity positively related to forest succession on the loess plateau. *CATENA* **2013**, *110*, 1–7. [CrossRef]
47. Wei, Z.C.; Huang, J.; Liu, Y.H.; Li, H.T.; Wu, P.F.; Liu, A.Q. Community characteristics of soil bacteria of *Cunninghamia lanceolata* plantations at different developmental stages. *J. Southwest For. Univ.* **2017**, *5*, 122–129.
48. Chen, G.S.; Yang, Z.J.; Gao, R.; Xie, J.S.; Guo, J.F.; Huang, Z.Q.; Yang, Y.S. Carbon storage in a chronosequence of Chinese fir plantations in southern China. *For. Ecol. Manag.* **2013**, *4*, 68–76. [CrossRef]
49. Liu, S.; Liu, X.S.; Zhu, X.Z.; Cheng, K.Y.; Guo, X.M.; Zhang, W.Y. Rhizosphere effects of nutrients and enzyme activities of Cunninghania lanceolata and soil fertility assessment. *J. Plant Nutr. Fertil.* **2017**, *2*, 492–501.
50. Yang, Z.J.; Chen, S.D.; Liu, X.F.; Xiong, D.C.; Xu, C.; Arthur, M.A.; McCulley, R.L.; Yang, Y.S. Loss of soil organic carbon following natural forest conversion to Chinese fir plantation. *For. Ecol. Manag.* **2019**, *449*, 117476. [CrossRef]

51. Wu, Z.Y.; Li, J.J.; Zheng, J.; Liu, J.F.; Liu, S.Y.; Lin, W.X.; Wu, C.Z. Soil microbial community structure and catabolic activity are significantly degenerated in successive rotations of Chinese fir plantations. *Sci. Rep.* **2017**, *1*, 6691. [CrossRef]
52. Lan, S.A.; Du, H.; Zeng, F.P.; Song, T.Q.; Peng, W.X.; Han, C.; Chen, L.; Su, L. Carbon storage and allocation in *Cunninghamia lanceolata* plantations with different stand ages. *China J. App. Ecol.* **2016**, *4*, 1125–1134.
53. Xu, L.; Shi, Y.J.; Fang, H.Y.; Zhou, G.M.; Xu, X.J.; Zhou, Y.F.; Tao, J.X.; Ji, B.Y.; Xu, J.; Li, C.; et al. Vegetation carbon stocks driven by canopy density and forest age in subtropical forest ecosystems. *Sci. Total Environ.* **2018**, *631–632*, 619–626. [CrossRef] [PubMed]
54. Zhang, Y.; Li, H.; Jiang, X.Y.; Dai, Q.S.; Zhang, H.C.; Song, H.L.; Zhang, J.Y. Soil factors influencing juvenile growth of *Liriodendron chinense* × *tulipifera*. *J. Zhejiang Agric. For. Univ.* **2016**, *1*, 94–101.
55. Li, R.; Zhou, D.M.; Wu, Y.; Zhou, G.Y.; Huang, P.F.; Deng, X.J. Selection and characteristics of phosphate-solubilizing bacteria in rhizosphere of *Cunninghaimia lanceolata*. *J. Cent. South Univ.* **2012**, *4*, 95–99.
56. Hao, Z.M.; Wu, S.R.; Qin, L.; Tan, L.; Guo, W.F. Soil physical and chemical properties of pure *Pinus massoniana* and its mixed forests in different ages in southern guangxi. *Asian Agric. Res.* **2018**, *8*, 53–57.
57. Joshi, B.; Pant, S.C. Pedology of mixed deciduous and plantation forests in Tarai and Bhawar of Kumaun Himalaya, Uttarakhand. *Indian For.* **2014**, *7*, 679–684.
58. Diakhate, S.; Gueye, M.; Chevallier, T.; Diallo, N.H.; Assigbetse, K.; Abadie, J.; Diouf, M.; Masse, D.; Sembène, M.; Ndour, Y.B.; et al. Soil microbial functional capacity and diversity in a millet-shrub intercropping system of semi-arid Senegal. *J. Arid Environ.* **2016**, *129*, 71–79. [CrossRef]
59. Debenport, S.J.; Assigbetse, K.; Bayala, R.; Chapuis-Lardy, L.; Dick, R.P.; Gardener, B.B.M. Association of shifting populations in the root zone microbiome of millet with enhanced crop productivity in the Sahel Region (Africa). *Appl. Environ. Microbiol.* **2015**, *8*, 2841–2851. [CrossRef]
60. Yarwood, S.A.; Hgberg, M.N. Soil bacteria and archaea change rapidly in the first century of Fennoscandian boreal forest development. *Soil Biol Biochem.* **2017**, *114*, 160–167. [CrossRef]
61. Ren, C.J.; Liu, W.C.; Zhao, F.Z.; Zhong, Z.K.; Deng, J.; Han, X.H.; Yang, G.H.; Feng, Y.Z.; Ren, G.X. Soil bacterial and fungal diversity and compositions respond differently to forest development. *CATENA* **2019**, *181*, 104071. [CrossRef]
62. Williams, M.A.; Jangid, K.; Shanmugam, S.G.; Whitman, W.B. Bacterial communities in soil mimic patterns of vegetative succession and ecosystem climax but are resilient to change between seasons. *Soil Biol. Biochem.* **2013**, *57*, 749–757. [CrossRef]
63. Zhou, Z.H.; Wang, C.K.; Luo, Y.Q. Meta-analysis of the impacts of global change factors on soil microbial diversity and functionality. *Nat. Commun.* **2020**, *1*, 3072. [CrossRef] [PubMed]
64. Li, Q. Effect of Neighboring Competition on Photosynthetic Characteristics and Biomass Allocation of Chinese Fir Seedlings under Low Phosphorus Stress. Master's Thesis, Fujian Agriculture and Forestry University, Fuzhou, China, 2014.
65. Zhuang, Z.; Li, Y.J.; Liu, Q.Q.; YU, Y.Y.; Liu, B.; Liu, A.Q. Sustained release of phosphorus by organic acid extraction from litter-derived red soil from Chinese fir plantation in southern China. *J. Fujian Agric. For. Univ.* **2017**, *5*, 569–575.
66. Kang, H.Z.; Gao, H.H.; Yu, W.J.; Tang, Y.; Wang, Y.; Ning, M.L. Changes in soil microbial community structure and function after afforestation depend on species and age: Case study in a subtropical alluvial island. *Sci. Total Environ.* **2018**, *625*, 1423–1432. [CrossRef]
67. Ogilvie, L.A.; Hirsch, P.R.; Johnston, A.W.B. Bacterial diversity of the broadbalk 'classical' winter wheat experiment in relation to long-term fertilizer inputs. *Microb. Ecol.* **2008**, *3*, 525–537. [CrossRef] [PubMed]
68. Kivlin, S.N.; Hawkes, C.V. Tree species, spatial heterogeneity, and seasonality drive soil fungal abundance, richness, and composition in neotropical rainforests. *Environ. Microbiol.* **2016**, *12*, 4662–4673. [CrossRef] [PubMed]
69. Zhang, J.Z.; Chen, X.R.; Yang, C.D.; Xue, L. A study on the diversity of soil cultured fungi in the alpine grassland of eastern Qilian mountains. *Acta Pratac. Sin.* **2010**, *2*, 124–132.
70. Eisenlord, S.D.; Zak, D.R. Simulated atmospheric nitrogen deposition alters actinobacterial community composition in forest soils. *Soil Sci. Soc. Am. J.* **2010**, *4*, 1157–1166. [CrossRef]
71. Verzeaux, J.; Alahmad, A.; Habbib, H.; Nivelle, E.; Roger, D.; Lacoux, J.M.; Decocq, G.; Hirel, B.; Catterou, M.; Spicher, F.; et al. Cover crops prevent the deleterious effect of nitrogen fertilisation on bacterial diversity by maintaining the carbon content of ploughed soil. *Geoderma* **2016**, *281*, 49–57. [CrossRef]
72. Debruyn, J.M.; Nixon, L.T.; Fawaz, M.N.; Johnson, A.M.; Radosevich, M. Global biogeography and quantitative seasonal dynamics of *Gemmatimonadetes* in Soil. *Appl. Environ. Microbiol.* **2011**, *17*, 6295–6300. [CrossRef]
73. Su, J.Q.; Wei, B.; Ou-Yang, W.Y.; Huang, F.Y.; Zhao, Y.; Xu, H.J.; Zhu, Y.G. Antibiotic resistome and its association with bacterial communities during sewage sludge composting. *Environ. Sci. Technol.* **2015**, *12*, 7356–7363. [CrossRef]
74. Yelle, D.J.; Ralph, J.; Lu, F.; Hammel, K.E. Evidence for cleavage of lignin by a brown rot basidiomycete. *Environ. Microbiol.* **2008**, *7*, 1844–1849. [CrossRef] [PubMed]
75. Liang, Y.; Guo, L.D.; Mark, P. Role of mycorrhizal fungi in ecosystems. *Chin. J. Plant Ecol.* **2002**, *6*, 739–745.
76. Nielsen, U.N.; Osler, G.H.R.; Campbell, C.D.; Burslem, D.F.R.P.; van der Wal, R. The influence of vegetation type, soil properties and precipitation on the composition of soil mite and microbial communities at the landscape scale. *J. Biogeogr.* **2010**, *7*, 1317–1328. [CrossRef]
77. Lauber, C.L.; Hamady, M.; Knight, R.; Fierer, N. Pyrosequencing-Based assessment of soil pH as a predictor of soil bacterial community structure at the continental scale. *Appl. Environ. Microbiol.* **2009**, *15*, 5111–5120. [CrossRef] [PubMed]

78. Griffiths, R.I.; Thomson, B.C.; James, P.; Bell, T.; Whiteley, A.S. The bacterial biogeography of British soils RID D-5237-2011 RID A-3395-2012. *Environ. Microbiol.* **2011**, *6*, 1642–1654. [CrossRef]
79. Li, L.; Li, S.M.; Sun, J.H.; Zhou, L.L.; Bao, X.G.; Zhang, H.G.; Zhang, F.S. Diversity enhances agricultural productivity via rhizosphere phosphorus facilitation on phosphorus-deficient soils. *Proc. Natl. Acad. Sci. USA* **2007**, *27*, 11192–11193. [CrossRef]
80. Liu, W.X.; Jiang, L.; Yang, S.; Wang, Z.; Tian, R.; Peng, Z.Y.; Chen, Y.L.; Zhang, X.X.; Kuang, J.L.; Ling, N.; et al. Critical transition of soil bacterial diversity and composition triggered by nitrogen enrichment. *Ecology* **2020**, *101*, e03053. [CrossRef]
81. Maisto, G.; Marco, A.D.; Meola, A.; Sessa, L.; Santo, A.V.D. Nutrient dynamics in litter mixtures of four Mediterranean maquis species decomposing in situ. *Soil Biol. Biochem.* **2011**, *3*, 520–530. [CrossRef]
82. Chen, Y.L.; Kang, L.H.; Malajczuk, N.; Dell, B. Selecting ectomycorrhizal fungi for inoculating plantations in south China: Effect of scleroderma on colonization and growth of exotic *Eucalyptus globulus, E. urophylla, Pinus elliottii*, and *P. radiata. Mycorrhiza* **2006**, *4*, 251–259. [CrossRef]
83. Wang, J.Q.; Shi, X.Z.; Zheng, C.Z.; Suter, H.; Huang, Z. Different responses of soil bacterial and fungal communities to nitrogen deposition in a subtropical forest. *Sci. Total Environ.* **2020**, *755*, 142449. [CrossRef]
84. Zhao, Y.G.; Zhang, F.H.; Yang, L.; Wang, D.; Wang, W.C. Response of soil bacterial community structure to different reclamation years of abandoned salinized farmland in arid China. *Arch. Microbiol.* **2019**, *9*, 1219–1232. [CrossRef] [PubMed]
85. Fierer, N.; Jackson, R.B. The diversity and biogeography of soil bacterial communities. *Proc. Natl. Acad. Sci. USA* **2006**, *3*, 626–631. [CrossRef] [PubMed]
86. Dennis, P.G.; Rushton, S.P.; Newsham, K.K.; Lauducina, V.A.; Ord, V.J.; Daniell, T.J.; O'Donnell, A.G.; Hopkins, D.W. Soil fungal community composition does not alter along a latitudinal gradient through the maritime and sub-Antarctic. *Fungal Ecol.* **2012**, *4*, 403–408. [CrossRef]
87. Lin, K.M.; Zhang, Z.Q.; Cao, G.Q.; He, Z.M.; Ma, X.Q. Decomposition characteristics and its nutrient dynamics of leaf litter mixtures of both Chinese fir and *Phoeba bournei. Acta Ecol. Sin.* **2006**, *8*, 2732–2738.
88. Xia, Z.C.; Yu, L.; He, Y.; Korpelainen, H.; Li, C.Y. Broadleaf trees mediate chemically the growth of Chinese fir through root exudates. *Biol. Fertil. Soils* **2019**, *7*, 737–749. [CrossRef]

forests

MDPI

Article

Characteristics of Microbial Abundance in Rhizosphere and Non-Rhizosphere Soils of Permafrost Peatland, Northeast China

Chao Gong [1,†], Xiuyan Ma [1,†], Yanyu Song [1,*], Dan Zhang [2,*], Mengyuan Zhu [1,3], Xianwei Wang [1], Siqi Gao [1,3], Jinli Gao [1] and Changchun Song [1,4]

1 Key Laboratory of Wetland Ecology and Environment, Northeast Institute of Geography and Agroecology, Chinese Academy of Sciences, Changchun 130102, China; gongchao@iga.ac.cn (C.G.); maxiuyan@iga.ac.cn (X.M.); zhumengyuan@iga.ac.cn (M.Z.); wangxianwei@iga.ac.cn (X.W.); sqgao_st@rcees.ac.cn (S.G.); gaojinli@iga.ac.cn (J.G.); songcc@iga.ac.cn (C.S.)
2 College of Landscape Architecture, Changchun University, Changchun 130022, China
3 University of Chinese Academy of Sciences, Beijing 100049, China
4 School of Hydraulic Engineering, Dalian University of Technology, Dalian 116024, China
* Correspondence: songyanyu@iga.ac.cn (Y.S.); zhangdan@ccu.edu.cn (D.Z.)
† These authors contributed equally to this work.

Abstract: The rhizosphere microenvironment is crucial to plant–soil physiological processes. The differences among microbial communities in the rhizosphere and non-rhizosphere peatland topsoil (0–15 cm) and subsoil (15–30 cm) in five plant communities dominated by *Carex schmidtii*, *Chamaedaphne calyculata*, *Ledum palustre*, *Betula fruticosa*, and *Vaccinium uliginosum*, as well as non-rhizosphere soil in discontinuous and continuous permafrost regions, were studied. We found that the bacteria and *nif*H gene abundances in the *C. calyculata* rhizosphere soil in the discontinuous permafrost region were higher than those in continuous permafrost region, while the *nir*K and *nif*H gene abundances in the non-rhizosphere soil of the discontinuous permafrost region were lower than those in the continuous permafrost region. The ratio of bacteria to fungi decreased and that of *nir*K to *nir*S increased significantly from the discontinuous to the continuous permafrost region, indicating that permafrost degradation can change soil microbial community composition. Fungal abundance was higher in the rhizosphere than the non-rhizosphere soils, suggesting that plant roots provide a more suitable environment for fungi. Moreover, the abundances of the topsoil bacteria; the fungi; and the *nir*K, *nir*S, and *nif*H genes were higher than those in the subsoil because of the organic matter from plant litter as a source of nutrients. The microbial abundance in the subsoil was also more affected by nutrient availability. To sum up, the microbial abundance varied among the different types of rhizosphere and non-rhizosphere soils, and the carbon and nitrogen cycling processes mediated by soil microorganisms may be greatly altered due to permafrost degradation under climate warming.

Keywords: bacteria; fungi; *nir*K and *nir*S genes; *nif*H gene; rhizosphere; permafrost peatland

1. Introduction

Plants play indispensable roles in maintaining the structure and functional stability of wetland ecosystems. Vegetation changes may vastly affect the quality and quantity of organic matter (OM) returned to soil and its microbes, which integrally influence the dynamics of soil organic carbon (SOC) [1,2]. The rhizosphere, an area of soil in the vicinity of 1–2 mm from plant roots, has rich microbial diversity, making it closely related to biogeochemical processes [3,4]. Root exudates are composed of carbohydrates, amino acids, fatty acids, enzymes, and other organic compounds that affect soil's physicochemical properties, creating differences in the microbial composition and activity in the vicinity of the roots [5–7]. The rhizosphere's function is an important subject of the study of plant–soil interactions in different ecosystems [8]. Previous studies have shown that microbial abundance is strongly associated with vegetation and depth in rhizosphere and non-rhizosphere soils [9,10]. Clarifying the differences in microbial composition characteristics between rhizosphere and non-rhizosphere soils is conducive to disentangling the mechanisms of the interaction effects between plants and soil microorganisms under global warming.

Microorganisms play a fundamental role in soil, regulating various biogeochemical processes, such as organic matter mineralization and nutrient cycling [11,12]. Soil microorganisms are repositories of nutrients, supplying energy for soil chemical and biological processes and those nutrients to plants [13–15]. The rhizosphere is critical to microorganism-driven ecological processes in terrestrial ecosystems, being characterized by high microbial activity and substrate availability [16]. Typically, the composition and abundance of soil microbial communities vary significantly depending on the plant species [17]. Plants have a certain impact on soil indigenous microbial populations, and each plant species selectively so for specific microbial communities. Plant root exudates and residue inputs are crucial factors that change bacterial community structures [18,19], directly affecting the decomposition of organic matter or indirectly affecting the soil C and N cycle through the food web [20]. Thus, the abundance and diversity of soil microorganisms is extremely important for understanding soil quality and nutrient status, and thus, their study has always been a hot topic in the contexts of both rhizosphere and non-rhizosphere soil [21,22]. Presently, only a few studies have examined the characteristics of microbial composition in rhizosphere and non-rhizosphere soil in the permafrost region [16,23].

The Great Xing'an Mountains, the only area in China with a zonal permafrost distribution, are located in the northeast of the country. However, the permafrost wetlands are gradually degrading to the north with global warming, which may lead to variations in the plant composition and thus altered microbial communities in the local permafrost peat fields [24]. The purpose of the current research was to determine the differences in microbial abundance among different types of permafrost rhizosphere and non-rhizosphere peatland soil in the Great Xing'an Mountains. Microbial abundance was estimated in two soil layers (0–15 cm, 15–30 cm). The vegetation was dominated by *Carex schmidtii*, *Chamaedaphne calyculata*, *Ledum palustre*, *Betula fruticosa*, and *Vaccinium uliginosum* in the discontinuous and continuous permafrost areas. The abundances of bacterial 16S rRNA; fungal ITS; and the *nir*K, *nir*S, and *nif*H genes were detected using a quantitative polymerase chain reaction (q-PCR). We hypothesized that (1) the topsoil microbial abundance would be higher than that of subsoil regardless of the sampling site, (2) the abundance and composition of the microorganisms would differ between rhizosphere and non-rhizosphere soil at different depths due to variations in plant root density, and (3) the soil microbial abundance in the discontinuous permafrost region would be higher than that in the continuous permafrost region due to an increased substrate availability upon permafrost degradation.

2. Materials and Methods

2.1. Site Description and Sampling

In August 2021, soil samples were collected in the discontinuous permafrost peatland in Xinlin (51°35′5.74″ N–51°36′18.22″ N, 124°16′41.13″ E–124°20′14.37″ E) and the continuous permafrost peatland in Amur (52°15′03″ N–53°33′15″ N, 122°38′30″ E–24°05′05″ E; Figure 1). The thawing depths were 45 cm and 55 cm, respectively. The average annual air temperature and precipitation were −3.6 °C and 498 mm, respectively, in the discontinuous permafrost region and −5 °C and 430 mm, respectively, in the continuous permafrost peatland [25].

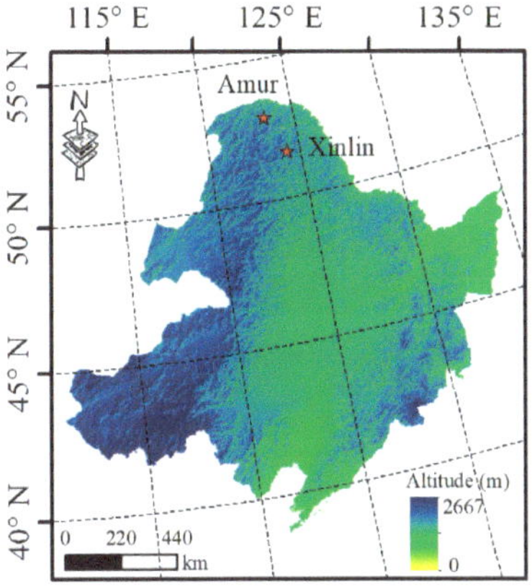

Figure 1. Sampling sites in Northeast China.

In each permafrost region, soil samples in three replications were collected at 0–15 and 15–30 cm depths from the rhizosphere of the dominant species *Carex schmidtii* (Cs), *Chamaedaphne calyculata* (Cc), *Ledum palustre* (Lp), *Betula fruticosa* (Bf), and *Vaccinium uliginosum* (Vu). The soil >2 mm away from the roots was carefully collected by hand and defined as non-rhizosphere (NR) soil. The rhizosphere soil was the 0–2 mm thick layer of soil attached to the fine roots. Samples were mixed separately from each soil layer of each plant species at each sampling site, stored in plastic bags, and then refrigerated with ice bags and transported to the laboratory as soon as possible. Part of the soil sample was frozen at −80 °C for DNA extraction, and the remaining part was air-dried for further determination of SOC content, total nitrogen (TN) content, total phosphorus (TP) content, and pH.

2.2. Soil Carbon, Nitrogen, Phosphorus, and pH Determination

The SOC content was measured via dry combustion using a Multi N/C 2100 analyzer (Analytik Jena, Jena, Germany). After wet digestion with sulfuric acid, the TN and TP contents were measured with an AA3 Continuous Flow Analyzer (Seal Analytical, Norderstedt, Germany) [26]. The soil pH was determined at a water-to-soil ratio of 10:1 using a calibrated pH meter (PHS-25, Shanghai, China).

2.3. DNA Extraction and Functional Gene Abundance Assay

DNA was extracted from 0.3 g of fresh soil using a FastDNA Spin Kit for Soil (MPbio, Santa Ana, CA, USA), following the manufacturer's instructions. The DNA concentrations were determined with a NanoDrop® ND-1000 UV-Vis spectrophotometer (Thermo Fisher Scientific, Inc., Waltham, MA, USA). Before functional gene analysis, the extracted DNA was stored inside a −20 °C freezer. The abundances of bacterial 16S rRNA; fungal ITS; and the *nir*K, *nir*S, and *nif*H genes were examined with real-time PCR (RT-PCR) using an ABI StepOne instrument (Applied Biosystems, San Francisco, CA, USA) with SYBR green dye. The RT-PCR analysis was replicated three times for each soil sample. The PCR amplification procedures and primers for each target gene are shown in Table 1. Each 25 μL PCR reaction mixture included 12.5 μL of SYBR Buffer (TaKaRa, Beijing, China), 0.4 μL of each primer (10 μM), 0.5 μL of ROXII (TaKaRa), 0.88 μL of 3% BSA, 0.63 μL of DMSO, and 10 ng of template DNA. Standard curves were generated by purifying phylogenetic and functionally labeled amplicon products using a Cyclic Purification Kit (OMEGA Bio-Tek, Norcross, GA, USA); they were then attached to the vector pMD18-T (TaKaRa) and converted to TOP10 *Escherichia coli*. competent cells. Plasmids were extracted using the Plasmid Mini Kit (OMEGA Bio-Tek). The plasmid specificity was detected with the Basic Local Alignment Search Tool [27], and the plasmid concentration was determined via Nanodrop 2000 (Thermo, Waltham MA, USA). We obtained the standard curve with successive dilution of the known-copy-number plasmids.

Table 1. Primers and amplification procedures for soil microbial functional genes.

Group	Primer	Sequence (5′—3′)	Amplification Details	Reference
Bacteria	Bacteria-338F Bacteria-518R	CCTACGGGAGGCAGCAG ATTACCGCGGCTGCTGG	95 °C, 2 min, 35 cycles; 95 °C, 30 s; 60 °C, 30 s; 72 °C, 30 s; 80 °C, 15 s	[28]
Fungi	ITSIF 5.8S	TCCGTAGGTGAACCTGCGG CGCTGCGTTCTTCATCG	94 °C, 15 min; 94 °C, 30 s; 59.4 °C, 30 s; 72 °C, 30 s; 80 °C, 30 s, 35 cycles	[29]
*nir*K	FlaCu R3Cu	ATCATGGTSCTGCCGCG GCCTCGATCAGRTTGTGGTT	95 °C, 10 min; 6 touchdown cycles: 95 °C, 15 s; 63 °C, 30 s (−1 °C); 72 °C, 30 s; and 95 °C, 15 s; 58 °C, 30 s; 72 °C, 30 s; 80 °C, 30 s, 35 cycles	[30]
*nir*S	cd3aF R3cd	GTSAACGTSAAGGARACSGG GASTTCGGRTGSGTCTTGA	95 °C, 10 min; 94 °C, 1 min; 57 °C, 1 min; 72 °C, 1 min; 83 °C, 30 s, 40 cycles	[31]
*nif*H	PolF PolR	TGCCAYCCSAARGCBGACTC ATSGCCATCATYTCRCCGGA	95 °C, 10 min, 40 cycles; 95 °C, 15 s; 60 °C, 30 s; 72 °C, 30 s; 80 °C, 15 s	[32]

2.4. Statistical Analyses

SPSS software (v. 16.0) was used to analyze the data. One-way analysis of variance (ANOVA) with LSD multiple comparison was used to determine the significant differences in the abundances of bacteria, fungi, and functional genes in the rhizospheres of the different plant species and the non-rhizosphere soil of each sampling site, and three-way ANOVA was performed to test the significant effects of the plant species, soil depth, and sampling site and their interactions on the SOC, TN, TP, and pH values and soil microbial abundances. The relationship between the soil chemical properties and the microbial abundances was analyzed using redundancy analysis (RDA). All data were tested for normality (S-W test) prior to the ANOVA and were distributed normally. The significance of the statistical analysis was considered to be at the $p < 0.05$ level. Origin8.5 was used for mapping.

3. Results

3.1. SOC, TN, TP, and pH Values

Figure 1 shows the SOC, TN, TP, and pH values at different depths of rhizosphere and non-rhizosphere soils from the two permafrost regions. These results indicate that the subsoil organic carbon content was lower than that of the topsoil at both sampling sites, and the SOC content in the discontinuous permafrost region was higher than that in the continuous permafrost region (Figure 2a,e). The subsoil TN content values under Cs and Bf were apparently higher than those of the topsoil in the discontinuous permafrost, and similar patterns were observed for Cs and Bf across the different sampling sites (Figure 2b,f). Overall, the soil TP content (Figure 2c,g) and pH values (Figure 2d,h) under the same species in the continuous permafrost region were clearly higher than those in the discontinuous permafrost region.

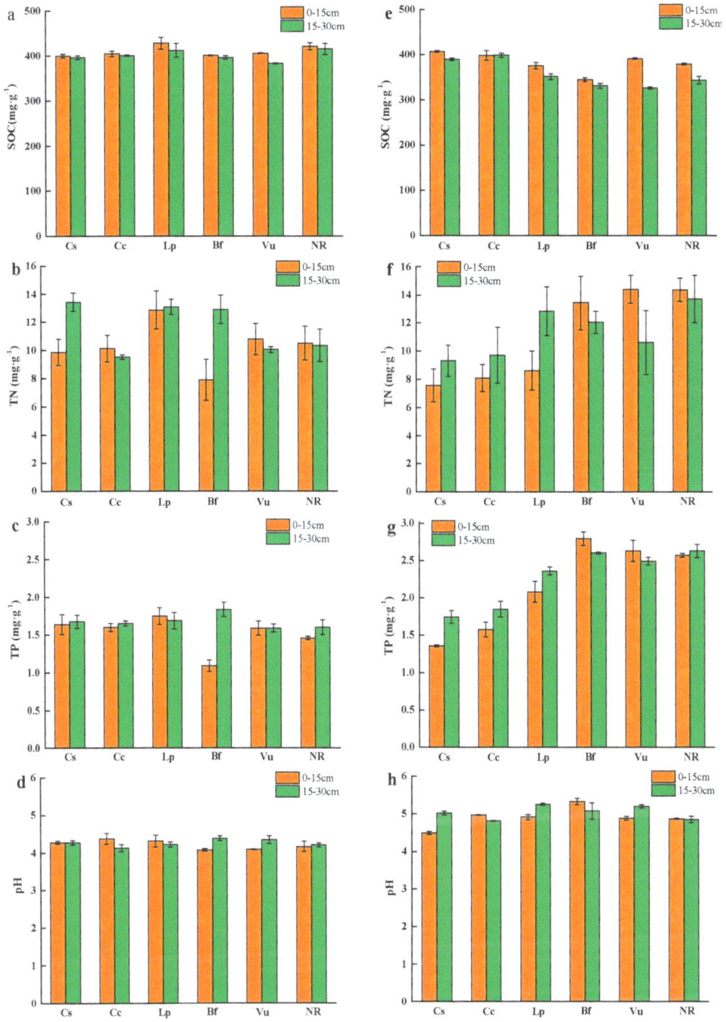

Figure 2. SOC, TN, TP, and pH values in rhizosphere and non-rhizosphere soils at different depths from (**a**–**d**) discontinuous permafrost peatland and (**e**–**h**) continuous permafrost peatland. Cs, *Carex schmidtii*; Cc, *Chamaedaphne calyculata*; Lp, *Ledum palustre*; Bf, *Betula fruticosa*; Vu, *Vaccinium uliginosum*; NR, non-rhizosphere soil. Error bars represent the standard error (n = 3).

3.2. Abundances of Bacteria and Fungi

The bacterial and fungal abundances in the topsoil were higher than those in the subsoil in both permafrost regions (Figure 3). In the discontinuous permafrost region, the Cc rhizosphere soil and non-rhizosphere soil had the highest and lowest bacterial and fungal abundances, respectively. Moreover, the bacterial abundance was significantly higher at both soil depths in the Cc rhizosphere soil in the discontinuous permafrost region compared with that in the continuous permafrost region ($p < 0.05$). The bacterial abundance in the Vu rhizosphere topsoil was higher than those in the Cs and Lp rhizosphere soils ($p < 0.05$), and the fungal abundance in the Vu rhizosphere subsoil was higher than those in the Cs and Lp rhizosphere soils ($p < 0.05$; Figure 3a,b). In the continuous permafrost region, the topsoil bacterial abundance in the Lp rhizosphere was the highest, while that in the Cs rhizosphere soil was the lowest (Figure 3d). The fungal abundances in the Lp rhizosphere soil at both depths were higher than those in the other investigated soils, while the non-rhizosphere soil had the lowest fungal abundance. The fungal abundance in the Vu rhizosphere topsoil was significantly lower than those in the Lp and Bf rhizosphere soils ($p < 0.05$). The subsoil fungal abundance under Cs was higher than those in the Bf and Vu rhizosphere soils ($p < 0.05$; Figure 3e). The ratio of bacteria to fungi decreased significantly from the discontinuous permafrost region to the continuous permafrost region (Figure 3c,f).

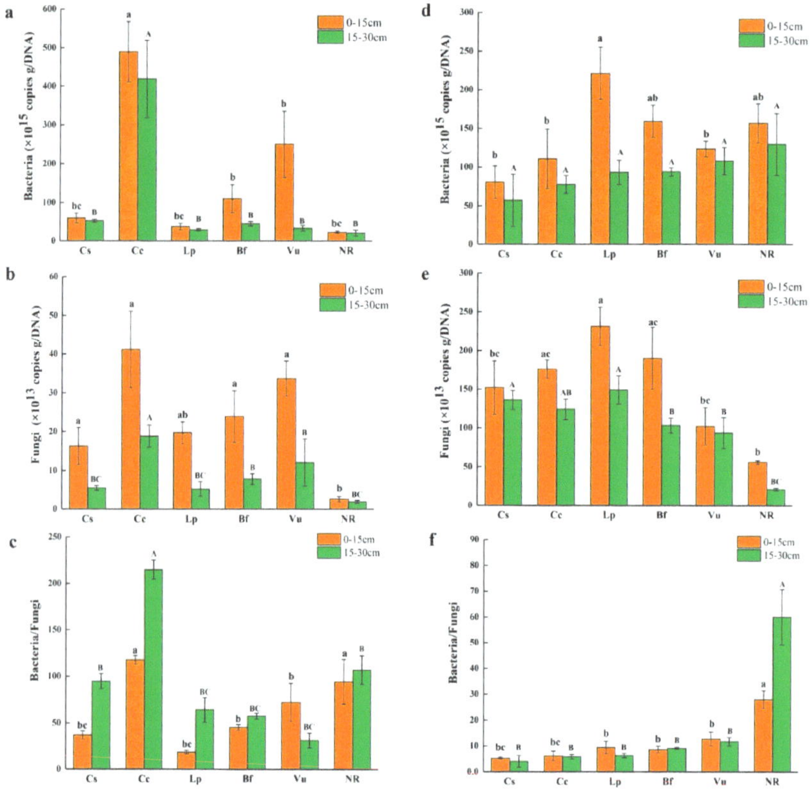

Figure 3. Bacterial and fungal abundances in rhizosphere and non-rhizosphere soils of (**a–c**) discontinuous permafrost peatland and (**d–f**) continuous permafrost peatland. Cs, *Carex schmidtii*; Cc, *Chamaedaphne calyculata*; Lp, *Ledum palustre*; Bf, *Betula fruticosa*; Vu, *Vaccinium uliginosum*; NR, non-rhizosphere soil. Different letters indicate significant differences at the $p < 0.05$ level. Error bars represent the standard error ($n = 3$).

3.3. Abundance of Denitrifying Bacteria

The numbers of *nir*K and *nir*S gene copies were used to express the abundance of denitrifying bacteria. The abundance of denitrifying genes was consistently higher in the topsoil compared to the subsoil in both permafrost regions (Figure 4). The Cs rhizosphere topsoil had the highest abundances of the *nir*K and *nir*S genes, and the lowest abundances of these genes were found in the Lp and Cc rhizosphere soils in the discontinuous permafrost region. The abundances of the *nir*K and *nir*S genes in the Cc and Vu rhizosphere subsoils were substantially lower than those in the other soils investigated (Figure 4a,b). In the continuous permafrost region, the abundance of the *nir*K gene in the non-rhizosphere soil was remarkably higher than in the other investigated rhizosphere soils, and the highest *nir*S gene abundance occurred in the Bf rhizosphere topsoil ($p < 0.05$; Figure 4d,e). The ratio of *nir*K to *nir*S significantly increased from the discontinuous permafrost region to the continuous permafrost region (Figure 4c,f).

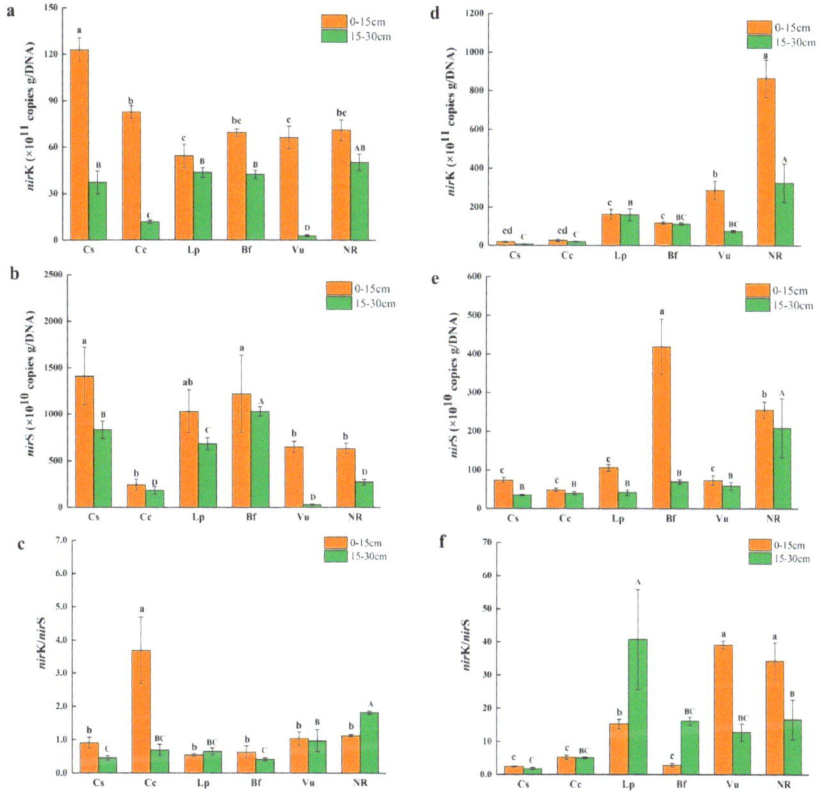

Figure 4. Abundances of *nir*K and *nir*S genes in rhizosphere and non-rhizosphere soils at different depths in (**a**–**c**) discontinuous permafrost peatland and (**d**–**f**) continuous permafrost peatland. Cs, *Carex schmidtii*; Cc, *Chamaedaphne calyculata*; Lp, *Ledum palustre*; Bf, *Betula fruticosa*; Vu, *Vaccinium uliginosum*; NR, non-rhizosphere soil. Different letters indicate significant differences at the $p < 0.05$ level. Error bars represent the standard error ($n = 3$).

3.4. Abundance of N-Fixing Bacteria

The number of *nif*H gene copies was used to express the abundance of N-fixing bacteria. We found that the *nif*H gene abundance in the topsoil was higher than that in the subsoil at the different sampling sites (Figure 5). In the discontinuous permafrost region, the *nif*H gene abundance in the Cc rhizosphere soil was clearly higher than those in the other soils investigated, and the same pattern was observed in the non-rhizosphere soils of the continuous permafrost region ($p < 0.05$; Figure 5a,b). The *nif*H gene abundance in the Cs rhizosphere soil was higher than those observed in the other rhizosphere soils (Cc excepted) in the discontinuous permafrost region, while it was the lowest in the continuous permafrost region ($p < 0.05$).

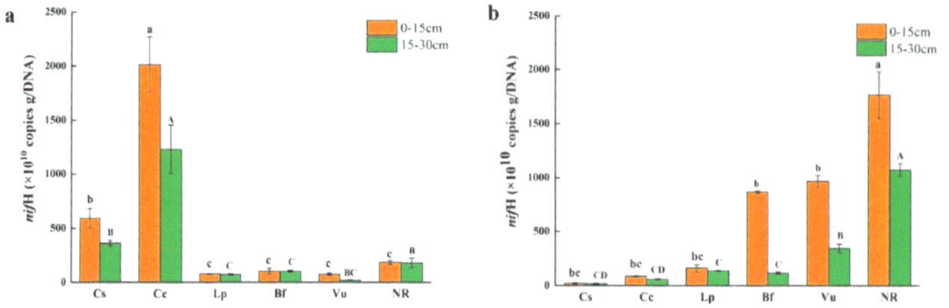

Figure 5. Abundance of the *nif*H gene in rhizosphere and non-rhizosphere soils in (**a**) the peatland of the discontinuous permafrost region and (**b**) the peatland of the continuous permafrost region. Cs, *Carex schmidtii*; Cc, *Chamaedaphne calyculata*; Lp, *Ledum palustre*; Bf, *Betula fruticosa*; Vu, *Vaccinium uliginosum*; NR, non-rhizosphere soil. Different letters indicate significant differences at the $p < 0.05$ level. Error bars represent the standard error ($n = 3$).

3.5. Effects of Vegetation Type, Soil Depth, and Permafrost Type on Soil Bacteria, Fungi, and Functional Gene Abundances

The results of the three-way ANOVA indicated that the bacterial abundance was significantly influenced by the vegetation type and the soil depth ($p < 0.01$). The interaction effects between the permafrost type and the soil depth significantly affected the bacterial abundance ($p < 0.01$). The interaction effects between the vegetation type and the soil depth also significantly affected the soil bacterial abundance ($p < 0.05$). The fungal abundance was significantly influenced by the vegetation type ($p < 0.01$). The interaction effects between the permafrost type and the soil depth and between the vegetation type and the soil depth significantly affected the fungal abundance ($p < 0.05$). Furthermore, the interaction effects among the vegetation type, permafrost type and soil depth had a significant impact on the abundance of *nir*K ($p < 0.01$; Table 2). The abundance of *nir*S was significantly influenced by the vegetation type and the permafrost type ($p < 0.05$). Furthermore, the interaction effects among the vegetation type, permafrost type, and soil depth had a significant impact on the abundance of *nir*S ($p < 0.01$). The interaction effects between the vegetation type and the permafrost type and among the vegetation type, permafrost type, and soil depth had a significant impact on the abundance of *nif*H ($p < 0.01$).

Table 2. Results of three-way ANOVA on individual and interactive effects of vegetation type, soil depth, and permafrost type on soil bacteria, fungi, and functional gene abundances (* $p < 0.05$, ** $p < 0.01$).

Factor	Bacteria	Fungi	*nir*K	*nir*S	*nif*H
Vegetation Type	3.814 **	3.819 **	13.470 **	2.671 *	0.611
Soil Depth	8.640 **	0.862	76.677 **	1.323	1.636
Permafrost Type	0.360	1.591	47.942 **	0.123	0.000
Vegetation Type × Permafrost Type	0.699	0.606	48.258 **	2.728 *	8.707 **
Permafrost Type × Soil Depth	9.520 **	6.22 *	29.046 **	0.001	0.376
Vegetation Type × Soil Depth	2.274 *	3.475 *	52.467 **	1.979	2.312 *
Vegetation Type × Permafrost Type × Soil Depth	1.005	1.756	17.412 **	5.283 **	4.101 **

The RDA results revealed that the SOC, TP, and pH were significant factors explaining the variations in the surface soil bacteria; fungi; and *nir*K, *nir*S, and *nif*H gene abundances at both sampling sites ($p < 0.05$; Figure 6a), in which the TP had the highest explanatory power (29.6%), followed by the pH (18.8%) and SOC (3.8%). For the subsoil, variations in the bacteria; fungi; and *nir*K, *nir*S, and *nif*H gene abundances were significantly explained by the pH, TP, TN, N/P, and SOC ($p < 0.05$; Figure 6b). Among them, pH had the highest explanatory power (39.7%), followed by the TP (20.8%) and the TN (6.2%).

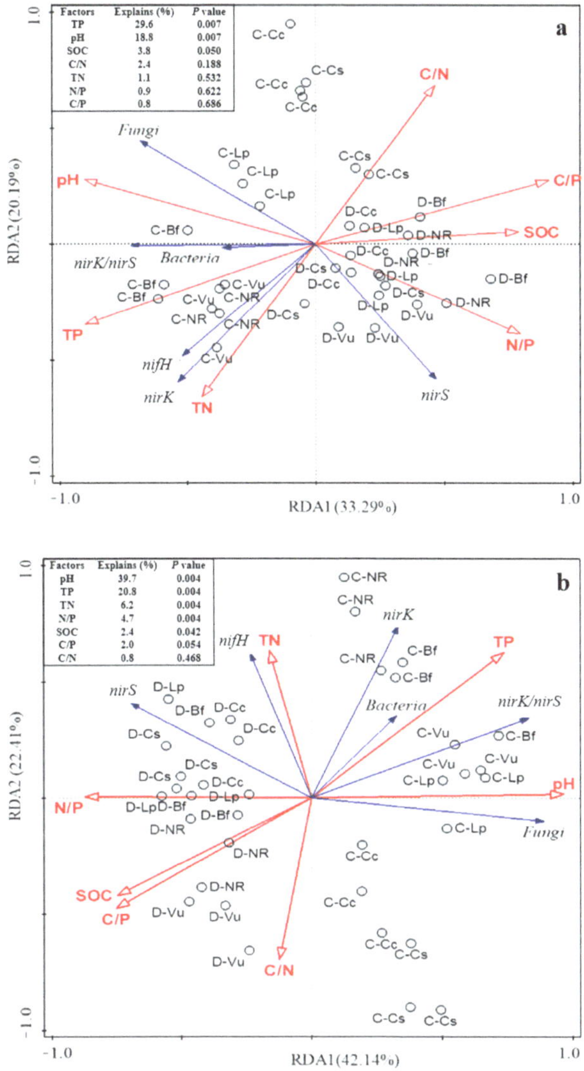

Figure 6. Redundancy analysis (RDA) of the relationships between (**a**) topsoil and (**b**) subsoil bacteria, fungi and functional gene abundances, and soil chemical characteristics in different permafrost regions. Note: open circles represent different plant species. C- and D- indicate continuous and discontinuous permafrost regions, respectively. Cs, *Carex schmidtii*; Cc, *Chamaedaphne calyculata*; Lp, *Ledum palustre*; Bf, *Betula fruticosa*; Vu, *Vaccinium uliginosum*; NR, non-rhizosphere soil.

4. Discussion

4.1. Differences in Soil Microbial Abundance between Rhizosphere and Non-Rhizosphere Soil

The results of our study have demonstrated that the bacterial and fungal abundance in the rhizosphere soil were higher than those in the non-rhizosphere soil in the discontinuous permafrost region, which is consistent with hypothesis 2. The same pattern was revealed for fungi in the continuous permafrost region, but not for bacteria in the rhizosphere and non-rhizosphere soils (Figure 3). Our results were in accordance with previous studies that have suggested that the number of bacteria in non-rhizosphere soil is lower than that in rhizosphere soil and this pattern is soil-environment-dependent [9,10]. Diverse plant species growing on the same soil host their own specific microbial communities, confirming that plants can shape their rhizosphere microbiomes [16,33–36]. Generally, bacteria are a substantial biological group of soil microorganisms due to their large amounts in soil and act as an indispensable driver that is greatly regulated by substrate availability in ecosystems [37]. The abundance of fungi in rhizosphere soil was greater than that of non-rhizosphere soil independently of region, indicating that rhizosphere soil is conducive to the growth and propagation of fungi given the high substrate availability from root debris and exudate. Additionally, the majority of the nutrient shift between plants and microbial communities took place directly in rhizosphere soil, which was different from non-rhizosphere soil, which induced higher bacterial and fungal abundances [38]. We also observed that the fungal abundance varied in the rhizosphere soils of different plant species, mainly due to the fact that different vegetation types have varying effects on soil physicochemical properties, leading to obvious differences in soil fungal community structure and quantity [39,40].

In this study, except for in the topsoil under *Betula fruticosa*, the abundances of the *nir*K and *nir*S genes in rhizosphere soil were clearly lower than those in non-rhizosphere soil in the continuous permafrost region. This is consistent with the research of Nie et al. (2014), who verified that denitrifying functional gene amplification resulted in fewer denitrifiers in rhizosphere soil [41]. The reasons may be that oxygen secretion from plant roots may induce an oxygen gradient in those roots, which is not conducive to denitrification [42], and non-rhizosphere soil provides more substrate sources for denitrifying microorganisms on account of plant competition shortage [43]. Bremer et al. (2007) revealed that plants could directly induce variations in the composition of the *nir*K-type denitrifier through root exudates [44]. We also observed that the abundances of the *nir*K and *nir*S genes in rhizosphere and non-rhizosphere soils under the same (both sampling sites) and different plant species were markedly different. This was relevant to the differences in the microenvironment and the root exudates in the plant rhizospheres, similarly to the results of previous research, suggesting that the *nir*K and *nir*S genes' abundances in rhizosphere and non-rhizosphere soils from different sites and plants differed remarkably [45].

4.2. Differences of Soil Microbial Abundance in Topsoil and Subsoil

In accordance with hypothesis 1, our results clarified that the abundances of bacteria; fungi; and the *nir*K, *nir*S, and *nif*H genes in the topsoil were greater than those in the subsoil in the permafrost region. The reason may be that the surface layer harbors more organic matter from plant litter and roots, which acts as a source of nutrients for soil microorganisms and increases the abundances of bacteria; fungi; and the *nir*K, *nir*S, and *nif*H genes. Plant roots can continuously produce organic matter into rhizosphere soil [46,47]. There are some researchers who have demonstrated that the primary food source of soil microbes closely associated with plants is root exudates, which are essential for microbial population enrichment in rhizosphere soil with high substrate availability [48]. Moreover, topsoil contains more nutrients than subsoil, which is beneficial to the reproduction of microorganisms. However, not only is the interaction between plants and microorganisms a prime factor in determining the composition of rhizosphere microbial communities, but the competitive interactions among microorganisms are the fundamental cause of differences in soil microbial abundance [49].

We discovered that vegetation type and soil depth had remarkable influences on soil microbial abundances, and the complex interactions between vegetation type and permafrost type or among vegetation type, permafrost type, and soil depth were of similar importance to the distribution of soil microbial abundance, which is in line with our hypotheses. Previous studies have found that soil microbial communities under diverse plant species could differ dramatically in their abundance and composition [17,50], and the magnitude and structure of each microbial community in rhizosphere soil is chiefly regulated by soil type [51,52].

4.3. Differences in Soil Microbial Abundances in Different Types of Permafrost

Previous studies have shown that constantly produced organic compounds are excreted into the soil through roots that modulate microbial growth and reproduction [46,47]. In our research, we only observed that the abundances of bacteria and the *nif*H gene in the *C calyculata* rhizosphere soil of the continuous permafrost region were lower than those of the discontinuous permafrost region, which is in accordance with hypothesis 3. Vegetation succession under the influence of global climate change may cause a series of variations in the connection between plants and microorganisms, which will subsequently alter nutrient cycles. The reasons for these discrepancies could be due to the facts that *C. calyculata* may be more sensitive to temperature and precipitation and that the higher air temperature and precipitation in the discontinuous permafrost region are conducive to the physiological activity of *C. calyculata*, inducing the releases of more soil-available nitrogen and other nutrients utilized by microorganisms.

Furthermore, the higher air temperature and precipitation in the discontinuous compared with the continuous permafrost region are conducive to increasing the soil temperature and moisture, which can directly increase the activity of soil microorganisms. The results of the RDA indicated that soil TP and pH significantly positively correlated with soil bacterial and fungal abundances in permafrost peatland soils, suggesting that the abundances of soil bacteria and fungi may be limited by soil phosphorus content and acidic condition, while increases in soil TP content and pH can stimulate their abundances. The abundances of the *nir*K and *nif*H genes in the non-rhizosphere soil in the continuous permafrost region were higher than those in the discontinuous permafrost region, as permafrost degradation is caused by climate warming. This could induce the nitrite reduction step of denitrification while limiting nitrogen fixation, resulting in reductions in the *nir*K and *nif*H genes. Previous studies have revealed that the mean annual precipitation in the discontinuous permafrost region is higher than that in the continuous permafrost region, leading to higher soil moisture content. This is not conducive to denitrifying or nitrogen-fixing bacteria [25]. We found that the ratios of bacteria to fungi increased and that of *nir*K to *nir*S significantly decreased from the continuous to the discontinuous permafrost region, indicating that permafrost degradation may shift the soil microbial community composition and the bacteria, specifically denitrifying bacteria (*nir*K) domination. This dominance of bacteria promotes soil organic matter decomposition, as labile substrates favor bacteria more than fungi, inducing more carbon to probably take part in cycling processes with permafrost degradation under climate warming [53,54].

5. Conclusions

The abundance of bacteria was more sensitive to permafrost degradation, leading to the release of more available nutrients that were conducive to bacteria growth and propagation in the rhizosphere soil. The characteristics of the *nif*H gene abundance in the varied permafrost region revealed that permafrost degradation is not conducive to soil microbial nitrogen fixation without the indirect effect of plants. The abundance of fungi in the rhizosphere soil was greater than that in the non-rhizosphere soil independently of the sites. The non-rhizosphere soil provided more substrate sources for denitrifying microorganisms as a lack of plant competition. Topsoil with more organic matter from plant litter induced higher abundances of bacteria; fungi; and the *nir*K, *nir*S, and *nif*H genes as compared to subsoil that is limited by nutrients in the permafrost region. Our study highlights the microbial abundances in different plant rhizosphere soils, varying with landscape changes, that may greatly impact carbon cycling under climate warming.

Author Contributions: Methodology, X.W.; Software, M.Z.; Formal analysis, J.G.; Resources, S.G.; Data curation, X.M.; Writing—review & editing, C.G.; Supervision, Y.S., D.Z. and C.S. All authors have read and agreed to the published version of the manuscript.

Funding: This work was supported by the National Natural Science Foundation of China (grant numbers 42271109, 41401106); the Professional Association of the Alliance of International Science Organizations (grant number ANSO-PA-2020-14); the Innovation Team Project of the Northeast Institute of Geography and Agroecology, the Chinese Academy of Sciences (grant number 2022CXTD02); and the Young Scientist Group Project of the Northeast Institute of Geography and Agroecology, Chinese Academy of Sciences (grant number 2022QNXZ01-01).

Data Availability Statement: Data are available on request.

Conflicts of Interest: The authors have no relevant financial or non-financial interest to disclose.

References

1. Templer, P.H.; Groffman, P.M.; Flecker, A.S.; Power, A.G. Land use change and soil nutrient transformations in the Los Haitises region of Dominican Republic. *Soil Biol. Biochem.* **2005**, *37*, 215–225. [CrossRef]
2. Kalbitz, K.; Solinger, S.; Park, J.H.; Michalzik, B.; Matzner, E. Controls on the dynamics of dissolved organic matter in soils: A review. *Soil Sci.* **2000**, *165*, 277–304. [CrossRef]
3. Dazzo, F.B.; Garoutte, A.; Hartmann, A. Rhizosphere. In *Encyclopedia of Microbiology*, 4th ed.; Academic Press: Cambridge, MA, USA, 2019; pp. 147–163.
4. Bashir, O.; Khan, K.; Hakeem, K.R.; Mir, N.A.; Rather, G.H.; Mohiuddin, R. Soil Microbe Diversity and Root Exudates as Important Aspects of Rhizosphere Ecosystem. In *Plant, Soil and Microbes*; Springer International Publishing: Berlin/Heidelberg, Germany, 2016.
5. Hütsch, B.W.; Augustin, J.; Merbach, W. Plant rhizodeposition–An important source for carbon turnover in soils. *J. Plant Nutr. Soil Sci.* **2002**, *165*, 397–407. [CrossRef]
6. Walker, T.S.; Bais, H.P.; Grotewold, E.; Vivanco, J.M. Root exudation and rhizosphere biology. *Plant. Physiol.* **2003**, *132*, 44–51. [CrossRef]
7. Turner, T.R.; Ramakrishnan, K.; Walshaw, J.; Heavens, D.; Alston, M.; Swarbreck, D.; Osbourn, A.; Grant, A.; Poole, P.S. Comparative metatranscriptomics reveals kingdom level changes in the rhizosphere microbiome of plants. *ISME J.* **2013**, *7*, 2248–2258. [CrossRef]
8. Lynch, J.P. Root architecture and plant productivity. *Plant Physiol.* **1995**, *109*, 7–13. [CrossRef]
9. He, Y.; Hu, W.G.; Ma, D.C.; Yang, Y.; Lan, H.Z.; Gao, Y. Diversity and abundance of ammonia-oxidizing microorganisms in relation to soil environment in rhizosphere soil of *Halocnemum strobilaceum* in Ebinur Lake wetland. *Acta Sci. Circumstantiae* **2017**, *37*, 1967–1975.
10. Wang, C.H.; Wu, F.; Hu, W.G.; Mo, C.; Zhang, X.H. Community diversity of ammonia-oxidizing bacteria of three plants rhizosphere in Ebinur Lake wetland. *Acta Microbiol. Sin.* **2015**, *55*, 1190–1200.
11. Liu, Y.; Shen, X.; Chen, Y.M.; Wang, L.F.; Chen, Q.M.; Zhang, J.; Xu, Z.F.; Tan, B.; Zhang, L.; Xiao, J.J.; et al. Litter chemical quality strongly affects forest floor microbial groups and ecoenzymatic stoichiometry in the subalpine forest. *Ann. Forest Sci.* **2019**, *76*, 106. [CrossRef]
12. Moreno-Espíndola, I.P.; Ferrara-Guerrero, M.J.; Luna-Guido, M.L.; Ramírez-Villanueva, D.A.; León-Lorenzana, A.S.D.; Gómez-Acata, S.; González-Terreros, E.; Ramírez-Barajas, B.; Navarro-Noya, Y.E.; Sánchez-Rodríguez, L.M.; et al. The bacterial community structure and microbial activity in a traditional organic milpa farming system under different soil moisture conditions. *Front. Microbiol.* **2018**, *9*, 2737. [CrossRef]
13. Damashek, J.; Francis, C.A. Microbial nitrogen cycling in estuaries: From genes to ecosystem processes. *Estuaries Coasts* **2018**, *41*, 626–660. [CrossRef]
14. Huang, L.; Riggins, C.W.; Rodríguez-Zas, S.; Zabaloy, M.C.; Villamil, M.B. Long-term N fertilization imbalances potential N acquisition and transformations by soil microbes. *Sci. Total Environ.* **2019**, *691*, 562–571. [CrossRef]
15. Linda, H.; Cécile, G.R.; Nicol, G.W.; Prosser, J.I. The consequences of niche and physiological differentiation of archaeal and bacterial ammon Aggravated phosphorus limitation on biomass production under increasing nitrogen loading: A meta-analysisia oxidisers for nitrous oxide emissions. *ISME J.* **2018**, *12*, 934–943.
16. Berg, G.; Smalla, K. Plant species and soil type cooperatively shape the structure and function of microbial communities in the rhizosphere. *FEMS Microbiol. Ecol.* **2009**, *68*, 1–13. [CrossRef] [PubMed]
17. Innes, L.; Hobbs, P.; Bardgett, R. The impacts of individual plant species on rhizosphere microbial communities in soils of different fertility. *Biol. Fertil. Soils* **2004**, *40*, 7–13. [CrossRef]
18. Weinert, N.; Meincke, R.; Gottwald, C.; Heuer, H.; Gomes, N.C.M.; Schloter, M.; Berg, G.; Smalla, K. Rhizosphere communities of genetically modified Zeaxanthin-accumulating potato plants and their parent cultivar differ less than those of different potato cultivars. *Appl. Environ. Microbiol.* **2009**, *75*, 3859–3865. [CrossRef]
19. Smalla, K.; Wieland, G.; Buchner, A.; Zock, A.; Parzy, J.; Kaiser, S.; Roskot, N.; Heuer, H.; Berg, G. Bulk and rhizosphere soil bacterial communities studied by denaturing gradient gel electrophoresis: Plant-dependent enrichment and seasonal shifts revealed. *Appl. Environ. Microbiol.* **2001**, *67*, 4742–4751. [CrossRef]
20. Wickings, K.; Grandy, A.S.; Reed, S.C.; Cleveland, C.C. The origin of litter chemical complexity during decomposition. *Ecol. Lett.* **2012**, *15*, 1180–1188. [CrossRef]

21. Xue, L.; Kuang, L.G.; Chen, H.Y.; Tan, S.M. Soil nutrients, microorganisms and enzyme activities of different stands. *Acta Pedol. Sin.* **2003**, *40*, 280–285.
22. Kuzyakov, Y.; Blagodatskaya, E. Microbial hotspots and hot moments in soil: Concept & review. *Soil Biol. Biochem.* **2015**, *83*, 184–199.
23. Song, Y.Y.; Jiang, L.; Song, C.C.; Wang, X.W.; Ma, X.Y.; Zhang, H.; Tan, W.W.; Gao, J.L.; Hou, A.X. Microbial abundance and enzymatic activity from tussock and shrub soil in permafrost peatland after 6-year warming. *Ecol. Indic.* **2021**, *126*, 107589.
24. Sun, L.; Li, X.Z.; Wang, X.W.; Lv, J.J.; Li, Z.M.; Hu, Y.M. Latitudinal pattern in species diversity and its response to global warming in permafrost wetlands in the Great Hing'an Mountains, China. *Russ. J. Ecol.* **2011**, *42*, 123–132.
25. Sun, G.Y.; Yu, S.P.; Wang, H.X. Causes, south borderline and subareas of permafrost in Da Hinggan Mountains and Xiao Hinggan Mountains. *Sci. Geogr. Sin.* **2007**, *27*, 68–74.
26. Bao, S.D. *Soil Agricultural Chemistry Analysis*, 3rd ed.; China Agricultural Press: Beijing, China, 2008.
27. Zhang, Z.; Schwartz, S.; Wagner, L.; Miller, W. A greedy algorithm for aligning DNA sequences. *J. Comput. Biol.* **2000**, *7*, 203–214. [CrossRef]
28. Wang, H.; Yang, J.P.; Yang, S.H. Effect of a 10 degrees C-elevated temperature under different water contents on the microbial community in a tea orchard soil. *Eur. J. Soil Biol.* **2014**, *62*, 113–120.
29. Gardes, M.; Bruns, T.D. Its primers with enhanced specificity for basidiomycetes-application of mycorrhizae and rusts. *Mol. Ecol.* **1993**, *2*, 113–118. [PubMed]
30. Hallin, S.; Lindgren, P.E. PCR detection of genes encoding nitrile reductase in denitrifying bacteria. *Appl. Environ. Microbiol.* **1999**, *65*, 1652–1657. [CrossRef]
31. Petersen, D.G.; Blazewicz, S.J.; Firestone, M.; Herman, D.J.; Turetsky, M.; Waldrop, M. Abundance of microbial genes associated with nitrogen cycling as indices of biogeochemical process rates across a vegetation gradient in Alaska. *Environ. Microbiol.* **2012**, *14*, 993–1008. [PubMed]
32. Fan, L. Response of Diazotrophic Microbial Community to Nitrogen Input and Glyphosate Application in Soils Cropped to Soybean. Ph.D. Thesis, Auburn University, Auburn, AL, USA, 2013.
33. Berendsen, R.L.; Pieterse, C.M.J.; Bakker, P. The rhizosphere microbiome and plant health. *Trends Plant Sci.* **2012**, *17*, 478–486.
34. Hartmann, A.; Schmid, M.; van Tuinen, D.; Berg, G. Plant-driven selection of microbes. *Plant Soil* **2009**, *321*, 235–257.
35. Patersen, E.; Gebbing, T.; Abel, C.; Sim, A.; Telfer, G. Rhizodeposition shapes rhizosphere microbial community structure in organic soil. *New Phytologist.* **2007**, *173*, 600–610.
36. Jiao, R.Z.; Yang, C.D. The changes of the soil microorganism in rhizosphere and outside in different developing stages of the Chinese fir plantation. *Sci. Silvae Sin.* **1999**, *35*, 53–59.
37. Elliott, D.R.; Caporn, S.J.M.; Nwaishi, F.; Nilsson, R.H.; Sen, R. Bacterial and Fungal Communities in a Degraded Ombrotrophic Peatland Undergoing Natural and Managed Re-Vegetation. *PLoS ONE* **2015**, *10*, e0124726. [CrossRef]
38. Kemp, J.; Lotter, D.; Meyer, A.; Kleinert, A.; Pérez-Fernández, M.; Valentine, A. Variation in rhizosphere nutrient cycling affects the source of nitrogen acquisition in wild and cultivated Aspalathus linearis (N.L.Burm.) R.Dahlgren plants. *Appl. Soil Ecol.* **2018**, *130*, 26–33.
39. Wang, Y.; Liu, S.; Guo, J.L.; Liu, B.B. Effects of different vegetation types on soil nutrients, enzyme activities and microorganisms on the Loess Plateau. *Bull. Soil Water Conserv.* **2018**, *38*, 62–68. (In Chinese)
40. Broeckling, C.D.; Broz, A.K.; Bergelson, J. Root exudates regulate soil fungal community composition and diversity. *Appl. Environ. Microbiol.* **2008**, *74*, 738–744.
41. Nie, S.A.; Xu, H.J.; Li, S.; Li, H.; Su, J.Q. Relationships between abundance of microbial functional genes and the status and fluxes of carbon and nitrogen in rice rhizosphere and bulk soils. *Pedosphere* **2014**, *24*, 645–651.
42. Zheng, Y.L.; Hou, L.J.; Liu, M.; Yin, G.Y.; Gao, J.; Jiang, X.F.; Lin, X.B.; Li, X.F.; Yu, C.D.; Wang, R. Community composition and activity of anaerobic ammonium oxidation bacteria in the rhizosphere of salt-marsh grass *Spartina alterniflora*. *Appl. Microbiol. Biot.* **2016**, *100*, 8203–8212.
43. Zhang, Z.X.; Zhang, W.Z.; Yang, H.C.; Sheng, R.; Wei, W.X.; Qin, H.L. Elevated N_2O emission by the rice roots: Based on the abundances of *nar*G and bacterial *amoA* genes. *Environ. Sci. Pollut. Res.* **2017**, *24*, 2116–2125.
44. Bremer, C.; Braker, G.; Matthies, D.; Reuter, A.; Engels, C.; Conrad, R. Impact of plant functional group, plant species, and sampling time on the composition of nirK-type denitrifier communities in soil. *Appl. Environ. Microbiol.* **2007**, *73*, 6876–6884. [PubMed]
45. Yang, Y. Characteristics of Denitrifying Microbial Community Structure and Niche Differentiation in Rhizosphere Soil of Dominant Plants in Ebinur Lake Wetland. Master's thesis, Shihezi University, Shihezi, China, 2018.
46. Uren, N.C. Types, amounts and possible functions of compounds released into the rhizosphere by soil grown plants. In *The Rhizosphere: Biochemistry, and Organic Substances at the Soil Interface*; Pinton, R., Varani, Z., Nanniperi, P., Eds.; Marcel Dekker Inc.: New York, NY, USA, 2000; pp. 19–40.
47. Rovira, A.D. Plant root excretions in relation to the rhizosphere effect. *Plant Soil* **2005**, *7*, 178–194.
48. Vogel, T.M.; Simonet, P.; Jansson, J.K.; Hirsch, P.R.; Tiedje, J.M.; Elsas, J.D.V.; Bailey, M.J.; Nalin, R.; Philippot, L. Terra Genome: A consortium for the sequencing of a soil metagenome. *Nat. Rev. Microbiol.* **2009**, *7*, 252–253.

49. Rasche, F.; Velvis, H.; Zachow, C.; Berg, G.; Elsas, J.D.V.; Sessitsch, A. Impact of transgenic potatoes expressing antibacterial agents on bacterial endophytes is comparable to effects of soil, wild type potatoes, vegetation stage and pathogen exposure. *Can. J. Microbiol.* **2006**, *43*, 555–566.
50. Schnitzer, S.A.; Klironomos, J.N.; HilleRisLambers, J.; Kinkel, L.L.; Reich, P.B.; Xiao, K.; Rillig, M.C.; Sikes, B.A.; Callaway, R.M.; Mangan, S.A.; et al. Soil microbes drive the classic plant diversity-productivity pattern. *Ecology* **2011**, *92*, 296–303.
51. Bulgarelli, D.; Rot, M.; Schlaeppi, K. Revealing structure and assembly cues for Arabidopsis root-inhabiting bacterial microbiota. *Nature* **2012**, *488*, 91–95. [PubMed]
52. Lundberg, D.S.; Lebeis, S.L.; Paredes, S.H.; Yourstone, S.; Gehring, J.; Malfatti, S.; Tremblay, J.; Engelbrektson, A.; Kunin, V.; Rio, T.G.D.; et al. Defining the core *Arabidopsis thaliana* root microbiome. *Nature* **2012**, *488*, 86–90.
53. Su, Y.G.; Huang, G.; Lin, Y.J.; Zhang, Y.M. No synergistic effects of water and nitrogen addition on soil microbial communities and soil respiration in a temperate desert. *Catena* **2016**, *142*, 126–133.
54. Keiblinger, K.M.; Hall, E.K.; Wanek, W.; Szukics, U.; Hämmerle, I.; Ellersdorfer, G.; Böck, S.; Strauss, J.; Sterflinger, K.; Richter, A.; et al. The effect of resource quantity and resource stoichiometry on microbial carbon-use-efficiency. *FEMS Microbiol. Ecol.* **2010**, *73*, 430–440.

MDPI AG
Grosspeteranlage 5
4052 Basel
Switzerland
Tel.: +41 61 683 77 34

Forests Editorial Office
E-mail: forests@mdpi.com
www.mdpi.com/journal/forests